Birkhäuser

Thomas Ernst

A Comprehensive Treatment of q-Calculus

Birkhäuser

Thomas Ernst
Department of Mathematics
Uppsala University
Uppsala
Sweden

ISBN 978-3-0348-0430-1 ISBN 978-3-0348-0431-8 (eBook)
DOI 10.1007/978-3-0348-0431-8
Springer Basel Heidelberg New York Dordrecht London

Library of Congress Control Number: 2012947523

Printed on acid-free paper

Springer is part of Springer Science+Business Media (www.springer.com)

To Neita Lundquist with affection and gratitude

Preface

In q-calculus we are looking for q-analogues of mathematical objects that have the original object as limits when q tends to 1. There are two types of q-addition, the Nalli-Ward-Al-Salam q-addition (NWA) and the Jackson-Hahn-Cigler q-addition (JHC). The first one is commutative and associative, while the second one is neither.

This is one of the reasons why sometimes more than one q-analogue exists. The two operators above form the basis of the method which unities hypergeometric series and q-hypergeometric series and which gives many formulas of q-calculus a natural form reminding directly of their classical origin. The method is reminiscent of Eduard Heine (1821–1881), who mentioned the case where one parameter in a q-hypergeometric series is $+\infty$. The q-addition is the natural way to extend addition to the q-case, as can be seen when restating addition formulas for q-trigonometric functions.

The history of q-calculus (and q-hypergeometric functions) dates back to the eighteenth century. It can in fact be taken as far back as Leonhard Euler (1707–1783), who first introduced the q in his *Introductio* [190] in the tracks of Newton's infinite series.

The formal power series were introduced by Christoph Gudermann (1798–1852) and Karl Weierstraß (1815–1897). In England, Oliver Heaviside (1850–1925) made yet another contribution to this subject.

In recent years the interest in the subject has exploded. Hardly a week goes by without a new paper on q. This is of course due to the fact that q-analysis has proved itself extremely fruitful in various fields and today has wide-ranging applications in such vital areas as computer science and particle physics, and also acts as an important tool for researchers working with analytic number theory or in theoretical physics.

The book has several aims. One is to give the student of q a basic insight and training in q-calculus or its equivalents, elliptic functions and theta functions. Another is to present the tools and methods that q-analysis requires and recount the history that has shaped the course of the q-discipline. Readers will find here a historical background, which has hitherto not been known to large parts of the mathematics and physics communities, since the treatments and theses containing the early q-history from the 18th and 19th centuries were written in Latin, German and French. An example is the book *Theory of Finite Differences* by Nørlund (1924) [403] which includes Bernoulli and Euler polynomials. The book is written in German and thus not easily accessible to modern English-oriented scientists. Some of these results can therefore be said to be rediscovered and represented as important foundations of the early q-calculus.

Our book is furthermore an attempt to bring the history and the overall insight into q-analysis and calculus up-to-date and hopefully into the future also. Various Schools in q-analysis have sprouted over the last decades, e.g. and foremost the Watson School and the Austrian School, making the subject somewhat confusing for a 'beginner' to study, and also making it difficult to find and define a 'common denominator' and a normal nomenclature for q. '*A comprehensive treatment of q-calculus*' (and later issues) therefore contains treatments and discussions of the very latest results and discoveries in the field of q, and furthermore presents a new and hopefully unifying logarithmic notation and an umbral method for studying q-hypergeometric series.

Acknowledgements

I thank *Cambridge University Press* for its support and for publishing the book [319]. Without this book by Jones and Singerman it would have been very difficult to penetrate so deeply into the field of meromorphic functions. The book [220] and its second edition [221] have also meant a great deal to the author; it enabled him to make the first steps in q-Analysis; without this book and the foreword by Tom Koornwinder, the important Chapter 7 of the present book would probably not have been in its current form.

I am particularly grateful to Neita Lundquist, Patrick Duncan (London) and Rolf Du Rietz for the linguistic processing. Christer Kiselman, Staffan Rodhe, Andreas Ruffing, Moritz Simon, Markus Gabrysch (CERN), Axel Riese (Linz), Leif Karlsson, Gunnar Ingelman, Per Karlsson, Mikael Rågstedt, Jonny Jansson, Ernst Dietrich and Karl-Heinz Fieseler have all contributed with sound advice.

Contents

Chapter 1
Introduction

1.1 A survey of the chapters

Each section in the following chapters has an inbuilt chronology.

The structure of the book is as follows:

In Section 1.2 we will describe the general nature of the different equations in q-calculus together with the character of the proofs. In Section 1.3 we will briefly list the available books on special functions (and q-analysis) and give a survey of relations to complex analysis.

Since the concept of infinity, ∞, is a central theme in q-analysis, we will briefly describe the relationship with nonstandard analysis in Section 1.4. We will make a comparison with physics concepts in Section 1.5, so that the reader is not lost in the many formulas. In Section 1.6 we will summarize the analogies between the q-difference and q-sum operators and the differentiation or integration operator in four tables. In Section 1.7 we will start with the first q-functions in order to facilitate the description of the various Schools. In the following sections we will sketch the connections to other subjects like analytic number theory and combinatorics.

In Chapter 2 we will give a survey of the different Schools in q-analysis, with special emphasis on the Watson and Austrian Schools. In Section 2.3 we talk for the first time about difference calculus and Bernoulli numbers in order to make a preparation for the important fourth chapter. In Section 2.5 we summarize the different attempts at elliptic and theta functions, both of which are intimately related to q-calculus.

We present the history of trigonometry, prosthaphaeresis and logarithms in Section 2.6 as a preparation for the introduction of q-complex numbers [175, 185], which were presented at the Conference in Honor of Allan Peterson, Novacella, Italy, in 2007. The development of calculus is treated in detail in Section 2.6, because we claim that Fermat introduced the precursor of the q-integral long before calculus was invented. We then continue with the historical development of symbols for derivative, formal equalities and finite sums. The reason is that we use different signs for definition, equality, formal equality and symbolic equality. J. Faulhaber's discrete mathematics is presented briefly in Section 2.8.

T. Ernst, *A Comprehensive Treatment of q-Calculus*,
DOI 10.1007/978-3-0348-0431-8_1, © Springer Basel 2012

The Hindenburg combinatoric School, which is presented in Section 2.9, provided a basis for the discovery of the Schweins q-binomial theorem. In Section 2.10 we briefly describe the so-called *Fakultäten*, the forerunner of the Γ function and q-factorial. We then come to the various German Schools of special functions that are described in Sections 2.11 to 2.17.

In Section 2.18 we go through the transition from fluxions to the Leibniz notation in Cambridge, which enabled the umbral calculus and the works of Jackson and Cigler on q-calculus.

In Sections 3.1 and 3.2 we present the duality between analytic number theory, combinatorial identities and q-series, to indicate the historical development of the allied disciplines. Then follows the history, in broad outline, of hypergeometric and special functions (Section 3.7). This is important, since we use a consistent nomenclature for hypergeometric identities in order to identify the different q-analogues. The important Γ function will be introduced in Section 3.7.1 as a preparation for the following subsection.

The concepts of well-poised and balanced series will be introduced in Section 3.7.2. The three Kummer summation formulas play here a special part; we will, e.g. find two similar summation formulas, like Bailey-Daum, accordingly q-analogues to Kummer's first summation formula.

Since the concept of infinity plays a certain part in this book, we discuss it and its relation to the conic sections in the Sections 3.7.7–3.7.9. Then follows the history of hypergeometric series, in Section 3.7.11 and further.

With the help of Carl Gustav Jacobi (1804–1851) [310, p. 145], we will show that the elliptic function sn$u$ (Section 3.8) can be written as a balanced quotient of infinite products.

In Chapter 4 we introduce the q-umbral calculus in the spirit of Rota. Chapter 4 is a mixture of different ideas and shows the unification of finite differences and differential calculus in the shape of q-Appell polynomials. In Section 4.1 we will present the algebraic rules for the two q-additions and the alphabet. This section is almost a prerequisite for the rest of the book. We will continue with Section 4.2, devoted to a q-analogue of Nørlund's finite difference calculus, which culminates in Section 4.3 with the q-Appell polynomials and their special cases, the Bernoulli and Euler polynomials. In Section 4.4 we will give a slight variation of the same theme. In the last Section 4.5 we will briefly treat the semiring of Ward numbers or q-natural numbers.

In Chapter 5 the q-Stirling numbers will be treated, and many beautiful operator formulas will be proved. It turns out that an almost mechanical calculation of the q-analogues of the formulas for q-Stirling numbers from Jordan [320] and the elementary textbooks by Johann Cigler [138] and Schwatt [456] is possible. To this end, various q-difference operators are used. A q-power sum of Carlitz [97, p. 994–995] plays a special role.

In Chapter 6 we are ready to give a thorough treatment of q-calculus. The notation is completely new and follows the publications of the author starting 2000.

After the introduction of the tilde operator, we can finally present q-analogues of formulas for the Pochhammer symbol. Originally the notation

$$\widetilde{a} \pm b = \widetilde{a \pm b} \tag{1.1}$$

for Eq. (6.28) for the tilde operator was used, but after criticism at the 2003 Rørås Conference, the congruence notation was chosen. This has not always been the case, since Alfred Pringsheim (1850–1941) [419, p. 920] used the notation $c \neq -n \pm \frac{2\pi mi}{\log q}$ for the requirement of a parameter in the denominator of a q-hypergeometric function.

The notation in Section 6.1 is sufficient for the majority of proofs related to q-hypergeometric functions. In Section 6.2 we define the q-derivative and present its basic properties.

Section 6.3 presents the q-integral together with its fundamental properties. In Section 6.4 two other tilde operators are given in order to be able to treat a root of unity and a certain geometric series. The second tilde operator is required for the important $\triangle$-operator.

This enables a q-analogue of the formula [425, p. 22, (2)]:

$$(a)_{kn} = k^{nk} \prod_{m=0}^{k-1} \left(\frac{a+m}{k}\right)_n. \tag{1.2}$$

In Section 6.5 we will present the q-binomial coefficients, which are fundamental for the following sections. This is followed by Cigler's equivalent operational method, in Section 6.6. Several Gould and Carlitz q-binomial coefficient identities are given in Section 6.7. We will present five q-exponential functions, limits, inequalities and the corresponding q-trigonometric functions in Section 6.8. Section 6.9 treats the Heine q-hypergeometric function together with the corresponding q-difference equations. In Section 6.10 we will briefly describe some oscillation properties in the q-analysis, which are connected with the q-exponential function.

In Sections 6.11 and 6.12 we continue with direct applications of the q-binomial coefficients, the Jackson-Hahn-Cigler (JHC) and Nalli-Ward-Al-Salam (NWA) q-additions. In each of these sections we will give the corresponding addition formulas for E_q and q-trigonometric functions.

In Section 6.12 we also present the Ward numbers and an equivalent definition of the q-difference operator (with NWA). In Section 6.13 we will discuss Weierstraß elliptic functions and theta functions, and their connection to the Γ_q function. We will conclude Chapter 6 with the parameter augmentation, the inverse of the limits $a \to \pm\infty$, in Section 6.14.

In Chapter 7 we treat more advanced q-calculus, starting from a general definition of q-hypergeometric series. This generalization of q-hypergeometric series is a compromise with Gasper and Rahman's notation, which always can be obtained by substitution from the generalized definition. The method we use is based on the tilde operator, which the author presented at the q-Series Conference in Urbana, 2000. The general tilde operator is used in the definition, but we do not need it until the next chapter. This chapter covers the theory of q-hypergeometric series until Gasper and Rahman [220, p. 39], i.e. to Rogers-Ramanujan identities and to Bailey's (1929) transformation formula for a terminating, balanced, very-well-poised ${}_{10}\phi_9$.

We also discuss Section 3.2 of [220], which is about q-hypergeometric ${}_4\phi_3$ and ${}_3\phi_2$ transformations; this is Section 7.12, on q-analogues of the Thomae transformations.

In Section 7.18.1 we will give three q-analogues of the summation formula [466, p. 56, 2.3.4.7], the proofs are small deviations from the proof of the q-Dixon formula. In Section 7.22, we will give three q-analogues of Euler's integral formula for the Γ function. The funny thing is that the integrands in the first two formulas are equal, but the upper limits of integration are different. Only one of these formulas was known earlier.

In Chapter 8, *Sundry topics*, we will finally use the general tilde operator. We will prove several summation and transformation formulas here, together with material on q-Gamma and theta functions. Some of these summation formulas can be found in Andrews's paper [25]. We will use a numbering of Kummer's summation formulas for easy reference. Most of the summation formulas are given both in q-hypergeometric form and in q-binomial coefficient form for convenience.

We provide a table showing the connection between the different summation formulas, ending with the q-Kummer's formulas.

We will find two quadratic q-hypergeometric transformations in Section 8.2. In Section 8.3 we find two further q-analogues of Kummer's first summation formula and use a q-analogue of Euler's mirror formula to prove a summation formula.

This formula, which will be the basis for a q-gamma function formula with a negative function value in the theory of multiple q-hypergeometric series, is very important. In Section 8.4 we give another proof of the q-Dixon formula by means of yet another q-analogue of Kummer's first summation formula, namely (8.40); this proof requires the complex method [180].

The finite version of the q-Dixon formula (Section 8.5) will be of great interest in coming articles. The 2-balanced, well-poised summation formula in Section 8.6 is also very interesting. In Section 8.7 we present the finite Jackson q-analogue of the Dixon formula. Section 8.8, *Further examples of q-special functions*, will treat modern q-special functions that will seldom be found in this book, but that may be of interest to those who are already familiar with this branch.

In Section 8.10 we find a q-analogue of Truesdell's function and its functional equation. The Bailey transformation for q-series (8.11) is of great theoretical interest. In Section 8.12 we find a q-Taylor formula with Lagrange remainder term. In Section 8.13 bilateral q-hypergeometric series are treated. Finally, in Section 8.14 we will briefly discuss fractional q-integrals. Sections 8.10, 8.13 and 8.14 are not required for the rest of the book.

Chapter 9 introduces three q-orthogonal polynomials and one q-polynomial, which is related to an orthogonal one. In Section 9.1 we will treat the Cigler q-Laguerre polynomials from an operational point of view. The generating function technique by Rainville [425] is used to prove recurrences for q-Laguerre polynomials. Several product operator formulas for these polynomials are deduced. The Al-Salam operator is used several times and is followed by its extension, the q-Laguerre

Rodriguez operator. Several q-analogues of Carlitz's formulas for this Rodriguez operator are found before embarking on the study of q-orthogonality. The q-Jacobi polynomials (Section 9.2) are treated in an analogous way, with the exception that here many equalities are purely formal, due to the slow convergence of the q-shifted factorial. We will prove orthogonality for both the above-mentioned polynomials using q-integration by parts. In Section 9.3.1 we will find q-orthogonality and q-difference equations for the q-Legendre polynomials. Finally, we briefly treat Carlitz-Al-Salam polynomials in Section 9.3.2.

In Chapter 10 we come to q-functions of many variables. Section 10.1 starts with the vector notation for the most important functions and q-Taylor formulas for power series and functions of inverse q-shifted factorials. This section can be read independently.

In Section 10.2 we will give a historical introduction to the rest of this long and interesting chapter and to the next chapter as well. We will also define q-Appell functions together with the normal form. Then we come to the two definitions of q-Kampé de Fériet functions due to Karlsson and Srivastava. In Section 10.3 the q-analogues of the Appell and Kampé de Fériet's transformation formulas require the Watson q-shifted factorial in the definition.

We will continue with Carlitz's Saalschützian formulas, Andrews's formal transformations and Carlson's transformations.

In Section 10.4 we will show the Jacksonian formula for the q-integral of the q-Appell function Φ_1 is equivalent to the first of Andrews's formal transformations.

In Section 10.5 we will give the first examples of multiple reduction formulas with general terms. These are used to find q-analogues of a reduction formula for the Humbert function, a relation for the product of two Bessel functions, summation formulas for the first Appell function and Lauricella function $\Phi_{\mathrm{D}}^{(n)}$. A relation for Γ_q functions with negative integer argument from Chapter 8 will be used in one of the proofs; this was really hard work!

Many summation formulas appear as doublets, a legacy of two q-Vandermonde summation formulas.

In Section 10.6 follow reduction formulas, transformation formulas and generating functions of Srivastava. Some of these formulas have general terms. In Section 10.6.3 (double-sum identities of Srivastava and Jain [481]), a second q-analogue of the Buschman and Srivastava formula [91, 3.10, p. 440] is proved. The four q-analogues of [91, 2.7–2.9, 3.10] can be found in the article [181].

In Section 10.7 we will find two reduction formulas with general terms, by Karlsson and Srivastava, which are proved by means of the Bailey-Daum theorem. The q-analogues of Karlsson's formulas in Section 10.8 follow the same pattern. A multiple reduction formula with general terms has as special case a reduction formula for a $\Phi_{\mathrm{D}}^{(3)}$ function. In Section 10.9 we will introduce the inverse pair of symbolic operators $\nabla_q(h)$ and $\Delta_q(h)$ due to Jackson.

We will first derive slightly improved versions of Jackson's expansions for q-Appell functions. Each of these expansions is equivalent to a combinatorial identity, which resembles a well-known q-summation formula. The q-Burchnall-

Chaundy operators will then be used to prove q-analogues of the Verma expansions, which are generalizations of the previous ones.

Another application of the q-Burchnall-Chaundy operators is an expansion formula for a certain q-Kampé de Fériet function with the help of Jackson's theorem.

In Section 10.11 we will use the q-derivative technique and summation formulas for $\Phi_{\mathrm{D}}^{(n)}$ to find two multiple extensions of the Rothe-von Grüson-Gauß formula. In Sections 10.11 and 10.12 we will conclude with an expansion formula in the spirit of Chaundy and formulas of the Burchnall-Chaundy and Jackson type.

In Chapter 11, we will begin with a Section 11.1 on linear partial q-difference equations as a preparation for the partial q-difference equations for q-functions of many variables at the end of this chapter.

In Section 11.3 we will give q-difference equations for the four q-Appell and q-Lauricella functions.

In Chapter 12 we will give a systematic summary of the applications of q-analysis in physics.

1.2 What is q-calculus?

The development of q-analysis started in the 1740s, when Euler initiated the theory of partitions, also called additive analytic number theory. Euler always wrote in Latin and his collected works were published only at the beginning of the 1800s, under the legendary Jacobi. In 1829 Jacobi presented his triple product identity (sometimes called the Gauß-Jacobi triple product identity), and his θ and elliptic functions, which in principle are equivalent to q-analysis.

The progress of q-calculus continued under C. F. Gauß (1777–1855), who in 1812 invented the hypergeometric series and their contiguity relations. Gauß would later invent the q-binomial coefficients and prove an identity for them, which forms the basis for q-analysis. This is the Rothe-von Grüson theorem, which in turn is a special case of the Schweins q-binomial theorem.

In q-analysis we usually consider formal power series (or inverse factorials). The Γ_q function also plays a prominent part. We can make a comparison with the historical development of these subjects, see Section 2.17.

We now proceed to characterize the proofs in q-calculus.

In q-analysis, we make formal proofs as follows. The validity of the obtained formula depends on the different parts of the proof. A typical proof in q-analysis reads:

Theorem 1.2.1

$$A = B \quad or \quad A \cong B. \tag{1.3}$$

Proof

$$A \overset{\text{by } (\star)}{=} \cdots \overset{\text{by } (\star\star)}{=} B. \tag{1.4}$$

□

Here, $\cong$ means formal equality. Now we ask: What are A and B? First, e.g. A can be an *operator expression* and B may be a sum of operators. Then the formula is generally valid for so-called q-holomorphic functions. For now it suffices to think of a q-holomorphic function as being a formal power series. Secondly, we have a *transformation formula* and thirdly, a *summation formula*. We now treat the last two cases. The proof goes on as follows: We begin with the left-hand side and show in n steps that it is equal to the right-hand side. When we are finished, we ask: When is the obtained formula valid? Answer: A finite summation formula is always valid.

For infinite sums or transformations, the result depends on the character of the n stopovers, i.e. $\star, \ldots, \star\star$. If we only use basic formulas such as (6.14)–(6.21), (6.31)–(6.34) or summation formulas like q-Vandermonde, q-Saalschütz, etc. as stopovers, then $A = B$ holds everywhere. If, however, we use formulas which do not always hold as intermediate stopovers, like the q-binomial theorem or the Jackson q-analogue of the Euler-Pfaff-Kummer transformation (7.62), it is likely that $A = B$ is only valid in a specific region. Examples of such formulas, which do not always hold for $q = 1$, are (10.90)–(10.94), (9.194). The Andrews formula (10.90) is equivalent to the q-analogue (10.104) of the Picard integral formula for the first q-Appell function.

In the spirit of Heine, [270, p. 306] we make the convention to use q^x as function argument for q-hypergeometric formulas, which have no hypergeometric counterpart.

Nor do the definitions of the q-integral (1.53), (6.55) seem to be valid for $q = 1$; a limit process shows, however, that the q-integral of a power function gives the same value as for the case $q = 1$. It is also possible that the corresponding formulas for $q = 1$ are very simple and well known, like in (9.192). Several examples will be given of q-formulas that do not correspond to known formulas for $q = 1$, e.g. the bilateral summation formula (8.103).

We will often speak of balanced series and balanced Γ_q quotients. These notions are fundamental, just as similar notions in mechanics. It turns out that the Jacobi elliptic function has a similar character and can be written as a balanced Γ_q function. When we write $\lim_{c \to \pm\infty} \Gamma_q(c)$ we mean a path in the complex plane that avoids all poles of $\Gamma_q(c)$. Usually c appears with the sign -1 in the Γ_q functions, this means that $\lim_{c \to \infty}$ could be problematic.

Finite hypergeometric summation formulas can often be written in two different forms, depending on whether N is even or odd. The expression for N odd is sometimes equal to 0. These two formulas correspond to one Γ quotient.

When we have a q-hypergeometric summation formula equal to a balanced Γ_q quotient, we can often obtain multiple q-analogues through the limit processes $a \to \pm\infty$, where a is a variable in the Γ_q quotient. For $q = 1$ this usually corresponds to a single formula.

In Chapter 4 we will treat the q-umbral calculus, which is very important for the function argument in q-hypergeometric series. In these formulas, we almost always consider formal power series, with variables called letters. It will be tacitly assumed that the q-umbral calculus will only be used for power functions of letters. Outside the alphabet, the notation x always means x^1, when x is a letter. A product like

$P_{n,q}(x,a)$ always alludes to the ordinary meaning. This is a duality like the wave particle duality.

A second duality is represented by the formulas that arise when we make an inversion of the basis, e.g. for the reduction formula (10.143). The duality between the NWA and JHC q-addition is also produced by an inversion of the basis. This construction is very natural, as is shown by Eqs. (4.174) and (4.232). Similarly, when we have a function $F(x,q)$, we can apparently, for shortness of notation, also write $F(x)$.

Another phenomenon is the slightly larger domain of convergence, e.g. for (7.62), we will often come back to this later. We also want to tell something about the various signs in q-analysis and their history. The sign for NWA q-addition $\oplus$ (excluding $\ldots_q$) occurred for the first time as *tet* in the Phoenician writing around 1000 BC. It was then called Θ in the Greek alphabet. A similar sign was also used in Paleo-Hebrew or Old Hebrew around 842 BC. This symbol $\oplus_q$ and its relative $\boxplus_q$ play an important role as operators in the infinite alphabet of the q-umbral calculus.

We use different signs for equality: $=$ is the usual equality. $\equiv$ means a definition. In the sense of Milne-Thomson [385] and Rainville [425], we write $\doteq$ to indicate the symbolic (umbral) nature.

The equivalence between letters in the alphabet (Section 4.1) is denoted by $\sim$.

The equality between equivalence classes in the alphabet is again denoted by $=$. Equalities between exponents (Sections 6.1 and 3.7.2) are denoted by $\equiv \mathrm{mod}(\frac{2\pi i}{\log q})$. Empty products are defined as 1 and empty sums are defined as 0. In equations RHS and LHS denote the right- and left-hand sides of the equation.

The hypergeometric functions are the basis of special functions, which are very important in theoretical physics. It turns out, however, that not all theoretical physicists are familiar with the concept of a hypergeometric function. We address this issue in Section 3.7.

In 1846 Heine introduced the q-hypergeometric function with notation ${}_2\phi_1(a,b;c;q,z)$. We denote this series by ${}_2\phi_1(a,b;c|q;z)$, to make it clear that the parameters to the left of $|$ are exponents. To the right of $|$ is the basis q and the function value z in the so-called alphabet. A q-analogue of a hypergeometric formula may in some cases contain an extra parameter, which will always be found to the right of the symbol $||$. During the OPSFA Conference in Copenhagen 2003, such q-analogues with an extra index in ϕ were presented. It turns out, however, that it is better to use k as default value for this parameter, as in Eq. (7.1) in Section 7.1. This is exactly like the default values i, j for the matrix elements. For double q-hypergeometric series, we use m, n as default values.

When we deal with q-special functions, the q-additions in the function argument (Section 4.1) will somehow get into the background, partly because of the convergence problem. We are not able to prove that a particular series converges with a particular q-addition. The special functions q-Appell, etc., play the main roles. But the q-umbral calculus will be used for other purposes.

Since the series in this book usually converge absolutely, they may be rearranged at will. We sometimes use some elementary series manipulations, a short list of them will be given in the next subsection.

1.2.1 Elementary series manipulations

The following double series transformations that we will occasionally use, are easy to prove [425, p. 56]:

$$\sum_{n=0}^{\infty}\sum_{k=0}^{\infty} A(k,n) = \sum_{n=0}^{\infty}\sum_{k=0}^{n} A(k,n-k), \tag{1.5}$$

[425, p. 57]

$$\sum_{n=0}^{\infty}\sum_{k=0}^{n} B(k,n) = \sum_{n=0}^{\infty}\sum_{k=0}^{\infty} B(k,n+k), \tag{1.6}$$

Carlitz (1960)

$$\sum_{m=0}^{\infty}\sum_{r=0}^{\min(m,n)} f(m,r,n) = \sum_{m'=0}^{\infty}\sum_{r=0}^{n} f\big(m'+r,r,n\big). \tag{1.7}$$

1.3 Update on q-calculus

The q-calculus is a part of the theory special functions. We therefore begin with this subject, with the beautiful name 33 in the Mathematical Sciences Classification system. A three-volume work serves as a tool for a variety of professionals in the field of special functions. This highly successful project took its name from the Caltech mathematician Harry Bateman (1882–1946), who on his death left behind a vast amount of information on special functions. Arthur Erdelyi (1908–1977) took on the task of editing his work. The result was *Higher Transcendental Functions* in three volumes [162] and two volumes covering *Tables of Integral Transforms*. The 1990'th saw the dawn of the new Askey-Bateman project, when Richard Askey found that "The Bateman project badly needs to be redone" on account of the multitude of discoveries in the decades since the Bateman project handbooks were published. Now it was also time for the q-functions; the present book serves just this purpose. Recently, Amédée Debiard and Bernard Gaveau have studied the partial differential equations of two-variable hypergeometric series and have corrected the differential equations for H_4, H_5 and H_7 in Erdelyi et al. [162] in the process. K. S. Kölbig (CERN) has published corrections of [162], [420] and other books about special functions.

Recently, there has been a lack of modern books on special functions. Although there is the book *Integrals and Series*, Vol. 3 by Prudnikov, Brychkov and Marichev [420], which contains a large amount of summation formulas for hypergeometric functions etc. But this book contains no proofs and also no references to most of the hypergeometric formulas.

Incidentally, Marichev now (2000) works for Mathematica in Champaign, Illinois. Last year the book *Handbook of Special Functions* by Brychkov [83] was published, which is obviously the best book on the market today. This book contains a long list of hypergeometric equations and formulas exemplified for special functions. It contains for the first time many operator formulas and mentions the name Toscano from the Italian School, which was previously forgotten. At the end a long list of Meijer G-function formulas appears, which however do not contain any $\triangle(l;\lambda)$ operators, which usually were customary in the sixties. This book contains no q-formulas and no proofs. Nor do many formulas have any references.

The book *Representations of Lie Groups and Special Functions*, Vol. 3, by Vilenkin and Klimyk [523] contains many formulas for special functions and shows the relationship to Lie groups.

In *Harmonic Analysis on Symmetric Spaces and Applications*, Vol. I (Springer Verlag, Heidelberg and New York, 1985) by Audrey Terras there are some observations about specific functions from the viewpoint of homogeneous spaces, a very interesting and extensive area that has been explored particularly by Helgasson. His books, however, have a different character.

The book *A Treatise on Generating Functions* by Srivastava and Manocha [484] covers specific functions from the standpoint of generating functions. There are no proofs, but many equations can be found there, as well a considerations about Lie algebras and multiple hypergeometric series (MHS).

The book *Multiple Gaussian Hypergeometric Series* by Srivastava and Karlsson [482] treats MHS, however, without proofs; the convergence conditions play an important role.

The book [275] by Henrici thoroughly treats the 24 Kummer solutions and quadratic transformations.

1.3.1 Current textbooks on this subject

We will briefly describe five textbooks on this theme that illustrate the present state. Slater's book [466] treats parts of Sections 3.7, 7.1 to 7.21 and part of Chapter 10.

The book [220] by Gasper and Rahman treats parts of Sections 3.7, 7.1 through 7.21 and parts of Chapter 8 to Section 9.2, partially with some other definitions.

For the French readers, there is the excellent textbook [41] *Fonctions Hypergéométriques et Hypersphériques* by Appell and Kampé de Fériet on the subject MHS.

Harold Exton (1928–2001) [193] has written a good textbook on MHS.

In his book *q-Hypergeometric Functions and Applications* [194], Exton may have tried to introduce the umbral notation. This book was read by many physicists. We are also going to make many comparisons with physics. Clearly, a refined textbook in the field of q-analysis is needed, since the earlier attempts were only partly successful.

1.3.2 Comparison with complex analysis

In the Russian counterpart of Mathematical Reviews, *Referativnyi Zhurnal Matematika*, and also in Zentralblatt from the sixties, special functions (Classification 33) are listed under Analysis. The way mathematical logic is the grammar of philosophy, special functions are the endgame of analysis. The special functions also play a large rôle in the Swedish textbook by V. J. Malmquist, V. Stenström and S. Danielson: *Matematisk Analys 2: Analytiska Funktioner och Lineära Differentialekvationer i det Komplexa Området*, Stockholm, 1952.

As we shall see, q-analysis has also much in common with algebra.

1.4 Comparison with nonstandard analysis

Nonstandard Analysis, a part of mathematical logic, was introduced in 1961 by Abraham Robinson (1918–1974). The nonstandard mathematics has been given a big boost in recent years through the many books that have appeared in this area.

Although q-analysis has more applications, for example, in physics, there is yet no book in German on this subject. We will compare these two types of analysis to show that q-analysis is at least as beautiful as nonstandard analysis.

The purpose of nonstandard analysis is to complete the real numbers with some missing elements. These missing elements are the infinitesimal elements and ∞. This defines a new field ${}^\star\mathbb{R}$,

$$
{}^\star\mathbb{R} \equiv \{\alpha\}, \tag{1.8}
$$

where α denotes all real numbers. There is an equivalence relation $\approx$ on ${}^\star\mathbb{R}$ with the following properties:

Theorem 1.4.1 [351, p. 23, 3.9(iv)] *Let $\alpha, \beta \in {}^\star\mathbb{R}$. Then α is infinitely close to β, $\alpha \approx \beta$, if $\alpha - \beta$ is infinitesimal.*

The infinitesimal elements correspond to our θ $(= 0)$, the α, β correspond to our letters and $\approx$ corresponds to our $\sim$ in Chapter 4.

This subject is about an analysis with a finite difference operator and an integral with finite sums. This is just like in works of Leibniz (1646–1716), Laplace (1749–1827), etc. There is a theorem which says that continuous functions can be approximated by polynomials.

Let $f(x)$ be a continuous function and $q(x)$ a polynomial. We can write this as

$$
f(x) \approx q(x). \tag{1.9}
$$

In nonstandard analysis $\approx$ is an equivalence relation.

The comparison with q-calculus is clear. Here we most likely consider formal power series. Two umbrae α and β are equivalent if $v(f, \alpha) = v(f, \beta)$. Here v is a linear functional. In q-Analysis there is an explicit representation for the q-integral and the inverse relation between the q-derivative and the q-integral. In nonstandard analysis, this is not highlighted. On the contrary, nonstandard analysis highlights on "Real Analysis", topology and measure theory.

1.5 Comparison with the units of physics

The theme of this book is extensive, but the distance to applications in sciences that depend on mathematics is often small. This proximity gives the reader the opportunity to control his knowledge in an interesting and visual way. The study of physics usually begins with a course in mechanics and in this section we will explain some general correlations, which are particularly important in this area.

The presentation is based on the use of SI units, which today are widely used.

Those who understand the whole SI system, should also be able to grasp q-analysis. We will arrange the symbolic comparisons in tabular form:

Quantity	SI-unit	q-terminology	Section
	SI	alphabet	Section 4.1
Length	1 m	Γ_q	Section 1.7
Time	1 s	ϵ	Section 1.7
Mass	1 kg	$\langle\alpha;q\rangle_n$	Section 1.7
Force	1 N	(6.14)–(6.21), (6.31)–(6.42)	Section 6.1
Speed	1 m/s	D_q	Sections 1.7, 6.2
Work	1 Nm	$\mathrm{D}_{\oplus_q}$	Section 4.1
Acceleration	1 m/s^2	$\langle\widetilde{\overset{m}{T}}a;q\rangle_n$, ${}_k\widetilde{\langle a;q\rangle_n}$	Section 6.4
Flux density	1 Gauß	hypergeometric series	Section 3.7
Flux density	1 Tesla	${}_2\phi_1(a,b;c\mid q;x)$	Section 6.9
Current	1 A	q-Laguerre polynomial	Section 9.1
Temperature	1 K	q-Legendre polynomial	Section 9.3.1
Chemical amount	1 mol	q-Jacobi polynomial	Section 9.2
Luminous intensity	1 cd	$\mathrm{E}_q(x)$	Sections 1.7, 6.8

The equation

$$m = \int dm_P \tag{1.10}$$

corresponds to the fact that for each equation, the index sum of $\langle\alpha;q\rangle_n$ must be equal.

The vector product can be compared to the tilde operator. This fits well, since these two operators have a fundamental significance for the respective disciplines.

$a\times b$	$\widetilde{\langle a;q\rangle_n}$	Section 6.1

The scalar triple product

$$(\vec{u} \times \vec{v}) \cdot \vec{w} = \begin{vmatrix} u_1 & v_1 & w_1 \\ u_2 & v_2 & w_2 \\ u_3 & v_3 & w_3 \end{vmatrix} \tag{1.11}$$

corresponds to (6.28).

The triple vector cross product

$$(\vec{u} \times \vec{v}) \times \vec{w} = \vec{v}(\vec{u} \cdot \vec{w}) - \vec{u}(\vec{v} \cdot \vec{w}) \tag{1.12}$$

corresponds to (6.29).

The Lagrange identity

$$(\vec{t} \times \vec{u}) \cdot (\vec{v} \times \vec{w}) = (\vec{t} \cdot \vec{v})(\vec{u} \cdot \vec{w}) - (\vec{t} \cdot \vec{w})(\vec{u} \cdot \vec{v}) \tag{1.13}$$

corresponds to (6.30).

The formula

$$\dot{\vec{u}} = \frac{(\vec{u} \times \dot{\vec{u}})}{\vec{u} \cdot \vec{u}} \times \vec{u} + \frac{\vec{u} \cdot \dot{\vec{u}}}{\vec{u} \cdot \vec{u}} \vec{u} \tag{1.14}$$

corresponds to (6.31).

The formula

$$\dot{\vec{e}}_i = \vec{\omega} \times \vec{e}_i \tag{1.15}$$

corresponds to (6.32).

The formula

$$\vec{M}_0 = \vec{M}_c + \vec{r}_c \times \vec{F} \tag{1.16}$$

gives the torque in a rigid body with respect to any point. This formula corresponds to (6.38) and (6.36).

The formula

$$\vec{L}_0 = \vec{L}_c + \vec{r}_c \times \vec{p} \tag{1.17}$$

gives the impulse with respect to any point in a rigid body. This formula corresponds to (6.42).

In the same way, the scalar product and the q-additions are connected to each other.

Scalar product	q-additions	Sections 6.11–6.12

The formula

$$\vec{a} \cdot \vec{b} = \vec{b} \cdot \vec{a} \tag{1.18}$$

corresponds to (4.23).

The formula

$$s(\vec{a} \cdot \vec{b}) = (s\vec{a}) \cdot \vec{b} \tag{1.19}$$

corresponds to (4.22).

The formula

$$\ddot{x} = \frac{d}{dx}\left(\frac{1}{2}\dot{x}^2\right) \tag{1.20}$$

corresponds to (6.71).

The formula

$$\ddot{\vec{r}} = \vec{e}_t\ddot{s} + \vec{e}_n\kappa\dot{s}^2 \tag{1.21}$$

gives the acceleration in the natural reference system. This formula corresponds to (6.72).

The formula for the curvature

$$|\kappa| = \left(\dot{\vec{r}}\cdot\dot{\vec{r}}\right)^{-\frac{3}{2}}\left(\left(\dot{\vec{r}}\cdot\dot{\vec{r}}\right)\left(\ddot{\vec{r}}\cdot\ddot{\vec{r}}\right) - \left(\dot{\vec{r}}\cdot\ddot{\vec{r}}\right)^2\right)^{\frac{1}{2}} \tag{1.22}$$

corresponds to (6.73) and (6.74).

The formula for the torsion

$$\tau = \left(\left|\dot{\vec{r}}\times\ddot{\vec{r}}\right|\right)^{-2}\left(\dot{\vec{r}}\times\ddot{\vec{r}}\right)\cdot\vec{r}^{(3)} \tag{1.23}$$

corresponds to (6.80). The differential equations correspond to the q-difference equations.

$A(D)=0$	$A(\mathrm{D}_q)=0$	Chapter 11

The Resonance corresponds to the Heine infinity.

Resonance	∞_{H}	Section 7.1

In physics there is a dimensional analysis saying that both sides in an equation must have the same dimension. This corresponds to the fact that the index sum for $\langle\alpha;q\rangle_n$ and for $\langle\widetilde{\overset{m}{\top} a};q\rangle_n$ must be the same in each factorial equation.

The oscillation corresponds to the complementary q-exponential function for $x<0$. As an example consider an RC-circuit.

Oscillation	$\mathrm{E}_{\frac{1}{q}}(x)$	Section 6.10

In mechanics, one begins with the statics to acquire the force equations. The equilibrium equations correspond to the balanced q-hypergeometric series and the balanced Γ_q functions. The equations for moment equilibrium correspond to the well-poised q-hypergeometric series.

When we arrive at the balanced Γ_q functions, it is necessary to pass on to the complex plane [180], because of the root of unity that appears in the exponent. This is equivalent to the generalized tilde operator. The $j\omega$-method thus corresponds to the balanced Γ_q functions. There is also the concept balanced hypergeometric series of two and three variables, which was studied in 1889 by J. Horn. Horn [286, p. 565] discovered that the sum of the Pochhammer symbols in the numerator and the denominator (with sign) must be equal to ± 1 for convergence.

Force equilibrium	balanced series	3.7.1, 7.2, 10.2
Moment equilibrium	well-poised series	3.7.1, 7.2
$j\omega$-method	balanced Γ_q functions	7.2

There are many finite and infinite (q-)summation formulas. The atomic number corresponds to the hypergeometric formula. The isotopes correspond to the various q-hypergeometric summation formulas. The neutrons then correspond to the factors $((-1)^k q^{\binom{k}{2}})^{\pm 1}$. An example is the formula (3.52) (atomic number), which has three q-summation formulas (isotopes), see Section 7.18. The q-hypergeometric summation formulas without apparent hypergeometric counterparts such as (10.90)–(10.94) and (9.194), correspond to the ions. The strange evaluation formulas of Gessel and Stanton correspond to the transition elements. The q-binomial coefficient formulas in Section 6.7 correspond to the electrons. In the same way the molecules correspond to the multiple q-hypergeometric summation formulas; symmetry plays here an important role, as is emphasized in Chapter 10.

The dynamics is treated in Sections 8.2, 10.3, 10.4, where there are many non-balanced formulas. Examples of dynamic formulas are (10.92) and (10.94). Thus, dynamics corresponds to the formal equalities.

As the most beautiful equations in mechanics are connected to the torque (and the angular momentum), the theta functions and the elliptic functions should be regarded as the most beautiful equations in calculus.

torque	theta functions, elliptic functions	Sections 3.8, 6.11

Ohm's law corresponds to the q-Laguerre Rodriguez formula (9.111). Kirchhoff's law corresponds to the q-orthogonality (9.120).

The equation

$$1K = 1C + 273.16 \tag{1.24}$$

corresponds to the q-orthogonality relation (9.185).

We now come to the unit 1 mol. We use the following symbols:

p	Pressure
V	Volume
n	Chemical amount
R	Universal gas constant
T	Temperature in Kelvin
N_A	Avogadro constant
F	Faraday constant
e	Elementary charge

The law

$$F = eN_A \tag{1.25}$$

corresponds to the Rodriguez formula (9.125). The law

$$pV = nRT \tag{1.26}$$

corresponds to the q-orthogonality relation (9.177).

We have now arrived to the last SI unit, 1 Candela. The q-exponential function is the light of q-analysis, which allows a large amount of operator formulas, see Section 9.1. The q-Appell polynomials in Chapter 4 as well as the pseudo-q-Appell polynomials have q-exponential functions in the generating function. The matrix pseudo-groups [179], which are q-analogues of the most famous Lie groups, are also associated with the q-exponential function. The q-Pascal matrix is a solution to the corresponding q-difference equation.

The q-Cauchy matrix is constructed by means of the corresponding matrix q-exponential function together with the NWA-q-additions. For the elliptic functions the corresponding elements are the Jacobi theta functions.

We will use the following symbols:

I	Luminous intensity (Candela)
Φ	Flux
Ω	Solid angle

The law

$$I = \frac{d\Phi}{d\Omega} \tag{1.27}$$

corresponds to the fundamental formula (6.233). The Gaussian lens formula (f denotes the focal length)

$$\frac{1}{f} = \frac{1}{a} + \frac{1}{b} \tag{1.28}$$

corresponds to (6.152).

The stress tensor (Matrices) corresponds to the generalized q-hypergeometric series.

Matrices	${}_p\phi_r((a);(b)\vert q;x)$	Section 7.1

The symmetries correspond to the q-functions of many variables.

$SO(3)$	q-Appell functions	Chapter 10
$SU(2)$	q-Lauricella functions	Chapter 10

The maximal torus in $SO(3)$ corresponds to the summation formula (10.126). The maximal torus in $SU(2)$ corresponds to the summation formula (10.268). In the area of summation and reduction formulas of q-functions of many variables, it is often possible to identify the origin of a formula. We therefore have a father-daughter relationship, or, as in nuclear physics, a daughter nucleus.

The summation formula (10.131) and the reduction formula (10.176) are daughter nuclei of the Bailey-Daum summation formula (7.80). The reduction formula (10.128) is a daughter nucleus of the q-Vandermonde formula.

The summation formula (10.273) is a daughter nucleus of the q-Gaussian summation formula.

1.6 Analogies between q-analysis and analysis

For the convenience of the reader we make tables with the relevant concepts exemplified for the relevant sections. In each table the continuous terms are always on the right.

The relevant sections are 4.2–4.3, 5.2, 5.3 and 1.6, 4.1 and Chapter 6. Sections 5.2 and 5.3 are somewhat related, because they use the same q-Stirling numbers. There is a certain similarity between Section 4.2 and Section 4.1, Chapter 6, the difference is, for example, in the definition of the q-derivative, which is the same for formal power series. Some formulas in the following tables stand in two rows.

Sections 4.2, 4.3	
$n \to (\overline{n}_q)$	$x \to f(x)$
$\mathrm{D}_{\oplus_q} f(x) \equiv \lim_{\delta x \to \theta} \frac{f(x \oplus_q \delta x) - f(x)}{(\delta x)^1}$	$\mathrm{D} f(x) \equiv \lim_{h \to 0} \frac{f(x+h)-f(x)}{h}$
$\triangle_{\mathrm{NWA},q}$; $\triangle_{\mathrm{JHC},q}$	$\mathrm{D} f(x) \equiv \lim_{h \to 0} \frac{f(x+h)-f(x)}{h}$
$\mathrm{E}(\oplus_q) = \mathrm{E}_q(\mathrm{D}_q)$; $\mathrm{E}(\boxplus_q) = \mathrm{E}_{\frac{1}{q}}(\mathrm{D}_q)$	$\mathrm{E} = e^{\mathrm{D}}$
$\int_0^{x \oplus_q 1} f(t)\, d_q(t)$	$\int_0^{x+1} f(t)dt$
$\sum_{k=0}^{n-1} f(\overline{k}_q) \equiv \sum_0^n f(\overline{x}_q)\delta_q(x)$	$\int_0^x f(t)dt$
$\sum_{k=0}^{n-1} \triangle_{\mathrm{NWA},q} f(\overline{k}_q) = f(\overline{x}_q)\vert_0^n$	$\int_0^x F'(t)dt = F(x) - F(0)$
$(n)_k = n(n-1)\cdots(n-k+1)$	x^k
$\binom{n}{k}$	$\frac{x^k}{k!}$
$f(\overline{n}_q) = \sum_{k=0}^n \binom{n}{k} \triangle_{\mathrm{NWA},q}^k f(0)$	$f(x) = \sum_{k=0}^\infty \frac{f^{(k)}(0)}{k!} x^k$
$\triangle_{\mathrm{NWA},q}^n (fg)$ $= \sum_{i=0}^n \binom{n}{i} \triangle_{\mathrm{NWA},q}^i f(\triangle_{\mathrm{NWA},q}^{n-i} \mathrm{E}(\oplus_q)^i) g$	$\mathrm{D}^n(fg)$ $= \sum_{k=0}^n \binom{n}{k} \mathrm{D}^k f \mathrm{D}^{n-k} g$
$\triangle_{\mathrm{NWA},q} \frac{f(x)}{g(x)} = \frac{g(x)\triangle_{\mathrm{NWA},q} f(x) - f(x)\triangle_{\mathrm{NWA},q} g(x)}{g(x) g(x \oplus_q 1)}$	$\mathrm{D}(\frac{f}{g}) = \frac{f'g - g'f}{g^2}$
$\sum_{k=0}^{n-1} f(\overline{k}_q)\triangle_{\mathrm{NWA},q} g(\overline{k}_q) = [f(\overline{x}_q) g(\overline{x}_q)]_0^n$ $- \sum_{k=0}^{n-1} \mathrm{E}(\oplus_q) g(\overline{k}_q) \triangle_{\mathrm{NWA},q} f(\overline{k}_q)$	$\int_a^b fg'dx$ $= [fg]_a^b - \int_a^b f'gdx$

Section 5.2	
$n \to \{n\}_q$	$x \to f(x)$
$\triangle_{\mathrm{H},q} f(x) \equiv \frac{f(qx+1) - f(x)}{1+(q-1)x}$	$\mathrm{D} f(x) \equiv \lim_{h \to 0} \frac{f(x+h)-f(x)}{h}$
$(x)_{k,q} \equiv \prod_{m=0}^{k-1} (x - \{m\}_q)$; $(\{n\}_q)_{k,q}$	x^k
$\binom{n}{k}_q$	$\frac{x^k}{k!}$
$\triangle_{\mathrm{H},q} (x)_{k,q} = \{k\}_q (x)_{k-1,q}$	$\mathrm{D}x^k = kx^{k-1}$
$f(x) = \sum_{k=0}^\infty \frac{(\triangle_{\mathrm{H},q}^k f)(0)}{\{k\}_q!} (x)_{k,q}$	$f(x) = \sum_{k=0}^\infty \frac{f^{(k)}(0)}{k!} x^k$
$\triangle_{\mathrm{H},q}^n (fg) = q^{-\binom{n}{2}}$ $\sum_{i=0}^n q^{\binom{i}{2} + \binom{n-i}{2}} \binom{n}{i}_q \triangle_{\mathrm{H},q}^i f(\triangle_{H,q}^{n-i} \mathrm{E}_{\mathrm{C},q}^i) g$	$\mathrm{D}^n(fg)$ $= \sum_{k=0}^n \binom{n}{k} \mathrm{D}^k f \mathrm{D}^{n-k} g$
$\sum_{i=0}^{n-1} (\{i\}_q)_{j,q} q^i = \frac{(\{n\}_q)_{j+1,q}}{\{j+1\}_q}$, $j < n$.	$\sum_{i=0}^{n-1} (i)_j = \frac{(n)_{j+1}}{j+1}$, $j < n$.

Section 5.3	
$x \to f(q^x)$	$x \to f(x)$
$\triangle_{\mathrm{CG},q} \equiv (\mathrm{E} \boxminus_q \mathrm{I})$	$\mathrm{D}f(x) \equiv \lim_{h\to 0} \frac{f(x+h)-f(x)}{h}$
$\sum_{k=0}^{n-1} f(k) \equiv \sum_0^n f(x)\delta_q(x)$	$\int_0^x f(t)dt$
$\sum_{k=0}^{n-1} \triangle_{\mathrm{CG},q} f(k) = f(n) - f(0)$	$\int_0^x F'(t)dt = F(x) - F(0)$
$\{n-k+1\}_{k,q}$	x^k
$\binom{n}{k}_q$	$\frac{x^k}{k!}$
$\triangle^m_{\mathrm{CG},q} \binom{x}{n}_q = \binom{x}{n-m}_q q^{m(x+m-n)},\ m \le n$	$\mathrm{D}^m x^n = \frac{n!}{(n-m)!} x^{n-m},\ m \le n$
$f(x) = \sum_{k=0}^{\infty} \frac{\triangle^k_{\mathrm{CG},q} f(0)}{\{k\}_q!} \{x-k+1\}_{k,q}$	$f(x) = \sum_{k=0}^{\infty} \frac{f^{(k)}(0)}{k!} x^k$
$\triangle^n_{\mathrm{CG},q}(fg) = \sum_{i=0}^n \binom{n}{i}_q \triangle^i_{\mathrm{CG},q} f \triangle^{n-i}_{\mathrm{CG},q} \mathrm{E}^i g$	$\mathrm{D}^n(fg) = \sum_{k=0}^n \binom{n}{k} \mathrm{D}^k f \mathrm{D}^{n-k} g$

Sections 1.6, 4.1, Chapter 6	
$x \to f(x)$	$x \to f(x)$
$(\mathrm{D}_q f)(x) \equiv \frac{\varphi(x)-\varphi(qx)}{(1-q)x}$	$\mathrm{D}f(x) \equiv \lim_{h\to 0} \frac{f(x+h)-f(x)}{h}$
$\int_0^x f(t,q)\, d_q(t) \equiv x(1-q) \sum_{n=0}^{\infty} f(xq^n, q) q^n$	$\int_0^x f(t)dt$
$\int_0^x \mathrm{D}_q F(t,q)\, d_q(t) = F(x) - F(0)$	$\int_0^x F'(t)dt = F(x) - F(0)$
x^k	x^k
$\frac{x^k}{\{k\}_q!}$	$\frac{x^k}{k!}$
$\mathrm{D}_q x^k = \{k\}_q x^{k-1}$	$\mathrm{D}x^k = kx^{k-1}$
$\int_0^x t^k\, d_q(t) = \frac{x^{k+1}}{\{k+1\}_q}$	$\int_0^x t^k dt = \frac{x^{k+1}}{k+1}$
$f(x \oplus_q y) = \sum_{k=0}^{\infty} \frac{y^k}{\{k\}_q!} \mathrm{D}^k_q f(x)$	$f(x+y) = \sum_{k=0}^{\infty} \frac{f^{(k)}(y)}{k!} x^k$
$f(x) = \sum_{k=0}^{\infty} \frac{\mathrm{D}^k_q f(y)}{\{k\}_q!} (x \boxminus_q y)^k$	$f(x) = \sum_{k=0}^{\infty} \frac{f^{(k)}(y)}{k!} (x-y)^k$
$\mathrm{D}^n_q(fg)(x) = \sum_{k=0}^n \binom{n}{k}_q \mathrm{D}^k_q(f)(xq^{n-k}) \mathrm{D}^{n-k}_q(g)(x)$	$\mathrm{D}^n(fg) = \sum_{k=0}^n \binom{n}{k} \mathrm{D}^k f \mathrm{D}^{n-k} g$
$\mathrm{D}_q(\frac{f(x)}{g(x)}) = \frac{g(x)\mathrm{D}_q f(x) - f(x)\mathrm{D}_q g(x)}{g(qx)g(x)}$	$\mathrm{D}(\frac{f}{g}) = \frac{f'g-g'f}{g^2}$
$\int_a^b (\mathrm{D}_q f(t,q)) g(t,q)\, d_q(t) =$ $= [f(t,q)g(t,q)]_a^b - \int_a^b f(qt,q) \mathrm{D}_q g(t,q)\, d_q(t)$	$\int_a^b f'g dx$ $= [fg]_a^b - \int_a^b fg' dx$

1.7 The first q-functions

We will now briefly describe the q-umbral method invented by the author [163–180].

This method is a mixture of ideas by Heine (in 1846) [269] and by Gasper and Rahman [220]. The advantages of this method have been summarized in [165, p. 495].

Four features of this q-umbral-calculus are the logarithmic behavior, the formal power series, the formal calculations and the frequent use of the infinity symbol.

The aim of this method is to combine the q-binomial theorem (2.19) with all hypergeometric formulas. This is achieved by means of the q-factorial and the associated q-hypergeometric series. The Γ_q function is closely associated with this.

Definition 1 The variables

$$a, b, c, a_1, a_2, \ldots, b_1, b_2, \ldots \in \mathbb{C}$$

denote certain parameters in hypergeometric series or q-hypergeometric series. The variables i, j, k, l, m, n, p, r will denote natural numbers, except for certain cases where it will be clear from the context that i denotes the imaginary unit. In connection with elliptic functions k always denotes the modulus.

Let $\delta > 0$ be an arbitrary small number. We will always use the following branch of the logarithm: $-\pi + \delta < \mathrm{Im}(\log q) \le \pi + \delta$. This defines a simply connected domain in the complex plane.

The power function is defined by

$$q^a \equiv e^{a \log(q)}. \tag{1.29}$$

Formula (1.29) means that $q^a q^b = q^{a+b}$.

1 kg, which is the unit for mass, can be compared to the q-factorial.

1 kg	$\langle \alpha; q \rangle_n$

Definition 2 The two q-factorials and the tilde operator are defined by

$$\langle a; q \rangle_n \equiv \prod_{m=0}^{n-1} \left(1 - q^{a+m}\right), \tag{1.30}$$

$$(a; q)_n \equiv \prod_{m=0}^{n-1} \left(1 - aq^m\right), \tag{1.31}$$

$$\langle \widetilde{a}; q \rangle_n \equiv \prod_{m=0}^{n-1} \left(1 + q^{a+m}\right). \tag{1.32}$$

A q-analogue of a complex number is also a complex number. The q-analogues of a complex number a, a natural number n and the factorial are defined as follows:

$$\{a\}_q \equiv \frac{1 - q^a}{1 - q}, \quad q \in \mathbb{C}\backslash\{0, 1\}, \tag{1.33}$$

$$\{n\}_q \equiv \sum_{k=1}^{n} q^{k-1}, \qquad \{0\}_q = 0, \quad q \in \mathbb{C}\backslash\{0, 1\}, \tag{1.34}$$

$$\{n\}_q! \equiv \prod_{k=1}^{n} \{k\}_q, \qquad \{0\}_q! \equiv 1, \quad q \in \mathbb{C}\backslash\{0, 1\}. \tag{1.35}$$

Definition 3 The q-Pochhammer-symbol $\{a\}_{n,q}$ is defined by

$$\{a\}_{n,q} \equiv \prod_{m=0}^{n-1} \{a+m\}_q. \tag{1.36}$$

An equivalent symbol is defined in Exton [194, p. 18] and is used throughout that book. This quantity can be very useful in some cases where we are looking for q-analogues and it is included in the new notation.

Since products of q-Pochhammer symbols and q-shifted factorials occur so often, to simplify them we shall frequently use the following more compact notation. Let $(a) = (a_1, \ldots, a_A)$ be a vector with A elements. Then

$$\{(a)\}_{n,q} \equiv \{a_1, \ldots, a_A\}_{n,q} \equiv \prod_{j=1}^{A} \{a_j\}_{n,q}, \tag{1.37}$$

$$\langle (a); q \rangle_n \equiv \langle a_1, \ldots, a_A; q \rangle_n \equiv \prod_{j=1}^{A} \langle a_j; q \rangle_n. \tag{1.38}$$

Let the Gaussian q-binomial coefficients [225] be defined by

$$\binom{n}{k}_q \equiv \frac{\langle 1; q \rangle_n}{\langle 1; q \rangle_k \langle 1; q \rangle_{n-k}}, \quad k = 0, 1, \ldots, n. \tag{1.39}$$

To justify the following three definitions of infinite products, we remind the reader of the following well-known theorem from complex analysis, see Rudin [443, p. 300]:

Theorem 1.7.1 *Let Ω be a region in the complex plane and let $H(\Omega)$ denote the holomorphic functions in Ω. Suppose $f_n \in H(\Omega)$ for $n = 1, 2, 3, \ldots$, no f_n is identically* 0 *in any component of Ω and the series*

$$\sum_{n=1}^{\infty} \left| 1 - f_n(z) \right| \tag{1.40}$$

converges uniformly on compact subsets of Ω. Then the product

$$f(z) = \prod_{n=1}^{\infty} f_n(z) \tag{1.41}$$

converges uniformly on compact subsets of Ω. Hence, $f \in H(\Omega)$.

Definition 4 The following functions are all holomorphic:

$$\langle a; q \rangle_\infty = \prod_{m=0}^{\infty} \left(1 - q^{a+m}\right), \quad 0 < |q| < 1, \tag{1.42}$$

$$(a;q)_\infty = \prod_{m=0}^{\infty}\left(1-aq^m\right), \quad 0<|q|<1. \tag{1.43}$$

Remark 1 If a in (1.42) is a negative integer the result will be zero, and in the following we have to be careful when these infinite products occur in denominators. Sometimes a limit process has to be used when two such factors occur in numerator and denominator.

We shall henceforth assume that $0<|q|<1$ whenever $\langle a;q\rangle_\infty$ or $(a;q)_\infty$ appears in a formula, since the infinite product in (1.42) diverges when

$$q^a \neq 0, \quad |q| \geq 1.$$

The following notation for the Γ function is often used:

Definition 5

$$\Gamma\begin{bmatrix} a_1,\ldots,a_p \\ b_1,\ldots,b_r \end{bmatrix} \equiv \frac{\Gamma(a_1)\cdots\Gamma(a_p)}{\Gamma(b_1)\cdots\Gamma(b_r)}. \tag{1.44}$$

The meter, the measure unit for the length, can be compared with the Γ_q function, a basis for q-calculus.

1 m	Γ_q

Definition 6 The Γ_q function is defined by

$$\Gamma_q(z) \equiv \begin{cases} \frac{\langle 1;q\rangle_\infty}{\langle z;q\rangle_\infty}(1-q)^{1-z}, & \text{if } 0<|q|<1, \\ \frac{\langle 1;q^{-1}\rangle_\infty}{\langle x;q^{-1}\rangle_\infty}(q-1)^{1-x}q^{\binom{x}{2}}, & \text{if } |q|>1. \end{cases} \tag{1.45}$$

Here we deviate from the usual convention $q<1$, because we want to work with meromorphic functions of several variables. The reason is that the q-analogue of the Euler reflection formula involves the first Jacobi theta function, which by construction is a complex function, not only real. The simple poles of Γ_q are located at $z=-n\pm\frac{2k\pi i}{\log q}$, $n,k\in\mathbb{N}$.

We find a q-analogue of [425, p. 23]:

$$\frac{\Gamma_q(\alpha+n)}{\Gamma_q(\alpha)} = \{\alpha\}_{n,q}. \tag{1.46}$$

The following equation holds:

$$\Gamma_q(1+n) = \{n\}_q!. \tag{1.47}$$

We denote the limit for $0<|q|<1$ by $\lim_{q\to 1^-}$ and the limit for $|q|>1$ by $\lim_{q\to 1^+}$.

We have $\lim_{q\to 1^-}\Gamma_q(x) = \Gamma(x)$ [343] and $\lim_{q\to 1^+}\Gamma_q(x) = \Gamma(x)$.
For $q < 1$ the residue at $x = -n$ is [220, p. 17]

$$\lim_{x\to -n}(x+n)\Gamma_q(x) = \frac{(1-q)^{n+1}}{\langle -n; q\rangle_n \log(q^{-1})}. \tag{1.48}$$

We find by l'Hospital's rule that $\lim_{q\to 1^-}$ of this is $\frac{(-1)^n}{n!}$.

We will later in Chapter 8 find a formula (8.42) involving a quotient of Γ_q functions evaluated at negative integers. The above formula serves as a kind of motivation for those computations.

In diagrams 1, 2, 3 we will show three graphs for the Γ_q function, which show the great resemblance to the Γ function. Everywhere we have $q = .5$.

Definition 7 Let S_r denote the additional poles of Γ_q, vertical if q is real and slanting if q is complex. Then the generalized Γ_q function, a function $(\mathbb{C}\setminus(\{\mathbb{Z}\le 0\}\cup S_r))^{p+r}\times\mathbb{C}\to\mathbb{C}$, is defined as follows:

$$\Gamma_q\begin{bmatrix} a_1,\ldots,a_p \\ b_1,\ldots,b_r\end{bmatrix} \equiv \frac{\Gamma_q(a_1)\cdots\Gamma_q(a_p)}{\Gamma_q(b_1)\cdots\Gamma_q(b_r)}. \tag{1.49}$$

In the same way as the Γ function plays a basic role in complex analysis, the Γ_q function is fundamental for q-calculus. The reason is the close connection to the q-factorial.

1 cd	$\mathrm{E}_q(x)$

Definition 8

$$\mathrm{E}_q(z) \equiv \sum_{k=0}^{\infty}\frac{1}{\{k\}_q!}z^k. \tag{1.50}$$

Diagram 1

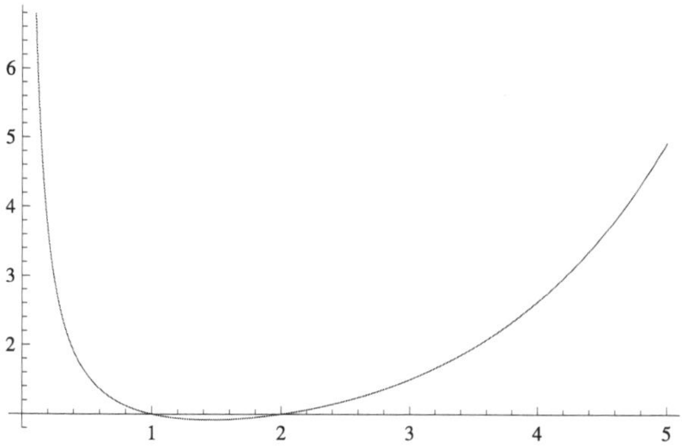

Diagram 2

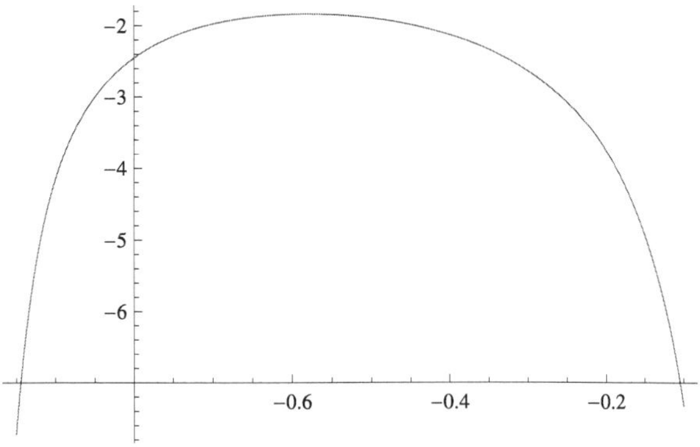

Diagram 3

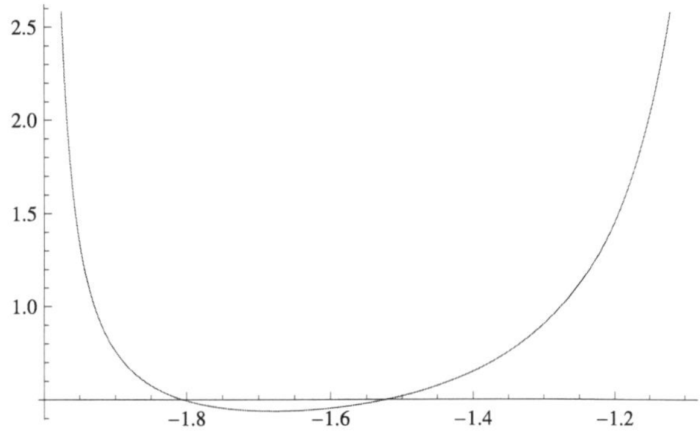

The speed can be compared to the q-derivative.

$\frac{ds}{dt}$	D_q	Section 6.2

Definition 9

$$(\mathrm{D}_q\varphi)(x) \equiv \begin{cases} \frac{\varphi(x)-\varphi(qx)}{(1-q)x}, & \text{if } q \in \mathbb{C}\setminus\{1\},\ x \neq 0, \\ \frac{d\varphi}{dx}(x), & \text{if } q = 1, \\ \frac{d\varphi}{dx}(0), & \text{if } x = 0. \end{cases} \tag{1.51}$$

Definition 10 The q-integral is the inverse of the q-derivative:

$$\int_a^b f(t,q)\,d_q(t) \equiv \int_0^b f(t,q)\,d_q(t) - \int_0^a f(t,q)\,d_q(t), \quad a, b \in \mathbb{R}, \tag{1.52}$$

where

$$\int_0^a f(t,q)\,d_q(t) \equiv a(1-q)\sum_{n=0}^{\infty} f\left(aq^n, q\right)q^n, \quad 0 < |q| < 1,\ a \in \mathbb{R}. \tag{1.53}$$

The second, which is the unit of measurement for the time, may be compared to the Cigler ϵ-operator. This operator is the simplest operator in q-analysis; it is called σ in French literature and η in the papers by Rogers and Jackson and in Hardy and Wright [268, p. 386].

$$\boxed{1\,\mathrm{s}}\boxed{\epsilon}$$

Definition 11 On the field of polynomials, we define

$$[\epsilon_x]f(x) \equiv \epsilon f(x) \equiv f(qx). \tag{1.54}$$

Let $q \in \mathbb{C}^\star$, $|q| \neq\neq 1$ and $\mathcal{M}er(\mathbb{C}^\star)$ be the field of meromorphic functions in $\mathbb{C}^\star$. The ϵ-operator is an automorphism on $\mathcal{M}er(\mathbb{C}^\star)$.

Definition 12 The operator M and the function $\mathrm{P}_{n,q}(x,a)$ are defined by

$$[\mathrm{M}_x]f(x) \equiv \mathrm{M}f(x) = xf(x), \tag{1.55}$$

$$\mathrm{P}_{n,q}(x,a) \equiv \prod_{m=0}^{n-1}\left(x + aq^m\right), \quad n = 1, 2, \ldots. \tag{1.56}$$

The following notation is often used when we have long exponents:

$$\mathrm{QE}(x) \equiv q^x. \tag{1.57}$$

When there are several q's we generalize this as follows:

$$\mathrm{QE}(x, q_i) \equiv q_i{}^x. \tag{1.58}$$

Theorem 1.7.2 *The Cigler q-binomial theorem* [133]. *Let A, B be elements of $\mathbb{C}[x]$ or belonging to an associative algebra, and satisfying*

$$BA = qAB. \tag{1.59}$$

Then

$$(A + B)^n = \sum_{k=0}^{n} \binom{n}{k}_q A^k B^{n-k}, \quad n = 0, 1, 2, \ldots. \tag{1.60}$$

Definition 13 Let a and b be elements of a commutative multiplicative semigroup. Then the NWA q-addition is given by

$$(a \oplus_q b)^n \equiv \sum_{k=0}^{n} \binom{n}{k}_q a^k b^{n-k}, \quad n = 0, 1, 2, \ldots. \tag{1.61}$$

Furthermore, we will put

$$(a \ominus_q b)^n \equiv \sum_{k=0}^{n} \binom{n}{k}_q a^k (-b)^{n-k}, \quad n = 0, 1, 2, \ldots. \tag{1.62}$$

We will now put $B = \mathrm{D}_q$ and $A = \epsilon$ in (1.60). Since the requirements are fulfilled, we conclude that we can also define the NWA q-addition for noncommuting A and B when (1.59) is valid. The NWA q-addition is a generalization of (1.60).

Example 1 We display the NWA as a function of the first $n = 1, \ldots, 9$ for two different values of a and b with $a + b = 1$ or 1.005.

$a = .45, b = .55, q = .99$	$(a \oplus_q b)^n$	$a = .455, b = .55, q = .99$	$(a \oplus_q b)^n$
$n = 1$	1.	$n = 1$	1.005
$n = 2$	0.997525	$n = 2$	1.00752
$n = 3$	0.9926	$n = 3$	1.00756
$n = 4$	0.985267	$n = 4$	1.0051
$n = 5$	0.975587	$n = 5$	1.00019
$n = 6$	0.963635	$n = 6$	0.992868
$n = 7$	0.949505	$n = 7$	0.983185
$n = 8$	0.933303	$n = 8$	0.971222
$n = 9$	0.915148	$n = 9$	0.95707

Apparently, the NWA quickly decreases for large n, and this is one of the reasons for the enlarged domain of convergence of the q-Appell functions in Section 10.2.

Mathematica calculations show that the function

$$F(n)_{a,b,q} : n \mapsto (a \oplus_q b)^n, \quad 0 < a < 1, 0 < b < 1 \tag{1.63}$$

first grows and then decreases and thus has an absolute maximum.

Example 2 As further confirmation of this hypothesis, we show the following table, which gives the n-values for the maximum of the function $F(n)_{a,a,.88}$:

a	n for maximum
.54	3
.55	4
.6	7
.65	11
.7	14
.8	24

Chapter 2
The different languages of q-calculus

The study of q is often considered to be one of the most difficult subjects to engage in mathematics. This is partly because of the many formulas involved, partly due to the wide range of different notations used. Below is an outline of how and why this wide and confusing variety of notations came into being, a situation which is closely connected to the 300-year old history of q. The history of the study of q may be illustrated by a tree.

This q-tree has 13 distinct roots:

1. the study of elliptic functions from the nineteenth century
2. the development of theta functions
3. additive analytic numbers theory, or theory of partitions
4. the field of hypergeometric functions
5. gamma function theory
6. Bernoulli and Euler polynomials
7. umbral calculus
8. theory of finite differences
9. combinatorics + q-binomial coefficient identities
10. theory of finite fields + primitive roots
11. Mock theta functions
12. multiple hypergeometric functions
13. elliptic integrals and Dedekind eta function

The main trunk of the tree is the subject q. The trunk is divided into two main branches, the two principal Schools in q-analysis: Watson and Austrian. Each of the two main branches bears several smaller twigs, smaller Schools which have sprouted—and still do—somewhat later from one of the principal branches.

Now, the following pages may remind the reader more of a study in languages and their interconnections than an exposition in mathematics. So be it for a while. To fully understand the development and current state of q-analysis, the pitfalls and the problems to be dealt with, it is imperative to establish a 'frame' within which to get a proper general view of the vast area and its many different and connected

T. Ernst, *A Comprehensive Treatment of q-Calculus*,
DOI 10.1007/978-3-0348-0431-8_2, © Springer Basel 2012

branches. We may indeed use the language metaphor to grasp q at its present, somewhat confusing state:

The various Schools of q to be outlined below have developed over a period of roughly 300 years since the Bernoullis and Euler. These Schools today use and communicate in such different languages, that they have problems understanding each other. The big School of Watson speaks—both literally and metaphorically—English. It has little exact consciousness of, indeed may not find it necessary, to study and know the roots of the vernacular. Still metaphorically speaking, the Watson School does not understand the development of old Icelandic—which was the beginnings of q—into a modern Scandinavian language. Thus, it is difficult for the Watson School practitioners to understand and indeed to communicate with Schools where completely different dialects have evolved. And we have in the q-area, in relation to other areas, a tendency to develop specific, new languages (e.g. notations), which surpasses this tendency in other fields of mathematics. This is either because practitioners find these more easy to speak, because they find them more beautiful or simply because they do not know and cannot pronounce the older forms.

2.1 Schools—traditions

It is today possible, indeed clarifying, to divide q-analysis/calculus into several Schools or traditions. What distinguishes these Schools is first and foremost

1. Their history (their roots).
2. Which specific modern language they write in.
3. Their different notations.

These Schools or traditions should not be seen as iron-fence surrounded exclusive units, but more as blocks with rather fluid and sometimes overlapping boundaries. It is thus sometimes convenient to place a q-scholar in two traditions or Schools. If we take Harold Exton as an example, it is possible to say that he was firmly rooted in the historic tradition of the Austrian School, but he nevertheless used (partly) the notation from the Watson School. He—to use a metaphor again—spoke Watsonian but with a strong Austrian accent. The result was, as is too often the case, a mishmash.

The two main Schools are, as already stated, the Austrian School and the Watson School. In this context, it might be of interest to tell something about the early development of the combinatorics in the German territory. The predecessor of the Austrian School was the combinatorial School of Karl-Friedrich Hindenburg (1741–1808) and Christoph Gudermann. The goal of the combinatorial School was to develop functions in power series by Taylor's formula.

The *Austrian School* is named in honour of one of its main figures, the Berliner Wolfgang Hahn (1911–98), who held a professorship in Graz, Austria, from 1964. Hahn was strongly influenced by Heine. The Austrian School is a continuation of the Heine q-umbral calculus from the mid nineteenth century, which at the time however met with little attention except for Rogers, who introduced the first q-Hermite polynomials and first proved the Rogers-Ramanujan identities.

The *Watson School* takes its name from the English mathematician George Neville Watson (1886–1965), who wrote the famous essay *Treatise on the Theory of Bessel Functions*, and furnished a rigorous proof for the Rogers-Ramanujan identities.

Both Schools or traditions recognize the early legacies of Gauß and Euler. Only the Austrian School, however, represents and incorporates the entire historical background which includes the pre-q mathematics, namely the Bernoulli and Euler numbers, the theta functions and the elliptic functions. The vast area of theta functions and elliptic functions, which will be dealt with in Sections 3.8 and 6.13, is in fact q-analysis before q was really introduced. Pre-q or q-analysis in disguise could perhaps be an appropriate term for this period and its practitioners. The Austrian School takes the development of q all the way from Jacob Bernoulli, Gauß and Euler in the 17th and 18th century, through the central European mathematicians of the nineteenth century: Heine, Thomae, Jacobi, and from the twentieth century: Pringsheim, Lindemann, Hahn, Lesky and Cigler, the Englishman Jackson, the Austrians Peter Paule, Hofbauer, Axel Riese and the Frenchman Appell. The present (Swedish) author of this book confesses to be firmly rooted in the Austrian School.

The Austrian School is little known in the English speaking world, e.g. the USA and the Commonwealth, for two main reasons: Immediately after the first world war, ca. 1920 until 1925, the German and Austrian mathematicians were barred from participation in the big mathematical conferences; they were simply deemed political pariahs by the French, unwanted, and the communications with the English speaking q-analysts were for a period limited, as was the exchange of ideas. The second and more important reason has to do with languages. Most of the mathematicians of the Austrian School wrote in either German or French and, as regards the oldest, namely Euler, Gauß and Jacobi, in Latin. Only few English-speaking mathematicians today master these languages (it is simply not part of the common American curriculum), and few are therefore familiar with the works of these early European scientists.

The Watson School is today the most widespread and influential of the two principal Schools/traditions. But again this is mainly due the language. If English—as is the case today for many mathematicians—is your primary and perhaps only approach to the study of mathematics, then adherence to the Watson School is almost automatic. The Watson School is nowadays the main highway to the study of q. Not necessarily, though, the best, the most correct or even the smoothest or most beautiful route to choose.

The Watson School, in this author's opinion—I hope and I will convince the reader as the book progresses—does not take the full and necessary steps to understand and incorporate the work early q and pre-q analysts into modern q-calculus.

Before Ramanujan, Cambridge had enjoyed quite a strong Austrian School representation with names such as James Glaisher (1848–1928) and Arthur Cayley (1821–95), from the nineteenth century.

F. H. Jackson (1870–1960), James Rogers (1862–1933) and Andrew Russell Forsyth (1858–1942) from the twentieth century represent a kind of transition to the Watson School.

The Watson School may be said to start in earnest with the Indian Cambridge mathematician Srinivasa Ramanujan's discoveries, namely the mock theta functions and the Rogers-Ramanujan identities. The Watson School goes, in a manner of speaking, straight from Gauß to the twentieth century Srinivasa Ramanujan (1887–1920) and thus skips and misses out a 150-years period of fruitful European studies. The early Watson School adherents and practitioners were the mathematicians at Cambridge who Ramanujan's short stay inspired, e.g. Eric Harold Neville (1889–1961) and Wilfrid Norman Bailey (1893–1961). Bailey devoted a lot of time to administration and sports during his academic career. He was an excellent teacher and among his students were L. J. Slater, Jackson, F. Dyson and Ernest Barnes (1874–1953). Interestingly enough, all of these turned into special functions. Barnes became a bishop and wrote a long series of papers on special functions and difference calculus. In a way, Barnes was a predecessor of Nørlund, but they studied different problems. Although Barnes was an excellent mathematician, far better than Jackson, he never returned to the academia after a few years as a teacher. In his youth, Bailey met Jackson in the Navy and they certainly discussed q-calculus already then. In a couple of papers around 1947 Bailey intended to simplify some of Rogers's proofs of generalizations of the Rogers-Ramanujan identities. He then invented a new notation and gave Dyson credit to some of the formulas. The famous Bailey's lemma comes from this time; there are several variations of this, and Wengchang Chu claims that Bailey's lemma is a special case of a generalization of the Carlitz q-Gould-Hsu inversion formula. In the author's opinion, the wisest way is to go back to Rogers's original proofs. After his retirement, in 1958, Bailey intended to write a major work on q-series. For some reasons this failed, although he moved to Eastbourne, where Jackson spent his last years.

2.2 Ramifications and minor Schools

The explanations in Sections 2.2, 12.7–12.9 are written for convenience and give a good account of the current state of affairs. These explanations are not in standard terms and cannot be cited. The study of q can be further divided into sub-groups, traditions or Schools. Some of these, e.g. the Chinese or Japanese ones, can hardly be said to form Schools as such; what they have in common however, is a cultural and linguistic background, which in a certain sense shapes their mathematical work.

The *Carlitz, Gould and Vandiver tradition* is perhaps better termed the American-Austrian School as its practitioners all tend to work in the European Watson School tradition. This School can be traced back to the 1930s.

Isaac Joachim Schwatt (1867–1934), Leonard Carlitz (1907–1999) and Harry Schultz Vandiver (1882–1973) were descendants of European immigrants to the USA and therefore read and worked in the pre-q, or q-in-disguise tradition. Henry

W. Gould (1928–) on the other hand simply enjoyed reading European mathematical literature. This group had a strong interest in Bernoulli numbers, finite differences and combinatorial identities. However, not all practitioners of the American-Austrian School wrote books on their subject—the largest part of their contribution is contained in their lectures and articles. Vandiver and Gould collected card-files on articles about Bernoulli numbers; this work was continued by Karl Dilcher and made available on the Web.

It is also possible that Carlitz was a student of Schwatt at University of Pennsylvania, Philadelphia. Carlitz was born in Philadelphia, and the two have worked on similar mathematical topics. Schwatt, who became PhD in 1893 in Philadelphia, remained in this city during his whole career (1897–1928). We will come back to the q-analogues of the Schwatt formulas in Chapter 5, where q-Stirling numbers are discussed. Carlitz spent a post-doc year 1930–31 with E. T. Bell in Pasadena, and we have come to the next School.

The so-called *E. T. Bell, Riordan, Rota-School* started in 1906 in San Francisco (the year of the big earthquake), when Eric Temple Bell (1883–1960) read some books on number theory [430, p. 109]. The first one was by Paul Bachmann (1837–1920) who enriched the whole theory with detailed proofs. Bachmann had a doctorate from Berlin 1862; his instructors were Ernst Kummer (1810–1893) and Martin Ohm (1792–1872) (Bachmann himself had no doctoral students). The second book Bell read was *Théorie des Nombres* [363] by E. Lucas (1842–1891) [363]; here Bell was initiated into the so-called umbral calculus. Morgan Ward (1901–1963) also belonged to this School; his supervisor was Bell. John Daum recognized the connection between q-series and hypergeometric functions; Daum belonged to the second mathematical generation after Bell.

John Riordan (1903–1988) and Gian-Carlo Rota (1932–1999) were also members of this group. In 1963, Riordan and Rota met in Boston and went to a restaurant where they discussed Riordan's new book *An Introduction to Combinatorial Analysis*; Riordan's book is actually dedicated to E. T. Bell. After 31 years, in the footsteps of Bell, Rota and his student Brian Taylor attempted a rigorous foundation for the umbral calculus in the excellent treatise *The Classical Umbral Calculus* [439]. Unfortunately, Rota learned late of Bell's combinatorial work [430, p. 227], so he could not find the q-analogue of [439], which is presented in this book.

Special functions have always been a major research topic in India, an inheritance from the old nineteenth century Austrian Cambridge School under Glaisher and Forsythe. The *Indian School* is made up of many different branches and can for convenience be divided into at least three different areas of interest: Srivastava, Hahn and Ramanujan. The Srivastavan branch was originally founded in the 1960s by Hari M. Srivastava in the footsteps of Shanti Saran. The city of Lucknow is a centre for experts, and will be of interest in the following. Saran got his PhD in 1955 in Lucknow on a treatise on hypergeometric functions of three variables; his supervisor was R. P. Agarwal (1925–2008). Srivastava received his PhD in 1965 in Jai Narain Vyas on a similar topic. The mathematical journal *Ganita* is printed in Lucknow since 1950; since the sixties, almost every issue contains an article about special functions. In the 1952 issue, there is a report on hypergeometric functions by H. M. Srivastava and A. M. Chak. In the 1956 issue there is a report on the hypergeometric functions of three variables by Krishna Ji Srivastava, which refers to Saran. These two mathematicians were probably unaware of the articles by Horn.

H. M. Srivastava was born in India, settled in Canada in the early seventies, but travelled frequently to India until 1985. He was a good friend of and therefore also inspired by Gould of the American–Austrian School. Srivastava left India permanently in the 1980s and without its founder the School quickly dwindled. The members of the Srivastava School were especially interested and produced works on hypergeometric functions of many variables, on generating functions and on different polynomials, e.g. Laguerre and/or differential operators. Recently (2009), Srivastava again visited Vijay Gupta in India and perhaps something interesting will arise from this cooperation.

H. L. Manocha (Polytechnic Institute of New York University) has written a book about generating functions [484]; one of his graduate students was Vivek Sahai.

Srinivasa Rao, who was earlier in Chennai, is since 2004 at the Ramanujan Center of Sastra University in Kumbakonam, one of his students was V. Rajeswari.

Subuhi Khan, Aligarh University, has done basic studies on connections between Lie algebras and special functions, which she has presented at conferences in Hong Kong and Decin.

Wolfgang Hahn of the Austrian School spent a year in India before taking up as professor at Graz and his Indian pupils, following in his footsteps, are especially interested in q-Laplace transformations and Hahn q-additions.

The Ramanujan branch was and is still a very active and productive group. They are inspired by the great Indian master Ramanujan and his mock theta functions. Brilliant works appear from time to time from this otherwise uneven group of Indian mathematicians.

The *Hungarian School* is yet another European branch with a strong affinity and connection to the Austrian School. This is also a q-School without the q, ... i.e. the work is done primarily on q-related topics such as the theory of orthogonal polynomials and recurrence relations. The Hungarian School emerged in the post WW1 years, the prominent figure being the Hungarian-born Gabor Szegő (1895–1985). Szegő is the father of the so-called Rogers-Szegő polynomials—q-polynomials with many similarities to Hermite polynomials. He later took up work at Stanford and wrote thence in the Watson tradition. One of his successors is Richard Askey; the Askey tableau of orthogonal polynomials stems from him. There is also a q-version of this. Another major contribution is the book *Calculus of Finite Differences* by Charles Jordan [320]. Other strong representatives of this School are Eugene Wigner (1902–1995) and John von Neumann (1903–1957), who also worked without q, but strongly influenced the development of quantum groups. Their works are of high quality, but rather difficult. Wigner introduced the 3-j coefficients, but his formulas could have been greatly improved by using hypergeometric functions in his formulas. This came only in our time, when Joris Van der Jeugt used multiple hypergeometric functions for 9-j coefficients.

The *Danish School*: This School is closely linked with the Austrian School by the close linguistic relationship with German and French. There are two branches: one deals with Stirling numbers and finite difference calculus, the other with multiple hypergeometric functions and gamma functions. In 1909 Thorvald N. Thiele (1838–1910) wrote a book about interpolation theory, containing a table of Stirling numbers. Johan Ludwig Jensen (1859–1925) wrote about gamma functions and thereby influenced Niels Nielsen (1865–1931) in the same area. Nielsen had a strong interest in special functions, in his own way, and introduced Stirling numbers in a paper in French. He has also written biographies on ancient French mathematicians. Nielsen has developed Fakultätenreihen by Bessel functions in his 1904 book on cylinder functions [396]. Here one can also find an excellent bibliography. In the footsteps of Nielsen, Niels Erik Nørlund (1885–1981) in the remarkable work [403] gave the first rigorous treatment of finite differences from the perspective of the mathematician. Nørlund gave lectures on hypergeometric series in Copenhagen until 1955. Nørlund also knew q-calculus; F. Ryde published a thesis on this subject under his supervision in Lund. The next link in the chain is perhaps Per Karlsson (1936–), the expert on multiple hypergeometric functions and friend, among other things, collaborator and model of the author. One can mention also Christian Berg, who works on moment problems and real analysis.

Russian School: Russian mathematicians have greatly influenced the above mentioned Hungarian School through studies of polynomials.

Russia has, in general, a strong tradition in mathematics, which dates back to Euler and the 18th century mathematicians who took up this heritage. Euler has very much contributed to the *Proceedings of the St. Petersburg Academy*. After his death, his successors could not keep the high level from before.

Euler himself died in St. Petersburg, and it is a well-established fact that both his direct influence and also his unpublished papers and work remained in Russia, which explains in part the high level of mathematics in Russia. Many Russian mathematicians from the nineteenth century did excellent work in the area of Bernoulli numbers and umbral calculus, among them Grigoriew, Chistiakov and Imchenetsky.

According to Grigoriew [247, p. 147], the generalized Bernoulli numbers, which Nørlund used in [403], were also used by Blissard (1803–1875) [75] and Imchenetsky [289]. L. Geronimus (1898–1984) wrote about certain Appell polynomials. Leading figures in the tradition of orthogonal polynomials and Bessel functions were P. L. Chebyschew (1821–1894) and Nikolay Yakovlevich Sonine (1849–1915). Their articles are, unfortunately, less accessible, being written in Russian, though some are translated into English. Sonine was one of the last representatives of this School, who could read Euler in the original language.

There is a connection to theoretical physics:

Valentine Bargmann (1908–1989) was born in Berlin to Russian parents. After studying in Berlin and Zurich, he went to Princeton, where he joined Einstein and Wigner. He is famous for the unitary irreducible representations of $SL_2(\mathbb{R})$ and the Lorentz group (1947) and for the Bargmann-Fock space. Russian mathematicians strive to develop, among other things, the theory of Heisenberg ferromagnetic equations and are also actively studying the connection between 3-j coefficients and hypergeometric functions.

The *Italian School* is quite strong. It started when Giuseppe Lauricella (1867–1913) studied hypergeometric functions of many variables in the nineteenth century.

Like Paul Appell (1855–1930), Lauricella focused on symmetric functions, because these provide the most beautiful formulas.

Pia Nalli (1886–1964), who made a highly interesting study of the so-called q-addition, was influenced by the Italian literature about elliptic functions. Nalli was the first one who used the NWA q-addition in her only article on q-calculus.

Letterio Toscano (Messina) (1905–1977), who at the beginning wrote only in Italian, published many interesting articles about Bernoulli, Euler and Stirling numbers in connection with the operator $x\mathrm{D}$.

Due to his friendship with Francesco Tricomi (1897–1978) (Torino), Toscano could write in publications of the Italian academy; the two belonged to the same generation. Tricomi published books on confluent hypergeometric functions and elliptic functions before joining the Bateman project.

In 2005, Donato Trigiante, a student of Tricomi, published elegant matrix representation for the Bernoulli and Euler polynomials.

In 2007, the author presented q-analogues of Trigiante's matrix formulas at the OPSFA Conference in Marseille.

Another Italian mathematician, who, among other things, published on Laguerre polynomials, was Giuseppe Palama (1888–1959) from Milano. In the last decades, Rota and Brini have published works on umbral calculus. We find a q-analogue of Rota's infinite alphabet in this book, that the author introduced (2005).

The *Scottish School* consisted of Thomas Murray MacRobert (1884–1962), who introduced the MacRobert E-function. In this context, MacRobert together with Meijer introduced the $\Delta(l;\lambda)$-operator of Srivastava [475]. Under his leadership began the *Proceedings of the Glasgow Mathematical Association* in 1951, in which many publications on special functions appeared. MacRobert had no doctoral students, but he helped many authors to write on special functions in British journals. After his death, many mathematicians, e.g. from Egypt or India, have followed in his footsteps.

2.2.1 Different notations

This large number of different Schools and their branches or off-springs has also resulted in a profusion of different notations. The scheme below is an attempt to bring some sort of order into the profusion and confusion of notations in the different areas.

The different notations:

1. q-hypergeometric functions

Watson: The q-shifted factorial is denoted by $(a;q)_n$, (1.31).

Austrian:

1. The q-shifted factorial is denoted by $\langle a;q\rangle_n$, (1.30).
2. Cigler only uses the Gaussian q-binomial coefficients, which are nearly equivalent to (1.30) above.

3. Yves André [22, p. 685] uses a notation that is equivalent to $\langle a; q\rangle_n$. André [22, p. 692] also denotes D_q by δ_σ.

Indian:

V. Rajeswari and K. Srinivasa Rao in 1991 [426] and in 1993 [470, p. 72] use my umbral notation in connection with the q-analogues of the 3-j and 6-j coefficients.

Russian:

1. Also Gelfand has used a similar notation in one of his few papers [228, p. 38] on q-calculus. His comment is the following: Let us assume that at first we use Watson's notation (1.31) for the q-hypergeometric series.
   If all α_i and β_i are non-zero, it is convenient to pass to the new parameters a_i, b_i, where $\alpha_i = q^{a_i}$, $\beta_i = q^{b_i}$.
2. Igor Frenkel (MIT) uses none of the above notations. He simply writes the definitions. In the book [186] the symbol $\{a\}$ is used instead of our $\{a\}_q$. In [186] there is a special notation for $(z;q)_a$, but no notation for D_q. Instead of q-factorials, Theta functions are used [186, p. 172].
3. Naum Vilenkin (1920–1991), and Anatoly Klimyk (1939–2008) [523] in their three volumes on representation theory for Lie groups use their own Watson-like notation, as does Boris Kupershmidt.

2.3 Finite differences and Bernoulli numbers

Finite differences and Bernoulli numbers are closely related to q-Analysis. The Bernoulli numbers were first used by Jacob Bernoulli (1654–1705) [70], who calculated the sum:

$$s_m(n) \equiv \sum_{k=0}^{n-1} k^m = \frac{1}{m+1}\sum_{k=0}^{m}\binom{m+1}{k} n^{m+1-k} B_k, \tag{2.1}$$

where the *Bernoulli numbers* B_n are defined by:

$$\frac{z}{e^z-1} = 1 - \frac{z}{2} + \sum_{n=1}^{\infty}\frac{\mathrm{B}_{2n} z^{2n}}{(2n)!}. \tag{2.2}$$

The *Bernoulli polynomials* are defined by:

$$\mathrm{B}_n(x) \equiv (\mathrm{B}+x)^n = \sum_{k=0}^{n}\binom{n}{k}\mathrm{B}_k x^{n-k}, \tag{2.3}$$

where B^n must be replaced by B_n on expansion.

In 1738 Euler used the generating functions to study the Bernoulli polynomials. The Bernoulli polynomials were also studied by J.-L. Raabe (1801–1859) [422] and Oskar Xaver Schlömilch (1823–1901) [453, p. 211].

We now write down the basic equations for finite differences where, E is the shift operator and $\triangle \equiv \mathrm{E} - \mathrm{I}$.

Theorem 2.3.1 [525, p. 200], [138, p. 26], [458, p. 9]. *Faulhaber-Newton-Gregory-Taylor series*

$$f(x) = \sum_{k=0}^{\infty} \binom{x}{k} (\triangle^k f)(0). \tag{2.4}$$

Theorem 2.3.2

$$\triangle^n f(x) = \sum_{k=0}^{n} (-1)^k \binom{n}{k} \mathrm{E}^{n-k} f(x). \tag{2.5}$$

This formula can be inverted.

Theorem 2.3.3 [460, p. 15, 3.1]

$$\mathrm{E}^n f(x) = \sum_{i=0}^{n} \binom{n}{i} \triangle^i f(x). \tag{2.6}$$

The Leibniz rule goes as follows:

Theorem 2.3.4 [320, p. 97, 10], [138, p. 27, 2.13], [385, p. 35, 2], [411, p. 19].

$$\triangle^n (fg) = \sum_{i=0}^{n} \binom{n}{i} \triangle^i f \, (\triangle^{n-i} \mathrm{E}^i) g. \tag{2.7}$$

In Chapter 4, we will retain the binomial coefficients in the corresponding q-formulas, whereas in Chapter 5, q-binomial coefficients for the corresponding formulas will be used.

Karl Weierstraß said that the calculus of finite differences will once play a leading role in mathematics. Two important elements of the calculus of finite differences are the Bernoulli numbers and the Γ function. Nørlund says in a letter to Mittag-Leffler 1919 [183] “Someone who is not an expert in these fields cannot be an expert on calculus of finite differences.” We will show that the Γ function can always be transferred to the Pochhammer-Symbol. The q-factorial and the Γ_q function are the corresponding q-terms. We infer that the (q-)hypergeometric function is also part of the calculus of finite differences. This is no accident, as Douglas Barker Sears and Hjalmar Mellin (1854–1933) have shown.

2.4 Umbral calculus, interpolation theory

The interpolation theory, which was often used by astronomers of the nineteenth century, like Gauß, Bessel, W. Herschel (1738–1822), J. Herschel (1792–1871), is essentially equivalent to the theory of finite differences.

Herschel [277] wrote: "*The want of a regular treatise, on the calculus of finite differences in English, has long been a serious obstacle to the progress of the enquiring student. The Appendix annexed to the translation of Lacroix's Differential and integral calculus, although from the necessity of studying compression it is not so complete as its author could have wished...*"

Computations on elliptic functions with finite differences were made by Jacobi, Weierstraß and Louis Melville Milne-Thomson (1891–1974).

In 1706, John Bernoulli (1667–1748) invented the difference symbol $\triangle$. Fifty years later, in 1755, Leonhard Euler used its inverse, the $\sum$ operator [189, Chapter 1]. Euler was John Bernoulli's student together with Bernoulli's two sons, Nicolaus II and Daniel. Even though John Bernoulli used the symbol $\triangle$ already in 1706, he had in mind not finite differences thereby, but differential quotients. Hence, Euler stands out as the one who devised the designation that has remained in use. Euler's proofs were however not entirely satisfactory from a modern point of view [277, p. 87].

The two symbols, sometimes also called the difference and sum calculus, correspond respectively to differentiation and integration in the continuous calculus. We will find 2 different q-analogues of the inverse operators $\triangle$ and $\sum$ in Sections 4.3 and 5.3.

Euler [189] and Joseph-Louis Lagrange (1736–1813) have introduced the umbral calculus, where operators like (4.126) were used. In 1812, in the footsteps of L. Arbogast (1759–1803), another Frenchman, Jacques-Frédéric Français, wrote E for the forward shift operator [345, p. 163] and reproved the Lagrange formula from 1772,

$$\mathrm{E} = e^{\mathrm{D}}. \tag{2.8}$$

A decade later, Augustin Cauchy (1789–1857) in his *Exercices de mathématiques* [345, p. 163] for the first time found operational formulas like

$$\mathrm{D}\big(e^{rx} f(x)\big) = e^{rx}(r + x) f(x). \tag{2.9}$$

Arbogast and Fourier regarded this umbral method as an elegant way of discovering, expressing, or verifying theorems, rather than as a valid method of proof [345, p. 172]. Cauchy had similar opinions. We will see that this sometimes also obtains for the method of the author.

Textbooks on the subject were written by Andreas von Ettingshausen (1796–1878), J. Herschel, J. Pearson and Augustus De Morgan (1806–1871).

Robert Murphy (1806–1843) was a forerunner of George Boole (1815–1864) and Heaviside, who among other things found nice formulas for derivatives in the spirit of Carlitz.

In 1854 Arthur Cayley introduced the concept of Cayley table in his article *On the theory of groups*. Cayley defined a group as a set of objects with a multiplication and a symbolic equation $\theta^n = 1$. Blissard was obviously influenced by Cayley's work, at least in what concerns his notation.

Another French entry came from E. Lucas, who invented a modern notation for umbral calculus. The Lucas umbral calculus was widespread in Russia, for example one finds the defining formula for the Bernoulli numbers in Chistiakov 1895 [127, p. 105]. The Blissard umbral calculus has attracted attention in Russia; in [127, p. 113] one finds the Blissard Bernoulli number formulas with sine and cosine.

By 1860 two textbooks on finite differences were published in England, one of them by Boole, which covered almost all the theorems that we know now. Heaviside was able to greatly simplify Maxwell's 20 equations in 20 variables to four equations in two variables. This and other articles about electricity problems, which appeared in 1892–98, were severely criticized for their lack of rigour by contemporary mathematicians.

It seems that Heaviside's contribution to mathematics was underestimated by his contemporaries, since in fact he both discussed formal power series and the rudiments of umbral calculus, which we present in this book.

Angelo Genocchi (1817–1889) and Salvatore Pincherle (1853–1936) contributed to the early Italian development of the subject. Alfred Clebsch (1833–1872) and Paul Gordan (1837–1912) continued the theory of invariants that had started with Sylvester and Cayley.

The Heine q-umbral calculus reached its peak in the thesis by Edwin Smith (1879–) [467] 1911, which was supervised by Pringsheim.

F. H. Jackson followed this path in the early twentieth century, and fully understood the symbolic nature of the subject in his first investigations of q-functions. Like Blissard, Jackson worked as a priest his whole life; both of them had studied in Cambridge. To honour Jackson, we will use his notation for $\mathrm{E}_q(x)$.

Steffensen [488], Jordan [320] and Milne-Thomson [385] wrote books about finite differences intended both for mathematicians and statisticians. Johann Cigler wrote an excellent book [138] on finite differences with a view to umbral calculus.

2.5 Elliptic and Theta Schools and notations, the oldest roots–the q-forerunners

Just as the nicest equations in mechanics are connected with the torque (and the angular momentum), the theta functions and the elliptic functions give the most beautiful equations in calculus.

moment	theta functions, elliptic functions

As was the case with q: Schools and traditions abound, so it is the case with the forerunners of q-analysis: different Schools, different notations. The early q-analysis may be defined as the mathematics done in this area even before the q was introduced and properly defined, namely the elliptic functions and theta functions from ca. 1750 and onwards. We may call these the q-forerunners or speak of the pre-q

mathematics or q-analysis in disguise. Euler had in fact already introduced q-series and Jacobi continued to use the letter q, which has survived until today.

Eight Schools stand out:

1. Jacobi theta functions and elliptic functions of one variable.
2. Riemann Theta functions of one or more variables or Abelian functions.
3. Weierstraß elliptic functions and sigma functions.
4. The Glaisher-Neville-School for Jacobi elliptic functions.
5. The Γ_q-School.
6. The Heine Ω function School.
7. The Weierstraß-Mellin School of Gamma functions and hypergeometric functions.
8. The Italian elliptic function School.

School 1 is about the Jacobi theta functions, so called after the German mathematician Carl Gustav Jacob Jacobi. Jacobi's development of the theta functions, of which we have four, was made in parallel to (but before) that of Weierstraß of School 3. This means that if we know all 4 Jacobi Theta functions, we can compute all Weierstraß σ functions by means of 4 linear equations and vice versa.

There were in fact originally three different notations for Jacobi elliptic functions:

Jacobi notation, Abel notation and finally the modern Gudermann notation $\mathrm{sn}u, \mathrm{cn}u, \mathrm{dn}u$ from 1844.

School 2 covers the Riemann Theta functions after G.F.B. Riemann (1826–66). These are functions of one or more variables. Prominent persons working in this School have been Krazer, Rauch and Lebowitz, Thomae, Göpel, Rosenheim and Forsyth.

School 3 deals with the Weierstraß elliptic functions, after Weierstraß. These are more general, defined on a general lattice in the complex plane.

School 4 is just a modern English version of School 1. By 1875 the theory of elliptic functions was very popular; the *Messenger of Mathematics* had since no. 4 of 1875 a separate Section on elliptic functions, where among others Glaisher and Cayley have contributed.

School 5 covers the q-Gamma function. There are, for instance, many modern papers with inequalities for quotients of these functions. Atakishiyev [50, p. 1326] has rediscovered the q-analogue of the Euler mirror formula.

The roots of the Heine Ω function (School 6) come from the Jacobi-Heine treatment of elliptic functions [270].

Johannes Thomae (1840–1921) [497, p. 262] claims that his teacher Heine was the first to find the q-analogue of the Euler mirror formula [270, p. 310].

The main difference between the two functions is that Ω has zeros, in contrast to the Γ_q function which has no zeros, and therefore $\frac{1}{\Gamma_q}$ is entire. In his thesis Ashton [45], supervised by Lindemann, showed its connection to elliptic functions. Daum [148] tried to find all the basic analogues of Thomae's ${}_3F_2$ transformation formula [501, Eq. 11], using a notation analogous to that used by Whipple [541] and by essentially replacing the Γ_q function by the Heine Ω function.

Daum [148] concludes his thesis by saying *It is hoped, however that the use of the modified Heine Ω function, will serve to emphasize the analogy between hypergeometric series and q-hypergeometric series and simplify the notation generally.*

Sonine wrote a book [469] on the Heine Ω function in Russian, which also treats the Γ function.

School 7 is in the area of the Mellin and Gamma functions and the nineteenth century work in this area further influenced the Danish School, featuring names such as Thiele, Nielsen and Nørlund.

School 8 started with the numerous Italian books about elliptic functions written by Bellacchi in 1894, E. Pascal (II) in 1895, Bianchi (1856–1928) in 1901, and Giulio Vivanti (1859–1949) in 1900 [524].

This influenced Pia Nalli to write her paper [391], where Theta functions and a q-integral formula for a q-Riemann zeta function were given.

One can note here that the Gudermann's notation was very quickly accepted in Japan, see [378, p. 87].

2.6 Trigonometry, prosthaphaeresis, logarithms

There has always been a strong connection between mathematics and physics.

The 1648 book *Mathematical Magick* by John Wilkins (1614–1672) contains basic mechanics, but no mathematics, by today's definition.

Textbooks on mathematics in the late eighteenth century contained a variety of subjects like mechanics, optics and astronomy. One example is the book *Die Elemente der Mathematik* by Johann Friedrich Lorenz (1737–1807) from 1797, which treated such diverse subjects as refraction, parallax, geography, the atmosphere of the moon. Elementary trigonometric formulas were given, and these trigonometric functions were used to treat the physics involved. This was natural since many earlier mathematicians, like the Bernoullis, were both physicians and physicists. John Bernoulli (1) has in fact even written a memoir about mathematical medicine.

We will start with a brief history about trigonometry and its relationship to spherical triangles and astronomy.

Probably Aristarchus of Samos, Greece (-260) [392, p. 108] used ratios similar to tangens. Menelaus of Alexandria ($+100$) in his treatise on spherical trigonometry introduced the concept of sine [392, p. 108].

Aryabhata ($+510$) has for the first time used special names for sine and computed tables for every angle [392, p. 108]. His contemporary Vara-Mihira has in the year 505 given formulas that are equivalent to sine and cosine [392, p. 108]. These Indian works were then taken over by the Arabs and transmitted to Europe [392, p. 108]. The Egyptian mathematician and astronomer Ibn Yunus (950–1009) demonstrated the product formula for cos and made many astronomical observations. During the doldrums of the dark ages not much happened in trigonometry until the renaissance with the rediscovery of the old Arabic culture. Prosthaphaeresis (the Greek word for addition and subtraction) is a technique for computing products quickly using trigonometric identities, which predated logarithms. Myriads of books have been written on trigonometry in Latin before the modern notations sin and cos were introduced by Euler. The Greek and Alexandrian mathematicians were prominent in proof theory and geometry, including conic sections. Certainly these ancient scientists had some notation for trigonometric functions and some of this great research has survived through Latin translations during the renaissance.

Georg Purbach (1423–1461), who was born near Linz, has enriched the trigonometry and astronomy with new tables and theorems. He actually combined the chords of Ptolemaios with the sines of the arabs and in that way introduced the first tables of sines with decimals. His student in Vienna Johannes Müller (1436–1476) continued Purbach's research and inter alia introduced the tangent in astronomy; Müller called it *foecundus* (Latin: fruitful) because of its great advantages. His work on flat and spherical trigonometry in five volumes was printed in Nuremberg in 1533.

The trigonometry of Regiomontanus (Müller) looked about the same as now, the biggest difference being that logarithms were not used. Many other mathematicians have continued his development of trigonometry, e.g. John Werner (1468–1528), who wrote a work in five volumes on triangles. Werner used the product formula for sin and also contributed to the development of instruments. In the footsteps of Werner, Georg Rheticus (1514–1574) used Prosthaphaeresis computations for instruments. Through his friendship with Duke Albert of Prussia, Rheticus agreed in 1541 to the printing of Copernicus work *De Revolutionibus*. Two years earlier, Rheticus visited Nicholaus Copernicus (1473–1543), whose great book could not yet be printed. Copernicus *De Revolutionibus* was finally printed in 1543 in Nuremberg; in an annex Rheticus adds tables for sin and cos. In 1596, Rheticus' student Valentin Otto (1550–1603) published several books on plane and spherical trigonometry, based on Rheticus' computations.

Erasmus Reinhold (1511–1553) and Franciscus Maurolycus (1494–1575) have also published tables of tangents or secants.

In 1591, Philippus Lansbergen (1561–1632) computed tables for sin, sec and tan by hand in *Geometria Triangularium*. This is a short, concise work.

The Danish physicist and mathematician Thomas Fincke (1561–1656) was born in Flensburg and later worked in Copenhagen. Fincke has introduced the concept of sec.

Probably the first western European work dealing with systematic computations in plane and spherical triangles was written 1579 by François Viète (1540–1603), called 'the father of modern algebraic notation'.

Viète used a certain notation for multiplication and found [230] formulas almost equivalent to multiplication and division for complex numbers as well as the de Moivre's theorem.

Joost Bürgi (1552–1632) was also an advocate for Prosthaphaeresis. At about the same time as Napier, he invented the logarithms, but unfortunately he did not dare to publish his invention until 1620 [110, p. 165]. Bürgi, who was a Swiss clockmaker was also a member of the so-called *Rosenkreuzer* society, to which we will come back later.

James VI of Scotland, who was on a journey to Norway in the year 1590 together with his entourage, including Dr. John Craig, visited the island Ven on his return to Scotland. There Tycho Brahe (1546–1601) had constructed a big machine for prosthaphaeresis computations to ease the burden of calculation. Craig told his friend John Napier (1550–1617) about the visit to the famous astronomer and that inspired Napier to develop his logarithms and the generation of his tables, a work to which he dedicated his remaining 25 years.

Johannes Kepler (1571–1630), who was collecting Tycho Brahe's immense data, read Napier's book on logarithms in 1616; he found that he could describe his laws for orbital periods and semimajor axes for planetary ellipses as a straight line in a log-log diagram.

Thanks to these laws Isaac Newton (1642–1727) was able to discover the gravitation law. At the request of Kepler, Bürgi finally brought himself to publish his important book on logarithms in 1620. Logarithms have been in great use ever since; even physicians and nurses have employed these tables for various long computations.

In Ulm (1627), Kepler then published his Rudolphine tables with tables for $\log \sin x$ together with so-called Antilogarithms, a forerunner of the exponential. In 1801, James Wilson in *scriptores Logarithmici* published a work on the use of logarithms in navigation. Wilson has also written a book on finite differences in 1820.

2.7 The development of calculus

The first calculating machine was built by the German astronomer Wilhelm Schickard (1592–1635) in 1623 [361, p. 48] and was designed for Kepler. The Schickard calculator could add, subtract, multiply and divide, but remained unknown for 300 years. In 1642, Blaise Pascal (1623–1662) constructed a mechanical calculator, capable of addition and subtraction, called Pascal's calculator or the *Pascaline*, in order to help his father with his calculations of taxes [361, p. 16]. Some of his calculators were also exhibited in museums both in Paris and Dresden, but they failed to be a commercial success. Although Pascal made further improvements and built fifty machines, the *Pascaline* became little more than a toy and status symbol for the very rich families in Europe, since it was extremely expensive. Also, people feared it might create unemployment, since it could do the work of six accountants.

Gottfried Wilhelm Leibniz made two attempts to build a calculating machine before he succeeded in 1673; his machine could do addition, subtraction, multiplication and possibly division [283, p. 79].

Since the q-difference operator is fundamental for our treatment, we will go through the historical development of the calculus in some detail. We will outline the development of the infinitesimal calculus and find that the q-integral occurs in geometric disguise. Euclid computed the volume of a pyramid by a geometric series in Elements XII 3/5 [65, p. 48], [459, p. 98]. Archimedes used a geometric series to do the quadrature of the parabola. In the 17th century each mathematician did his own proofs in calculus. Pierre de Fermat (1601–1665), Gilles Roberval (1602–75) and Evangelista Torricelli (1608–47) had great success in the theory of integration. All three, independent of each other, found the integral and derivative of power functions, but in a geometric way. Roberval kept his post as professor in Paris by winning every contest that was set up. Thus he could not publish his discoveries since then he would reveal the secret's of his methods [527, p. 21]. Roberval's win of the competition in 1634 was probably due to his knowledge of indivisibles [527, p. 21].

Between 1628 and 1634 Roberval invented his method of infinitesimals [527, p. 59]. Roberval plotted graphs of trigonometric functions before 1637 in connection with a volume calculation [527, p. 67]. He was also the first to compute certain trigonometric integrals [527, p. 72].

In 1635 Bonaventura Cavalieri (1598–1647) was the first to publish integrals of power functions x^n in his book *Geometria indivisibilis continuorum*; but he proved the result explicitly only for the first few cases, including $n = 4$, while, as he stated, the general proof which he published was communicated to him by a French mathematician Jean Beaugrand (1590–1640), who quite probably had got it from Fermat. Beaugrand made a trip to Italy in 1635 to tell Cavalieri about Fermat's achievements [366, p. 51]. Cavalieri's method was much like Roberval's, but mathematically inferior [527, p. 21].

Fermat was a famous mathematician who founded modern number theory, analytic geometry (together with Descartes), and introduced the precursor of the q-integral.

In that time period, father Marin Mersenne (1588–1648) kept track on science in France and knew about all important discoveries, a kind of human internet. Mersenne had made the work of Fermat, René Descartes (1596–1650) and Roberval known in Italy, both through correspondence with Galilei (1564–1642) dating from 1635 and during a pilgrimage to Rome in 1644.

Fermat's contribution became known through a translation of *Diophantus Arithmetica* by Claude Gaspard de Bachet (1591–1639) in the year 1621. Fermat adhered to the algebraic notation of Viète and relied heavily on Pappus in his development of calculus. Like Kepler, Fermat uses the fact that extreme values of polynomials are characterized by multiple roots of the function put equal to zero [284, p. 63].

Originally Fermat put $f(x+h) = f(x-h)$ for extreme values [292], then developed the expression in terms of powers of h, and finally decided the type of the extreme value from the sign. Later in 1643–44 he even talked of letting $h \to 0$ [284, p. 63]. The leading mathematician of the first part of the seventeenth century was Fermat, who was very talented in languages and handwritings [284, p. 62]. Few results in the history of science have been so closely examined as Fermat's method of maxima and minima [245, p. 24].

Laplace acclaimed Fermat as the discoverer of the differential calculus, as well as a codiscoverer of analytic geometry. Fermat was the first to consider analytic geometry in $\mathbb{R}^3$.

Descartes contended himself with $\mathbb{R}^2$ [66, p. 83]. According to Moritz Cantor (1829–1920) [96, p. 800], Descartes and Fermat were the greatest mathematicians mentioned here.

At that time Fermat and then also Torricelli, had already generalized the power formula to rational exponents $n \neq -1$. Fermat determined areas under curves which he called general parabolas and general hyperbolas. This was equivalent to calculating integrals for fractional powers of x.

Vincenzo Viviani (1622–1703), another prominent pupil of Galilei, determined the tangent to the cycloid.

In 1665, the first two scientific journals were published, *Philosophical Transactions of the Royal Society* in England and *Journal des Savants* in France. The idea of private, for-profit, journal publishing was already established during this time. Fermat however, chose not to publish, probably for political reasons. It is also possible that Fermat was influenced by the old Greek habit of not publishing one's own proofs [366, p. 31].

Roberval was the first to (q-)integrate certain trigonometric functions. At about the same time important contributions were made by the English mathematician John Wallis (1616–1703) in Oxford, who in his book *the Arithmetic of Infinitesimals* (1655) stressed the notion of the limit, see Section 3.7.8.

It was in mathematics that Wallis became an outstanding scientist in his country, although he engaged himself in a wide range of interests.

Blaise Pascal associated himself with his contemporaries in Paris, like Roberval and Mersenne. He learned a method similar to q-integration from his masters and also corresponded with Fermat. Their short correspondence in 1654 [66, p. 107] founded the theory of probability. During several months from 1658 to 1659, Pascal summed infinite series, calculated derivatives of the trigonometric functions geometrically and found power series for sine and cosine. However, in Pascal's time there were no signs for sine and cosine. During the last years of his life Pascal also published the philosophical work *Lettres provinciales* under the pseudonym Louis de Montalte.

Trigonometric tables were published in great numbers in the early seventeenth century; e.g. Mathias Bernegger (1582–1640) published such tables 1612 and 1619 in Strasbourg.

During his life, Jacob Heinlein (1588–1660) had mostly worked as a priest; when he was inaugurated by Kepler in the mathematical world at the time, Heinlein was allowed to lecture on mathematics in Thübingen for a time after Schickard's death in 1635.

Heinlein published *Synopsis mathematica universalis* posthumously in Thübingen in 1663 and 1679, where some trigonometric functions appear. This book was then translated into English and issued in three editions in London, in 1702, 1709 and 1729. In the English expanded edition you will also find trigonometric tables.

The power series for sin and arcsin were communicated by Henry Oldenburg (1626–1678) to Leibniz in 1675 [10]. These series were also known to Georg Mohr (1640–1697) and John Collins (1625–1683). Isaac Barrow (1630–77), who was Isaac Newton's teacher in Cambridge [66, p. 117], published his geometrical lectures in 1664. Barrow was familiar to the concept of drawing tangents and curves, probably from the works of Cavalieri and Pascal. Barrow developed a kind of calculus in a geometrical way, but did not have a suitable algebraic notation for it.

J. M. Child claims that Barrow was the first to give a rigorous demonstration of the derivative for fractional powers of x. But probably Fermat forestalled him. Barrow also touched upon logarithmic differentiation.

Leibniz was a child prodigy in Leipzig, where he learned Latin and Greek by studying books in his father's library. At the age of 15 Leibniz explained the theorems of Euclidean geometry to his fellow students at the university. Leibniz also studied philosophy and was fascinated by Descartes's ideas. Leibniz had found remarks on combinations of letters in a book by Christopher Clavius (1537–1612) [283, p. 3]. Leibniz could decipfer both letter- and number codes and in 1666 published a thesis [357], where the mathematical foundations of combinations were given.

He was also interested in alchemy and became a member of an alchemical society in Leipzig [3]. It was a secret society named after Christian Rosenkreuz (1378–1484), and founded in the seventeenth century. Its aim was to further knowledge, in particular mathematics and alchemy. Despite his outstanding qualifications in law, Leibniz was not given his doctor's degree in Leipzig, so he turned to another city… He stayed in Nuremberg for several months 1667 to learn the secrets of the *Rosenkreuzer* and their scientific books. Because of his outstanding knowledge in alchemy, he was elected as secretary of the society [346, p. 109]. He remembers that he had Cavalieri's book about indivisibles in his hands during this stay [283, p. 5]. During this time, Leibniz was completely under the spell of the concept of indivisibles and had no clear idea of the real nature of infinitesimal calculus [283, p. 8].

Bored by the trivialities of the alchemists and realizing that the world's scientific centre was in France, Leibniz then entered the diplomatic service for several German royal families. As a diplomat, he first went to Paris in 1672, where he mingled in scientific and mathematical circles for four years. He was advised by the Dutch mathematician Huygens to read Pascal's work of 1659 *A Treatise on the Sines of a Quadrant of a Circle.*

In January and February 1673 he visited the Royal Society and was elected to membership [540, p. 260]. He wished to display one of the first models of his calculating machine [283, p. 24]. An English calculating machine was also shown to him, which used Napier's bones [283, p. 25]. On 6 April 1673 Oldenburg mailed Leibniz a long report, which Collins had drawn up for him, on the status of British mathematics. James Gregory's (1638–1675) and Newton's work dominated the report, which included a number of series expansions, although no suggestion of the method of proof was given [540, p. 260]. At the end of 1675, Leibniz had received only some of Newton's results, all confined to infinite series [540, p. 262]. At Leibniz' second visit to London, it appeared that Leibniz' method proved to be more general, but that left Collins wholly unmoved. Even before Leibniz' visit, Collins had been impressed enough to urge Newton anew to publish his method. But since Newton at the time was engrossed in other interests, he did not respond to Collins' suggestion and Oldenburg, who was engaged in Newton's research, unfortunately died two years later [540, p. 264].

Leibniz rewrote Pascal's proof of $\sin' x = \cos x$ in terms of increment in y/increment in x, using finite differences.

The idea of a limit in the definition of the derivative was introduced by Jean d'Alembert (1717–83) and Augustin Louis Cauchy in 1821.

Earlier, there was a suspicion that Leibniz got many of his ideas from the unpublished works of Newton, but nowadays there is agreement that Leibniz and Newton have arrived at their results independently. They have both contributed successfully to the development of calculus; Leibniz was the one who started with integration while Newton started with differentiation.

It was Leibniz, however, who named the new discipline. In fact, it was his symbolism that built up European mathematics. Leibniz kept an important correspondence with Newton, where he introduced and forcefully emphasized his own ideas on the subject of tangents and curves. Newton says explicitly that he got the hint of the method of the differential calculus from Fermat's method of drawing tangents [66, p. 83].

Leibniz also continued the use of infinity that had been used by Kepler and Fermat. In fact Kepler started with an early calculus for the purpose of calculating the perimeter of an ellipse and the optimal shape of wine boxes. When we, according to Leibniz, speak of infinitely great or infinitely small quantities, we mean quantities that are infinitely large or infinitely small, i.e. as large or small as you please. Leibniz said: It will be sufficient if, when we speak of infinitely large (or more strictly unlimited), or of infinitely small quantities, it is understood that we mean quantities that are indefinitely large or indefinitely small, i.e. as large as you please, or as small as you please… These notions were then further used in the works of Euler, Gauß, Heine, and the present author.

Newton called his calculus the "the science of fluxions". Newton used the notation $\dot{x}$ for his fluxions; this symbol is much used in mechanics today. He wrote his *Methodus fluxionum et serium infinitorum* in 1670–1671, but this work was not published until 1736, nine years after his death. It was the astronomer Edmond Halley (1656–1742) who paid for the printing of the masterpiece *Principia* in 1687, where the first rigorous theory of mechanics and gravitation was given. At that time there was only one journal in England where Newton could publish, *Philosophical Transactions of the Royal Society*, dating back to 1665. Newton's theory of light was published there 1672. The *Transactions of the Cambridge Philosophical Society* did not appear until 1821–1928.

On the other hand, the infinity symbol was often used in the circle around Leibniz. In 1682 the first scientific journal of the German lands, *Acta Eruditorum*, was founded in Leipzig, with Otto Mencke (1644–1707) as editor. This journal was very broad in scope and at that time mathematics and astronomy were considered to be one subject. Leibniz published his first paper on calculus in this journal in 1684. One other important author was Ehrenfried Walter von Tschirnhaus (1651–1708), who published on tangents under the pseudonym D. T. The Bernoulli brothers, Jakob and John, soon also started to publish in *Acta Eruditorum*. The result was that almost all of the elementary calculus that we now know was published here before the end of the eighteenth century. For instance, the well-known Taylor formula was published in a different form by Johann Bernoulli in 1694 [210].

We will stop from time to time and note some connections in order to motivate the q-umbral calculus introduced here. Our first concern is of computational nature. Today there are several programs used by scientists for computations. The q-calculus computations are especially hard because of their symbolic nature; there are only two programs which can do the job well. We are only going to grade these two programs on a relative scale, since an absolute evaluation is not possible; we will take the ability of drawing colour graphs into consideration. To describe the relative grades we write a table which compares Maple and Mathematica on the one hand, and prosthaphaeresis and logarithms on the other hand. The astute reader will immediately observe that logarithms have the higher grade, since it shortens down the calculations considerably. The present paper continues the logarithmic method for q-calculus which enables additions instead of multiplications in computations. The resemblance to hypergeometric formulas is also appealing. The q-addition corresponds to the Viète formula for cos and to the de Moivre theorem. We try to use a uniform notation for q-special functions, like the Eulerian notations sin and cos which are now generally accepted. All this is called renaissance [170], [168], [165] in the table.

For each row, the relative merit is higher to the right.

Viète	Descartes
prosthaphaeresis	logarithms
congruences	index calculus
Pascaline	Babbage computer
fluxions	Arbogast notation
Basic	Fortran
Maple	Mathematica
Gasper/Rahman	Renaissance

2.8 The Faulhaber mathematics

In this section we refer to Schneider's biography [455] of Faulhaber. Johannes Faulhaber (1580–1635) was an outstanding Ulmian mathematician, who early was apprenticed as reckoner. He spoke, however, with a few exceptions, neither Latin nor French [455, p. 180]. He could, therefore, not access the large Latin literature, which was available at the time. The complex numbers of Rafael Bombelli (1526–1572) [76] were also inaccessible to him. Faulhaber was a representative of discrete mathematics and we will briefly summarize his contributions.

1. The first 16 Bernoulli numbers (1631) [161, p. 128].
2. A formula equivalent to (2.4).
3. Pascal's triangle.
4. Power sums similar to Section 5.3.
5. The formula (compare (5.89))

$$\sum_{s=k}^{n-1} \binom{s}{k} = \binom{n}{k+1}. \tag{2.10}$$

2.9 Descartes, Leibniz, Hindenburg, Arbogast

Descartes had travelled a lot in his youth and perhaps met Faulhaber in the year 1620 in Ulm [11, p. 223]. Faulhaber was a member of the *Rosenkreuzer*. The *Rosenkreuzer* liked to use unusual signs in their texts; e.g. Faulhaber showed the astrological sign of Jupiter to Descartes. The *Rosenkreuzer* also liked to collect scientific books; Faulhaber had collected several books about algebra and geometry in his home. According to [373, p. 50], Descartes attempted in vain to contact the brotherhood during his travels in Germany. Around that time mathematics in France was not prosperous, due to the dominance of the Catholic Church (Richelieu, Mazarin). As we have seen, the most talented scientists had to keep a secret correspondence: they knew what had happened to Galilei. In 1666, the French Academy of Sciences was founded at the suggestion of Colbert. The situation further improved after the translation of *Principia* into French in 1759 by Émilie du Châtelet (1706–1749). In Germany the situation was different, some states had changed to Protestantism. The *Rosenkreuzer* were opposed to the Catholic Church and preferred a reform of the religious system in continental Europe. They disliked the opposition of the Catholic Church to scientific ideas and wanted a change. This may have been the most important reason why the *Rosenkreuzer* brotherhood was a secret society. There were also English mathematicians, like John Dee (1527–1609), who were *Rosenkreuzer*. Dee had written a long preface to the English translation of *Elements*. He was in the service of Queen Elizabeth and had visited many people on his long trips, including Tycho Brahe. The famous scientist Robert Hooke (1635–1703), contemporary of Newton, wrote in codes and Newton was a dedicated alchemist who virtually gave up science during the last third of his life [64, p. 196]. When Descartes returned to Paris in October 1628, there were rumours that he had become a *Rosenkreuzer*. According to [3], Descartes started to keep a secret notebook with signs used by the *Rosenkreuzer*. Then at the end of 1628, Descartes definitely moved to Holland, where he stayed for 21 years.

The infinite series were introduced by Newton. The formal power series were conceptually introduced by Euler [190]. We will see that this concept prevailed for quite a while until the introduction of modern analysis by Cauchy in 1821.

Leibniz did not have many pupils in Germany, since he worked a lot as a librarian and travelled frequently. While in France mathematical physics was flourishing (Laplace, Lagrange, Legendre (1752–1833), Biot (1774–1862)), German mathematics had a weak scientific position in the time between Frederick the great (1712–1786) and the Humboldt education reform [312, p. 173], [95, p. 256]. One exception was Georg Simon Klügel (1739–1812), who introduced a relatively modern concept of trigonometric functions in [339] (1770) and [338] (1805).

Combinatorial notions such as permutation and combination had been introduced by Pascal and by Jacob Bernoulli in [70].

In the footsteps of Leibniz, Hindenburg, a professor of physics and philosophy in Leipzig founded the first modern School of combinatorics with the intention of promoting this subject to a major position in mathematics. One of the ways to achieve this aim was through journals. The *Acta Eruditorum* continued until 1782, then Hindenburg and John Bernoulli (III) (1744–1807) edited *Leipziger Magazin für reine und angewandte Mathematik* (1786–1789). This was followed by *Archiv der reinen und angewandten mathematik* (1794–1799) with Hindenburg as sole editor. Hindenburg used certain complicated notations for binomial coefficients and powers; one can feel the influence of the *Rosenkreuzer* secret codes here. The main advantages of the Hindenburg combinatorial School was the use of combinatorics in power series and the partial transition from the Latin language of Euler. The disadvantages were the limitation to formal computations and the old-fashioned notation.

A partial improvement was made by Bernhard Friedrich Thibaut (1775–1832) [312, p. 193], [496], who expressed the multinomial theorem, a central formula in the Hindenburg School, in a slightly more modern way. Thibaut also made a strict distinction between equality of formal power series and equality in finite formulas [312, p. 196], a central theme in q-calculus.

A similar discussion about formal power series was made by Gudermann in 1825 [253].

The Hindenburg combinatorial School can be divided into two phases [312, p. 171].

1. 1780–1808.
2. 1808–1840.

The first important book of Hindenburg [280] began with a long quotation from Leibniz [357]; this quotation would be normal in Hindenburg combinatorial School publications until 1801 [312, p. 178].

One of the main reasons for this is the so-called multinomial expansion theorem, a central formula in the Hindenburg combinatorial School, which was first mentioned in a letter 1695 from Leibniz to John Bernoulli, who proved it.

Hindenburg thought that a mechanical calculation would be an important objective for his School [312, p. 222] and that is exactly what we do in this book.

The bricks are the basic formulas for the q-*factorials*. The general formulas and developments can often be expressed through the q-binomial coefficients as (4.74) and (4.75).

Heinrich August Rothe (1773–1842) introduced a sign for sums, which was used by Gudermann. In 1793 Rothe found a formula for the inversion of a formal power series, improving on a formula found, with no proof provided, by Hieronymus Eschenbach (1764–1797) in 1789 [312, p. 200]. This invention gave the combinatorial School a rise in Germany, as can be seen from the list of its ensuing publications [312, p. 201].

Rothe's presentation was marked by clarity, order and completeness, according to a report in the newspaper *Jenaische Allgemeine Literaturzeitung* in 1804. Rothe was the teacher of Martin Ohm, of whom we shall hear much later.

In the year 1800, Louis Arbogast suggested [42] to substitute a capital D for the little $\frac{dy}{dx}$ of Leibniz to simplify the computations. Arbogast [42, p. v] writes: *The Leibniz theory of combinatorics has been improved by Hindenburg, who has studied the development of functions of one variable and given the multinomial theorem. The procedure and notation of Hindenburg is not familiar. I don't know his writings except for the title; I have followed my proper ideas. The procedure I will give is very analytical.*

In [42, p. 127] Arbogast gives Taylor's formula for a function of two variables. The use of signs for sums would significantly increase the clarity of this long formula... At any rate, Arbogast's formula had a long-lasting influence on the development of calculus in Germany and in England.

This can be seen from the different notations in two publications by Hindenburg. In 1795, the Journal *Archiv der Reinen und Angewandten Mathematik* received some papers on the Taylor formula, in which it was expressed through a difference operator. In 1803, however, Hindenburg [279, p. 180] also used the symbol D and apparently noticed the difference between the two. The above-mentioned journal also received some military reports, among others from Johann Heinrich Lambert (1728–1777), mathematician and physicist. Hindenburg, who was also a physicist, read Lambert's articles in physics with interest, which in turn had a strong influence on Gudermann.

From about this time onwards, the theory of special functions according to Euler can be said to have started. This development was parallel to the theory of the Bernoulli numbers and the Stirling numbers.

2.10 The Fakultäten

In 1730 James Stirling (1692–1770), by a remarkable numerical analysis, confirmed a result which would now be written in the form [159, p. 18]:

$$\Gamma\left(\frac{1}{2}\right) = \sqrt{\pi}.$$

The Frenchman Alexandre-Théophile Vandermonde (1735–1796) found the same formula [516]. This article was translated into German and the formula appeared in [515, p. 77].

In 1772 Vandermonde introduced the following notation [515]:

Definition 14 The falling factorial is defined by

$$(x)_{n-} \equiv x(x-1)(x-2)\cdots(x-n+1). \tag{2.11}$$

We will now explain how the development of the Fakultäten occurred in parallel with the development of Newton's binomial theorem and the fluxions. To avoid the metaphysical difficulties of the fluxions, John Landen (1719–1790) suggested to use a purely algebraic method, which could be compared with Lagrange's operational method. That's why Landen was called the English d'Alembert.

The Italian mathematician A. M. Lorgna (1730–1796) had similar ideas. In the period 1791–1806 a series of books entitled *Scriptores Logarithmici* [371] were published in London. In the second volume Landen gives a proof of Newton's binomial theorem with positive quotient-exponent. In volume five he gives the proof for exponent $-\frac{1}{n}$. The name and the method has many similarities with the so-called *Fakultäten*. As we shall see, the fluxions disappeared in the English teaching about twenty years later.

In 1798 Christian Kramp (1760–1826) introduced the *Fakultäten*, a function similar to the Γ function. In 1800 Arbogast used the word Faktoriell and in 1808 Kramp introduced the notation $n!$.

Vandermonde [515] used the function

$$f(u,x,y) \equiv \prod_{m=0}^{y-1} (u+mx). \tag{2.12}$$

When $y \in \mathbb{N}$, the following relations hold:

$$f\big(u,x,y+y'\big) = f(u,x,y) f\big(u+yx,x,y'\big), \tag{2.13}$$

$$f(u,x,1) = u, \tag{2.14}$$

$$f(ku,kx,y) = k^y f(u,x,y), \tag{2.15}$$

$$f(u,x,y) = f(u+yx-x,-x,y), \tag{2.16}$$

$$f(u,0,y) = u^y. \tag{2.17}$$

This combinatorial School of Vandermonde and Kramp enjoyed a certain popularity in the period 1772–1856. The aim of this School was to divide the Fakultäten into four classes: positive, negative, whole and fractional exponents.

Each class had its own laws, similar as for the q-factorial. The *Fakultäten* were also considered by Ettingshausen [525, p. 190]. Arbogast also writes a lot about the *Fakultäten* in his book [42, p. 364]. Vandermonde and Kramp [347] tried to extend this function to all $y \in \mathbb{R}_+$ by the definitions (2.13)–(2.17). This however did not turn out well, as was shown by Friedrich Wilhelm Bessel (1784–1846) in 1812 [71, p. 241]. Bessel tried to mend this theory by defining the factorial $f(u,x,y)$ in another way. In this connection, Bessel [71, p. 348] published a formula very similar to the Euler reflection formula

$$\Gamma(x)\Gamma(1-x) = \frac{\pi}{\sin \pi x}. \tag{2.18}$$

This formula was expressed in terms of *Fakultäten*. Two other similar formulas were published in the same paper. In 1824, M. Ohm and Ludwig Öttinger (1797–1869) made further attempts in *Crelle Journal* to rescue the theory of the *Fakultäten* without the introduction of complex numbers and of the principal branch of the logarithm. In 1843 and 1856 Weierstraß [539] published a long paper, which put the theory of analytic factorials on a solid mathematical foundation by connecting the formulas to the Γ function. Weierstraß' investigations also contained Bessel's formula (2.18). Thereafter, the Γ function was mainly used. Nevertheless, the *Fakultäten* have continued to live in the Russian textbook by Guelfond, *Calcul des Différences Finies* [257, p. 25]; this book was translated from Russian into many languages.

2.11 Königsberg School

The Königsberg School was able to develop quickly, as there were not so many visitors and one could therefore concentrate on research. It included many celebrities, such as Immanuel Kant (1724–1804) and Jacobi. The proximity of St. Petersburg was also noticeable, e.g. for the Bernoulli numbers. The Bessel function, which was introduced in 1824 by Bessel, was investigated earlier by Jacob Bernoulli, Daniel Bernoulli (1700–1782), Euler and Lagrange. There is also a link to differential equations: the Bessel differential equation is related to the Riccati differential equation, which was introduced in 1724 by Francesco Riccati (1676–1754).

This paved the way for the development of the hypergeometric function, the basis of special functions and the prerequisite for q-hypergeometric functions.

Jacobi discovered the theta functions by a brilliant formal derivation. After Jacobi's death (1851) his School was continued by his favourite student Friedrich Julius Richelot (1808–1875).

Franz Ernst Neumann (1798–1895) and his doctoral student Louis Saalschütz (1835–1913) have both contributed to the theory of special functions. Neumann, who formulated the law of electric induction in 1845 and 1847, but with very different notation, has also contributed to the theory of Bessel functions (Neumann function).

In his book [446] Saalschütz summarized the present knowledge in the field of Bernoulli numbers, and has [446, p. 54–116], according to Gould, given 38 explicit formulas for Bernoulli numbers.

Saalschütz also republished the Euler-Pfaff-Saalschütz summation formula for hypergeometric functions.

Reinhold Hoppe (1816–1900), mathematical physicist and professor in Berlin, also belonged to the Königsberg School.

2.12 Viennese School

This is a School without q-calculus.

Ettingshausen introduced the notation $\binom{m}{n}$ for the binomial coefficients in his book about combinatorics [525, p. 195] published in 1826 and used the so-called Stirling numbers. At that time, Euler's *Vollstaendige Anleitung zur Integralrechnung* was also translated from Latin into German and published in Vienna.

Jozef Petzval (1807–1891) [227] was a world famous optician and good mathematics teacher, who wrote an excellent textbook on differential equations.

Leopold Gegenbauer (1849–1903) wrote excellent articles about orthogonal polynomials; the Gegenbauer polynomials are named after him.

Lothar Koschmieder (1890–1974) worked in Wroclaw, Brno, Aleppo, Graz and Tübingen. In the forties Koschmieder published, among other journals, in Austrian journals. He worked a great deal with differential operator computations on polynomials and (multiple) hypergeometric series.

2.13 Göttingen School

In 1808 Gauß wrote his only article on q-analysis [442], where he introduced the q-binomial coefficients and computed some sums of these. This has evidently influenced Rothe and Johann Philipp von Grüson (1768–1857) to find the following important theorem. According to Ward [531, p. 255] and Kupershmidt [349, p. 244], the identity (2.19) was already known to Euler. Gauß in 1876 [222] proved this formula.

Theorem 2.13.1 *Rothe* (1811) [441], [526, p. 36] (1814) *von Grüson. Fundamental Theorem of q-calculus*:

$$\sum_{n=0}^{m}(-1)^n\binom{m}{n}_q q^{\binom{n}{2}}u^n = (u;q)_m. \tag{2.19}$$

We note that the names of two students of Gauß, Grünert and Gudermann, can be found in the next section, and this is no coincidence.

Johannes Friedrich Posselt (1794–1823) used *Fakultäten* in his Göttingen thesis *Dissertatio analytica de functionibus quibusdam symmetricis* (1818); he spent his last years at the observatory in Jena.

Without ever having passed through a baccalaureate degree, Moritz Stern (1807–1894), born in Frankfurt, received in 1829 the Doctorate for a treatise on the theory of continued fractions. Stern then became interested in the field of number theory, to which he would devote by far the greater part of his life. In the period until the appointment to professor in 1859, Stern also pursued the teaching, which Gauß did not like: the two were good friends.

In 1847 Stern wrote the book *Zur Theorie der Eulerschen Integrale*. Stern had a considerable role in the massive reform of university mathematics teaching that took place in this period.

2.14 The combinatorial School: Gudermann, Grünert

The goal of the combinatorial School was to develop functions in power series by Taylor's formula. Taylor's formula was originally formulated with finite difference-quotients, so-called fluxions. Earlier, there were two names for finite differences, after Brook Taylor (1685–1731) and Jasques Cousin (1739–1800). Instead, in 1795, Hindenburg introduced the notation [281, p. 94] $\overset{k}{\underset{y}{}}$ in *Archiv der Reinen and Angewandten Mathematik.*

Christoph Gudermann, encouraged by his close friendship with August Crelle (1780–1855), was first school teacher in Cleve, later professor in Münster. He first wrote in German and later alternately turned to Latin, the common scientific language of that time and could therefore reach international recognition.

Gudermann wrote an excellent Latin in a time when the Latin in Europe had declined.

Crelle was very interested in the contemporary mathematical questions and was able to find publishers for Gudermann's textbooks.

Gudermann even had a crucial role as the teacher of Weierstraß. It is reported that thirteen students came to the first lecture of Gudermann about elliptic functions. At the end of the semester only one remained, namely Weierstraß. Gudermann was one of the first who discovered Weierstraß' extraordinary talent for mathematics. Weierstraß was inspired by Gudermann's theories on series expansions and often expressed his great gratitude for his old teacher. Weierstraß, for his part, developed and modified the Gudermann combinatorial School.

Bernard Riemann was probably also influenced by Gudermann. In his textbook on elliptic functions [432] one finds the series (7.32)–(7.34) with the Gudermann notation. This book contains the Riemann lectures of 1855–56 and 1861–62.

Influenced by Lambert, who introduced the hyperbolic functions, Gudermann, among other things, developed the function $\frac{1}{\cosh(x)}$ in powers of x, using the work of Scherk on the so-called Euler numbers. The Gudermann names for the trigonometric functions had many followers until 1908 [94, p. 173].

Gudermann used the sign for sums of Rothe and also a product symbol for $\sin x$ and $\cos x$ in the form of infinite products [255, p. 68]. Gudermann has often developed his functions by Taylor's formula; he then used a precursor of the Pochhammer symbol – in the disguise of Kramp's notation.

One could say that the circle around Gudermann formed a School of its own. This School consisted, among others, of Johann August Grünert (1797–1872), editor of *Archiv der Mathematik und Physik*, which started in 1841, and Oscar Schlömilch, editor of *Zeitschrift für Mathematik and Physik*, which started in 1856. These two journals differed from Crelle's Journal, which had a more purely mathematical character. Grünert, mathematician and physicist, also took over the completion of the dictionary of Klügel and the preparation of two supplementary volumes. He was a student of Pfaff and Gauß, wrote, among other things, early on *Fakultäten*, and thus composed tables of Stirling numbers [250, p. 71], [252, p. 279]. Stirling numbers were later used in series expansions for Bernoulli functions [545, p. 210]. Grünert worked almost 39 years as a full professor in Greifswald, where he founded a mathematical seminar in 1825, and also made his private library available to his students.

In Sweden, Malmsten and Björling both contributed to the *Grünert Archiv*. This magazine also included publications about hyperbolic functions and spherical trigonometry. This last name is a modern term for analytical Sphaerics, which was treated in [254]. We will later return to contributions about elliptic functions in Grünert Archiv.

You could say that this was an earlier start of the Mathematical Reviews and Centralblatt Mathclass section 33 (Special functions with applications) in Europe.

Grünert had a conflict with Grassmann in 1862, and perhaps this is why his name is not mentioned in Klein's eminent book [336]; Klein also treats Gudermann unfairly.

In 1853, when Grünert was 56, *Archiv der Mathematik und Physik* began its decline. After the death of Grünert in 1872, Hoppe took over the editorial.

2.15 Heidelberg School

Franz Ferdinand Schweins (1780–1856) taught mathematics 46 years at the University of Heidelberg. Heidelberg became the centre of the combinatorial School under Schweins. Öttinger was director of the Pädagogicum in Durlach (near Karlsruhe) in 1820, professor at the Gymnasium in Heidelberg in 1822 and next lecturer at the University of Heidelberg from 1831 to 1836.

One might therefore speak of a Heidelberg School. In the textbook by Öttinger of 1831 [408, p. 26] one finds Euler's special case of the formula (2.20) below; the proof has been carried out by using partitions, just like Euler did. The textbook by Schweins [458, p. 613] contains the *Fakultäten*. He often referred to Jozef Maria Hoene Wronski (1778–1853). Schweins then proves the following theorem (2.20) with the help of the *Fakultäten*, that together with (2.19) forms the basis of q-analysis.

Theorem 2.15.1 *The Schweinsian q-binomial theorem* [457, p. 294, (497)]:

$$\sum_{n=0}^{\infty} \frac{\langle a;q\rangle_n}{\langle 1;q\rangle_n} z^n = \frac{(zq^a;q)_\infty}{(z;q)_\infty},$$

$$|z| < 1, \ 0 < |q| < 1. \tag{2.20}$$

In the following, we call this relation simply the q-binomial theorem; it is more general than the two Euler formulas (6.188), (6.189).

Remark 2 The notation of Schweins leaves much to be desired, yet it is better than Hindenburg's notation.

Remark 3 In his book [458, p. 317–] Schweins gives a theory of "Producte mit Versetzungen", which according to Muir [389] is equivalent to a deep theory of determinants. His colleague Öttinger has written a similar book *Differential und Differenzenrechnung* in 1831. It is clear that the two have worked together.

2.16 Weierstraß, formal power series and the Γ function

Karl Weierstraß was a colleague of M. Ohm, Kummer and Hoppe in Berlin. In the footsteps of Gudermann, Weierstraß introduced the modern Analysis.

A famous student of Weierstraß, Gösta Mittag-Leffler (1846–1927) tells [491, p. 214] that formal power series were always used. The Euler and Gaussian formulas for the Γ function were here brought together in modern notation.

Earlier Adrien-Marie Legendre had denoted the Γ function by Γ and computed the Euler integrals for $\Gamma(x)$. As we have already mentioned, in 1856 Weierstraß replaced the *Fakultäten* by the complex Γ function. In the same year Alfred Enneper's dissertation *Über die Funktion* Γ *von Gauß mit komplexem Argument* appeared. Based on the Legendre integral representation of the Gamma function, Enneper proceeds purely formal and arrives through long logarithmic computations to a series of profound results.

These two works by Weierstraß and Enneper (1830–1885) thus have virtually extended the Γ function to the complex plane. In the present book the Γ_q function is extended to the complex plane, compare [180]. The elliptic functions in a more general form than Gudermann's also played an important role in Weierstraß work. The number of Weierstraß students was high, and we list here only those who mainly dealt with special functions. Nicolai Bugaev (1837–1903), doctorate in 1866, had a gifted student Sonine, who worked with Laguerre polynomials and Bessel functions. Mathias Lerch (1860–1922) has written some interesting articles on Theta functions, which are similar to q-analysis. The works of Lerch in q-analysis have many similarities with those of Leopold Schendel [450], who has also written a book on this subject [451]. Schendel published work on the q-Gaussian Taylor series and pointed out an expansion of the logarithmic integral.

2.17 Halle q-analysis School

In 1844 Gudermann published his famous book on elliptic functions [256]. Two years later, in 1846, E. Heine, a docent in Bonn, who had studied with Gauß, Lejeune Dirichlet (1805–1859) and Jacobi, wrote the following letter to his professor Dirichlet. This letter, which is a natural continuation of the work of Gudermann, was published in the same year [269] in the Crelle Journal.

> Sehr viele Reihen, darunter auch solche, auf welche die elliptischen Funktionen führen, sind in der allgemeinen Reihe
>
> $$1+\sum_{k=1}^{\infty}\frac{\prod_{m=0}^{k-1}(q^{a+m}-1)\prod_{m=0}^{k-1}(q^{b+m}-1)}{\prod_{m=0}^{k-1}(q^{1+m}-1)\prod_{m=0}^{k-1}(q^{c+m}-1)}z^k$$
>
> enthalten, die ich zur Abkürzung mit
>
> $${}_2\phi_1(a,b;c;q,z)$$
>
> bezeichne, gerade so wie es bei der hypergeometrischen Reihe zu geschehen pflegt, in welche unser ϕ für $q=1$ übergeht. Es scheint mir nicht uninteressant, die ϕ ganz ähnlich zu behandeln, wie **Gauß** the ${}_2\mathrm{F}_1$ in den 'Disquisitiones generales' untersucht hat. Ich

> will hier nur flüchtige **Andeutungen** zu einer solchen Übertragung geben. Es entspricht jedem F im §5 der Disquisitiones genau eine Reihe ϕ. Einigen Formeln entsprechen zwei oder mehr verschiedene ϕ, nämlich denen, in welchen a oder b unendlich werden. So hat die Reihe für e^t zwei Analoga.

Heine worked as a professor in Halle, he often went to Berlin, as his sister, who was married to Felix Mendelssohn, lived there.

Heine introduced the q-hypergeometric functions and proved their transformation formulas formally by continued fractions [270]. This was the first time that q-equations had to be corrected; certainly it would have been better for Heine (already in 1847) to use the improved notation (1.57).

Heine translated the work of the Swedish mathematician Göran Dillner (1832–1906) about quaternions into German. Heine published his famous book on spherical harmonics [272] in 1861. In the same year, Thomae began his studies at the Universität Halle, near his home. Heine had the greatest influence on Thomae, who thus developed his great liking for function theory.

The astronomer Ernst Schering (1833–1897) was tutor of Thomae in Göttingen 1864, and wrote a treatise called *Allgemeine Transformation der Thetafunctionen*. In 1867 Thomae became a docent in Halle, where he was a colleague of Heine and Georg Cantor (1845–1918). Together with the Reverend Jackson, Thomae has developed the so-called q-integral, the inverse of the q-derivative. Thomae also wrote important works about hypergeometric series and in fact many years these two subjects have developed together.

The cooperation with Heine in q-analysis lasted till 1879.

Karl Heun (1859–1929), a student of Schering and Enneper in the period 1878–80 in Göttingen, did not stay long in Halle between April and October 1880. Heun went back to Göttingen and began his doctoral work, which was inspired by Heine; his supervisor was again Schering. In 1881 Heun defended his doctoral dissertation *Kugelfunctions and Lamésche functions als Determinanten*. The Heun equation is a linear differential equation of second order of Fuchsian type with four singular points.

2.18 Jakob Friedrich Fries, Martin Ohm, Babbage, Peacock and Herschel

In the following we frequently quote Elaine Koppelman [345] and E. P. Ozhigova [409].

Robert Woodhouse (1773–1827), George Peacock (1791–1858), Charles Babbage (1791–1871) and John Herschel (1792–1871) were all from Cambridge.

Woodhouse discussed at length the importance of good notation [345, p. 177], since the development of calculus in Cambridge had been slow until 1820 [345, p. 155]. His conclusion was that the notation of Arbogast is by far superior.

Cajori [93] has drawn a similar conclusion.

Already in 1803 Woodhouse had tried to put calculus on a rigorous algebraic basis by a formal power series development, similar to Lagrange, in his important work *Principles of Analytic Calculation*, which had a lasting influence on Babbage [345, p. 178]. Another attempt by Woodhouse to bring mathematics at Cambridge up-to-date was in 1804, when he published a paper on elliptic integrals in the *Philosophical Transactions of the Royal Society*. He fully realised the significance of the topic, which earlier had received little attention in Cambridge.

The first immediate reaction to Woodhouse's book from 1803 came when Rev. John Brinkley (1763–1835), in a paper in *Phil. Trans. Royal Soc. London* in 1807, started the first symbolic calculus in England. Brinkley's paper contained some abbreviations for expressions like $\frac{x^n}{n!}$ and $\frac{\dot{x}}{n!}$. Here $\dot{x}$ is the fluxion of x. Brinkley also calculated with expressions for "differences of nothing", the precursor of Stirling numbers. Of course the Stirling numbers had been known already to Thomas Harriot (1560–1621), but Brinkley probably was not aware of this. Although Brinkley knew about Arbogast, he writes: *My publication has hitherto been delayed by my unwillingness to offer a fluxional notation different from either that of Newton or Leibniz, each of which is very inconvenient as far as regards the application of the theorems for finding fluxions.*

Brinkley's work became widely known in Russia, partly due to his fame as an astronomer. His work was also published in France by the mathematician and astronomer Dominique François Jean Arago (1786–1853) in 1827 [409, 138]. After receiving the chair of astronomy at Trinity College, Dublin, in 1790, Brinkley had to wait 18 years until the new telescope was erected, and still stands. He had eighteen years more in which to use it. During the first of these periods Brinkley devoted himself to mathematical research; during the second he became a celebrated astronomer.

Jakob Friedrich Fries (1773–1843) was a German philosopher, who valued mathematics very highly. In 1822 Fries's work on the mathematical philosophy of nature appeared.

Fries says: Every philosophy that matches the exact sciences may be true, any that contradicts them must necessarily be false.

Among others, Gauß highly valued the philosophy of Fries, and Schlömilch was a student of Fries.

The fluxion concept was dominant in England until 1820, when the four people from Cambridge managed to recognize the notation of Leibniz and Arbogast in England [345, p. 156]. This led shortly to the introduction of operator calculus [345, p. 156] or umbral calculus. The fluxion notation was cumbersome, an expression could have many meanings [409, p. 139]. Other mathematicians could not understand the fluxions [409, p. 139].

The astronomical tables had a great importance for the navigation.

Herschel and Babbage have pointed to the many errors in the astronomical tables and insisted that an automatic calculating machine is needed [409, p. 139]. Many computations were then made using logarithms, see the book by Wilson from 1820 about calculus of finite differences.

In the translation of Lacroix's book by Babbage, Peacock and Herschel (1816) it is claimed in the preface that calculus was discovered by Fermat, made analytical by Newton and enriched with a powerful and comprehensive notation by Leibniz [345, p. 181]. Before Babbage dropped this subject he once again stressed the importance of a good notation for calculus [345, p. 184]. E. T. Bell (adviser of Morgan Ward) wrote that operational mathematics, which was developed in England during the period 1835–1860, despite its obvious utility, was scarcely reputable mathematics, because no validity condition or validity region accompanied the formulas obtained [345, p. 188]. In his book *Treatise on Algebra* (1830) Peacock studied the relationship between algebra and natural numbers; he called it symbolic algebra.

One of Peacock's students was De Morgan [409, p. 143]. England thus became the centre for symbolic calculations. Combinatorial analysis (Germany) and symbolic computations (England, France, Italy) developed in two different directions [409, p. 121]. The symbolic calculus was introduced in 1880 in Italy [115]; in the footsteps of Koschmieder, Johann Cigler reintroduced the symbolic method in Austria 1979 [133].

In 1837 the *Cambridge Mathematical Journal* was founded by, among others, Duncan Gregory (1813–1844), to provide a place for publication of short mathematical research papers and thus encourage young researchers. In an 1845 letter to John Herschel [345, p. 189], De Morgan described the journal and Gregory's contributions as full of very original communications, very full of symbols. In Gregory's first paper on the separation of symbols, the linear differential equation with constant coefficients was treated. Similar studies had already been published by Cauchy in France; Gregory was familiar with Cauchy's and Brisson's works on this subject [345, p. 190]. As pointed out by De Morgan in 1840 [345, p. 234], the symbolic algebra method gives a strong presumption of truth, not a method of proof. Gregory correctly claimed that the operations of multiplication and function differentiation obey the same laws [345, p. 192]. Gregory's methods only applied to differential operators with constant coefficients.

A generalization to non-commutative operators was given in 1837 [345, p. 195] by Robert Murphy. The studies of non-commutative operators were continued by, among others, George Boole, William Donkin (1814–1869) and Charles Graves (1812–1899). More general functional operators were studied by W. H. L. Russell [345, p. 204], William Spottiswoode (1825–1883), William Hamilton (1805–1865) and William Clifford (1845–1879).

Let's summarize: the calculus of operations, imported from France and extended by Babbage and Herschel, was an important mathematical research area in England between 1835 and 1865 [345, p. 213]. Most of these articles were published in the *Cambridge Mathematical Journal* and its successor, the *Cambridge and Dublin Mathematical Journal* (CDMJ) (1845–1854).

The CDMJ handled all kinds of subjects such as physics and astronomy. In CDMJ 1 De Morgan summarized the work of Arbogast.

In CDMJ 3, Rev. Brice Bronwin wrote an article about umbral calculus. This description is typical of the many so-called mathematicians in England who were clergymen and were not familiar with the modern Analysis of Cauchy. Formal power series were introduced in England in the 1880s by Oliver Heaviside. Soon after England took over the lead in umbral calculus.

The *Quarterly Journal of Pure and Applied Mathematics* (QJPAM) rose from the ashes of the CDMJ in April 1855 [144]. The first two British editors were Ferrers and J. J. Sylvester (1814–1897), who was editor until 1878. It was here that the first works on umbral calculus were published by Horner in 1861, John-Charles Blissard in the years 1861–68 [144] and Glaisher. It was Sylvester who coined the name umbral calculus.

F. H. Jackson, the first master of q-calculus in the twentieth century, who was also a priest like Blissard, published many of his papers in QJPAM.

From about 1860 the calculus of operations split into different areas, some of which are:

1. umbral calculus.
2. q-calculus.
3. theory of linear operators [345, p. 214].
4. algebra.

These different subjects are however far from disjoint.

The theory of formal power series within the Hindenburg combinatorial School continued under Martin Ohm, who under the influence of Cauchy's *Cours d'analyse*, obtained convergence criteria for the known elementary transcendental functions in certain regions. Ohm defined mathematical analysis using seven basic operations and built a rather involved theory, which clearly was a forerunner of later attempts to base all of mathematics on the integers [345, p. 226].

Some students of M. Ohm were

1. Eduard Heine, who introduced the q-hypergeometric series.
2. Leo Pochhammer (1841–1920), known for the Pochhammer symbol and the Pochhammer integral.
3. Friedrich Prym (1841–1915), who made many investigations about special functions and even founded a mathematical School in Würzburg.

Martin Ohm made mathematics into a clean, accurate subject, without physics. He also wrote physics books. In a textbook of 1862 (short guide) Ohm writes on page 62 about *Fakultäten*. In 1848 Ohm unsuccessfully tried to merge the Γ function, which after 1856 was universal and the *Fakultäten*.

In 1833 Hamilton read a paper expressing complex numbers as algebraic couples, and in 1837 he presented an article on the arithmetization of analysis [372]. This was a careful, detailed and logical criticism of the foundations of algebra, and it represents an important step in the development of modern abstract algebra (C. C. Macduffee, 1945) [345, p. 222].

In 1846 Hamilton published a series of papers with the title *On symbolic geometry*. Again he cited Peacock and Martin Ohm as the authors who had inspired him to a deeper appreciation of the new School of algebra [345, p. 222]. Like De Morgan, Hamilton wanted, at first, a system which would form an associative and commutative division algebra over the reals [345, p. 228]. Out of this finally grew the famous quaternions.

2.19 Different styles in q-analysis

In mathematics there are different styles. There is the so-called real analysis, in modern literature often connected with Sobolev spaces, where one the main purposes is to find different norms and inequalities for integrals. Then there is the field of formal computations with special functions, which is closely connected with number theory and induction.

In q-analysis, there is a similar layout. Here in the real analysis case, often the Γ_q function is used to formulate inequalities. Also q-integrals occur, but here far from all possibilities are exhausted.

The formal computations within q-analysis play, as we will see, a big role. In most cases, a real analysis article contains few formal calculations and vice versa. But this need not always be the case, both styles can learn from each other.

Chapter 3
Pre q-Analysis

3.1 The early connection between analytic number theory and q-series

Euler started analytic number theory by inventing the *Euler-Maclaurin summation formula* 1732 [187] and 1736 (proof) [188], which expresses the average value of an arithmetical function using integrals and the greatest integer function [201], [37].

Then came the *Euler product* in 1737, which expresses the infinite sum of a multiplicative arithmetical function as the infinite product over the primes of an infinite sum of function values of prime powers. If the function is *completely multiplicative*, the product simplifies.

Euler developed the theory of partitions, also called *additive number theory* in the 1740s. The *partition function* $p(n)$ is the number of ways to write n as a sum of integers. There are also other partition functions, but they have some restriction on the parts, e.g. only odd parts are allowed. Euler found all the generating functions for these partitions. In 1748 Euler considered the infinite product $\prod_{k=0}^{\infty}(1-q^{k+1})^{-1}$ as a generating function for $p(n)$.

In the same year Euler discovered the first two q-exponential functions, a prelude to the q-binomial theorem, and at the same time introduced an operator which would over a hundred years later lead to the q-difference operator.

In 1750, by using induction, Euler proved the *pentagonal number theorem*

$$1+\sum_{m=1}^{\infty}(-1)^m\left(q^{\frac{m(3m-1)}{2}}+q^{\frac{m(3m+1)}{2}}\right)=\prod_{m=1}^{\infty}(1-q^m),\quad 0<|q|<1; \tag{3.1}$$

this was the first example of a q-series, and at the same time the first example of a theta function.

Yet another example of an early q-series is the following result of Gauß, published in 1866, 11 years after his death in 1855. The reason for the late publication was Gauß's work on geodesy and magnetism. Probably he found the formula during

T. Ernst, *A Comprehensive Treatment of q-Calculus*,
DOI 10.1007/978-3-0348-0431-8_3, © Springer Basel 2012

his youth.

$$1+\sum_{m=1}^{\infty} q^{\binom{m+1}{2}}=\prod_{m=1}^{\infty}\frac{1-q^{2m}}{1-q^{2m-1}},\quad |q|<1. \tag{3.2}$$

In many areas of mathematics, Gauß generalized the results of Euler. Gauß published little during his life, he had few students, but many mathematicians have been strongly influenced by him, among them August Ferdinand Möbius (1790–1868), Gudermann, Grünert and Riemann.

Assume that we have a congruence

$$a \equiv b \pmod{m}. \tag{3.3}$$

Now pick out a primitive root r for m.

According to the so-called index calculus, this is equivalent to studying the congruence

$$i(a) \equiv i(b) \big(\operatorname{mod} \varphi(m)\big), \tag{3.4}$$

where $i(a)$ and $i(b)$ are the so-called indices of a and b for the root r. The index calculus obeys the logarithm law [37].

$$i(ab) \equiv \big[i(a)+i(b)\big] \big(\operatorname{mod} \varphi(m)\big). \tag{3.5}$$

We will later introduce a similar calculus for q-analysis, which goes back to Heine.

The idea of indices was thus introduced by Euler, but the first systematic treatment of the subject was given by Gauß in the third Section of his *Disquisitiones*, published in 1801.

3.2 Some aspects of combinatorical identities

The vast subject of combinatorics is concerned with enumeration problems, a field which has recently developed explosively. For several reasons, Richard Stanley's (1944–) book on Enumerative Combinatorics from 1986 was the first comprehensive treatment of the subject and it greatly helped spark the interest. It also gave lecturers on the subject the possibility to actually teach combinatorics in a meaningful way.

China has in recent years become a centre for the study of combinatorics—American and European professors gladly travel far to be in the thick of things. Furthermore, combinatorics today has applications in such diverse fields as chemistry, biology, finance and economics, scheduling, forestry, logistics, location, airlines, telecommunication, transportation, production planning—to name but a few....

q-calculus forms one of the languages for algebraic combinatorics. That is, many results in combinatorics may be expressed by q-series.

This Section (3.2) and Section 3.3 introduce the necessary notation with regard to the combinatorial identities and Bernoulli numbers. Section 3.7 deals with hypergeometric series, and Sections 3.11–3.11.4 treat orthogonal polynomials. Combinatorical identities, hypergeometric functions and orthogonal polynomials are in fact part and parcel of the same, or, differently stated, three different approaches to binomial coefficient identities. They are intimately connected, there are even direct links between algebraic combinatorics and orthogonal polynomials. In order to fully treat a problem in this field, to comprehend the scope of it and be able to communicate across the 'language' barriers that separate the fields, it is necessary to at least keep in mind the three approaches. The Venn diagram shows the intersections between the areas. q-calculus—outside the Venn diagram encircles the entire field, and several topics, e.g. topology or algebra, tie up with one or the other area though with no immediate connection. In the very centre of the diagram, you achieve, of course, the highest understanding and outlook on the different areas.

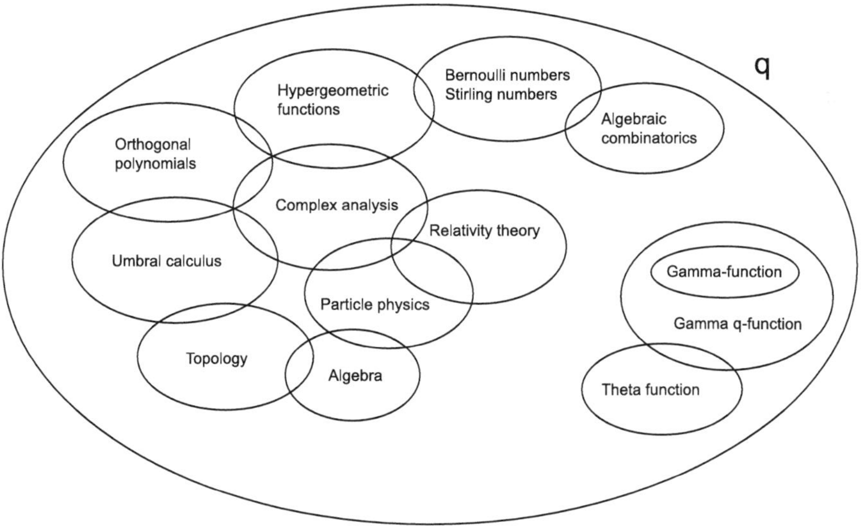

Now, the historic mathematical literature in general contains large numbers of combinatorial identities. Combinatorial identities can be expressed either as hypergeometric formulas—the canonical form—or as binomial coefficient identities. Combinatorial results are often expressed as q-formulas. Here we have a similar duality: q-hypergeometric formulas—the canonical form—or q-binomial coefficient identities. Now q-binomial coefficient computations are easier to handle by hand, or by computer. Some mathematicians, like for example Catalan, Bateman and Gould, have tried (in vain) to provide systematic treatments of binomial coefficient identities. Nonetheless, q-hypergeometric formulas and q-binomial coefficient identities belong together. Accordingly, we represent many theorems in Chapter 8 both ways. These unfortunately do not appear in a 'canonical' form and are therefore often difficult to recognize. Many 'new' discoveries in the area of combinatorial identities therefore turn out, by a closer inspection, to be not so new and revolutionary—they have already been found, perhaps a hundred or two hundred years ago, but have been forgotten, not recognized at the time of their discovery, and are today hidden and forgotten in the numerous classical articles. Understanding the classics and thus the roots of the modern day combinatorics may make it easier for the present-day scientist, the mathematical archaeologist and scientists generally interested in scientific history to find, decipher and use the research already done.

A step on the way to this is taken in the following subsections, where we attempt to introduce and define e.g. the Bernoulli numbers, their historical background and different notations, in order to follow the development up to the present. The chronological account of (3.6) thus presents hitherto forgotten mathematicians and their valuable contributions to the subject.

There is also a link to hypergeometric transformation formulas of several variables. Any such transformation corresponds to a combinatorial identity, as explained in Section 10.9.

3.2.1 Faà di Bruno formula

The Faà di Bruno formula gives the nth derivative of a composite function.

Theorem 3.2.1 *If g and f are functions with a sufficient number of derivatives, then*

$$\frac{d^n}{dx^n} g\big(f(x)\big) = \sum \frac{n!}{k_1!k_2!\cdots k_n!} g^{(k)}\big(f(x)\big) \prod_{m=1}^{n} \left(\frac{f^{(m)}(x)}{m!}\right)^{k_m}, \tag{3.6}$$

where the sum goes over all positive k_i, $k_1 + 2k_2 + \cdots + nk_n = n$ and where $k = \sum_i k_i$.

The Faà di Bruno formula was once considered a part of analysis [318]. Later, Riordan [433] and Comtet [141] put the formula into algebraic combinatorics. These books were the forerunners of [486]. The Faà di Bruno formula is a hot research topic, and has links e.g. to partitions. The coefficients in (3.6) are the Stirling numbers of the second kind. Francesco Faà di Bruno (1825–1888) had studied in the period 1850–1859 with Cauchy in Paris [547].

3.3 The duality between Bernoulli and Stirling numbers

To develop a meaningful language for combinatorial identities, we shall first define a series of numbers and related polynomials. Some of their q-analogues will be presented in Chapters 4 and 5.

Let the *Pochhammer symbol* $(a)_n$ be defined by

$$(a)_n = \prod_{m=0}^{n-1}(a+m), \qquad (a)_0 = 1. \tag{3.7}$$

Since *products* of Pochhammer symbols occur very often, we shall frequently use the more compact notation

$$(a_1, a_2, \ldots, a_m)_n \equiv \prod_{j=1}^{m}(a_j)_n. \tag{3.8}$$

James Stirling, a contemporary of Euler, expressed Maclaurin's formula in a different form by using the Stirling numbers of the second kind [233, p. 102].

The Stirling numbers of the first kind are the coefficients in the expansion

$$(x)_{n-} = \sum_{k=0}^{n} \mathrm{s}(n,k)x^k. \tag{3.9}$$

The Stirling numbers of the second kind are defined by

$$x^n = \sum_{k=0}^{n} \mathrm{S}(n,k)(x)_{k-}. \tag{3.10}$$

In combinatorics, the unsigned Stirling numbers of the first kind $|\mathrm{s}(n,k)|$ count the number of permutations of n elements with k disjoint cycles.

The Stirling numbers of the second kind $\mathrm{S}(n,k)$ give the number of partitions of n objects into k non-empty subsets.

Tables of the first $\mathrm{S}(n,k)$ were given by De Morgan [151, p. 253] and by Grünert [250], [252, p. 279]. These tables were subsequently extended by Cayley [114].

Gould [238] remarks that many sums involving binomial coefficients greatly benefit from the use of Bernoulli, Euler, or Stirling numbers.

One of the reasons Bernoulli and Stirling numbers have such wide-ranging applications in computer technology [244] and in numerical analysis [218] is because computers use difference operators rather than derivatives, and these numbers are used in the transformation process.

The Bernoulli and Stirling numbers are intimately connected by a well-known formula; they complement each other. The sum $\mathcal{B}_n \equiv \sum_{k=1}^{n} \mathrm{S}(n,k)$ is the nth Bell number [240, p. 456, 4.12], with applications in partition theory [30].

In 1877 Dobinski [156] found the following formula:

$$\mathcal{B}_n = \frac{1}{e}\sum_{k=0}^{\infty}\frac{k^n}{(k)!}. \tag{3.11}$$

q-Bernoulli and q-Euler numbers are discussed in Chapter 4; in Chapter 5 we come back to Stirling and Bell numbers. In Section 5.2 we will find a q-analogue for the Schwattian generalization of Dobinski's formula.

3.4 Tangent numbers, Euler numbers

The *tangent numbers* T_{2n+1} are integers defined by André 1879 [21], [34, p. 380]:

$$\tan x = \sum_{n=0}^{\infty} \frac{\mathrm{T}_{2n+1} x^{2n+1}}{(2n+1)!}. \tag{3.12}$$

The following equation is valid [108, p. 413]:

$$\mathrm{T}_{2n+1} = \left((1+x^2)D\right)^{2n+1} x\big|_{x=0}. \tag{3.13}$$

The *Euler numbers* or *secant numbers* $\mathcal{S}_{2n}$ are numbers defined by André 1879 [21], [108, p. 414]:

$$\frac{1}{\cos x} = \sum_{n=0}^{\infty} \frac{\mathcal{S}_{2n} x^{2n}}{(2n)!}. \tag{3.14}$$

The following equation holds [108, p. 414]:

$$\mathcal{S}_{2n} = \left(x + (1+x^2)D\right)^{2n} 1\big|_{x=0}. \tag{3.15}$$

3.5 The occurrence of binomial coefficient identities in the literature

According to primarily Andrews, Gould and Riordan, the mass of binomial coefficient identities cannot be organized in a coherent manner. These identities are both inexhaustible and unpredictable. Riordan took the most pessimistic view: "The age-old dream of putting order in this chaos is doomed to failure".

Some have however tried putting order in the chaos. H. W. Gould tabulated 555 combinatorial identities, but as he himself stresses, there still remains the problems of finding one's own identity amongst the huge number of formulas; one also has to consider changes of variables and introduction of redundant factors.

The Caltech mathematician Harry Bateman at his death in 1946, left an unpublished manuscript of 520 pages on binomial coefficient identities.

3.6 Nineteenth century: Catalan, Grigoriew, Imchenetsky

The Catalan numbers $C_n \equiv \frac{1}{n+1}\binom{2n}{n}$ [136] were first discovered in China about 1730 by J. Luo [352] and were studied by Euler in the middle of the eighteenth century. They became famous through the 1838 paper by Catalan. Catalan numbers are very important in combinatorics. Cigler and other authors have introduced q-Catalan numbers in several papers.

In the nineteenth century, the work in the field of generalized Bernoulli polynomials by Mikhail Egorovich Vashchenko-Zakharchenko (1825–1912), Grigoriew and Imchenetsky began. Vashchenko-Zakharchenko (Ukraine) had studied in Paris during the period 1847–48 and continued the tradition of symbolic calculus.

One of his many students has written a book on Bernoulli polynomials [127]. It is, however, difficult to assess the importance and contributions of the Russian mathematicians in this field as their works have been published only in Russian. We know that they were influenced by Blissard and they themselves had some impact on the Dane Niels Nørlund and, via his work, the present author's q-analogues of generalized Bernoulli polynomials; this influence can be seen from the correspondence between Nørlund and Mittag-Leffler [183].

3.7 A short history of hypergeometric series

We now come to the special functions. The *magnetic flux density* corresponds to the hypergeometric series.

1 Gauß	hypergeometric series

The theory of hypergeometric functions is fundamental for mathematical physics, since almost all elementary functions can be expressed as either hypergeometric or ratios of hypergeometric functions, and many nonelementary functions in this field can be expressed as hypergeometric functions. The general theory of hypergeometric functions is very powerful, and it is worthwhile to check if a given series is hypergeometric, because we may gain a lot of insight into a function by first recognizing that it is hypergeometric, then identifying its parameters and finally by using known results about such functions.

This chapter deals with some well-known formulas from complex analysis, most of them related to hypergeometric series and Γ functions.

In later sections we will prove their q-analogues.

Now follow some definitions, the q-analogues look very similar.

Definition 15 The *generalized hypergeometric series* ${}_p\mathrm{F}_r$ is defined by

$$
\begin{aligned}
&{}_p\mathrm{F}_r(a_1,\ldots,a_p;b_1,\ldots,b_r;z)\\
&\equiv {}_p\mathrm{F}_r\left[\begin{matrix}a_1,\ldots,a_p\\ b_1,\ldots,b_r\end{matrix};z\right]=\sum_{n=0}^{\infty}\frac{(a_1,\ldots,a_p)_n}{n!(b_1,\ldots,b_r)_n}z^n. \qquad (3.16)
\end{aligned}
$$

For hypergeometric series with function value equal to 1, z is not written.

Denote the series (3.16) by $\sum_{n=0}^{\infty} A_n z^n$. The quotient $\frac{A_{n+1}}{A_n}$ is a rational function $R(n)$ (Askey).

The Oxford professor John Wallis first used the word 'hypergeometric' (from the Greek) meaning thus, a geometric series for the series

$$\sum_{k=0}^{\infty}(a)_k$$

in connection with the calculation of certain 'Beta functions' in his work *Arithmetica Infinitorium* (1655).

Wallis, however, had used a geometric representation of the integral and no variables; for this he was strongly criticized by Fermat, who had found many of Wallis's results several decades earlier. Newton has calculated various surfaces by extrapolation of the infinite series of Wallis.

3.7.1 The **Γ** *function*

As we know, for a long time there was no definitive notation for the Γ function, which stems from Legendre. In the year 1729, de Moivre and Stirling [160, p. 228] discussed asymptotic expansions for the factorial.

Then, in 1730, Stirling found the asymptotic expansion of the Γ function; in modern notation:

$$\Gamma(\lambda+k) \sim k^{\lambda+k-\frac{1}{2}}e^{-k}\sqrt{2\pi}. \tag{3.17}$$

Euler used a predecessor of the Pochhammer symbol to find the value $\Gamma(\frac{1}{2}) = \sqrt{\pi}$. The proof was obtained by Wallis's formula for π.

The formula (1.44) is called balanced, if

$$\sum_{k=1}^{p} a_k = \sum_{k=1}^{r} b_k, \quad p = r. \tag{3.18}$$

Some examples of balanced Γ functions formulas are:

1. The Legendre duplication formula.
2. The Euler reflection formula.
3. The associated formula

$$\Gamma\begin{bmatrix} \frac{1}{2}+x, \frac{1}{2}-x \\ \frac{1}{2}, \frac{1}{2} \end{bmatrix} = \frac{1}{\cos \pi x}. \tag{3.19}$$

We will see that the balanced property remains after the q-deformation and is an important characterization also for the Γ_q function.

3.7.2 Balanced and well-poised hypergeometric series

In mechanics, one begins with the statics to acquire the force equations. The force equilibrium equations correspond to the balanced hypergeometric series and the balanced Γ functions. The moment equilibrium equations correspond to the well-poised hypergeometric series.

force equilibrium	balanced series
moment equilibrium	well-poised series

We will now introduce the terms balanced and well-poised series.

Definition 16 An ${}_{r+1}\mathrm{F}_r$ series is called *k-balanced* if

$$b_1 + \cdots + b_r = k + a_1 + \cdots + a_{r+1}, \tag{3.20}$$

and a 1-balanced series is called balanced, after L. Saalschütz.

Remark 4 In some books, e.g. [220], a balanced hypergeometric series is defined by the extra condition $z = 1$. The above definition is in accordance with [26, p. 475].

The hypergeometric series

$${}_{r+1}\mathrm{F}_r(a_1, \ldots, a_{r+1}; b_1, \ldots, b_r; z) \tag{3.21}$$

is called *well-poised* if its parameters satisfy the relations

$$1 + a_1 = a_2 + b_1 = a_3 + b_2 = \cdots = a_{r+1} + b_r. \tag{3.22}$$

The hypergeometric series (3.21) is called *nearly-poised* [544] if its parameters satisfy the relations

$$1 + a_1 = a_{j+1} + b_j \tag{3.23}$$

for all values of j, $1 \leq j \leq r$, with one exception. If the series (3.21) is well-poised and $a_2 = 1 + \frac{1}{2}a_1$, then it is called *very-well-poised.*

3.7.3 Fractional differentiation

The concept of fractional differentiation (and integration) appeared first in a letter from Leibniz to l'Hôpital dated September 30, 1695 and in another letter dated May 28, 1697, from Leibniz to Wallis [484]. Leibniz writes: *Itaque judicavi, præter affectiones quantitatis hactenus receptas*

$$y,\ y^2,\ y^3,\ y^{\frac{1}{2}},\ y^{\frac{1}{3}},\ \text{etc.},$$

vel generaliter y^e, *sive* $[p^e]$ y, *vel potentiæ ipsius y secundum exponentem e; posse adhiberi novas Differentiarum vel Fluxionum affectiones,* dy, d^2y, *(seu* ddy*),* d^3y, *(seu* $dddy$*), imo utiliter etiam occurrit* $d^{\frac{1}{2}}y$, *et similiter generaliter que* $d^e y$. For an extensive bibliography on occurrences of fractional derivatives see [404].

In Section 8.14 we briefly treat q-fractional integrals.

3.7.4 Newton, Taylor, Stirling, Montmort

The *binomial series*, first studied by Newton, is defined by

$$(1-z)^{-\alpha} \equiv \sum_{n=0}^{\infty} \frac{(\alpha)_n}{n!} z^n, \quad |z| < 1, \ \alpha \in \mathbb{C}. \tag{3.24}$$

In 1715 Brook Taylor used calculus of finite differences in his monumental work *Methodus Incrementorum Directa et Inversa*. Taylor travelled to France to see Pierre Rémond de Montmort (1678–1719), who, in 1711, had published a transformation formula for series with inverse factorial function argument. Montmort's formula was generalized by Euler, and these two formulas are q-deformed in (5.82) and (5.86).

After Newton and Taylor many other mathematicians studied similar series; one of them was James Stirling, who in his most important work *Methodus Differentialis* [490] gave a treatise on the calculus of finite differences. Stirling [490, p. 40, Prop. VII] also found the *Euler-Pfaff-Kummer transformation formula* (3.28) for the special case $a = 1$, when solving a difference equation [511, p. 35]. In Stirling's book there are very many difference equations, which are recurrences for the coefficients of the "hypergeometric series". Many of Stirling's series can be written in hypergeometric form. Stirling [510, p. 210] has found many special cases of the Gaussian summation formula (3.31).

Stirling has generalized summation formulas of Halley and William Brouncker (1620–1684). Brouncker, a graduate student of Wallis, has found a formula for the inverse of π in the form of continued fraction. We have thus described the first seeds of the hypergeometric function and the Γ function in this period. Both of these functions have been presented as Fakultäten; the hypergeometric function came only a century later.

3.7.5 Euler's contribution

Leonhard Euler introduced the concept of a function together with several formulas for Γ- and hypergeometric functions written in different notation. Euler proved the following four equations [220, p. 10, 19]:

The *beta integral*

$$\int_0^1 x^{s-1}(1-x)^{t-1}\,dx = \frac{\Gamma(s)\Gamma(t)}{\Gamma(s+t)} \equiv B(s,t), \quad \Re(s)>0,\ \Re(t)>0; \tag{3.25}$$

the integral representation of the hypergeometric series ${}_2F_1(a,b;c;z)$ (1748):

$${}_2F_1(a,b;c;z) = \frac{\Gamma(c)}{\Gamma(b)\Gamma(c-b)} \int_0^1 x^{b-1}(1-x)^{c-b-1}(1-zx)^{-a}\,dx, \tag{3.26}$$

$$\left|\arg(1-z)\right| < \pi, \quad \Re(c) > \Re(b) > 0.$$

The *Euler transformation formula*:

$${}_2F_1(a,b;c;z) = (1-z)^{c-a-b}{}_2F_1(c-a,c-b;c;z). \tag{3.27}$$

The *Euler-Pfaff-Kummer transformation formula* [348], [425, p. 60]:

$${}_2F_1(a,b;c;z) = (1-z)^{-a}{}_2F_1\big(a,c-b;c;(z/z-1)\big), \tag{3.28}$$

$|z|<1$ and $|\frac{z}{z-1}|<1$.

3.7.6 Vandermonde and Pfaffian summation formulas

In this section we continue the work on the function of one real variable. This is according to the Papperitz description of 1889. We have seen how the theory of the Fakultäten was founded in the eighteenth century.

One of the results turned out to be the *Chu-Vandermonde summation formula* [129], [392], [516], [515, p. 72]:

$${}_2F_1(-n,b;c) \equiv \sum_{k=0}^{n} \frac{(-1)^k (b)_k}{(c)_k}\binom{n}{k} = \frac{(c-b)_n}{(c)_n}, \quad n=0,1,\ldots. \tag{3.29}$$

Johann Friedrich Pfaff was perhaps the leading mathematician in Germany in the latter part of the eighteenth and early nineteenth centuries and is known for his work on differential equations [159, p. 26]. Beginning in the late 1790s he was a teacher and friend of C. F. Gauß in Helmstedt. Gauß stayed four months in Pfaff's big house and was able to use his library [46, p. 4]. Pfaff [414, p. 46–48] introduced the name hypergeometric series and the contiguous functions to Gauß [159, p. 28]. Pfaff's method of proof has recently received attention in some articles by George Andrews [29], where the Pfaffian method is used for some other proofs. In 1797 Pfaff [414, pp. 51–52], [445] stated another summation formula, which is now called the Pfaff-Saalschütz summation formula for a terminating balanced hypergeometric series:

$${}_3F_2(a,b,-n;c,1+a+b-c-n) = \frac{(c-a,c-b)_n}{(c,c-a-b)_n}, \quad n=0,1,\ldots. \tag{3.30}$$

Pfaff [414] also proved the transformation formula (3.28). This formula together with (3.27) was then proved by Jacobi and Gudermann in various ways.

The relevant work of Euler and Pfaff was mentioned by Jacobi and in a well-known textbook by Heine [273, pp. 357–398], [159, p. 29].

3.7.7 Conic sections in the seventeenth century

In Richelieu's France the work of the old masters was not very widespread, but Girard Desargues (1591–1661) published a book about projective geometry in 1639 [361, p. 16]. This work was highly praised by Fermat [96, p. 1618], though sadly enough it was destroyed after a few years for political reasons. Only one person managed to complete Desargues's geometrical achievements: in 1640, Blaise Pascal wrote a paper on conic sections, which Leibniz reviewed in a letter to Pascal's nephew written in 1676. In the next subsection we will see how the conic sections have indirectly contributed to hypergeometric series.

3.7.8 The infinity in England

The work on the quadrature of the parabola by Archimedes, where the infinity is used, was translated into English at the end of the sixteenth century. Wallis also built on the work of Kepler, Cavalieri, Roberval, Torricelli and Descartes (and indirectly of Fermat and Desargues). The infinity was introduced by Wallis in his work *De Sectionibus Conicis* (1655) in England [487, p. 6]. In his work, [528] Wallis has given the first examples of hypergeometric series. In the two tables [528, Prop. 184, Prop. 189], [487, p. 6] we find some quotients. Every other line and every second column begin with the Latin words *Monadici*, *Laterales*, *Triangulares* and *Pyramidales*. The first item in the table is ∞.

Wallis also introduced the induction in England. The induction was also used by Plato, Jacob Bernoulli, Fermat and Pascal. The first examples of infinite series with one variable were given by Newton [487, p. 20], in the footsteps of Wallis. Therefore, there were no proofs of the infinite series which Leibniz received from Oldenburg in 1675 (Section 2.7).

3.7.9 The infinity in the hands of Euler

Euler's *Introductio* [190] was published in 1748. In the German translation from 1788 [192] the considerations of ∞ begin in Chapter 15, "Von den Reihen, die aus der Entwicklung der Faktoren entspringen". In Chapter 16, "Von der Teilung der Zahlen", there are intuitive considerations showing the equivalence between partition and infinite products.

One finds the Euler equation (6.188) for the q-exponential function $e_q(z)$ in [192, pp. 336–337]. One finds the Euler equation (6.189) for the q-exponential function $e_{\frac{1}{q}}(z)$ in [192, pp. 338–339]. Euler does not write down the general terms in the sums, but in the spirit of Hindenburg capitals.

Euler's *Introductio* is now considered his best book; he also wrote other books, see Section 2.12.

3.7.10 The infinity, the binomial coefficients

W. J. G. Karsten (1732–1787) early on had close contact with L. Euler. In 1765, Johann Karsten ordered the release of Euler's *Theoria motus corporum rigidorum* in Greifswald. Karsten's first employment was in Bützow in 1760–1778. Through his friendship with Euler, Karsten was in 1765 appointed professor at the Academy in St. Petersburg, but he refused. Instead, he dedicated his life to his family and the writing of mathematical and physical papers. Unlike Euler, who published mainly in *Abh. der Petersburger Akademie*, Karsten preferred the *Abh. Münchener Akademie*. Like Gudermann, Karsten published alternately in Latin and German. Karsten published a lot, his greatest contribution being *Von den Logarithmen vermeinter Größen*, written in 1768, where he gave geometric interpretation of logarithms of complex numbers. In 1778, Karsten became a Professor of Mathematics and Natural Philosophy in Halle. His last book, *Mathematische Abhandlungen* (1786), begins with the chapter *Vom mathematischen Unendlichen* and shows the great influence of Euler's *Introductio* [190].

Abraham Gotthelf Kästner (1719–1800) taught as a docent in Leipzig until 1746 and as a professor until 1756, when he moved to Göttingen. He wrote manuals and books about math history. In [329] the title of the first Chapter reads: *Gründe der Lehre vom Unendlichen*. There is also a Chapter *Vergleichung der Binomialkoeffizienten mit den Koeffizienten der Differentiale*. Possibly Gauß did not visit his lectures, but may have read his books in the libraries at Göttingen and Helmstedt. Thus we see that Gauß's use of the infinity in his definition of the hypergeometric series is not his own invention. Certainly Gauß had realized that the binomial coefficients belong together with the Pochhammer symbol.

3.7.11 Gauß' contribution

The unit for the *magnetic flux density* corresponds to the hypergeometric series.

1 Gauß	hypergeometric series

The Swede Anders Lexell (1740–1784) had shown (together with Laplace), that Uranus, which was discovered in 1781 by Herschel, was not a comet, but a planet. Gauß calculated the elliptical orbits of Ceres and achieved great fame for it. Gauß asked the Duke of Brunswick for a double reward in Helmstedt, he had offers from both Berlin and from St. Petersburg. But no answer came, and the Napoleonic War prevented travel. In the period 1807–1813, Göttingen belonged to Westphalia under Jérôme Bonaparte (1784–1860). Due to economic difficulties, the university in Helmstedt was decommissioned in 1810, and Gauß was professor of astronomy in Göttingen. At the end of the eighteenth century, von Grüson had translated some of Euler's works into German and introduced some of Hindenburg's names ... Gauß discovered the transformation (3.28) in 1809 or 1805 [46].

In early 1797 [452] Gauß began a systematic treatment of elliptic functions, starting with lemniscate functions. When Gauß had explored elliptic functions and the complex plane, he could continue his investigations on the functions satisfying (3.32). Gauß also discovered the equations for the division of periods. The inverse of the arithmetic-geometric mean is a hypergeometric series, which Gauß had investigated thoroughly. In a letter to Bessel [452, p. 85], he had expressed the desire to explore the Γ function for very large values, to learn more about the connection to the primes. Gauß writes: *I am working on a paper with the Kramp Fakultäten.* Euler had dealt with a differential equation $\star$, that is the subject of a Section in Pfaff's *Disquisitiones Analyticae* [413]. Formula (3.32) was first found by Euler, we give two quotations:

1. Institutiones Calculi Integralis II (1769), Opera Omnia: Series 1, Volume 12 Kapitel 8–11 [35]. Here Euler finds a more general differential equation of second order with certain solutions.
2. (1778) (Eneström 710) Paragraph 6 in *Specimen transformationis singularis serierum* Nova Acta Academiae Scientarum Imperialis Petropolitinae 12, 1801 (1794), pp. 58–70 Opera Omnia: Series 1, Volume 16, pp. 41–55. Beweis einer einzigartigen Transformation von Reihen übersetzt von: Alexander Aycock, im Rahmen des Projektes Eulerkreis Mainz. Euler derives equation (3.32) for the hypergeometric series on page 60.

Then on January 20, 1812, C. F. Gauß [224] proved the famous *Gaußian summation formula*

$$ {}_2\mathrm{F}_1(a,b;c) = \frac{\Gamma(c)\Gamma(c-a-b)}{\Gamma(c-a)\Gamma(c-b)}, \quad \Re(c-a-b) > 0, \tag{3.31} $$

a generalization of the Vandermonde summation formula.

Gauß also gave the ratio of two F functions by means of continued fractions, compare with the works of Stern and Heine, who made important contribution to this theory. In the following year Gauß found that the series ${}_2\mathrm{F}_1(a,b;c;z)$ satisfies the following second order differential equation:

$$ z(1-z)\frac{d^2y}{dz^2} + \{c-(1+a+b)z\}\frac{dy}{dz} - aby = 0, \tag{3.32} $$

which has *regular singular points* at 0, 1, ∞ and at no other singular points [73, p. 224]. Gauß also found contiguous relations for the hypergeometric series. Gauß viewed the function ${}_2\mathrm{F}_1(a,b;c;z)$ as a function of four variables.

Gauß [223, p. 127, (2)], [41, p. 23] found the following formula:

$$(1+x)^n + (1-x)^n = 2\,{}_2F_1\left(\frac{-n}{2}, \frac{1-n}{2}; \frac{1}{2}; x^2\right). \tag{3.33}$$

Gauß also proved the following equation (γ denotes Euler's constant)

$$\Psi(x) \equiv \frac{\Gamma(x)'}{\Gamma(x)} = -\frac{1}{x} - \gamma + \sum_{n=1}^{\infty}\left(\frac{1}{n} - \frac{1}{x+n}\right), \tag{3.34}$$

which can be used to give a simple proof of an advanced version of the Euler-Maclaurin summation formula. Nowadays this function is called the digamma function.

Furthermore Gauß continued Euler's work on the Γ function and together with Legendre found the *Legendre duplication formula*, which says that

$$\Gamma(2x)\Gamma\left(\frac{1}{2}\right) = 2^{2x-1}\Gamma(x)\Gamma\left(x+\frac{1}{2}\right). \tag{3.35}$$

3.7.12 After Gauß; Clausen, Jacobi

This section continues the work on hypergeometric functions of real variables. In the year 1828, Clausen [140] found a formula which expresses the square of a ${}_2F_1$ series as a ${}_3F_2$ series. Thomas Clausen (1801–1885) was born in a small Danish village, and alternately worked as an astronomer in Altona and Munich, and finally in Dorpat. Clausen is also known for the von Staudt-Clausen theorem on Bernoulli numbers and for numerous data on comet orbits. Jacobi contributed to the topic in an article about the orthogonality and the Rodriguez formula for Jacobi polynomials.

3.7.13 Kummer's contribution

In the year 1834, Ernst Kummer was a gymnasium teacher in Liegnitz, and published a book in Latin about a differential equation of third order. A special case of this equation is the formula for the transformation of hypergeometric functions. In his long and important 1836 paper [348], Kummer derived all the 24 solutions of the hypergeometric differential equation (3.32) by the application of a general Möbius transformation.

It turned out that the only solutions were certain unimodular transformations, such as the previously known Euler and Euler-Pfaffian transformations.

The method which Kummer had used occurred already in the work of Euler [191, p. 58]. Through a sophisticated calculation, which used the rational functions of the solutions of (3.32) with various number parameters, Kummer was able to write down a list of quadratic transformations.

The complete theory of Möbius transformations, named after the mathematician and astronomer Möbius, was not yet fully developed at the time.

In modern notation, these transformations have a somewhat different form. Three examples in a neighbourhood of $x = 0$ are:

Theorem 3.7.1 [425, p. 65, (1)]:

$$ {}_2\mathrm{F}_1(\lambda, a; 2a; 2x) = (1-x)^{-\lambda}\, {}_2\mathrm{F}_1\left(\frac{\lambda}{2}, \frac{\lambda+1}{2}; a+\frac{1}{2}; \frac{x^3}{(1-x)^2}\right). \tag{3.36} $$

Theorem 3.7.2 [35, p. 125, (3.1.4)], [275, p. 169], [56, p. 9, (2)], [348, p. 78, (53)], [425, p. 67, (3)], [220, p. 59, (3.1.4)]:

$$ {}_2\mathrm{F}_1(a, b; 1+a-b; x) = (1-x)^{-a}\, {}_2\mathrm{F}_1\left(\frac{a}{2}, \frac{a+1}{2} - b; 1+a-b; \frac{-4x}{(1-x)^2}\right). \tag{3.37} $$

Theorem 3.7.3 *Gauß* [223, p. 225, (101)], [348, p. 78, (51)], [220, p. 68, (3.5.6)], [425, p. 65, (12)]:

$$ {}_2\mathrm{F}_1\big(a, b; 1+a-b; x^2\big) = (1-x)^{-2a}\, {}_2\mathrm{F}_1\left(a, a+\frac{1}{2} - b; 1+2a-2b; \frac{-4x}{(1-x)^2}\right). \tag{3.38} $$

Kummer found three fundamental summation formulas, which we will call Kummer's first summation formula, Kummer's second summation formula and Kummer's third summation formula.

Theorem 3.7.4 [348, p. 134, (1)] *Kummer's first formula for the sum of a well-poised* ${}_2\mathrm{F}_1$ *function with function value* -1:

$$ {}_2\mathrm{F}_1(a, b; 1+a-b; -1) = \Gamma\left[\begin{matrix} 1+a-b, 1+\frac{a}{2} \\ 1+a, 1+\frac{a}{2}-b \end{matrix}\right]. \tag{3.39} $$

Theorem 3.7.5 [348, p. 134, (2)] *Kummer's second summation formula*:

$$ {}_2\mathrm{F}_1\left(a, b; \frac{1}{2}(1+a+b); \frac{1}{2}\right) = \Gamma\left[\begin{matrix} \frac{1+a+b}{2}, \frac{1}{2} \\ \frac{1+b}{2}, \frac{1+a}{2} \end{matrix}\right]. \tag{3.40} $$

This formula is also called *Gauß's second summation theorem* after Gauß [223, p. 227, (105)].

Theorem 3.7.6 [348, p. 134, (3)] *Kummer's third summation formula*:

$$
{}_2\mathrm{F}_1\left(a, 1-a; c; \frac{1}{2}\right) = \Gamma\left[\begin{matrix} \frac{c}{2}, \frac{1+c}{2} \\ \frac{1+c-a}{2}, \frac{a+c}{2} \end{matrix}\right]. \tag{3.41}
$$

Proof [466] We prove Kummer's third summation formula. We put $b = 1 - a$ and $z = \frac{1}{2}$ in (3.26), make the substitution $y \to (1-x)^2$.
The Beta function identity gives

$$
{}_2\mathrm{F}_1\left(a, 1-a; c; \frac{1}{2}\right) = 2^{a-1} \frac{\Gamma(c)\Gamma(\frac{c+a-1}{2})}{\Gamma(c+a-1)\Gamma(\frac{c+1-a}{2})}. \tag{3.42}
$$

Finally, we apply the Legendre duplication formula. □

3.7.14 *Cauchy, Riemann, Heine, Thomae, Papperitz*

Cauchy developed the theory of a complex variable; his textbooks were translated into German by Grünert and Schlömilch.

In the footsteps of Cauchy and Gauß, Riemann introduced the Riemann sphere and the Riemann-Papperitz equation, a generalization of the hypergeometric equation. The Riemann-Papperitz equation describes the 24 Kummer solutions of (3.32) by a very abstract matrix notation, by the method of Frobenius. The notation ∞ is already assigned to the real infinity, so we need a new notation.

Definition 17 We denote the infinity point on the Riemann sphere $\hat{\mathbb{C}}$ by ∞_{Rie}.

The solutions of a homogeneous differential (or q-difference) equation of degree n form an n-dimensional vector space, and any solution of the homogeneous equation can be written as a linear combination of fundamental solutions.

Kummer derived all the 24 solutions of the hypergeometric differential equation (3.32). In accordance with [275], we denote these by $\{u_i(z)\}_{i=1}^{24}$.

Theorem 3.7.7 [275, p. 155]. *If $c \notin \mathbb{Z}$, the functions*

$$
\begin{aligned}
u_1(z) &\equiv {}_2\mathrm{F}_1(a, b; c; z), \\
u_2(z) &\equiv z^{1-c}_2\mathrm{F}_1(1+a-c, 1+b-c; 2-c; z)
\end{aligned} \tag{3.43}
$$

form a fundamental system for (3.32) *around* $z = 0$.

Theorem 3.7.8 [275, p. 155]. *The functions*

$$\begin{aligned} u_3(z) &\equiv z^{-a}{}_2\mathrm{F}_1\left(a-c+1, a; a-b+1; \frac{1}{z}\right), \\ u_4(z) &\equiv z^{-b}{}_2\mathrm{F}_1\left(b-c+1, b; b-a+1; \frac{1}{z}\right) \end{aligned} \tag{3.44}$$

form a fundamental system for (3.32) *around* $z = \infty_{\mathrm{Rie}}$.

Theorem 3.7.9 [275, p. 156]. *The functions*

$$\begin{aligned} u_5(z) &\equiv {}_2\mathrm{F}_1(a, b; a+b+1-c; 1-z), \\ u_6(z) &\equiv (1-z)^{c-a-b}\ {}_2\mathrm{F}_1(c-a, c-b; 1+c-a-b; 1-z) \end{aligned} \tag{3.45}$$

form a fundamental system for (3.32) *around* $z = 1$.

We will find more results in the wake of Thomae, the successor of Jacobi and Heine, who loved long computations and complicated formulas. In 1879, Thomae [501] found a transformation formula for a ${}_3\mathrm{F}_2$ series with function value equal to unity. Thomae used a different notation for the hypergeometric function. In the same year, he [501] found all relationships between Kummer's 24 solutions.

3.7.15 1880–1914; Sonine, Goursat, Stieltjes, Schafheitlin, Pochhammer, Mellin

We have now reached the third period according to Papperitz description, which starts with the work of Weierstraß, in which the beginning of a systematic revision of the function theory is clearly seen.

If the parameters a_i, b_j in (3.16) are chosen in such a way that the series does not end and is not undefined, it can be proved that the series converges for all z if $p \le r$, converges for $|z| < 1$ if $p = r+1$ and diverges for all $z \neq 0$ if $p > r+1$. The differential equation of the generalized hypergeometric series (3.16) is [382]

$$\left(z\frac{d}{dz}+a_1\right)\cdots\left(z\frac{d}{dz}+a_p\right) - \frac{d}{dz}\left(z\frac{d}{dz}+b_1-1\right)\cdots\left(z\frac{d}{dz}+b_r-1\right) = 0. \tag{3.46}$$

For example, the hypergeometric series

$$y(z) \equiv {}_3\mathrm{F}_2(a, b, c; d, e; z) \tag{3.47}$$

satisfies the differential equation of third order

$$z^2(z-1)\frac{d^3y}{dz^3} + z\big((a+b+c+3)z - (d+e+1)\big)\frac{d^2y}{dz^2}$$

$$+\big((bc+ac+ab+a+b+c+1)z-de\big)\frac{dy}{dz}+abcy=0, \qquad (3.48)$$

as shown by Thomae in 1870 [498], Goursat in 1883 [243] and Pochhammer in 1888 [417].

Pochhammer called the generalized hypergeometric function *Gaußsche Reihe angrenzende Funktion*, labeled H_n. He introduced the so-called Pochhammer integrals for hypergeometric series.

In the footsteps of Riemann, Pincherle and Mellin considered similar integrals. Paul Schafheitlin (1861–1924), doctorate in Halle-Wittenberg in 1885, has published a lot about special functions. The highlight was probably the book *Die Theorie der Besselschen Funktionen* (1908). But we must not forget that Schafheitlin published the Dixon and Watsonian summation formulas for hypergeometric series already in 1912 [449, p. 24]. Schafheitlin used the Gaussian terms for Γ functions.

Mittag-Leffler introduced what is known as the Mittag-Leffler function E_α (1903 and 1905):

$$\mathrm{E}_\alpha(z) \equiv \sum_{k=0}^{\infty} \frac{z^k}{\Gamma(\alpha k+1)}, \quad \alpha > 0. \qquad (3.49)$$

The Mittag-Leffler function has recently found applications in fractional calculus.

3.7.16 First half of the twentieth century; England, USA

At the beginning of the twentieth century the main scene moved to England and America. One of the reasons was of course Ramanujan, who caused quite a sensation in Cambridge. Ramanujan himself found many hypergeometric formulas. But the main reason is the general increase of the level of mathematics that followed after the Cambridge Palace revolution by Babbage and his fellows.

Again the history of q-calculus took a great leap forward with the help of hypergeometric series. Several equations for products of hypergeometric series have been proved in 1899 by Orr [405].

In 1907 John Dougall (1867–1960) [158] found the following summation formula for a very-well-poised 2-balanced series, i.e.

$$1+2a+n=b+c+d+e,$$

$$\begin{aligned}&{}_7\mathrm{F}_6\left[\begin{matrix} a, 1+\frac{1}{2}a, b, c, d, e, -n \\ \frac{1}{2}a, 1+a-b, 1+a-c, 1+a-d, 1+a-e, 1+a+n \end{matrix}\right]\\ &= \frac{(1+a, 1+a-b-c, 1+a-b-d, 1+a-c-d)_n}{(1+a-b, 1+a-c, 1+a-d, 1+a-b-c-d)_n}. \qquad (3.50)\end{aligned}$$

This formula was found three years later by Ramanujan [69, p. 9].

Corollary 3.7.10 [466, p. 56, 2.3.4.5]:

$$
{}_5F_4\left[\begin{matrix} a, b, c, d, 1+\frac{1}{2}a \\ 1+a-b, 1+a-c, 1+a-d, \frac{1}{2}a \end{matrix}\right]
$$
$$
=\Gamma\left[\begin{matrix} 1+a-b, 1+a-c, 1+a-d, 1+a-b-c-d \\ 1+a, 1+a-b-c, 1+a-b-d, 1+a-c-d \end{matrix}\right]. \tag{3.51}
$$

Proof Let $n \to +\infty$ in (3.50). □

Corollary 3.7.11 [466, p. 56, 2.3.4.7]:

$$
{}_4F_3\left[\begin{matrix} a, b, c, 1+\frac{1}{2}a \\ 1+a-b, 1+a-c, \frac{1}{2}a \end{matrix} \middle| -1\right]
$$
$$
=\Gamma\left[\begin{matrix} 1+a-b, 1+a-c \\ 1+a, 1+a-b-c \end{matrix}\right]. \tag{3.52}
$$

Proof Let $d \to +\infty$ in (3.51). □

A. C. Dixon (1865–1936) was a student of Cayley in Cambridge, who proved the following summation formula, which was also proved by Schafheitlin [449, p. 24, (22)].

Theorem 3.7.12 *The Dixon-Schafheitlin* [155], [449, p. 24, (22)] *summation formula for a well-poised series*:

$$
{}_3F_2(a, b, c; 1+a-b, 1+a-c)
$$
$$
=\Gamma\left[\begin{matrix} 1+\frac{1}{2}a, 1+a-b, 1+a-c, 1+\frac{1}{2}a-b-c \\ 1+a, 1+\frac{1}{2}a-b, 1+\frac{1}{2}a-c, 1+a-b-c \end{matrix}\right], \tag{3.53}
$$

provided that the series is convergent, i.e. $\Re(1+\frac{1}{2}a-b-c) > 0$.

In the year 1908, Barnes [63] found a line integral for a ${}_2F_1$ hypergeometric series:

$$
{}_2F_1(a, b; c; z) = \frac{\Gamma(c)}{\Gamma(a)\Gamma(b)} \frac{1}{2\pi i} \int_{-i\infty}^{i\infty} \frac{\Gamma(a+s)\Gamma(b+s)\Gamma(-s)}{\Gamma(c+s)} (-z)^s ds, \tag{3.54}
$$

where $|\arg(-z)| \le \pi - \delta$, $\delta > 0$. The q-analogue of the Barnes line integral (7.133) is to be found in Section 7.21.

F. J. W. Whipple (1876–1943) was a mathematician and a meteorologist educated at Trinity College, Cambridge (Second Wrangler, 1897). In 1923, Whipple [541] showed that by iterating Thomae's ${}_3F_2$ transformation formula [501, (11)], one obtains a set of 120 such series, and he tabulated the parameters of these 120 series [514]. Whipple also developed the concepts well-poised and balanced series in a series of papers [542], [543].

In 1925 Watson [534] [466], [59, (1.2), p. 237], together with Schafheitlin in 1912 [449, p. 24], proved the following generalization of Kummer's second summation formula:

Theorem 3.7.13 *The Watson-Schafheitlin summation formula*:

$$
{}_3F_2\left(a,b,c;\frac{1}{2}(1+a+b),2c\right)
= \Gamma\left[\begin{matrix}\frac{1}{2},\frac{1}{2}+c,\frac{1}{2}+\frac{1}{2}a+\frac{1}{2}b,\frac{1}{2}-\frac{1}{2}a-\frac{1}{2}b+c\\ \frac{1}{2}+\frac{1}{2}a,\frac{1}{2}+\frac{1}{2}b,\frac{1}{2}-\frac{1}{2}a+c,\frac{1}{2}-\frac{1}{2}b+c\end{matrix}\right]. \tag{3.55}
$$

Corollary 3.7.14

$$
{}_2F_1\left(\frac{c}{2},-N;c;2\right)=\begin{cases}\frac{(\frac{1}{2})_{\frac{N}{2}}}{(\frac{1+c}{2})_{\frac{N}{2}}}, & \textit{if } N \textit{ is even}\\ 0, & \textit{if } N \textit{ is odd.}\end{cases} \tag{3.56}
$$

Proof Let $a\to\infty$ in (3.55). □

This finite form is chosen in agreement with some q formulas in Section 8.1.

Theorem 3.7.15 *The Whipple summation formula* [541, p. 114], [59, (1.3), p. 237] *reads*:

$$
\begin{cases}{}_3F_2(\frac{c}{2},2N,-2N+1;c+1-e,e)=\frac{(-1)^N(\frac{e-2N+1}{2},\frac{e-c}{2})_N}{(\frac{e}{2},\frac{1-e+c}{2})_N},\\ {}_3F_2(\frac{c}{2},1+2N,-2N;c+1-e,e)=\frac{(-1)^N(\frac{e-2N}{2},\frac{e+1-c}{2})_N}{(\frac{e+1}{2},\frac{2-e+c}{2})_N}.\end{cases} \tag{3.57}
$$

For clarity we make a table of the main hypergeometric summation formulas and their context. We have divided these formulas into three classes.

Dougall (3.50)	–	–
Slater (3.51)	–	–
Dixon-Schafheitlin	Watson-Schafheitlin	Whipple
$c\to-\infty$	$c\to\infty;\ a\to\infty$	$c\to\infty$
Kummer1	Kummer2; (3.56)	Kummer3

The first class contains, from top to bottom, two very-well-poised 2-balanced series, two well-poised series.

Theorem 3.7.16 *The important Whipple formula from* 1926 [543, 7.7], [35, p. 145] *transforms a finite, very-well-poised* ${}_7\mathrm{F}_6$ *series into a balanced* ${}_4\mathrm{F}_3$ *series*:

$$
\begin{aligned}
&{}_7\mathrm{F}_6\left[\begin{matrix} a, b, c, d, 1+\frac{1}{2}a, e, -n \\ 1+a-b, 1+a-c, 1+a-d, 1+a-e, \frac{1}{2}a, 1+a+n \end{matrix}\right] \\
&\quad = \frac{(1+a, 1+a-d-e)_n}{(1+a-d, 1+a-e)_n}\, {}_4\mathrm{F}_3\left[\begin{matrix} d, e, 1+a-b-c, -n \\ 1+a-b, 1+a-c, d+e-n-a \end{matrix}\right]. \qquad (3.58)
\end{aligned}
$$

Finally, we have the following example of Bailey:

Theorem 3.7.17 *Bailey's transformation formula from* 1929 *for a finite* 2*-balanced, very-well-poised* ${}_9\mathrm{F}_8$ *hypergeometric series. With the notations*

$$
(\alpha) \equiv \left(a, b, c, d, e, f, 1+\frac{1}{2}a, \lambda+a+n+1-e-f, -n\right), \qquad (3.59)
$$

$$
\begin{aligned}
(\beta) \equiv &\left(1+a-b, 1+a-c, 1+a-d, 1+a-e,\right. \\
&\left.\frac{1}{2}a, a+1-f, e+f-n-\lambda, a+n+1\right), \qquad (3.60)
\end{aligned}
$$

$$
\begin{aligned}
(\gamma) \equiv &\left(\lambda, \lambda+b-a, \lambda+c-a, \lambda+d-a, e, f,\right. \\
&\left.1+\frac{1}{2}\lambda, \lambda+a+n+1-e-f, -n\right), \qquad (3.61)
\end{aligned}
$$

$$
\begin{aligned}
(\delta) \equiv &\left(1+a-b, 1+a-c, 1+a-d, 1+\lambda-e,\right. \\
&\left.\frac{1}{2}\lambda, \lambda+1-f, e+f-n-a, \lambda+n+1\right), \qquad (3.62)
\end{aligned}
$$

the formula reads [55]:

$$
{}_9\mathrm{F}_8\left[\begin{matrix} (\alpha) \\ (\beta) \end{matrix}\right] = \frac{(1+a, 1+a-e-f, 1+\lambda-e, 1+\lambda-f)_n}{(1+a-e, 1+a-f, 1+\lambda-e-f, 1+\lambda)_n}\, {}_9\mathrm{F}_8\left[\begin{matrix} (\gamma) \\ (\delta) \end{matrix}\right], \qquad (3.63)
$$

$$
\textit{where} \quad n = 0, 1, 2, \ldots, \quad \textit{and} \quad 2a+1 = \lambda+b+c+d. \qquad (3.64)
$$

Remark 5 In Bailey's paper, a slightly different version of this equation was given, which however is equivalent to the above equation.

3.7.17 *Special functions defined by integrals*

We now turn to generalizations of the hypergeometric series. The first attempt was the MacRobert E-function from the end of the 1930s, which is defined by an integral. MacRobert was a famous Scottish mathematician. In the sixties, R. P. Agarwal introduced a q-analogue of the E-function. Then came the Meijer G-function from 1936, named after Cornelis Simon Meijer (1904–1974). Today the Foxian H-function has found some applications in fractional calculus. Almost all special functions in applied mathematics are special cases of the Foxian H-function, introduced by Charles Fox (1897–1977). For the last two functions a comparison can be made with the formula (7.134) in Section 7.21, which shows certain similarities. Formula (7.134) indeed contains the Γ_q function and all the three functions above contain similar expressions for Γ functions.

3.7.18 *Second half of the twentieth century*

We now come to a period in which the number of special polynomials and special functions completely explodes. The majority of these polynomials are actually q-polynomials, which are presented in Section 8.8. We content ourselves with one example of hypergeometric polynomial.

Definition 18 The Wilson polynomials are given by the balanced series

$$p_n(y) \equiv (a+b, a+c, a+d)_n \times {}_4F_3(-n, a+b+c+d+n-1, a-iy, a+iy; a+b, a+c, a+d). \tag{3.65}$$

3.8 The Jacobi theta functions; different notations; properties

Torque	theta functions, elliptic functions

In the nineteenth century, the elliptic functions and the theta functions were developed, they being closely linked. No other area has so many nice mathematical formulas. The so-called Jacobi elliptic functions can be developed in a series of q-hypergeometric functions [177]. Jacobi's theta functions have series that converge very rapidly when q is small. The theta functions were known in special cases to Euler and Gauß, but were completely and systematically explored only by Jacobi in 1829. Euler had started with q-series, and Jacobi continued to use this notation; he put $q \equiv \exp(-\pi \frac{K'}{K})$, $\Re(\frac{K'}{K}) > 0$. Here K and K' are the constants from elliptic function theory.

This notation has survived until today. With the aid of his famous transformation formula, and together with his pupils in Königsberg, Jacobi was able to compute his

Theta functions with rapidly converging series. This enabled him to compile highly accurate tables of his elliptic functions and hence, of course, of the elliptic integral $F(z)$. Much more importantly, it led to a beautiful theory, which is still valid. The use of theta and elliptic functions is varied and broad. They are very important in connection with representation theory for Lie groups.

Sonia Kowalevsky has used elliptic functions in solving the three-body problem. Another, modern application is the Kadomtsev-Petviashvili (KP) hierarchy.

Previously there were at least four notations for Jacobi theta functions and elliptic functions:

1. The four Jacobi small theta functions (as in the present text), given by

$$\begin{aligned}
\theta_1(z,q) &\equiv 2\sum_{n=0}^{\infty}(-1)^n \mathrm{QE}\left(\left(n+\frac{1}{2}\right)^2\right)\sin(2n+1)z,\\
\theta_2(z,q) &\equiv 2\sum_{n=0}^{\infty}\mathrm{QE}\left(\left(n+\frac{1}{2}\right)^2\right)\cos(2n+1)z,\\
\theta_3(z,q) &\equiv 1+2\sum_{n=1}^{\infty}\mathrm{QE}\left(n^2\right)\cos(2nz),\\
\theta_4(z,q) &\equiv 1+2\sum_{n=1}^{\infty}(-1)^n\mathrm{QE}\left(n^2\right)\cos(2nz),
\end{aligned} \tag{3.66}$$

where $q \equiv \exp(\pi i t)$, $t \in \mathfrak{U}$. Jacobi was then able to express his elliptic functions in terms of his four theta functions.

Theorem 3.8.1 *The Jacobi theta functions can be represented as products in the following way*:

$$\theta_1(z,q) = 2q^{\frac{1}{4}}\sin z\left\langle 1, 1+\frac{iz}{\log q}, 1-\frac{iz}{\log q}; q^2\right\rangle_\infty, \tag{3.67}$$

$$\theta_2(z,q) = 2q^{\frac{1}{4}}\cos z\left\langle 1, \widetilde{\frac{1}{2}+\frac{iz}{\log q}}, \widetilde{\frac{1}{2}-\frac{iz}{\log q}}; q^2\right\rangle_\infty, \tag{3.68}$$

$$\theta_3(z,q) = \left\langle 1, \widetilde{\frac{1}{2}+\frac{iz}{\log q}}, \widetilde{\frac{1}{2}-\frac{iz}{\log q}}; q^2\right\rangle_\infty, \tag{3.69}$$

$$\theta_4(z,q) = \left\langle 1, \frac{1}{2}+\frac{iz}{\log q}, \frac{1}{2}-\frac{iz}{\log q}; q^2\right\rangle_\infty. \tag{3.70}$$

2. The following four functions are written according to Hancock [p. 220].

$$\Theta_1(z,q) \equiv \sum_{m=-\infty}^{\infty}\mathrm{QE}\left(m^2\right)\mathrm{Exp}\left(\frac{\pi i m z}{K}\right), \tag{3.71}$$

$$\mathrm{H}_1(z,q) \equiv \sum_{m=-\infty}^{\infty} \mathrm{QE}\left(\left(m+\frac{1}{2}\right)^2\right)\mathrm{Exp}\left((2m+1)\frac{\pi i z}{2K'}\right), \tag{3.72}$$

$$\mathrm{H}(z,q) \equiv \mathrm{H}_1(z-K,q), \tag{3.73}$$

$$\Theta(z,q) \equiv \Theta_1(z-K,q). \tag{3.74}$$

3. The Thomae [500] notation is equivalent to the notation of Krazer, Rauch and Lebowitz [428]. It corresponds to (in one variable) the Göpel notation for double-Theta functions and to the notation for the Riemannian Theta functions.

The notation is

$$\Theta_{11}(z,q) = \mathrm{H}(z,q), \tag{3.75}$$

$$\Theta_{01}(z,q) = \Theta(z,q), \tag{3.76}$$

$$\Theta_{10}(z,q) = \mathrm{H}_1(z,q), \tag{3.77}$$

$$\Theta(z,q) = \Theta_{00}(z,q) = \Theta_1(z,q)(z,q). \tag{3.78}$$

The functions

$$\Theta(2Kz,q),\ \mathrm{H}(2Kz,q),\ \mathrm{H}_1(2Kz,q),\ \Theta_1(2Kz,q),$$

correspond to

$$\theta_1(z,q),\ \theta_2(z,q),\ \theta_3(z,q),\ \theta_4(z,q),$$

respectively.

However, we have [428]

$$\theta_1(0,q) = -\theta\begin{bmatrix}1\\0\end{bmatrix}. \tag{3.79}$$

4. Around 1881, Glaisher introduced a uniform notation containing the same number of letters for each of the twelve functions

$$\begin{aligned} &\mathrm{sn}u,\ \mathrm{cn}u,\ \mathrm{dn}u,\ \frac{1}{\mathrm{sn}u},\ \frac{1}{\mathrm{cn}u},\ \frac{1}{\mathrm{dn}u},\\ &\frac{\mathrm{sn}u}{\mathrm{cn}u},\ \frac{\mathrm{sn}u}{\mathrm{dn}u},\ \frac{\mathrm{cn}u}{\mathrm{dn}u},\ \frac{\mathrm{cn}u}{\mathrm{sn}u},\ \frac{\mathrm{dn}u}{\mathrm{sn}u},\ \frac{\mathrm{dn}u}{\mathrm{cn}u}. \end{aligned} \tag{3.80}$$

He found it convenient to denote the quotient functions by pairs of letters consisting of the first letter of the numerator followed by the first letter of the denominator and the reciprocals of $\mathrm{sn}u$, $\mathrm{cn}u$, $\mathrm{dn}u$ by $\mathrm{ns}(u)$, $\mathrm{nc}(u)$, $\mathrm{nd}(u)$, the order of the letters being thus reversed so that we have

$$\mathrm{ns}u,\mathrm{nc}u,\mathrm{nd}u,\mathrm{sc}u,\mathrm{sd}u,\mathrm{cd}u,\mathrm{cs}u,\mathrm{ds}u,\mathrm{dc}u. \tag{3.81}$$

A similar notation was introduced by Neville 1944 [395]. He defined twelve elliptic functions with the following properties:

Definition 19 $f_j(z)$, $g_j(z)$ and $h_j(z)$ behave like $1/z$ in the vicinity of $z=0$. The function $f_j(z)$ is doubly periodic, with periods $2\omega_f$ and $4\omega_g$.

Today there are some small variations from the usual notation (3.66).

1. Chandrasekharan.
2. Mumford.
3. Nonnenmacher-Voros.

There are two parallel theories for

1. The four Jacobi theta functions.
2. The four Weierstraß σ functions.

If we know all Jacobi theta functions, we can compute all Weierstraß σ functions by means of 4 linear equations and vice versa. The Weierstraß elliptic functions, which are defined on a general lattice, are more general.

The Weierstraß ζ function $\zeta(z)$ is equal to the logarithmic derivative of $\sigma(z)$ (compare Section 6.13.1) and the same for $\zeta_i(z)$, $\sigma_i(z)$, $i=1,2,3$.

Jacobi used the corresponding function

$$Z(z,q) \equiv \frac{\Theta'(z,q)}{\Theta(z,q)}, \tag{3.82}$$

where $'$ denotes derivative with respect to z. The corresponding Thomae notation is

$$Z_{11}(z,q) \equiv \frac{\Theta'_{11}(z,q)}{\Theta_{11}(z,q)}, \tag{3.83}$$

$$Z_{01}(z,q) \equiv \frac{\Theta'_{01}(z,q)}{\Theta_{01}(z,q)}, \tag{3.84}$$

$$Z_{10}(z,q) \equiv \frac{\Theta'_{10}(z,q)}{\Theta_{10}(z,q)}, \tag{3.85}$$

$$Z_{00}(z,q) \equiv \frac{\Theta'_{00}(z,q)}{\Theta_{00}(z,q)}. \tag{3.86}$$

Another (equivalent) way to define the theta functions is via the *theta characteristic*, which is a two by one matrix of integers, written

$$\begin{bmatrix} \epsilon \\ \epsilon' \end{bmatrix}.$$

Next, given a complex number ζ and another complex number τ, satisfying $\operatorname{Im}\tau > 0$ and characteristic, we define the *first order theta function* with argument ζ and *theta period* τ, by

$$\theta(\zeta,\tau) \equiv \sum_{n=-\infty}^{+\infty} \exp\left(i\pi\left(\tau\left(n+\frac{\epsilon}{2}\right)^2 + 2\left(n+\frac{\epsilon}{2}\right)\left(\zeta+\frac{\epsilon'}{2}\right)\right)\right). \tag{3.87}$$

This function is analytic with respect to both variables. Up to sign this theta function is completely determined by the residue classes, mod 2, of ϵ and ϵ'; this is expressed by the *reduction formula.* Now we define the *reduced characteristics* to be those characteristics in which ϵ and ϵ' are either 0 or 1 and the *reduced representative* of an arbitrary characteristic to be that reduced characteristic whose entries are the least nonnegative residues mod 2 of ϵ and ϵ' respectively. We define a characteristic, not necessarily reduced, to be even or odd depending on whether $\epsilon\epsilon'$ is even or odd. The theta function is an even or odd function of ζ, depending on whether the characteristic is even or odd.

The theta function considered as a function only of ζ is entire and quasi-double periodic, with periods 1 and τ.

The following table shows the zeros of this function in one period.

Char	Zero
00	$\frac{1}{2}+\frac{\tau}{2}$
01	$\frac{\tau}{2}$
10	$\frac{1}{2}$
11	0

The function sn$u$ can now be expressed as a quotient of theta functions. The theta constant is the function value for $\zeta = 0$. By using the concept of *half-period* it is possible to express the *modulus* of sn$u$ as a quotient of Theta constants. We show by a detailed analysis that the elliptic function sn$u$ can be expressed as a balanced ratio of infinite products.

We used throughout the notation of Rauch and Lebowitz [428, p. 108–109].

Definition 20

$$\begin{aligned} Q_0 &\equiv \langle 1; q^2\rangle_\infty, \qquad Q_1 \equiv \langle \widetilde{1}; q^2\rangle_\infty, \\ Q_2 &\equiv \left\langle \frac{\widetilde{1}}{2}; q^2\right\rangle_\infty, \qquad Q_3 \equiv \left\langle \frac{1}{2}; q^2\right\rangle_\infty. \end{aligned} \tag{3.88}$$

This gives the following values for the theta constants:

$$\begin{aligned} \theta\begin{bmatrix}0\\0\end{bmatrix} &= \langle 1; q^2\rangle_\infty \left(\left\langle \frac{\widetilde{1}}{2}; q^2\right\rangle_\infty\right)^2, \\ \theta\begin{bmatrix}1\\0\end{bmatrix} &= 2q^{\frac{1}{4}}\langle 1; q^2\rangle_\infty \left(\langle \widetilde{1}; q^2\rangle_\infty\right)^2, \\ \theta\begin{bmatrix}0\\1\end{bmatrix} &= \langle 1; q^2\rangle_\infty \left(\left\langle \frac{1}{2}; q^2\right\rangle_\infty\right)^2. \end{aligned} \tag{3.89}$$

We get

Theorem 3.8.2 [310, p. 145]:

$$\begin{aligned} \mathrm{sn}u &= \sin x \, \Gamma_{q^2}\begin{bmatrix} 1+b, 1+b, a+\frac{1}{2}, -a+\frac{1}{2} \\ a+1, -a+1, \frac{1}{2}+b, \frac{1}{2}+b \end{bmatrix}, \\ q^{2a} &\equiv e^{2ix}, \qquad b \equiv \frac{\log(-1)}{2\log q}. \end{aligned} \tag{3.90}$$

This implies

Corollary 3.8.3

$$\frac{1}{\sqrt{k}} = \frac{\langle \widetilde{\tfrac{1}{2}}, \widetilde{\tfrac{1}{2}}; q^2\rangle_\infty}{\langle \widetilde{1}, \widetilde{1}; q^2\rangle_\infty} \frac{1}{2q^{\frac{1}{4}}}. \tag{3.91}$$

An nth order theta function may be expressed as the product of n first-order theta functions. Its characteristic is given by the matrix sum of the n characteristics. There is a formula for the number of linear independent even and odd theta functions of nth order with a certain characteristic.

The *addition theorem for* sn$u$ can be proved by using the *theta identities*.

The first period of sn$u$ can be expressed in terms of a theta constant by using the product expansions for the theta functions.

The *theta group* is the group generated by $z \mapsto z+2$ and $z \mapsto -1/z$. It has a fundamental domain bounded by the vertical lines $z=1$, $z=-1$ and the unit circle.

3.9 Meromorphic continuation and Riemann surfaces

Until about 1920 Abelian functions was the name of certain complex functions. The theory of Abelian functions and Abelian integrals is a generalization of the theory of elliptic functions and elliptic integrals. These are prerequisites for an understanding of the Riemann surfaces. All q-functions are meromorphic, i.e. Abelian complex functions with poles. q-analysis is thus a special case of these more general Abelian/complex functions. In this section focus is on a simple description of meromorphic continuation and Riemann surfaces, which form part of complex functions on a higher, more abstract level. We now, so to speak, move to a higher level of abstraction.

Definition 21 A region is a non-empty open connected subset of the Riemann sphere Σ.

Definition 22 Jones and Singerman [319, p. 123]. A function element is a pair (D, f), where D is a region and $f : D \to \Sigma$ a meromorphic function on D.

Definition 23 [319, p. 138] Let (D, f) be a *function element*, $a \in D$ and γ be a path in Σ from a to some point b in Σ. Then a *meromorphic continuation of* (D, f) *along* γ is a finite sequence of direct meromorphic continuations $(D, f) \sim (D_1, f_1) \sim (D_2, f_2) \sim \cdots \sim (D_m, f_m)$ such that:

1. each *region* D_i is an open disc in Σ with $a \in D_1 \subseteq D$;
2. there is a subdivision $0 = s_0 < s_1 < \cdots < s_m = 1$ of I such that $\gamma([s_{i-1}, s_i]) \subseteq D_i$ for $i = 1, 2, \ldots, m$, where $I = [0, 1]$.

When we do a q-deformation, a branching singularity is replaced by an infinite number of poles and zeroes.

This process is described briefly by an example, because the subject is vast.

According to Johannes Thomae (1869) [497, p. 262] and Smith (1911) [467, p. 51], the solution of the difference equation

$$(1-z)U(z) = \big(1-q^a z\big)U(qz) \tag{3.92}$$

is the function

$$p(a,q,z) = \lim_{n\to\infty} \big(1-q^n z\big)^{-a} \prod_{m=0}^{n-1} \frac{1-zq^{a+m}}{1-zq^m}, \tag{3.93}$$

which makes sense both for $0 < |q| < 1$ and $|q| > 1$. The first statement is obvious by (7.27). The last statement is proved as follows:

$$p(a,q,z) = \lim_{n\to\infty} (-z)^{-a} \left(1 - \frac{1}{zq^n}\right)^{-a} \prod_{m=0}^{n-1} \frac{1-\frac{1}{zq^{a+m}}}{1-\frac{1}{zq^m}} \tag{3.94}$$

or

$$p(a,q,z) = (-z)^{-a} p\left(a, \frac{1}{q}, \frac{1}{z}\right), \tag{3.95}$$

so that this follows from the previous case. For $q = 1$, the infinite product is equal to one and the result is the multiply valued function $(1-z)^{-a}$.

We infer that $p(a,q,z)$ is meromorphic in q and z and for $0 < |q| < 1$, $p(a,q,z) \equiv {}_1\phi_0(a; -|q, z)$. This means that both functions and their meromorphic continuations coincide everywhere.

3.10 Wave equation

The wave equation

$$\frac{\partial^2 V}{\partial x^2} + \frac{\partial^2 V}{\partial y^2} + \frac{\partial^2 V}{\partial z^2} = 0 \tag{3.96}$$

has the so-called spherical harmonics as its solutions, which for the first time were studied by Gauß. After separation of variables, the angular part contains the associated Legendre polynomials, which are treated in Section 3.11.1. The wave equation (3.96) was also solved by d'Alembert in 1752, who then arrived at the Cauchy-Riemann equations. Gauß expanded the Earth's magnetic potential in terms of the first four spherical harmonics to express the distribution of the Earth's magnetism.

3.11 Orthogonal polynomials

As Fujiwara [219] showed, the most important property of the Jacobi, Laguerre and Hermite polynomials (JLH) is the generalized Rodriguez formula. There are basically three different kinds of methods to treat orthogonality.

1. Three term recurrence theory (Favard theorem).
2. Rodriguez formula and integration by parts.
3. Self-adjoint differential equations with boundary conditions (A. Krall and P. Lesky).

The classical JLH polynomials have many things in common, as was beautifully explained by Ervin Feldheim (1912–1944) [200]. If we have an equation for the Jacobi polynomials, we automatically get a corresponding formula for Laguerre and Hermite polynomials.

3.11.1 Legendre-d'Allonville-Murphy polynomials

In the physics world, attributes with many names are common. In mathematics this is less frequent; one example is the article by V. I. Lebedev: *A unified formula for the phase function of extremal Chebyshev-Markov-Bernstein-Szegő polynomials*. We will use this kind of attribute only in the titles and keep the usual names in the text.

In 1722 in France Jasques Eugene d'Allonville (1671–1732) studied the polynomials defined by a certain generating function in connection with the isochronal property of the cycloid curve.

Except for a simple change of variables, he thus arrived at the Legendre polynomials, which were officially introduced 1784. The Legendre polynomials $\mathrm{P}_l(x)$ satisfy the differential equation

$$\left(1-x^2\right)f'' - 2xf' + l(l+1)f = 0. \tag{3.97}$$

The Rodriguez formula says that

$$\mathrm{P}_l(x) = \frac{1}{2^l l!}\frac{d^l}{dx^l}\left(x^2-1\right)^l. \tag{3.98}$$

The associated Legendre functions $\mathrm{P}_l^m(x)$ satisfy the differential equation

$$\left(1-x^2\right)f''-2xf'+\left[l(l+1)-\frac{\mu^2}{1-x^2}\right]f=0. \tag{3.99}$$

The associated Legendre functions are given by the formula

$$\mathrm{P}_l^m(x)=\left(1-x^2\right)^{\frac{m}{2}}\frac{d^m}{dx^m}P_l(x). \tag{3.100}$$

Equation (3.99) arises after separation of variables in the theory of the hydrogen atom.

In 1833 and 1835, Robert Murphy expressed the Legendre polynomials in the following hypergeometric form:

$$\mathrm{P}_l(x)={}_2\mathrm{F}_1\left(-n,1+n;1;\frac{1-x}{2}\right). \tag{3.101}$$

There are two types of Legendre polynomials:

1. $\mathrm{P}_l(x)$, orthogonal on $[-1,1]$;
2. $\mathrm{P}_l(x)^\star$, orthogonal on $[0,1]$.

We will find q-analogues of $\mathrm{P}_l(x)$ in Section 9.3.1. The n-dimensional versions of Legendre polynomials are called Lamé functions after Gabriel Lamé (1795–1870). The geometry of the n-dimensional sphere was very popular in France during that time. It was also the inspiration to the invention of the Hermite polynomials. The 1878 book *Kugelfunktionen* by Heine represented a breakthrough for this subject in the rest of the Western world.

Heine's work was a bit uneven, but he influenced many contemporaries to work in this area.

Rogers was strongly influenced by Heine to introduce the first q-Hermite polynomials, and to prove the Rogers-Ramanujan identities. Rogers was also very interested in elliptic functions.

3.11.2 Laguerre-Abel-Sonine-Murphy-Chebyshev-Halphen-Szegő polynomials

In 1833 and 1835 Robert Murphy studied the orthogonality of Laguerre polynomials.

These polynomials were later studied by Laguerre 1879 and 1883 for the case $\alpha=0$.

Laguerre's contribution was to find physical applications of these polynomials. The Laguerre polynomials $\mathrm{L}_n(x)$ satisfy the differential equation

$$xf''+(1-x)xf'+nf=0. \tag{3.102}$$

Laguerre used the notation $f_n(x) = n!\mathrm{L}_n(-x)$.

The associated Laguerre functions $\mathrm{L}_n^k(x)$, which arise after separation of variables in the Schrödinger equation for the hydrogen atom, satisfy the differential equation

$$xf'' + (k+1-x)f' + (n-k)f = 0. \tag{3.103}$$

The associated Laguerre functions are given by the formula

$$\mathrm{L}_n^k(x) = \frac{d^k}{dx^k}\mathrm{L}_n(x). \tag{3.104}$$

In 1880, Sonine [468, p. 41] generalized the Laguerre polynomials in a long paper about Bessel functions.

Definition 24 The generalized Laguerre polynomials are defined by

$$\mathrm{L}_n^{(\alpha)}(x) \equiv \frac{(\alpha+1)_n}{n!}\,{}_1\mathrm{F}_1(-n;\alpha+1;x). \tag{3.105}$$

The generalized Laguerre polynomials $\mathrm{L}_n^{(\alpha)}(x)$, which will form the basis for our study of the Cigler q-Laguerre polynomials in Section 9.1, should not be confused with the similar associated Laguerre functions $\mathrm{L}_n^k(x)$.

Abel's contribution was published posthumously in 1881.

In the year 1925, Pierre Humbert (1891–1953) [287] published a paper about so-called Sonine polynomials where reference to earlier contributions to the subject by Abel (1802–1829), Milne (1914), Vaney (1921) were given. The name Laguerre polynomials was often used after Szegő's two important 1926 papers. In the first paper, an expansion of a function in a series of Laguerre polynomials was given.

Erdelyi proved many formulas for Laguerre polynomials by Laplace transform. Laguerre polynomials are confluent hypergeometric functions. Other important functions are the confluent Whittaker function and the error function.

In 1939, Erdelyi wrote about the importance of the Hille-Hardy formula for computations with Laguerre polynomials. The *Hille-Hardy formula* (Hille [278], Hardy [265] and Watson [536]) for the generalized Laguerre polynomials $\mathrm{L}_n^{(\alpha)}(x)$ reads

$$\sum_{n=0}^{\infty} \frac{n!}{\Gamma(\alpha+n+1)}\mathrm{L}_n^{(\alpha)}(x)\mathrm{L}_n^{(\alpha)}(y)t^n$$
$$= (1-t)^{-1}(xyt)^{-(\frac{1}{2})\alpha}\exp\left(-\frac{(x+y)t}{1-t}\right)\mathrm{I}_\alpha\left(\frac{2\sqrt{xyt}}{1-t}\right), \tag{3.106}$$

where $\mathrm{I}_\alpha(z)$ denotes the modified Bessel function.

We will find a q-analogue of the Hille-Hardy formula in Section 9.1.5.

The idea of expanding products of Laguerre polynomials in series of Laguerre polynomials stirred a lot of attention in the thirties.

Bailey, Howell, Erdelyi and Watson have found several such expansions.

3.11.3 Jacobi polynomials

The Jacobi polynomials are defined as follows:

$$\mathrm{P}_n^{(\alpha,\beta)}(x) \equiv \frac{(1+\alpha)_n}{n!}\,{}_2\mathrm{F}_1(-n,\beta+n;1+\alpha;x). \tag{3.107}$$

The q-analogues of the Jacobi polynomials are found in Section 9.2. There are also the Szegő-Jacobi polynomials, big Jacobi polynomials, which we will not discuss.

3.11.4 Hermite polynomials

There are two types of Hermite polynomials in the literature; we will concentrate on the first one, which appeared first in Appell's book.

The Hermite polynomials $\mathrm{He}_n(x)$ satisfy the differential equation

$$f'' - xf' + nf = 0 \tag{3.108}$$

and the recurrence relation

$$\mathrm{He}_{n+1}(x) = x\mathrm{He}_n(x) - n\mathrm{He}_{n-1}(x). \tag{3.109}$$

Mehler [374] first stated [537] the famous Mehler formula for the Hermite polynomials $\mathrm{H}_n(x)$. In modern notation Mehler's formula takes the following form:

$$\sum_{n=0}^{\infty} \mathrm{He}_n(x)\mathrm{He}_n(y)\frac{t^n}{n!} = \left(1-t^2\right)^{-\frac{1}{2}} \exp\left(\frac{xyt - (x^2+y^2)\frac{t^2}{2}}{1-t^2}\right). \tag{3.110}$$

A formula for the product of two Hermite polynomials was proved in 1938 by Feldheim [196] and Watson [538] and in 1940 by Burchnall [84].

Chapter 4
The q-umbral calculus and semigroups. The Nørlund calculus of finite differences

In this chapter we discuss how the two q-additions lead to q-analogues of many results from Nørlund's investigations of difference analysis [402], [403]. We will use Milne-Thomson [385] as a basis for the notation of the various polynomials, which are defined by the generating function in Sections 4.3 and 4.4.

The main topics in these sections are the sets of almost parallel formulas, anticipated by Ward [531], for the q-Bernoulli, q-Euler, q-Lucas and q-G numbers and polynomials. These numbers will belong to $\mathbb{Z}(q)$. In number theory there is also a notion of Lucas number, which is not to be confused with this one. The reason for the name G-number is that E. T. Bell and after him his graduate student Ward, for some reason called these numbers Genocchi numbers.

The reason for introducing the second or JHC polynomials is that they are needed in the q-analogues of complementary argument formulas. The term second or JHC polynomials will be used throughout. The equations in Sections 4.3 and 4.4 are more systematically presented here than in [403], which makes this presentation an amplification and a complement to [403] even for the case $q = 1$. Despite the "telegraphic style" of the proofs, which assume that the reader knows the basic technical tools and various q-identities, it is likely that readers with a taste for q-analogues will find much here to enjoy. For example, the q-analogues of the Euler-Maclaurin summation formula might even be of some general interest.

We have added q-analogues of some formulas from Szegő's review [493] of [402]. The operational q-additions make the formulas remarkably pretty.

It is well-known that there are at least two types of q-Bernoulli numbers. Let us consider the first one, i.e. NWA. Its complement is JHC.

In the spirit of Milne-Thomson [385] and Rainville [425], we replace the $=$ by $\doteqdot$ to indicate the symbolic nature. Lucas [363] used a different symbol.

Example 3 The following recurrence holds [531, p. 265], [12, p. 245, 4.3]:

$$\mathrm{B}_{\mathrm{NWA},0,q} = 1, \qquad (\mathrm{B}_{\mathrm{NWA},q} \oplus_q 1)^k - \mathrm{B}_{\mathrm{NWA},k,q} \doteqdot \delta_{1,k}, \tag{4.1}$$

where $\mathrm{B}^n_{\mathrm{NWA},q}$ is replaced by $\mathrm{B}_{\mathrm{NWA},n,q}$ on power expansion. We see immediately that $\mathrm{B}_{\mathrm{NWA},n,q} \in \mathbb{Z}(q)$.

T. Ernst, *A Comprehensive Treatment of q-Calculus*,
DOI 10.1007/978-3-0348-0431-8_4, © Springer Basel 2012

Remark 6 The different Carlitz [97] q-Bernoulli numbers of 1948 are not included. These numbers are, however, the basis of the work of T. Kim.

4.1 The q-umbral calculus and semigroups

Those who understand the whole SI system (International System of Units) also have good chances to grasp q-analysis.

SI system	q-umbral calculus

Cigler considered a special case of the following q-umbral calculus; the case $q = 1$ was treated in [138].

Definition 25 Almost a q-analogue of Rota, Taylor [439, p. 696]. A pair $(A, \star)$, which consists of a set A together with a binary, associative and commutative operation $\star$ (multiplication) is called a commutative semigroup.

Definition 26 A q-umbral calculus contains a set A, called the alphabet, with elements called letters or umbrae. These letters are denoted α, β etc.

The most common alphabets are

1. $\mathbb{R}$.
2. $\mathbb{C}$ [185].

In each case, we adhere to one alphabet, each letter must belong to it.

If α, β are different umbrae, then a new umbra is created by $\alpha * \beta$, where $* \in \{\oplus_q, \boxplus_q, \ominus_q, \boxminus_q\}$. For convenience, the operators $\boxplus_q$ and $\boxminus_q$ should be as far right as possible.

Let M be a subset of A. Then $\langle M\rangle$ denotes the set generated by M together with the four operations above.

The set $\mathbb{R}_q$ is defined by

$$\mathbb{R}_q \equiv \langle\mathbb{R}\rangle. \tag{4.2}$$

There is a certain linear q-functional v, $\mathbb{C}[x] \times \mathbb{R}_q \to \mathbb{R}$, with $v(0) = a_0$, called the evaluation. In the following, an arbitrary $v \in \mathbb{C}[x]$ will be used.

Examples of evaluation are

1. The NWA q-addition:

$$(\alpha \oplus_q \beta)^n \equiv \sum_{k=0}^{n} \binom{n}{k}_q \alpha^k \beta^{n-k}, \quad n = 0, 1, 2, \ldots. \tag{4.3}$$

2. The NWA q-subtraction:

$$(\alpha \ominus_q \beta)^n \equiv \sum_{k=0}^{n} \binom{n}{k}_q \alpha^k (-\beta)^{n-k}, \quad n = 0, 1, 2, \ldots. \tag{4.4}$$

3. The JHC q-addition:

$$(\alpha \boxplus_q \beta)^n \equiv \sum_{k=0}^{n} \binom{n}{k}_q q^{\binom{k}{2}} \beta^k \alpha^{n-k}$$
$$= \alpha^n \left(-\frac{\beta}{\alpha}; q\right)_n, \quad n = 0, 1, 2, \ldots. \tag{4.5}$$

4. The JHC q-subtraction:

$$(\alpha \boxminus_q \beta)^n \equiv \sum_{k=0}^{n} \binom{n}{k}_q q^{\binom{k}{2}} (-\beta)^k \alpha^{n-k}$$
$$= \alpha^n \left(\frac{\beta}{\alpha}; q\right)_n, \quad n = 0, 1, 2, \ldots. \tag{4.6}$$

Let $\alpha \in \mathbb{R}_q$ and $z \in \mathbb{R}$. We assume that z is not a letter. We can now introduce a q-scalar product as follows:

$$\beta \equiv \alpha z, \tag{4.7}$$

on condition that we must operate immediately with v, see for example formula (4.80).

We can also introduce a q-interval as follows: $[0, a]$, $a \in \mathbb{R}_q$. Two umbrae $\alpha \in \mathbb{R}_q \vee \mathbb{R}$ and $\beta \in \mathbb{R}_q \vee \mathbb{R}$ are said to be equivalent, denoted $\alpha \sim \beta$, if $v(f, \alpha) = v(f, \beta)$. Equivalent umbrae form an equivalence class and $\sim$ is an equivalence relation i.e.,

$$\alpha \sim \alpha, \tag{4.8}$$

$$\alpha \sim \beta, \beta \sim \gamma \Rightarrow \alpha \sim \gamma. \tag{4.9}$$

If α and β are not equivalent, we write

$$\alpha \nsim \beta. \tag{4.10}$$

The equivalence class that contains α is denoted by $[\alpha]$.

The equality between equivalence classes is denoted by $=$. Thus $[\alpha] = [\beta] \Leftrightarrow \alpha \sim \beta$.

There is a distinguished element $\theta \in \mathbb{R}$, called the zero, such that for all $x \in \mathbb{R}$ and all f,

$$x \boxminus_q x \sim \theta, \quad v(f, \theta) = a_0. \tag{4.11}$$

In the case of matrices we interpret a_0 as a constant times the unit matrix.

The elements α and $\beta \in \mathbb{R}$ are said to be inverse to each other, if $\alpha \boxplus_q \beta \sim \theta$. There is also the concept of equation in the alphabet.

If

$$\alpha \oplus_q \beta \sim \gamma \tag{4.12}$$

or

$$\alpha \boxplus_q \beta \sim \gamma, \tag{4.13}$$

we can compute α explicitly. The solutions are

$$\alpha \sim \gamma \boxminus_q \beta \quad \text{or} \quad \alpha \sim \gamma \ominus_q \beta. \tag{4.14}$$

We can calculate β explicitly from (4.13) as

$$-\beta \sim \alpha \boxminus_q \gamma. \tag{4.15}$$

This fact is related to the concept of normal multiple hypergeometric series (in the broad sense), equivalent to the NWA q-addition and abnormal multiple hypergeometric series (in the broad sense), equivalent to the JHC q-addition of Jackson [307, p. 70]. These concepts are introduced in Section 10.2.

However, we can use a symmetry argument to show that

$$\alpha \boxminus_q \beta \sim \gamma \tag{4.16}$$

is equivalent to

$$\alpha \boxminus_q \gamma \sim \beta. \tag{4.17}$$

Definition 27 Let $\psi, \varphi \in \mathbb{R}_q$. The q-Kronecker-delta is defined by

$$\delta_{\psi,\varphi} \equiv \begin{cases} 1, & \text{if } \psi \sim \varphi, \\ 0, & \text{otherwise.} \end{cases} \tag{4.18}$$

We now go over to a closer examination of the q-addition.

Definition 28 There is a Ward number $\overline{n}_q \in \mathbb{R}_q$

$$\overline{n}_q \sim 1 \oplus_q 1 \oplus_q \cdots \oplus_q 1, \tag{4.19}$$

where the number of 1's in the RHS is n.

There is a corresponding Jackson number $\tilde{n}_q \in \mathbb{R}_q$

$$\tilde{n}_q \sim 1 \boxplus_q 1 \boxplus_q \cdots \boxplus_q 1, \tag{4.20}$$

where the number of 1's in the RHS is n.

If $\alpha_1, \ldots, \alpha_n \in \mathbb{R}$, $\alpha \sim \alpha_i$, $i = 1, \ldots, n$, then we have $\alpha_1 \oplus_q \cdots \oplus_q \alpha_n \sim \overline{n}_q \alpha$.

The last condition is a q-analogue of [402, p. 125, (13)], [403, p. 132, (49)].

$$F\big(x \oplus_q \mathrm{B}_{\mathrm{NWA},q}(y)\big) \doteq \sum_{n=0}^{\infty} \frac{\mathrm{B}_{\mathrm{NWA},n,q}(y)}{\{n\}_q!} \mathrm{D}_q^n F(x). \tag{4.21}$$

Here B can be changed to any q-polynomial sequence.

Theorem 4.1.1 [391, p. 345]. *The q-addition* (1.61) *has the following properties, for $\alpha, \beta, \gamma \in A$:*

$$(\alpha \oplus_q \beta) \oplus_q \gamma \sim \alpha \oplus_q (\beta \oplus_q \gamma), \tag{4.22}$$

$$\alpha \oplus_q \beta \sim \beta \oplus_q \alpha. \tag{4.23}$$

Proof The first property (associativity) is proved as follows: It suffices to prove that

$$\left[(a \oplus_q b) \oplus_q c\right]^n = \left[a \oplus_q (b \oplus_q c)\right]^n. \tag{4.24}$$

But this is equivalent to

$$\begin{aligned}&\sum_{k=0}^{n} \binom{n}{k}_q \sum_{l=0}^{k} \binom{k}{l}_q a^l b^{k-l} c^{n-k}\\ &\quad = \sum_{k'=0}^{n} \binom{n}{k'}_q a^{k'} \sum_{l'=0}^{n-k'} \binom{n-k'}{l'}_q b^{l'} c^{n-k'-l'}.\end{aligned} \tag{4.25}$$

Now put $l = k'$ and $l' = k - l$ to conclude the proof. The proof of the commutativity law is obvious. □

Equation (4.22) applies generally to compositions of $\oplus_q$ and $\boxplus_q$, if no $\oplus_q$ are to the right of $\boxplus_q$. Compare with Eqs. (4.72) and (4.75), where only $\boxplus_q$ occur.

In Section 10.1 we will extend the q-umbral calculus to n variables. There we will also introduce the Ward q-addition for q-factorials.

Definition 29 Limit for letters. Let $f \in \mathbb{C}[\![z]\!]$ and z_0 be a letter. Then limits of letters are defined as follows:

$$\lim_{z \to z_0} f(z) = \alpha \tag{4.26}$$

means that for every positive ϵ there is a positive δ, such that

$$0 < \left|(z \boxminus_q z_0)^1\right| < \delta \quad \text{implies} \quad \left|f(z) - \alpha\right| < \epsilon. \tag{4.27}$$

The work corresponds to the Ward q-derivative.

$F \cdot s$	$\mathrm{D}_{\oplus_q}$

In 1994, K. S. Chung, W. S. Chung, S. T. Nam and H. J. Kang [132] rediscovered the NWA together with a new form of the q-derivative:

Definition 30 If $f(x) \in \mathbb{C}[\![x]\!]$, we define the Ward q-derivative as follows:

$$\mathrm{D}_{\oplus_q} f(x) \equiv \lim_{\delta x \to \theta} \frac{f(x \oplus_q \delta x) - f(x)}{(\delta x)^1}. \tag{4.28}$$

Theorem 4.1.2 *This q-derivative $\mathrm{D}_{\oplus_q}$ satisfies the following rules*:

$$\mathrm{D}_{\oplus_{q,x}}(x \oplus_q a)^n = \{n\}_q (x \oplus_q a)^{n-1}, \tag{4.29}$$

$$\mathrm{D}_{\oplus_q}\mathrm{E}_q(x) = \mathrm{E}_q(x). \tag{4.30}$$

We will give three examples of other scientists who have used these q-additions in other contexts.

In 1954 A. Sharma and A. M. Chak [463] constructed q-Appell sequences for the JHC. The JHC has also been used by Goulden and D. M. Jackson [242], who used the notation

$$Q_n(-y, x) \equiv \mathrm{P}_{n,q}(x, y).$$

Ward explained how the NWA can be used as the function argument in a formal power series. The theory of formal power series is outlined in Niven [401], see also Hofbauer [282]. The formal power series form a vector space with respect to term-wise addition and multiplication by complex scalars. In the rest of this chapter, as in [531, p. 258], unless otherwise stated, we assume that functions $f(x), g(x), F(x), G(x) \in \mathbb{C}[\![x]\!]$.

Definition 31 Ward [531, p. 258]. If

$$F(x) = \sum_{k=0}^{\infty} a_k x^k, \tag{4.31}$$

put

$$F(x \oplus_q y) \equiv \sum_{n=0}^{\infty} a_n \sum_{k=0}^{n} \binom{n}{k}_q x^k y^{n-k} \tag{4.32}$$

and [307, p. 78]

$$F(x \boxplus_q y) \equiv F[x+y]_q \equiv \sum_{n=0}^{\infty} a_n \sum_{k=0}^{n} \binom{n}{k}_q q^{\binom{k}{2}} y^k x^{n-k}. \tag{4.33}$$

Before we embark on q-Taylor theorems, the following remark of Pearson [411] might be of interest. The differential calculus is a particular case of the direct method of finite differences and the integral calculus is a particular case of the inverse method of finite differences. In fact Taylor published his formula in terms of finite differences.

In the literature one finds at least three q-analogues of the Taylor formula for formal power series which we will list here.

Theorem 4.1.3 *The Nalli–Ward q-Taylor formula.* [391, p. 345], [531, p. 259]:

$$F(x \oplus_q y) = \sum_{n=0}^{\infty} \frac{y^n}{\{n\}_q!} \mathrm{D}_q^n F(x). \tag{4.34}$$

Theorem 4.1.4 *The first Jackson q-Taylor formula* [301, p. 63]:

$$F(x) = \sum_{n=0}^{\infty} \frac{(x \boxminus_q y)^n}{\{n\}_q!} \mathrm{D}_q^n F(y). \tag{4.35}$$

Theorem 4.1.5 *The second Jackson q-Taylor formula* [307, (51, p. 77)]:

$$F(x \boxplus_q y) = \sum_{n=0}^{\infty} \frac{y^n}{\{n\}_q!} q^{\binom{n}{2}} \mathrm{D}_q^n F(x). \tag{4.36}$$

Remark 7 Wallisser [529] found a criterion for the validity of the q-Taylor series (4.35) for an entire function $F(x)$ with $y = 1$ and $0 < q < 1$. Put $M_{\mathrm{E}_{\frac{1}{q}}}(r) = \max_{|x|=r} |\mathrm{E}_{\frac{1}{q}}(x)|$.

If the maximum of the absolute value of an entire function F satisfies on $|x| = r$ the inequality

$$M_F(r) \le C M_{\mathrm{E}_{\frac{1}{q}}}(r\tau), \quad q \in \mathbb{R},\ 0 < q < 1,\ \tau < \left(\frac{1}{q} - 1\right)^{-1}, \tag{4.37}$$

then $F(x)$ can be expanded in the q-Taylor series (4.35) for the special case $y = 1$.

Remark 8 Schendel [450] first proved (4.35), possibly influenced by Gauß.

The following general inversion formula will be useful.

Theorem 4.1.6 *Gaussian inversion* [9, p. 96], *a corrected version of* [232, p. 244]. *A q-analogue of* [433, p. 4]. *The following two equations for arbitrary sequences* a_n, b_n *are equivalent.*

$$a_n = q^{-f(n)} \sum_{l=0}^{n} (-1)^l q^{\binom{l}{2}} \binom{n}{l}_q b_{n-l}, \tag{4.38}$$

$$b_n = \sum_{i=0}^{n} q^{f(i)} \binom{n}{i}_q a_i. \tag{4.39}$$

Proof It will suffice to prove that

$$a_n = q^{-f(n)} \sum_{l=0}^{n} (-1)^l \binom{n}{l}_q q^{\binom{l}{2}} \sum_{i=0}^{n-l} q^{f(i)} \binom{n-l}{i}_q a_i. \tag{4.40}$$

The first sum is zero except for $i = n$ and $l = 0$. □

We give one last example that needs a definition:

Definition 32 If α denotes a letter $x \oplus_q y$, we can define the inverse of the q-factorial as follows:

$$\frac{1}{(x \oplus_q y; q)_a} \equiv \sum_{k=0}^{\infty} \frac{\langle a; q\rangle_k}{\langle 1; q\rangle_k} \sum_{l=0}^{k} \binom{k}{l}_q x^l y^{k-l}. \tag{4.41}$$

Then

Example 4

$$\mathrm{D}_{q,x} \frac{1}{(xs \oplus_q yt; q)_a} = s\{a\}_q \frac{1}{(xs \oplus_q yt; q)_{a+1}}. \tag{4.42}$$

Proof

$$\mathrm{D}_{q,x} \frac{1}{(xs \oplus_q yt; q)_a} = \mathrm{D}_{q,x} \sum_{k=0}^{\infty} \frac{\langle a; q\rangle_k}{\langle 1; q\rangle_k} \sum_{l=0}^{k} \binom{k}{l}_q (xs)^l (yt)^{k-l}$$

$$= \sum_{k=1}^{\infty} \frac{\langle a; q\rangle_k}{\langle 1; q\rangle_k} \sum_{l=1}^{k} \sum_{l=1}^{k} \frac{s\{k\}_q!(xs)^{l-1}(yt)^{k-l}}{\{l-1\}_q!\{k-l\}_q!} = s\{a\}_q \frac{1}{(xs \oplus_q yt; q)_{a+1}}. \tag{4.43}$$

□

4.2 Finite differences

In this section we will find, for example, q-analogues of the Nørlund and Jordan difference operators, which lie at the basis of the computations in the next two sections. We have retained the Nørlund notation $\omega \in \mathbb{C}$.

Definition 33 The Ward-Al-Salam q-forward shift operator $\mathrm{E}(\oplus_q)^\omega$ [12, p. 242, 3.1], a q-analogue of [79, p. 16], [138, p. 18], [403, p. 3], looks as follows:

$$\mathrm{E}(\oplus_q)^\omega (x^n) \equiv (x \oplus_q \omega)^n. \tag{4.44}$$

We denote the corresponding operator for the JHC by $\mathrm{E}(\boxplus_q)$, i.e.

$$\mathrm{E}(\boxplus_q)^\omega (x^n) \equiv (x \boxplus_q \omega)^n. \tag{4.45}$$

If $\omega = 1$, we denote these operators by $\mathrm{E}(\oplus_q)$ and $\mathrm{E}(\boxplus_q)$.

Definition 34 The invertible linear difference operator for NWA, a q-analogue of [403, p. 3], [385, p. 23], is defined by:

$$\underset{\omega\,\mathrm{NWA},q}{\triangle} \equiv \frac{\mathrm{E}(\oplus_q)^{\omega} - \mathrm{I}}{\omega}, \tag{4.46}$$

where I is the identity operator.

If $\omega = 1$, we denote this operator by $\triangle_{\mathrm{NWA},q}$ [12, p. 243, 3.5], compare [531, p. 264, 15.1]. We have $\triangle_{\mathrm{NWA},q} \in \mathbb{C}[\![\mathrm{D}_q]\!]$.

Remark 9 In contrast to what is claimed in [385], the formula $\lim_{\omega \to 0} \underset{\omega\,\mathrm{NWA},q}{\triangle}$ does not corresponds to a q-difference operator.

Definition 35 The linear difference operator for JHC, a q-analogue of [403, p. 3], is defined by:

$$\underset{\omega\,\mathrm{JHC},q}{\triangle} \equiv \frac{\mathrm{E}(\boxplus_q)^{\omega} - \mathrm{I}}{\omega}. \tag{4.47}$$

If $\omega = 1$, we denote this operator $\triangle_{\mathrm{JHC},q}$. We have $\triangle_{\mathrm{JHC},q} \in \mathbb{C}[\![\mathrm{D}_q]\!]$.

Definition 36 If ω is a Ward number $\overline{n}_q$, the difference operator for NWA is defined by

$$\underset{\overline{n}_q\,\mathrm{NWA},q}{\triangle} \equiv \frac{\mathrm{E}(\oplus_q)^{\overline{n}_q} - \mathrm{I}}{n}. \tag{4.48}$$

Definition 37 If ω is a Jackson number $\tilde{n}_q$, the difference operator for the JHC is defined by

$$\underset{\tilde{n}_q\,\mathrm{JHC},q}{\triangle} \equiv \frac{\mathrm{E}(\boxplus_q)^{\tilde{n}_q} - \mathrm{I}}{n}. \tag{4.49}$$

The formulas (4.59) and (4.85) show that the minus between $\mathrm{E}(\oplus_q)$ and I is not a q-subtraction.

We are now going to present an operational equation, which was first found by Lagrange 1772 for $q = 1$. It played a major role in the theory of finite differences, for example in Lacroix's treatise on differences from 1800, [42], [79, p. 18], [276, p. 26], [277, p. 66]; and later in the first umbral calculus by Blissard [75]. The following dual q-analogues of [138, p. 28], see [12, p. 242, 3.3, p. 243, 3.9], [531, p. 264] hold:

$$\mathrm{E}(\oplus_q)^{\omega} = \mathrm{E}_q(\omega \mathrm{D}_q), \tag{4.50}$$

$$\mathrm{E}(\boxplus_q)^{\omega} = \mathrm{E}_{\frac{1}{q}}(\omega \mathrm{D}_q). \tag{4.51}$$

Definition 38 A q-analogue of the mean value operator of Jordan [320, p. 6] ($\omega = 1$), Nørlund [403, p. 3] and [385, p. 30].

$$\underset{\omega\,\mathrm{NWA},q}{\nabla} \equiv \frac{\mathrm{E}(\oplus_q)^{\omega} + \mathrm{I}}{2}. \tag{4.52}$$

If $\omega = 1$, we denote this operator $\nabla_{\mathrm{NWA},q}$.

Definition 39 A JHC q-analogue of the mean value operator of Jordan [320, p. 6] ($\omega = 1$), Nørlund [403, p. 3] and [385, p. 30].

$$\underset{\omega\,\mathrm{JHC},q}{\nabla} \equiv \frac{\mathrm{E}(\boxplus_q)^{\omega} + \mathrm{I}}{2}. \tag{4.53}$$

If $\omega = 1$, we denote this operator $\nabla_{\mathrm{JHC},q}$.

In the following definition, the q-additions are written first in additive, then in multiplicative form. In the first case, we assume that the function arguments operate from left to right when using the two q-additions. In the second case, we assume that the function argument operates from right to left in accordance with (4.50) and (4.51). So do not forget that the following two equations are not associative.

Definition 40 If $f(x) = \sum_{k=0}^{\infty} a_k x^k$, we define

$$f(x \oplus_q y \boxminus_q z) \equiv \sum_{k=0}^{\infty} a_k \sum_{l=0}^{k} \binom{k}{l}_q y^{k-l} \sum_{m=0}^{l} (-1)^m \binom{l}{m}_q q^{\binom{m}{2}} z^m x^{l-m}, \tag{4.54}$$

$$f\big(\mathrm{E}(\boxminus_q)^{-z}\mathrm{E}(\oplus_q)^{y} x\big) \equiv \sum_{k=0}^{\infty} a_k \sum_{l=0}^{k} \binom{k}{l}_q y^{k-l} \sum_{m=0}^{l} (-1)^m \binom{l}{m}_q q^{\binom{m}{2}} z^m x^{l-m}. \tag{4.55}$$

We will now give a number of theorems for arbitrary letters which illustrate certain symmetry properties of this umbral calculus.

Theorem 4.2.1 *The NWA and the JHC are dual operators*:

$$f(x \oplus_q \alpha \boxminus_q \alpha) \equiv f\big(\mathrm{E}(\boxminus_q)^{-\alpha}\mathrm{E}(\oplus_q)^{\alpha} x\big) = f(x), \quad \alpha \in \mathbb{R}. \tag{4.56}$$

Proof Use (4.50) and (4.51). □

By Goulden and D. M. Jackson [242], we obtain two further formulas of this type.

Theorem 4.2.2 [242, p. 228]

$$(\alpha \boxminus_q \beta) \oplus_q (\gamma \boxminus_q \delta) \sim (\alpha \boxminus_q \delta) \oplus_q (\gamma \boxminus_q \beta), \quad \alpha, \beta, \gamma, \delta \in \mathbb{R}. \tag{4.57}$$

Proof Expand both sides. The powers of β and δ are denoted i and j. The corresponding terms are $\beta^i q^{\binom{i}{2}}$ and $\delta^j q^{\binom{j}{2}}$. Compare with (4.74) ($a_j = 0$, $j \geq 4$). □

Theorem 4.2.3 [242, p. 228]

$$(\alpha \boxminus_q \gamma) \sim (\alpha \boxminus_q \beta) \oplus_q (\beta \boxminus_q \gamma), \quad \alpha, \beta, \gamma \in \mathbb{R}. \tag{4.58}$$

The two Leibniz theorems go as follows. Notice the binomial coefficient on the right.

Theorem 4.2.4 *A q-analogue of* [138, p. 27, 2.13], [320, p. 97, 10], [385, p. 35, 2]. *Let $f(x)$ and $g(x)$ be formal power series. Then*

$$\triangle_{\text{NWA},q}^n (fg) = \sum_{i=0}^{n} \binom{n}{i} \triangle_{\text{NWA},q}^i f \left(\triangle_{\text{NWA},q}^{n-i} \text{E}(\oplus_q)^i\right) g. \tag{4.59}$$

Proof Analogous to [320, p. 96–97]. □

Theorem 4.2.5 *A q-analogue of* [138, p. 27, 2.13], [320, p. 97, 10], [385, p. 35, 2]. *Let $f(x)$ and $g(x)$ be two formal power series. Then*

$$\triangle_{\text{JHC},q}^n (fg) = \sum_{i=0}^{n} \binom{n}{i} \triangle_{\text{JHC},q}^i f \left(\triangle_{\text{JHC},q}^{n-i} \text{E}(\boxplus_q)^i\right) g. \tag{4.60}$$

The q-difference of a quotient of functions can be computed as

Theorem 4.2.6 *A q-analogue of* [411, p. 2], [79, p. 29].

$$\triangle_{\text{NWA},q} \frac{f(x)}{g(x)} = \frac{g(x)\triangle_{\text{NWA},q} f(x) - f(x)\triangle_{\text{NWA},q} g(x)}{g(x)g(x \oplus_q 1)}. \tag{4.61}$$

Theorem 4.2.7 *A q-analogue of* [411, p. 2], [79, p. 29].

$$\triangle_{\text{JHC},q} \frac{f(x)}{g(x)} = \frac{g(x)\triangle_{\text{JHC},q} f(x) - f(x)\triangle_{\text{JHC},q} g(x)}{g(x)g(x \boxplus_q 1)}. \tag{4.62}$$

The important q-binomial coefficients are presented in Sections 1.7 and 6.5. The q-multinomial coefficients are products of q-binomial coefficients.

Definition 41 The q-multinomial coefficient is defined by

$$\binom{n}{k_1, k_2, \ldots, k_m}_q \equiv \frac{\langle 1; q\rangle_n}{\langle 1; q\rangle_{k_1} \langle 1; q\rangle_{k_2} \cdots \langle 1; q\rangle_{k_m}}, \tag{4.63}$$

where $k_1+k_2+\cdots+k_m=n$. If the number of k_i is unspecified for $m=\infty$ in (4.63), we denote the q-multinomial coefficients by

$$\binom{n}{\vec{k}}_q, \quad \sum_{i=1}^{\infty} k_i = n. \tag{4.64}$$

We give some examples of q-multinomial coefficients:

Example 5

$$\binom{3}{1,1,1}_q = 1+2q+2q^2+q^3, \tag{4.65}$$

$$\binom{3}{1,1,1,1}_q = 1+3q+5q^2+6q^3+5q^4+3q^5+q^6, \tag{4.66}$$

$$\binom{3}{2,1,1}_q = 1+2q+3q^2+3q^3+2q^4+q^5, \tag{4.67}$$

$$\binom{3}{2,2}_q = 1+q+2q^2+q^3+q^4. \tag{4.68}$$

Definition 42 The symbol $\sum_{\vec{m}}$ denotes a multiple summation with the indices $m_1,\ldots,m_n$ running over all non-negative integer values. In this connection we will put $m \equiv \sum_{j=1}^{n} m_j$.

If $\vec{m}$ and $\vec{k}$ are two arbitrary vectors with n elements, their q-binomial coefficient is defined as

$$\binom{\vec{m}}{\vec{k}}_{\vec{q}} \equiv \prod_{j=1}^{n} \binom{m_j}{k_j}_{q_j}. \tag{4.69}$$

In order to present Ward's q-analogue of De Moivre's formula [531], (6.234) and (6.235) (Chapter 6) and (6.235), we need a new notation. Compare with Al-Salam [12, p. 244, 3.16], where the notation $P_k(n)$ was used.

Definition 43 [531, p. 258] Let

$$(\overline{n}_q)^k \equiv \sum_{m_1+\cdots+m_n=k} \binom{k}{m_1,\ldots,m_n}_q, \tag{4.70}$$

where each partition of k is multiplied by its number of permutations. We have the following special cases:

$$(\overline{0}_q)^k = \delta_{k,0}; \qquad (\overline{n}_q)^0 = 1; \qquad (\overline{n}_q)^1 = n. \tag{4.71}$$

The following definition shows how Jackson numbers usually appear in applications.

Definition 44

$$(\tilde{n}_q)^k = \sum_{m_1+\cdots+m_n=k} \binom{k}{m_1,\ldots,m_n}_q q^{\binom{\vec{m}}{2}}, \quad \vec{m} = (m_2,\ldots,m_n), \tag{4.72}$$

where each partition of k is multiplied by its number of permutations.

We have the following special cases:

$$(\tilde{0}_q)^k = \delta_{k,0}; \qquad (\tilde{n}_q)^0 = 1; \qquad (\tilde{n}_q)^1 = n. \tag{4.73}$$

The following table lists some of the first $(\overline{n}_q)^k$. Compare with [141, p. 309], where a long list of multinomial coefficients is given. The reader can convince himself that the results coincide with the definition (4.70) for the case $q = 1$.

	$k=2$	$k=3$	$k=4$
$n=1$	1	1	1
$n=2$	$3+q$	$4+2q+2q^2$	$5+3q+4q^2+3q^3+q^4$
$n=3$	$6+3q$	$10+8q+8q^2+q^3$	$3(5+5q+7q^2+6q^3+3q^4+q^5)$

and for $n=4$

$k=2$	$k=3$	$k=4$
$10+6q$	$4\,(5+5q+5q^2+q^3)$	$5\,(7+9q+13q^2+12q^3+7q^4+3q^5)+q^6$

The following table lists some of the first $(\tilde{n}_q)^k$:

	$k=2$	$k=3$	$k=4$
$n=1$	1	1	1
$n=2$	$2+2q$	$2(1+q+q^2+q^3)$	$3+2q+3q^2+3q^3+2q^4+q^5+2q^6$
$n=3$	$4+5q$	$6+8q+8q^2+5q^3$	$(2+2q+2q^2+3q^3)^2$

and for $n=4$

$k=2$	$k=3$	$k=4$
$7+9q$	$4(2+5q+5q^2+4q^3)$	$8+24q+41q^2+63q^3+56q^4+39q^5+25q^6$

According to Netto [394], the so-called multinomial expansion theorem was first mentioned in a letter (1695) from Leibniz to John Bernoulli, who proved it. In 1698 De Moivre first published a paper about multinomial coefficients in England [161, p. 114].

Two natural q-analogues are given by

Definition 45 If $f(x)$ is the formal power series $\sum_{l=0}^{\infty} a_l x^l$, its k'th NWA-power is given by

$$\left(\bigoplus_{q,l=0}^{\infty} a_l x^l\right)^k \equiv (a_0 \oplus_q a_1 x \oplus_q \cdots)^k \equiv \sum_{|m|=k} \prod_{l=0}^{\infty} (a_l x^l)^{m_l} \binom{k}{\vec{m}}_q. \tag{4.74}$$

Definition 46 If $f(x)$ is a formal power series $\sum_{l=0}^{\infty} a_l x^l$, its k'th JHC-power is given by

$$\left(\boxplus_{q,l=0}^{\infty} a_l x^l\right)^k \equiv (a_0 \boxplus_q a_1 x \boxplus_q \cdots)^k \equiv \sum_{|m|=k} \prod_{l=0}^{\infty} (a_l x^l)^{m_l} \binom{k}{\vec{m}}_q q^{\binom{\vec{n}}{2}}, \tag{4.75}$$

where $\vec{n} = (m_2, \ldots, m_n)$.

Definition 47 If $f(x)$ is a formal power series $\sum_{k=0}^{\infty} a_k x^k$, the (Ward) q-sum is defined by

$$\sum_{k=n}^{m} f(\overline{k}_q) \equiv \sum_{k=n}^{m} \sum_{l=0}^{\infty} a_l (\overline{k}_q)^l, \quad n, m \in \mathbb{N},\ n \le m, \tag{4.76}$$

where for each k the function value for the corresponding Ward number is computed. If $n > m$, the sum is equal to 0. Similarly, we define

$$(-1)^{\overline{n}_q} \equiv (-1)^n. \tag{4.77}$$

Definition 48 If $f(x)$ is a formal power series $\sum_{k=0}^{\infty} a_k x^k$, the (Jackson) q-sum is defined by

$$\sum_{k=n}^{m} f(\tilde{k}_q) \equiv \sum_{k=n}^{m} \sum_{l=0}^{\infty} a_l (\tilde{k}_q)^l, \quad n, m \in \mathbb{N},\ n \le m. \tag{4.78}$$

Similarly, we define

$$(-1)^{\tilde{n}_q} \equiv (-1)^n. \tag{4.79}$$

Definition 49 Let $f(x) \in \mathbb{R}[\![x]\!]$. We generalize the definition of q-integral (1.53) to

$$\int_0^a f(t)\, d_q(t) \equiv a^1 (1-q) \sum_{n=0}^{\infty} f(aq^n) q^n, \quad 0 < q < 1,\ a \in \mathbb{R}_q. \tag{4.80}$$

Definition 50 Let I be the product interval $[0,a]\times(0,1)$. Then $\mathrm{L}_q^1(I)$ is the space of all functions $f(x,q)\in\mathbb{R}[\![x]\!]$ on I such that

$$\int_0^a f(t,q)\,d_q(t) \tag{4.81}$$

converges.

Theorem 4.2.8 $\mathrm{L}_q^1(I)$ *is a vector space with respect to term-wise addition and multiplication by real scalars.*

If we have a function $G(\overline{n}_q)=F(x\oplus_q\overline{n}_q)$ or $G(\tilde{n}_q)=F(x\oplus_q\tilde{n}_q)$, like in (4.82) and (4.93), $G(0)$ always means $F(x)$.

Theorem 4.2.9 *A* q*-analogue of the Newton-Gregory series* [138, p. 21, 2.7], [320, p. 26], [385, 2.5.1], [363, p. 243].

$$f(\overline{n}_q)=\sum_{k=0}^{n}\binom{n}{k}\triangle_{\mathrm{NWA},q}^{k}f(0). \tag{4.82}$$

This can be generalized to

Theorem 4.2.10 *A* q*-analogue of* [403, p. 4, (7)]

$$f(\omega\overline{n}_q)=\sum_{k=0}^{n}\binom{n}{k}\omega^k\Big(\underset{\omega\,\mathrm{NWA},q}{\triangle}\Big)^k f(0). \tag{4.83}$$

Theorem 4.2.11 *A* q*-analogue of* [403, p. 4, (8)]

$$f(\omega\overline{n}_q)=\sum_{k=0}^{n}(-1)^{n-k}\binom{n}{k}2^k\Big(\underset{\omega\,\mathrm{NWA},q}{\nabla}\Big)^k f(0). \tag{4.84}$$

The formula (4.82) can be inverted as follows.

Theorem 4.2.12 *A corrected form of* [531, p. 264, (iii)] *and a* q*-analogue of* [244, p. 188, (5.50)], [363, p. 136, (3)], [385, 2.5.2]:

$$\triangle_{\mathrm{NWA},q}^{n}f(x)=\sum_{k=0}^{n}(-1)^{n-k}\binom{n}{k}f(x\oplus_q\overline{k}_q). \tag{4.85}$$

We will use the following simplified notations, where $k_l\in\mathbb{N}$, $l=1,\ldots,n$.

$$\Omega\equiv(k_1\omega_1\oplus_q k_2\omega_2\oplus_q\cdots\oplus_q k_n\omega_n),\quad k\equiv\sum_{l=1}^{n}k_l, \tag{4.86}$$

$$\Phi \equiv (k_1 m_1 \oplus_q k_2 m_2 \oplus_q \cdots \oplus_q k_n m_n). \tag{4.87}$$

The symbol $\sum_{\vec{k}}$ denotes a multiple summation with each of the indices $k_1, \ldots, k_n$ running between 0, 1. The formula (4.85) can be generalized to

Definition 51 Two q-analogues of [403, p. 4]:

$$\underset{\omega_1,\ldots,\omega_n \mathrm{NWA},q}{\triangle^n} f(x) \equiv (\omega_1 \cdots \omega_n)^{-1} \sum_{\vec{k}} (-1)^{n-k} f(x \oplus_q \Omega), \tag{4.88}$$

$$\underset{\overline{m}_{1_q},\ldots,\overline{m}_{n_q} \mathrm{NWA},q}{\triangle^n} f(x) \equiv (m_1 \cdots m_n)^{-1} \sum_{\vec{k}} (-1)^{n-k} f(x \oplus_q \Phi). \tag{4.89}$$

There is a similar formula for $\nabla_{\mathrm{NWA},q}$:

$$\nabla^n_{\mathrm{NWA},q} f(x) \equiv 2^{-n} \sum_{k=0}^{n} \binom{n}{k} f(x \oplus_q \overline{k}_q). \tag{4.90}$$

In a similar way, the formula (4.90) can be generalized to

Definition 52 A q-analogue of [403, p. 4].

$$\underset{\omega_1,\ldots,\omega_n \mathrm{NWA},q}{\nabla^n} f(x) \equiv 2^{-n} \sum_{\vec{k}} f(x \oplus_q \Omega). \tag{4.91}$$

Theorem 4.2.13 *A q-analogue of* [320, (12), p. 114].

$$\nabla^{-1}_{\mathrm{NWA},q} = \sum_{m=0}^{\infty} \frac{(-1)^m}{2^m} \triangle^m_{\mathrm{NWA},q}. \tag{4.92}$$

We now write down the corresponding formulas for JHC.

Theorem 4.2.14 *A q-analogue of the Newton-Gregory series* [138, p. 21, 2.7], [320, p. 26], [385, 2.5.1], [363, p. 243]:

$$f(\tilde{n}_q) = \sum_{k=0}^{n} \binom{n}{k} \triangle^k_{\mathrm{JHC},q} f(0). \tag{4.93}$$

This can be generalized to

Theorem 4.2.15 *A q-analogue of* [403, p. 4, (7)]:

$$f(\omega \tilde{n}_q) = \sum_{k=0}^{n} \binom{n}{k} \omega^k \Big(\underset{\omega\, \mathrm{JHC},q}{\triangle} \Big)^k f(0). \tag{4.94}$$

Theorem 4.2.16 *A q-analogue of* [403, p. 4, (8)]:

$$f(\omega\tilde{n}_q) = \sum_{k=0}^{n}(-1)^{n-k}\binom{n}{k}2^k\Big(\underset{\omega\,\mathrm{JHC},q}{\nabla}\Big)^k f(0). \tag{4.95}$$

The formula (4.93) can be inverted as follows.

Theorem 4.2.17 *A q-analogue of* [244, p. 188, (5.50)], [363, p. 136, (3)], [385, 2.5.2]:

$$\Delta^n_{\mathrm{JHC},q} f(x) = \sum_{k=0}^{n}(-1)^{n-k}\binom{n}{k} f(x \boxplus_q \tilde{k}_q). \tag{4.96}$$

We will use the following abbreviations, where $k_l \in \mathbb{N}$:

$$\Omega \equiv (k_1\omega_1 \boxplus_q k_2\omega_2 \boxplus_q \cdots \boxplus_q k_n\omega_n), \quad k \equiv \sum_{l=1}^{n} k_l. \tag{4.97}$$

$$\Phi \equiv (k_1 m_1 \boxplus_q k_2 m_2 \boxplus_q \cdots \boxplus_q k_n m_n). \tag{4.98}$$

The symbol $\sum_{\vec{k}}$ denotes a multiple summation with each of the indices $k_1, \ldots, k_n$ running between 0, 1.

Formula (4.96) can be inverted as follows.

Definition 53 Two q-analogues of [403, p. 4]:

$$\underset{\omega_1,\ldots,\omega_n\,\mathrm{JHC},q}{\Delta^n} f(x) \equiv (\omega_1\cdots\omega_n)^{-1}\sum_{\vec{k}}(-1)^{n-k} f(x \boxplus_q \Omega), \tag{4.99}$$

$$\underset{\tilde{m}_{1q},\ldots,\tilde{m}_{nq}\,\mathrm{JHC},q}{\Delta^n} f(x) \equiv (m_1\cdots m_n)^{-1}\sum_{\vec{k}}(-1)^{n-k} f(x \boxplus_q \Phi). \tag{4.100}$$

There is a similar formula for $\nabla_{\mathrm{JHC},q}$:

$$\nabla^n_{\mathrm{JHC},q} f(x) \equiv 2^{-n}\sum_{k=0}^{n}\binom{n}{k} f(x \boxplus_q \tilde{k}_q). \tag{4.101}$$

In a similar way, formula (4.101) can be generalized to

Definition 54 A q-analogue of [403, p. 4]:

$$\underset{\omega_1,\ldots,\omega_n\,\mathrm{JHC},q}{\nabla^n} f(x) \equiv 2^{-n}\sum_{\vec{k}} f(x \boxplus_q \Omega). \tag{4.102}$$

Theorem 4.2.18 *A q-analogue of* [320, (12), p. 114]:

$$\nabla_{\mathrm{JHC},q}^{-1} = \sum_{m=0}^{\infty} \frac{(-1)^m}{2^m} \triangle_{\mathrm{JHC},q}^{m}. \tag{4.103}$$

4.3 q-Appell polynomials

We will now describe the q-Appell polynomials. In the spirit of Milne-Thomson [385, p. 125–147], which we will follow closely, we will call these q-polynomials Φ_q polynomials and express them by a certain generating function. Examples of q-Appell polynomials or Φ_q polynomials are $\mathrm{B}_{\mathrm{NWA},\nu,q}^{(n)}(x)$ and $\mathrm{F}_{\mathrm{NWA},\nu,q}^{(n)}(x)$, i.e. q-Bernoulli and q-Euler polynomials.

Definition 55 A q-analogue of [385, p. 124]. For every power series $f_n(t)$, with $f_n(0) \neq 0$, the Φ_q polynomials of degree ν and order n have the following generating function:

$$f_n(t)\mathrm{E}_q(xt) = \sum_{\nu=0}^{\infty} \frac{t^\nu}{\{\nu\}_q!} \Phi_{\nu,q}^{(n)}(x). \tag{4.104}$$

Putting $x = 0$, we have

$$f_n(t) = \sum_{\nu=0}^{\infty} \frac{t^\nu}{\{\nu\}_q!} \Phi_{\nu,q}^{(n)}, \tag{4.105}$$

where $\Phi_{\nu,q}^{(n)}$ is called a Φ_q number of degree ν and order n.

It will be convenient to fix the value for $n = 0$ and $n = 1$:

$$\Phi_{\nu,q}^{(0)}(x) \equiv x^\nu; \qquad \Phi_{\nu,q}^{(1)}(x) \equiv \Phi_{\nu,q}(x); \qquad \Phi_{\nu,q}^{(0)} \equiv 0. \tag{4.106}$$

By (4.104) we obtain

Theorem 4.3.1 *A q-analogue of* [40], [385, p. 125, (4), (5)]:

$$\mathrm{D}_q \Phi_{\nu,q}^{(n)}(x) = \{\nu\}_q \Phi_{\nu-1,q}^{(n)}(x), \tag{4.107}$$

$$\int_a^x \Phi_{\nu,q}^{(n)}(t)\, d_q(t) = \frac{\Phi_{\nu+1,q}^{(n)}(x) - \Phi_{\nu+1,q}^{(n)}(a)}{\{\nu+1\}_q}. \tag{4.108}$$

By (4.34), (4.36) we obtain the two q-Taylor formulas

Theorem 4.3.2

$$\Phi_{\nu,q}^{(n)}(x \oplus_q y) = \sum_{k=0}^{\nu} \binom{\nu}{k}_q \Phi_{\nu-k,q}^{(n)}(x) y^k, \tag{4.109}$$

$$\Phi_{\nu,q}^{(n)}(x \boxplus_q y) = \sum_{k=0}^{\nu} \binom{\nu}{k}_q q^{\binom{k}{2}} \Phi_{\nu-k,q}^{(n)}(x) y^k. \tag{4.110}$$

Pay attention to the slight difference of polynomials of q-binomial type in (4.109).

The first formula (or [15, p. 33, 2.5]) gives the symbolic equality

Theorem 4.3.3 *A q-analogue of* [385, p. 125, (3)]:

$$\Phi_{\nu,q}^{(n)}(x) \doteq \left(\Phi_q^{(n)} \oplus_q x\right)^\nu. \tag{4.111}$$

Theorem 4.3.4 *A q-analogue of* [385, p. 125]:

$$\left(\mathrm{E}_q(t) - 1\right) f_n(t) \mathrm{E}_q(xt) = \sum_{\nu=0}^{\infty} \frac{t^\nu}{\{\nu\}_q!} \triangle_{\mathrm{NWA},q} \Phi_{\nu,q}^{(n)}(x). \tag{4.112}$$

Proof Operate on (4.104) with $\triangle_{\mathrm{NWA},q}$. □

Theorem 4.3.5 *A q-analogue of* [385, p. 125]:

$$\frac{(\mathrm{E}_q(t) + 1)}{2} f_n(t) \mathrm{E}_q(xt) = \sum_{\nu=0}^{\infty} \frac{t^\nu}{\{\nu\}_q!} \nabla_{\mathrm{NWA},q} \Phi_{\nu,q}^{(n)}(x). \tag{4.113}$$

Proof Operate on (4.104) with $\nabla_{\mathrm{NWA},q}$. □

Theorem 4.3.6 *A q-analogue of* [385, p. 125]:

$$\left(\mathrm{E}_{\frac{1}{q}}(t) - 1\right) f_n(t) \mathrm{E}_q(xt) = \sum_{\nu=0}^{\infty} \frac{t^\nu}{\{\nu\}_q!} \triangle_{\mathrm{JHC},q} \Phi_{\nu,q}^{(n)}(x). \tag{4.114}$$

Proof Operate on (4.104) with $\triangle_{\mathrm{JHC},q}$. □

Theorem 4.3.7 *A q-analogue of* [385, p. 125]:

$$\frac{(\mathrm{E}_{\frac{1}{q}}(t) + 1)}{2} f_n(t) \mathrm{E}_q(xt) = \sum_{\nu=0}^{\infty} \frac{t^\nu}{\{\nu\}_q!} \nabla_{\mathrm{JHC},q} \Phi_{\nu,q}^{(n)}(x). \tag{4.115}$$

Proof Operate on (4.104) with $\nabla_{\mathrm{JHC},q}$. □

The simplest example of a Φ_q polynomial is the Rogers-Szegő polynomial [434, 494], [133, p. 90, (11)]:

$$\mathrm{H}_{n,q}(x,a) \equiv (x \oplus_q a)^n. \tag{4.116}$$

The following recurrence [135, p. 35, (5)] holds:

$$\mathrm{H}_{n+1,q}(x,a) = (x+a)\mathrm{H}_{n,q}(x,a) + a\big(q^n - 1\big)x\mathrm{H}_{n-1,q}(x,a). \tag{4.117}$$

A special case of the Φ_q polynomials are the β_q-polynomials of degree ν and order n, which are obtained by putting $f_n(t) = \frac{t^n g(t)}{(\mathrm{E}_q(t)-1)^n}$ in (4.104):

Definition 56

$$\frac{t^n g(t)}{(\mathrm{E}_q(t)-1)^n}\mathrm{E}_q(xt) \equiv \sum_{\nu=0}^{\infty} \frac{t^\nu \beta_{\nu,q}^{(n)}(x)}{\{\nu\}_q!}. \tag{4.118}$$

We now give the first examples of umbral equations in this section. The proofs are a good exercise for the reader.

Theorem 4.3.8 [12, p. 255, 10.8], *a q-analogue of* [385, (2), p. 126], [364, p. 21], [439, p. 704], [363, p. 240]:

$$\triangle_{\mathrm{NWA},q}\beta_{\nu,q}^{(n)}(x) = \{\nu\}_q \beta_{\nu-1,q}^{(n-1)}(x) = \mathrm{D}_q \beta_{\nu,q}^{(n-1)}(x). \tag{4.119}$$

Proof Use (4.112). □

By (4.111), the following symbolic relations apply:

Theorem 4.3.9 *A q-analogue of* [385, p. 126]. *The second equation implies formula* (4.1):

$$\big(\beta_q^{(n)} \oplus_q x \oplus_q 1\big)^\nu - \big(\beta_q^{(n)} \oplus_q x\big)^\nu \doteq \{\nu\}_q \big(\beta_q^{(n-1)} \oplus_q x\big)^{\nu-1}, \tag{4.120}$$

$$\big(\beta_q^{(n)} \oplus_q 1\big)^\nu - \beta_{\nu,q}^{(n)} \doteq \{\nu\}_q \beta_{\nu-1,q}^{(n-1)}. \tag{4.121}$$

Theorem 4.3.10 *A q-analogue of* [402, (20), p. 163]:

$$\triangle_{\mathrm{NWA},q} f\big(\beta_{\nu,q}^{(n)}(x)\big) \equiv f\big(\beta_{\nu,q}^{(n)}(x) \oplus_q 1\big) - f\big(\beta_{\nu,q}^{(n)}(x)\big) \doteq \mathrm{D}_q f\big(\beta_{\nu,q}^{(n-1)}(x)\big). \tag{4.122}$$

Theorem 4.3.11 *Almost a q-analogue of* [485, p. 378, (26)]:

$$\sum_{k=1}^{\nu} \binom{\nu}{k}_q \beta_{\nu-k,q}^{(n)}(x) = \{\nu\}_q \beta_{\nu-1,q}^{(n-1)}(x). \tag{4.123}$$

Proof Use (4.109) and (4.120). □

4.3.1 The generalized q-Bernoulli polynomials

A particular case of β_q polynomials are the generalized q-Bernoulli polynomials of degree ν and order n, $\mathrm{B}^{(n)}_{\mathrm{NWA},\nu,q}(x)$, which were defined for $q=1$ in [385, p. 127], [402] and for complex order in [12, p. 254, 10.3].

These polynomials and the generalized q-Euler polynomials are immediate generalizations of $\mathrm{B}_{\mathrm{NWA},q}(x)$ and $\mathrm{F}_{\mathrm{NWA},q}(x)$ and we shall call them in an extended sense q-Bernoulli and q-Euler polynomials of higher order. The study of these polynomials has been motivated by several facts. The $\mathrm{B}_{\mathrm{NWA},q}(x)$ and the $\mathrm{F}_{\mathrm{NWA},q}(x)$ are indispensable in the study of q-difference equations of the first order. Likewise, the new polynomials are needed in the study of q-difference equations of any order.

Definition 57 [12, p. 254, 10.3], [403, (36), p. 132], [485]. The generating function for $\mathrm{B}^{(n)}_{\mathrm{NWA},\nu,q}(x)$ is a q-analogue of [439, p. 704], [418, p. 1225, ii]:

$$\frac{t^n}{(\mathrm{E}_q(t)-1)^n}\mathrm{E}_q(xt)=\sum_{\nu=0}^{\infty}\frac{t^\nu \mathrm{B}^{(n)}_{\mathrm{NWA},\nu,q}(x)}{\{\nu\}_q!},\quad |t|<2\pi. \tag{4.124}$$

This can be generalized to:

Definition 58 Let $\{\omega_i\}_{i=1}^n\subset\mathbb{C}$. The generating function for

$$\mathrm{B}^{(n)}_{\mathrm{NWA},\nu,q}(x|\omega_1,\ldots,\omega_n)$$

is the following q-analogue of [403, (77), p. 143]:

$$\frac{t^n\omega_1\cdots\omega_n}{\prod_{k=1}^n(\mathrm{E}_q(\omega_k t)-1)}\mathrm{E}_q(xt)=\sum_{\nu=0}^{\infty}\frac{t^\nu \mathrm{B}^{(n)}_{\mathrm{NWA},\nu,q}(x|\omega_1,\ldots,\omega_n)}{\{\nu\}_q!},$$
$$|t|<\min\big(|2\pi/\omega_1|,\ldots,|2\pi/\omega_n|\big). \tag{4.125}$$

Remark 10 The values for t above are for $q=1$. For general q, the domain of convergence can be different. The above definitions are mostly formal.

Corollary 4.3.12 *A q-analogue of* [356, p. 639]. *Compare with* [488, p. 192].

$$\mathrm{E}_q(t\mathrm{B}_{\mathrm{NWA},q})\doteq\frac{t}{\mathrm{E}_q(t)-1}. \tag{4.126}$$

In the literature there are many different definitions of Bernoulli polynomials. One example is [320], which has an extra $n!$ in the denominator.

Obviously, $\mathrm{B}^{(n)}_{\mathrm{NWA},\nu,q}(x|\omega_1,\ldots,\omega_n)$ is symmetric in $\omega_1,\ldots,\omega_n$ and in particular

$$\mathrm{B}^{(1)}_{\mathrm{NWA},\nu,q}(x|\omega)=\omega^\nu \mathrm{B}_{\mathrm{NWA},\nu,q}\left(\frac{x}{\omega}\right), \tag{4.127}$$

$$\Delta^n_{\omega_1,\ldots,\omega_n \mathrm{NWA},q} \mathrm{B}^{(n)}_{\mathrm{NWA},\nu,q}(x|\omega_1,\ldots,\omega_n) = \frac{\{\nu\}_q!}{\{\nu-n\}_q!} x^{\nu-n}. \qquad (4.128)$$

Theorem 4.3.13 *The successive differences of q-Bernoulli polynomials can be expressed as q-Bernoulli polynomials. A q-analogue of* [403, (46), p. 131]:

$$\Delta^p_{\omega_1,\ldots,\omega_p \mathrm{NWA},q} \mathrm{B}^{(n)}_{\mathrm{NWA},\nu,q}(x|\omega_1,\ldots,\omega_n) = \frac{\{\nu\}_q!}{\{\nu-p\}_q!} \mathrm{B}^{(n-p)}_{\mathrm{NWA},\nu-p,q}(x|\omega_{p+1},\ldots,\omega_n), \qquad (4.129)$$

$$\mathrm{B}^{(n)}_{\mathrm{NWA},\nu,q}(x|\omega_1,\ldots,\omega_n) = \left(\mathrm{B}^{(n)}_{\mathrm{NWA},\nu,\omega_1,\ldots,\omega_n,q} \oplus_q x\right)^{\nu}. \qquad (4.130)$$

Example 6 An explicit formula for generalized q-Bernoulli polynomials:

$$\mathrm{B}^{(n)}_{\mathrm{NWA},\nu,q}(x) = \sum_{k=0}^{\nu} \binom{\nu}{k}_q x^k D^{\nu-k}_{q,t}\left(\frac{t^n}{(E_q(t)-1)^n}\right)\bigg|_{t=0}. \qquad (4.131)$$

Proof Operate with $D^{\nu}_{q,t}$ on both sides of (4.124), use the q-Leibniz theorem and finally put $t=0$. □

4.3.2 The Ward q-Bernoulli numbers

The following special case is often used.

Definition 59 The Ward q-Bernoulli numbers [531, p. 265, 16.4], [12, p. 244, 4.1] are given by

$$\mathrm{B}_{\mathrm{NWA},n,q} \equiv \mathrm{B}^{(1)}_{\mathrm{NWA},n,q}. \qquad (4.132)$$

The following table lists some of the first Ward q-Bernoulli numbers:

$n=0$	$n=1$	$n=2$	$n=3$	$n=4$
1	$-(1+q)^{-1}$	$q^2(\{3\}_q!)^{-1}$	$(1-q)q^3(\{2\}_q)^{-1}$ $\times (\{4\}_q)^{-1}$	$q^4(1-q^2-2q^3-q^4+q^6)$ $\times (\{2\}_q^2\{3\}_q\{5\}_q)^{-1}$

Theorem 4.3.14 *We have the following operational representation, a q-analogue of Lucas* [362, p. 83], *Grigoriew* [247, (4), p. 147] *and Szegő* [493]:

$$\mathrm{B}^{(n)}_{\mathrm{NWA},\nu,q}(\omega_1,\ldots,\omega_n) \stackrel{..}{=} \left(\bigoplus_{q,l=1}^{n} \omega_l \mathrm{B}_{\mathrm{NWA},q}\right)^{\nu}. \qquad (4.133)$$

The following operator will be useful in connection with $\mathrm{B}^{(n)}_{\mathrm{NWA},\nu,q}(x)$.

Definition 60 Compare with [138, p. 32] ($n = 1$). The invertible operator $\mathrm{S}^n_{\mathrm{B,N},q} \in \mathbb{C}(\mathrm{D}_q)$ is given by

$$\mathrm{S}^n_{\mathrm{B,N},q} \equiv \frac{(\mathrm{E}_q(\mathrm{D}_q) - \mathrm{I})^n}{\mathrm{D}^n_q}. \tag{4.134}$$

This implies

Theorem 4.3.15

$$\triangle^n_{\mathrm{NWA},q} = \mathrm{D}^n_q \mathrm{S}^n_{\mathrm{B,N},q}. \tag{4.135}$$

Theorem 4.3.16 *A q-analogue of* [418, p. 1225, i]. *The q-Bernoulli polynomials of degree ν and order n can be expressed as*

$$\mathrm{B}^{(n)}_{\mathrm{NWA},\nu,q}(t) = \mathrm{S}^{-n}_{\mathrm{B,N},q} t^\nu. \tag{4.136}$$

Proof

$$\mathrm{LHS} = \sum_{k=0}^{\nu} \binom{\nu}{k}_q \mathrm{B}^{(n)}_{\mathrm{NWA},k,q} t^{\nu-k} = \sum_{k=0}^{\infty} \frac{\mathrm{B}^{(n)}_{\mathrm{NWA},k,q}}{\{k\}_q!} \mathrm{D}^k_q t^\nu \overset{\text{by (4.124)}}{=} \mathrm{RHS}. \tag{4.137}$$

□

Example 7 A q-analogue of a generalization of [138, p. 43, 3.3]:

$$\sum_{k=0}^{n} (-1)^{n-k} \binom{n}{k} \mathrm{B}^{(n)}_{\mathrm{NWA},\nu,q}(x \oplus_q \overline{k}_q) = \{\nu - n + 1\}_{n,q} x^{\nu-n}. \tag{4.138}$$

Proof

$$\begin{aligned}\triangle^n_{\mathrm{NWA},q} \mathrm{B}^{(n)}_{\mathrm{NWA},\nu,q}(x) &= \mathrm{D}^n_q \mathrm{S}^n_{\mathrm{B,N},q} \mathrm{B}^{(n)}_{\mathrm{NWA},\nu,q}(x) = \mathrm{D}^n_q \mathrm{S}^n_{\mathrm{B,N},q} \mathrm{S}^{-n}_{\mathrm{B,N},q} x^\nu \\ &= \mathrm{D}^n_q x^\nu = \{\nu - n + 1\}_{n,q} x^{\nu-n}.\end{aligned} \tag{4.139}$$

□

Corollary 4.3.17

$$\sum_{k=0}^{n} (-1)^{n-k} \binom{n}{k} \mathrm{B}^{(n)}_{\mathrm{NWA},\nu,q}(\overline{k}_q) = \{n\}_q!\, \delta_{0,\nu-n}. \tag{4.140}$$

Proof Put $x = 0$ in (4.138). □

The following is an example for $\mathrm{B}_{\mathrm{NWA},q}$ or first q-Bernoulli numbers.

Theorem 4.3.18 [12, p. 253, 9.5] *A q-analogue of* [363, (1), p. 240], [439, p. 699]:

$$f(x \oplus_q \mathrm{B}_{\mathrm{NWA},q} \oplus_q 1) - f(x \oplus_q \mathrm{B}_{\mathrm{NWA},q}) \doteqdot \mathrm{D}_q f(x), \tag{4.141}$$

where here and in the sequel, we have abbreviated the umbral symbol by $\mathrm{B}_{\mathrm{NWA},q}$.

In the literature we will find a very large number of recurrences between Bernoulli numbers and polynomials. The authors often arrive at these formulas in a rather roundabout way. But the common source of all these relations is formula (4.141) ($q = 1$).

We will also state the corresponding equation for $\mathrm{B}^{(n)}_{\mathrm{NWA},\nu,q}$ written in two different forms.

Theorem 4.3.19 *A q-analogue of* [247, (7), p. 152], [402, (11), p. 124], [403, (36), p. 132]:

$$f\big(x \oplus_q \mathrm{B}^{(n)}_{\mathrm{NWA},q} \oplus_q 1\big) - f\big(x \oplus_q \mathrm{B}^{(n)}_{\mathrm{NWA},q}\big) \doteqdot \mathrm{D}_q f\big(x \oplus_q \mathrm{B}^{(n-1)}_{\mathrm{NWA},q}\big), \tag{4.142}$$

$$f\big(\mathrm{B}^{(n)}_{\mathrm{NWA},q}(x) \oplus_q 1\big) - f\big(\mathrm{B}^{(n)}_{\mathrm{NWA},q}(x)\big) \doteqdot \mathrm{D}_q f\big(\mathrm{B}^{(n-1)}_{\mathrm{NWA},q}(x)\big). \tag{4.143}$$

Theorem 4.3.20 *Compare* [138, 3.15, p. 51], *where the corresponding formula for Euler polynomials was given*:

$$\mathrm{B}_{\mathrm{NWA},\nu,q}(x) \equiv \frac{\{\nu\}_q}{\mathrm{E}_q(\mathrm{D}_q) - \mathrm{I}} x^{\nu-1} = \frac{\{\nu\}_q}{\mathrm{E}(\oplus_q) - \mathrm{I}} x^{\nu-1} \doteqdot (x \oplus_q \mathrm{B}_{\mathrm{NWA},q})^{\nu}. \tag{4.144}$$

We will now follow Cigler [138] and give a few equations for the first q-Bernoulli polynomials. The first two of these equations are well-known in the literature ($q = 1$).

Definition 61 A q-analogue of the power sum [277, p. 87], [138, p. 13], [517, p. 575]:

$$s_{\mathrm{NWA},m,q}(n) \equiv \sum_{k=0}^{n-1} (\overline{k}_q)^m, \qquad s_{\mathrm{NWA},0,q}(1) \equiv 1. \tag{4.145}$$

Next, following [12, p. 248, 5.13], [531, p. 265, 16.5], is a q-analogue of [138, p. 13, p. 17: 1.11, p. 36], [363, p. 237]:

Theorem 4.3.21 *An expression for the q-analogue of the power sum in terms of q-Bernoulli polynomials*:

$$\begin{aligned} s_{\mathrm{NWA},m,q}(n) &= \frac{\mathrm{B}_{\mathrm{NWA},m+1,q}(\overline{n}_q) - \mathrm{B}_{\mathrm{NWA},m+1,q}}{\{m+1\}_q} \\ &\equiv \frac{1}{\{m+1\}_q} \sum_{k=1}^{m+1} \binom{m+1}{k}_q (\overline{n}_q)^k \mathrm{B}_{\mathrm{NWA},m+1-k,q} \\ &\equiv \frac{1}{\{m+1\}_q} \sum_{k=0}^{m} \binom{m+1}{k}_q (\overline{n}_q)^{m+1-k} \mathrm{B}_{\mathrm{NWA},k,q}. \end{aligned} \tag{4.146}$$

Proof We use the generating function technique.

$$\begin{aligned} \sum_m s_{\mathrm{NWA},m,q}(n) \frac{x^m}{\{m\}_q!} &= \sum_m \frac{x^m}{\{m\}_q!} \sum_{k=0}^{n-1} (\overline{k}_q)^m \\ &= \sum_{k=0}^{n-1} \sum_m \frac{x^m (\overline{k}_q)^m}{\{m\}_q!} = \sum_{k=0}^{n-1} \mathrm{E}_q\big(x(\overline{k}_q)\big) \\ &= \frac{\mathrm{E}_q(x\overline{n}_q) - 1}{\mathrm{E}_q(x) - 1} = \sum_{k=1}^{\infty} \frac{x^{k-1}(\overline{n}_q)^k}{\{k\}_q!} \sum_{j=0}^{\infty} \frac{x^j \mathrm{B}_{\mathrm{NWA},j,q}}{\{j\}_q!} \\ &= \sum_m \frac{x^m}{\{m\}_q!} \sum_{k+j-1=m} \frac{(\overline{n}_q)^k \mathrm{B}_{\mathrm{NWA},j,q}}{\{k\}_q!\{j\}_q!} \frac{\{m+1\}_q!}{\{m+1\}_q}. \end{aligned} \tag{4.147}$$

In the end, we equate the coefficients of $\sum_m \frac{x^m}{\{m\}_q!}$. □

Theorem 4.3.22 *A q-analogue of* [138, p. 45], [402, p. 127, (17)]:

$$x^n = \int_x^{x \oplus_q 1} \mathrm{B}_{\mathrm{NWA},n,q}(t)\, d_q(t) = \frac{\mathrm{B}_{\mathrm{NWA},n+1,q}(x \oplus_q 1) - \mathrm{B}_{\mathrm{NWA},n+1,q}(x)}{\{n+1\}_q}. \tag{4.148}$$

Proof q-integrate (4.107) for $n = 1$ and use (4.119). □

This can be rewritten as a q-analogue of the well-known identity [239, p. 496, 8.2]

$$x^n = \frac{1}{\{n+1\}_q} \sum_{k=0}^{n} \binom{n+1}{k}_q \mathrm{B}_{\mathrm{NWA},k,q}(x). \tag{4.149}$$

Cigler has given some examples of translation invariant operators. One of them is the Bernoulli operator.

Definition 62 The first q-Bernoulli operator is given by the following q-integral, a q-analogue of [138, p. 91], [146, p. 154], [440, p. 59], [439, p. 701, 703], [418, p. 1217]:

$$\mathrm{J}_{\mathrm{B,N},q} f(x) \equiv \int_x^{x \oplus_q 1} f(t)\, d_q(t). \tag{4.150}$$

We have $\mathrm{J}_{\mathrm{B,N},q} \in \mathbb{C}(\mathrm{D}_q)$.

Theorem 4.3.23 *A q-analogue of* [138, p. 44–45], [418, p. 1217]:
The first q-Bernoulli operator can be expressed in the form

$$\mathrm{J}_{\mathrm{B,N},q} f(x) = \frac{\triangle_{\mathrm{NWA},q}}{\mathrm{D}_q} f(x). \tag{4.151}$$

Proof Use (4.136) and (4.148). □

Theorem 4.3.24 *A q-analogue of* [138, p. 44–45]. *We can expand a given formal power series in terms of the* $\mathrm{B}_{\mathrm{NWA},k,q}(x)$ *as follows*:

$$f(x) = \sum_{k=0}^{\infty} \int_0^{\overline{1}_q} \mathrm{D}_q^k f(t)\, d_q(t) \frac{\mathrm{B}_{\mathrm{NWA},k,q}(x)}{\{k\}_q!}. \tag{4.152}$$

Proof Assume that

$$f(x) = \sum_{k=0}^{\infty} \frac{a_k}{\{k\}_q!} \mathrm{B}_{\mathrm{NWA},k,q}(x). \tag{4.153}$$

We have

$$x^k = \mathrm{S}_{\mathrm{B,N},q} \mathrm{B}_{\mathrm{NWA},k,q}(x), \tag{4.154}$$

$$f(x) = \sum_{k=0}^{\infty} \frac{a_k}{\{k\}_q!} \mathrm{S}_{\mathrm{B,N},q}^{-1} x^k, \tag{4.155}$$

$$\mathrm{S}_{\mathrm{B,N},q} f(x) = \sum_{k=0}^{\infty} \frac{a_k}{\{k\}_q!} x^k. \tag{4.156}$$

This implies

$$a_k = \mathrm{D}_q^k \mathrm{S}_{\mathrm{B,N},q} f(x)\big|_{x=0} = \mathrm{D}_q^k \frac{\triangle_{\mathrm{NWA},q}}{\mathrm{D}_q} f(x)\bigg|_{x=0} = \int_0^{\overline{1}_q} \mathrm{D}_q^k f(t)\, d_q(t). \tag{4.157}$$

□

A special case of Φ_q polynomials are the γ_q-polynomials of degree ν and order n, which are obtained by putting $f_n(t) = \frac{t^n g(t)}{(\mathrm{E}_{\frac{1}{q}}(t)-1)^n}$ in (4.104).

Definition 63

$$\frac{t^n g(t)}{(\mathrm{E}_{\frac{1}{q}}(t)-1)^n}\mathrm{E}_q(xt) \equiv \sum_{\nu=0}^{\infty} \frac{t^\nu \gamma_{\nu,q}^{(n)}(x)}{\{\nu\}_q!}. \tag{4.158}$$

Theorem 4.3.25 *A q-analogue of* [385, (2), p. 126], [364, p. 21], [439, p. 704]:

$$\triangle_{\mathrm{JHC},q}\gamma_{\nu,q}^{(n)}(x) = \{\nu\}_q \gamma_{\nu-1,q}^{(n-1)}(x) = \mathrm{D}_q \gamma_{\nu,q}^{(n-1)}(x). \tag{4.159}$$

Proof Use (4.114). □

By (4.115), the following symbolic relations hold

Theorem 4.3.26 *A q-analogue of* [385, p. 126]. *The second equation implies* (4.162).

$$\left(\gamma_q^{(n)} \oplus_q x \boxplus_q 1\right)^\nu - \left(\gamma_q^{(n)} \oplus_q x\right)^\nu = \{\nu\}_q \left(\gamma_q^{(n-1)} \oplus_q x\right)^{\nu-1}, \tag{4.160}$$

$$\left(\gamma_q^{(n)} \boxplus_q 1\right)^\nu - \gamma_{\nu,q}^{(n)} = \{\nu\}_q \gamma_{\nu-1,q}^{(n-1)}, \tag{4.161}$$

$$\mathrm{B}_{\mathrm{JHC},0,q} = 1, \qquad (\mathrm{B}_{\mathrm{JHC},q} \boxplus_q 1)^k - \mathrm{B}_{\mathrm{JHC},k,q} \doteq \delta_{1,k}, \tag{4.162}$$

where $\mathrm{B}_{\mathrm{JHC},q}^n$ *is replaced by* $\mathrm{B}_{\mathrm{JHC},n,q}$ *on expansion.*

Theorem 4.3.27 *A q-analogue of* [402, (20), p. 163]:

$$\triangle_{\mathrm{JHC},q} f\left(\gamma_{\nu,q}^{(n)}(x)\right) \equiv f\left(\gamma_{\nu,q}^{(n)}(x) \boxplus_q 1\right) - f\left(\gamma_{\nu,q}^{(n)}(x)\right) = \mathrm{D}_q f\left(\gamma_{\nu,q}^{(n-1)}(x)\right). \tag{4.163}$$

Theorem 4.3.28 *Almost a q-analogue of* [485, p. 378, (26)]:

$$\sum_{k=1}^{\nu} \binom{\nu}{k}_q \gamma_{\nu-k,q}^{(n)}(x) q^{\binom{k}{2}} = \{\nu\}_q \gamma_{\nu-1,q}^{(n-1)}(x). \tag{4.164}$$

Proof Use (4.109) and (4.120). □

4.3.3 The generalized JHC q-Bernoulli polynomials

A special case of γ polynomials are the second generalized q-Bernoulli polynomials $\mathrm{B}_{\mathrm{JHC},\nu,q}^{(n)}(x)$ of degree ν and order n, which were defined for $q = 1$ in [385, p. 127], [402].

Definition 64 The generating function for $\mathrm{B}_{\mathrm{JHC},\nu,q}^{(n)}(x)$ is a q-analogue of [439, p. 704], [418, p. 1225, ii]:

$$\frac{t^n}{(\mathrm{E}_{\frac{1}{q}}(t)-1)^n}\mathrm{E}_q(xt) = \sum_{\nu=0}^{\infty} \frac{t^\nu \mathrm{B}_{\mathrm{JHC},\nu,q}^{(n)}(x)}{\{\nu\}_q!}, \quad |t| < 2\pi. \tag{4.165}$$

This can be generalized to:

Definition 65 The generating function for $B_{JHC,\nu,q}^{(n)}(x|\omega_1,\ldots,\omega_n)$ is the following q-analogue of [403, (77), p. 143]:

$$\frac{t^n \omega_1 \cdots \omega_n}{\prod_{k=1}^n (E_{\frac{1}{q}}(\omega_k t) - 1)} E_q(xt) = \sum_{\nu=0}^{\infty} \frac{t^\nu B_{JHC,\nu,q}^{(n)}(x|\omega_1,\ldots,\omega_n)}{\{\nu\}_q!},$$

$$|t| < \min(|2\pi/\omega_1|,\ldots,|2\pi/\omega_n|). \tag{4.166}$$

Corollary 4.3.29 *Compare with* [488, p. 192]:

$$E_q(tB_{JHC,q}) \stackrel{\cdot\cdot}{=} \frac{t}{E_{\frac{1}{q}}(t) - 1}. \tag{4.167}$$

Obviously, $B_{JHC,\nu,q}^{(n)}(x|\omega_1,\ldots,\omega_n)$ is symmetric in $\omega_1,\ldots,\omega_n$ and in particular

$$B_{JHC,\nu,q}^{(1)}(x|\omega) = \omega^\nu B_{JHC,\nu,q}\left(\frac{x}{\omega}\right), \tag{4.168}$$

$$\Delta_{\omega_1,\ldots,\omega_n\,JHC,q}^n B_{JHC,\nu,q}^{(n)}(x|\omega_1,\ldots,\omega_n) = \frac{\{\nu\}_q!}{\{\nu-n\}_q!} x^{\nu-n}. \tag{4.169}$$

Theorem 4.3.30 *The successive differences of the second q-Bernoulli polynomials can be expressed as q-Bernoulli polynomials. A q-analogue of* [403, (46), p. 131]:

$$\Delta_{\omega_1,\ldots,\omega_p\,JHC,q}^p B_{JHC,\nu,q}^{(n)}(x|\omega_1,\ldots,\omega_n) = \frac{\{\nu\}_q!}{\{\nu-p\}_q!} B_{JHC,\nu-p,q}^{(n-p)}(x|\omega_{p+1},\ldots,\omega_n), \tag{4.170}$$

$$B_{JHC,\nu,q}^{(n)}(x|\omega_1,\ldots,\omega_n) = \left(B_{JHC,\nu,\omega_1,\ldots,\omega_n,q}^{(n)} \oplus_q x\right)^\nu. \tag{4.171}$$

Example 8 An explicit formula for the second generalized q-Bernoulli polynomials:

$$B_{JHC,\nu,q}^{(n)}(x) = \sum_{k=0}^{\nu} \binom{\nu}{k}_q x^k D_{q,t}^{\nu-k}\left(\frac{t^n}{(E_{\frac{1}{q}}(t)-1)^n}\right)\bigg|_{t=0}. \tag{4.172}$$

Proof Operate with $D_{q,t}^\nu$ on both sides of (4.165), use the q-Leibniz theorem and finally put $t=0$. □

The following special case is often used.

Definition 66 The Jackson q-Bernoulli numbers [531, p. 265, 16.4], [12, p. 244, 4.1] are given by

$$B_{JHC,n,q} \equiv B_{JHC,n,q}^{(1)}. \tag{4.173}$$

The following table lists some of the first Jackson q-Bernoulli numbers:

$n=0$	$n=1$	$n=2$	$n=3$	$n=4$
1	$-q(1+q)^{-1}$	$q^2(\{3\}_q!)^{-1}$	$(q^4-q^3)(\{2\}_q)^{-1} \times (\{4\}_q)^{-1}$	$q^4(1-q^2-2q^3-q^4+q^6) \times (\{2\}_q^2\{3\}_q\{5\}_q)^{-1}$

We know that the uneven Bernoulli numbers are equal to zero. There is a similar rule for the q-case.

Theorem 4.3.31

$$\begin{aligned} &\textit{For } \nu \textit{ even}, \quad \mathrm{B}_{\mathrm{NWA},\nu,q} = \mathrm{B}_{\mathrm{JHC},\nu,q}, \\ &\textit{For } \nu \textit{ uneven}, \quad \mathrm{B}_{\mathrm{NWA},\nu,q} = -\mathrm{B}_{\mathrm{JHC},\nu,q}, \quad \nu > 1. \end{aligned} \tag{4.174}$$

Proof Use the generating function (4.165) with $-t, n=1$ and $x=0$, multiply with $\mathrm{E}_q(t)$ in denominator and numerator. Finally compare with the generating function (4.124). □

Theorem 4.3.32 *We have the following operational representation, a q-analogue of* [247, (4), p. 147], [493]:

$$\mathrm{B}_{\mathrm{JHC},\nu,q}^{(n)}(\omega_1,\ldots,\omega_n) \doteqdot \left(\bigoplus_{l=1}^{q,n} \omega_l \mathrm{B}_{\mathrm{JHC},q} \right)^{\nu}. \tag{4.175}$$

The following operator will be useful in connection with $\mathrm{B}_{\mathrm{JHC},\nu,q}^{(n)}(x)$.

Definition 67 Compare with [138, p. 32] ($n=1$). The invertible operator $\mathrm{S}_{\mathrm{B,J},q}^n \in \mathbb{C}(\mathrm{D}_q)$ is given by

$$\mathrm{S}_{\mathrm{B,J},q}^n \equiv \frac{(\mathrm{E}_{\frac{1}{q}}(\mathrm{D}_q)-\mathrm{I})^n}{\mathrm{D}_q^n}. \tag{4.176}$$

With this definition, we have

Theorem 4.3.33

$$\triangle_{\mathrm{JHC},q}^n = \mathrm{D}_q^n \mathrm{S}_{\mathrm{B,J},q}^n. \tag{4.177}$$

Theorem 4.3.34 *A q-analogue of* [418, p. 1225, i]. *The second q-Bernoulli polynomials of degree ν and order n can be expressed as*

$$\mathrm{B}_{\mathrm{JHC},\nu,q}^{(n)}(t) = \mathrm{S}_{\mathrm{B,J},q}^{-n} t^{\nu}. \tag{4.178}$$

Proof

$$\mathrm{LHS} = \sum_{k=0}^{\nu} \binom{\nu}{k}_q \mathrm{B}_{\mathrm{JHC},k,q}^{(n)} t^{\nu-k} = \sum_{k=0}^{\infty} \frac{\mathrm{B}_{\mathrm{JHC},k,q}^{(n)}}{\{k\}_q!} \mathrm{D}_q^k t^\nu \overset{\text{by (4.165)}}{=} \mathrm{RHS}. \tag{4.179}$$

□

Example 9 A q-analogue of a generalization of [138, p. 43, 3.3]:

$$\sum_{k=0}^{n} (-1)^{n-k} \binom{n}{k} \mathrm{B}_{\mathrm{JHC},\nu,q}^{(n)}(x \boxplus_q \tilde{k}_q) = \{\nu - n + 1\}_{n,q} x^{\nu-n}. \tag{4.180}$$

Proof

$$\begin{aligned}\triangle_{\mathrm{JHC},q}^{n} \mathrm{B}_{\mathrm{JHC},\nu,q}^{(n)}(x) &= \mathrm{D}_q^n \mathrm{S}_{\mathrm{B,J},q}^n \mathrm{B}_{\mathrm{JHC},\nu,q}^{(n)}(x) = \mathrm{D}_q^n \mathrm{S}_{\mathrm{B,J},q}^n \mathrm{S}_{\mathrm{B,J},q}^{-n} x^\nu \\ &= \mathrm{D}_q^n x^\nu = \{\nu - n + 1\}_{n,q} x^{\nu-n}.\end{aligned} \tag{4.181}$$

□

Corollary 4.3.35

$$\sum_{k=0}^{n} (-1)^{n-k} \binom{n}{k} \mathrm{B}_{\mathrm{JHC},\nu,q}^{(n)}(\tilde{k}_q) = \{n\}_q! \delta_{0,\nu-n}. \tag{4.182}$$

Proof Put $x = 0$ in (4.180). □

Theorem 4.3.36 *A q-analogue of* [363, p. 240, (1)], [439, p. 699]:

$$f(x \oplus_q \mathrm{B}_{\mathrm{JHC},q} \boxplus_q 1) - f(x \oplus_q \mathrm{B}_{\mathrm{JHC},q}) \doteq \mathrm{D}_q f(x). \tag{4.183}$$

Let us also state the corresponding equation for $\mathrm{B}_{\mathrm{JHC},\nu,q}^{(n)}$ written in two different forms.

Theorem 4.3.37 *A q-analogue of* [402, (11), p. 124], [403, (36), p. 132]:

$$f\big(x \oplus_q \mathrm{B}_{\mathrm{JHC},q}^{(n)} \boxplus_q 1\big) - f\big(x \oplus_q \mathrm{B}_{\mathrm{JHC},q}^{(n)}\big) \doteq \mathrm{D}_q f\big(x \boxplus_q \mathrm{B}_{\mathrm{JHC},q}^{(n-1)}\big), \tag{4.184}$$

$$f\big(\mathrm{B}_{\mathrm{JHC},q}^{(n)}(x) \boxplus_q 1\big) - f\big(\mathrm{B}_{\mathrm{JHC},q}^{(n)}(x)\big) \doteq \mathrm{D}_q f\big(\mathrm{B}_{\mathrm{JHC},q}^{(n-1)}(x)\big). \tag{4.185}$$

Theorem 4.3.38 *Compare to* [138, 3.15, p. 51], *where the corresponding formula for Euler polynomials was given*:

$$\mathrm{B}_{\mathrm{JHC},\nu,q}(x) \equiv \frac{\{\nu\}_q}{\mathrm{E}_{\frac{1}{q}}(\mathrm{D}_q) - \mathrm{I}} x^{\nu-1} = \frac{\{\nu\}_q}{\mathrm{E}(\boxplus_q) - \mathrm{I}} x^{\nu-1} \doteq (x \oplus_q \mathrm{B}_{\mathrm{JHC},q})^\nu. \tag{4.186}$$

We will now follow Cigler [138] and give a few equations for second q-Bernoulli polynomials. The first two of these equations are well-known in the literature $(q = 1)$.

Definition 68 A q-analogue of the power sum [277, p. 87], [138, p. 13], [517, p. 575]:

$$s_{\mathrm{JHC},m,q}(n) \equiv \sum_{k=0}^{n-1} (\tilde{k}_q)^m, \qquad s_{\mathrm{JHC},0,q}(1) \equiv 1. \tag{4.187}$$

Theorem 4.3.39 *A q-analogue of* [138, p. 13, p. 17: 1.11, p. 36], [363, p. 237]:

$$\begin{aligned} s_{\mathrm{JHC},m,q}(n) &= \frac{\mathrm{B}_{\mathrm{JHC},m+1,q}(\tilde{n}_q) - \mathrm{B}_{\mathrm{JHC},m+1,q}}{\{m+1\}_q} \\ &\equiv \frac{1}{\{m+1\}_q} \sum_{k=1}^{m+1} \binom{m+1}{k}_q (\tilde{n}_q)^k \mathrm{B}_{\mathrm{JHC},m+1-k,q} \\ &\equiv \frac{1}{\{m+1\}_q} \sum_{k=0}^{m} \binom{m+1}{k}_q (\tilde{n}_q)^{m+1-k} \mathrm{B}_{\mathrm{JHC},k,q}. \end{aligned} \tag{4.188}$$

Theorem 4.3.40 *A q-analogue of* [138, p. 45], [402, p. 127, (17)]:

$$x^n = \int_x^{x \boxplus_q 1} \mathrm{B}_{\mathrm{JHC},n,q}(t)\, d_q(t) = \frac{\mathrm{B}_{\mathrm{JHC},n+1,q}(x \boxplus_q 1) - \mathrm{B}_{\mathrm{JHC},n+1,q}(x)}{\{n+1\}_q}. \tag{4.189}$$

Proof q-integrate (4.107) for $n = 1$ and use (4.159). □

This can be rewritten as a q-analogue of the well-known identity [239, p. 496, 8.2]

$$x^n = \frac{1}{\{n+1\}_q} \sum_{k=0}^{n} \binom{n+1}{k}_q \mathrm{B}_{\mathrm{JHC},k,q}(x) q^{\binom{n+1-k}{2}}. \tag{4.190}$$

Definition 69 The second q-Bernoulli operator is given by the following q-integral, a q-analogue of [138, p. 91], [146, p. 154], [440, p. 59], [439, p. 701, 703], [418, p. 1217]:

$$\mathrm{J}_{\mathrm{B,J},q} f(x) \equiv \int_x^{x \boxplus_q 1} f(t)\, d_q(t). \tag{4.191}$$

We have $\mathrm{J}_{\mathrm{B,J},q} \in \mathbb{C}(\mathrm{D}_q)$.

Theorem 4.3.41 *A q-analogue of* [138, p. 44–45], [418, p. 1217]. *The second q-Bernoulli operator can be expressed in the form*

$$\mathrm{J}_{\mathrm{B,J},q} f(x) = \frac{\triangle_{\mathrm{JHC},q}}{\mathrm{D}_q} f(x). \tag{4.192}$$

Proof Use (4.178) and (4.189). □

Theorem 4.3.42 *A q-analogue of* [138, p. 44–45]. *We can expand a given formal power series in terms of the $\mathrm{B}_{\mathrm{JHC},k,q}(x)$ as follows*:

$$f(x) = \sum_{k=0}^{\infty} \int_0^{\tilde{1}_q} \mathrm{D}_q^k f(t)\, d_q(t) \frac{\mathrm{B}_{\mathrm{JHC},k,q}(x)}{\{k\}_q!}. \tag{4.193}$$

Proof Analoguous to the case of NWA. □

We will now turn to q-Appell polynomials with a different type of generating function. It turns out that the formulas obtained are quite similar in character to the previous ones. A special case of the Φ_q polynomials are the η_q polynomials of order n, which are obtained by putting $f_n(t) = \frac{g(t)2^n}{(\mathrm{E}_q(t)+1)^n}$ in (4.104).

Definition 70 A q-analogue of [385, p. 142, (1)]:

$$\frac{2^n}{(\mathrm{E}_q(t)+1)^n} g(t) \mathrm{E}_q(xt) = \sum_{\nu=0}^{\infty} \frac{t^\nu \eta_{\nu,q}^{(n)}(x)}{\{\nu\}_q!}. \tag{4.194}$$

By (4.113), we get a q-analogue of [385], [384, p. 519]:

$$\nabla_{\mathrm{NWA},q} \eta_{\nu,q}^{(n)}(x) = \eta_{\nu,q}^{(n-1)}(x). \tag{4.195}$$

4.3.4 NWA q-Euler polynomials

We will now define the first q-Euler polynomials, a special case of the η_q polynomials. There are many similar definitions of these, but we will follow the one in [385, p. 143–147], [138, p. 51], because it is equivalent to the q-Appell polynomials from [15].

Definition 71 The generating function for the first q-Euler polynomials of degree ν and order n, $\mathrm{F}_{\mathrm{NWA},\nu,q}^{(n)}(x)$, is the following q-analogue of [437, p. 102], [385, p. 309], [504, p. 345]:

$$\frac{2^n \mathrm{E}_q(xt)}{(\mathrm{E}_q(t)+1)^n} = \sum_{\nu=0}^{\infty} \frac{t^\nu}{\{\nu\}_q!} \mathrm{F}_{\mathrm{NWA},\nu,q}^{(n)}(x), \quad |t| < \pi. \tag{4.196}$$

This can be generalized to:

Definition 72 Let $\{\omega_i\}_{i=1}^n \in \mathbb{C}$. The generating function for the first q-Euler polynomials of degree ν and order n, $\mathrm{F}^{(n)}_{\mathrm{NWA},\nu,q}(x|\omega_1,\ldots,\omega_n)$, is the following q-analogue of [403, p. 143, (78)]:

$$\frac{2^n \mathrm{E}_q(xt)}{\prod_{k=1}^n(\mathrm{E}_q(\omega_k t)+1)} = \sum_{\nu=0}^{\infty} \frac{t^\nu}{\{\nu\}_q!}\mathrm{F}^{(n)}_{\mathrm{NWA},\nu,q}(x|\omega_1,\ldots,\omega_n),$$
$$|t| < \min\big(|\pi/\omega_1|,\ldots,|\pi/\omega_n|\big). \tag{4.197}$$

Corollary 4.3.43

$$\mathrm{E}_q(t\mathrm{F}_{\mathrm{NWA},q}) \doteq \frac{2}{\mathrm{E}_q(t)+1}. \tag{4.198}$$

Obviously, $\mathrm{F}^{(n)}_{\mathrm{NWA},\nu,q}(x|\omega_1,\ldots,\omega_n)$ is symmetric in $\omega_1,\ldots,\omega_n$ and, in particular,

$$\mathrm{F}^{(1)}_{\mathrm{NWA},\nu,q}(x|\omega) = \omega^\nu \mathrm{F}_{\mathrm{NWA},\nu,q}\left(\frac{x}{\omega}\right). \tag{4.199}$$

From

$$\underset{\omega_1,\ldots,\omega_n \mathrm{NWA},q}{\nabla^n} \mathrm{F}^{(n)}_{\mathrm{NWA},\nu,q}(x|\omega_1,\ldots,\omega_n) = x^\nu \tag{4.200}$$

we obtain

$$\underset{\omega_1,\ldots,\omega_p \mathrm{NWA},q}{\nabla^p} \mathrm{F}^{(n)}_{\mathrm{NWA},\nu,q}(x|\omega_1,\ldots,\omega_n) = \mathrm{F}^{(n-p)}_{\mathrm{NWA},\nu,q}(x|\omega_{p+1},\ldots,\omega_n). \tag{4.201}$$

Theorem 4.3.44 *A q-analogue of* [385, p. 144, (7)], [403, (7), p. 121], [485, p. 378, (28)].

$$\sum_{k=0}^{\nu}\binom{\nu}{k}_q \mathrm{F}^{(n)}_{\mathrm{NWA},\nu-k,q}(x) + \mathrm{F}^{(n)}_{\mathrm{NWA},\nu,q}(x) = 2\mathrm{F}^{(n-1)}_{\mathrm{NWA},\nu,q}(x). \tag{4.202}$$

With this formula we can compute all first q-Euler polynomials of order n in terms of the polynomials of order $n-1$.

4.3.5 The NWA generalized q-Euler numbers

Definition 73 A q-analogue of [402, p. 139], [363, p. 252]. The first generalized q-Euler numbers are given by

$$\mathrm{F}^{(n)}_{\mathrm{NWA},\nu,q} \equiv \mathrm{F}^{(n)}_{\mathrm{NWA},\nu,q}(0). \tag{4.203}$$

Furthermore, we put

$$\mathrm{F}_{\mathrm{NWA},k,q} \equiv \mathrm{F}^{(1)}_{\mathrm{NWA},k,q}, \qquad \mathrm{F}_{\mathrm{NWA},\nu,q}(x) \equiv \mathrm{F}^{(1)}_{\mathrm{NWA},\nu,q}(x). \tag{4.204}$$

Remark 11 The numbers $\mathrm{F}_{\mathrm{NWA},k,q}$ are q-analogues of the numbers in [384, p. 520], [138, p. 51], which are multiples of the tangent numbers [447, p. 296]. Lucas [363, p. 250] called them G after Genocchi, but I disagree with this. The q-analogues of the original integral Euler numbers (secant numbers), see Salié [447], appear in [33].

It was J.-L. Raabe who for the first time used the name Euler numbers. They were then used again by Sylvester, Catalan, Glaisher, Lucas, and from 1877 on the name was used in Germany. Also see J. Herschel [277, p. 79].

Remark 12 If $q = 1$, we have $\mathrm{F}_\nu(x) = \nu!\mathrm{F}_{J,\nu}(x)$, where $\mathrm{F}_{J,\nu}(x)$ are the Euler polynomials used in [320].

Theorem 4.3.45 *The following operator expression is a q-analogue of* [138, 3.15, p. 51]:

$$\mathrm{F}_{\mathrm{NWA},\nu,q}(x) \equiv \frac{2}{\mathrm{E}_q(\mathrm{D}_q)+\mathrm{I}} x^\nu = \frac{2}{\mathrm{E}(\oplus_q)+\mathrm{I}} x^\nu \doteq (x \oplus_q \mathrm{F}_{\mathrm{NWA},q})^\nu. \tag{4.205}$$

The next two recursion formulas are quite useful for the computations of the first q-Euler polynomials.

Theorem 4.3.46 *A q-analogue of* [403, (27), p. 24], [485, p. 378, (29)]:

$$\mathrm{F}_{\mathrm{NWA},\nu,q}(x) + \sum_{k=0}^{\nu} \binom{\nu}{k}_q \mathrm{F}_{\mathrm{NWA},k,q}(x) = 2x^\nu. \tag{4.206}$$

Theorem 4.3.47 *A q-analogue of* [138, 3.16, p. 51], [363, p. 252]:

$$(1 \oplus_q \mathrm{F}_{\mathrm{NWA},q})^n + (\mathrm{F}_{\mathrm{NWA},q})^n \doteq 2\delta_{0,n}. \tag{4.207}$$

Theorem 4.3.48 *A q-analogue of* [98, p. 6, (4.3)], [402, (19), p. 136], *a corrected version of* [363, p. 261]:

$$f(x \oplus_q \mathrm{F}_{\mathrm{NWA},q} \oplus_q 1) + f(x \oplus_q \mathrm{F}_{\mathrm{NWA},q}) \doteq 2f(x). \tag{4.208}$$

We will also state the corresponding equation for $\mathrm{F}^{(n)}_{\mathrm{NWA},\nu,q}$ written in two different forms.

Theorem 4.3.49 *A q-analogue of* [402, (19), p. 150, p. 155], [403, (29), p. 126]:

$$\begin{aligned}\nabla_{\mathrm{NWA},q} f\big(x \oplus_q \mathrm{F}^{(n)}_{\mathrm{NWA},q}\big) &\doteq f\big(x \oplus_q \mathrm{F}^{(n-1)}_{\mathrm{NWA},q}\big)\\ &\doteq \nabla_{\mathrm{NWA},q} f\big(\mathrm{F}^{(n)}_{\mathrm{NWA},q}(x)\big) \doteq f\big(\mathrm{F}^{(n-1)}_{\mathrm{NWA},q}(x)\big).\end{aligned} \tag{4.209}$$

The following table lists some of the first q-Euler numbers $\mathrm{F}_{\mathrm{NWA},n,q}$:

$n=0$	$n=1$	$n=2$	$n=3$	$n=4$
1	-2^{-1}	$2^{-2}(-1+q)$	$2^{-3}(-1+2q + 2q^2-q^3)$	$2^{-4}(q-1)\{3\}_q! \times (q^2-4q+1)$

Theorem 4.3.50 *We have the following operational representation, a q-analogue of* [493]:

$$\mathrm{F}^{(n)}_{\mathrm{NWA},\nu,q}(\omega_1,\dots,\omega_n)=\left(\bigoplus_{q,l=1}^{n}\omega_l \mathrm{F}_{\mathrm{NWA},q}\right)^{\nu}. \tag{4.210}$$

Theorem 4.3.51 *The first q-Euler polynomials can be expressed as a finite sum of difference operators on x^n. Almost a q-analogue of* [320, p. 289]:

$$\mathrm{F}_{\mathrm{NWA},n,q}(x)=\sum_{m=0}^{n}\frac{(-1)^m}{2^m}\Delta^m_{\mathrm{NWA},q}x^n. \tag{4.211}$$

Theorem 4.3.52 *A generalization of* (4.206):

$$2^{-n}\sum_{k=0}^{n}\binom{n}{k}\mathrm{F}^{(n)}_{\mathrm{NWA},\nu,q}(x\oplus_q \overline{k}_q)=x^{\nu}. \tag{4.212}$$

Proof Expand $\nabla^n_{\mathrm{NWA},q}\mathrm{F}^{(n)}_{\mathrm{NWA},\nu,q}(x)$. □

Definition 74 A q-analogue of [277, p. 88]. The notation of Nielsen [398, p. 401] is a slightly modified variant of that in the original paper by Lucas [363]:

$$\sigma_{\mathrm{NWA},m,q}(n)\equiv\sum_{k=0}^{n-1}(-1)^k(\overline{k}_q)^m. \tag{4.213}$$

Theorem 4.3.53 *A q-analogue of* [138, p. 53], [385, p. 307], [233, p. 136]:

$$\sigma_{\mathrm{NWA},m,q}(n)=\frac{(-1)^{n-1}\mathrm{F}_{\mathrm{NWA},m,q}(\overline{n}_q)+\mathrm{F}_{\mathrm{NWA},m,q}}{2}. \tag{4.214}$$

Proof

$$\begin{aligned}\mathrm{LHS}&=\sum_{k=0}^{n-1}(-1)^k\nabla_{\mathrm{NWA},q}\mathrm{F}_{\mathrm{NWA},m,q}(\overline{k}_q)\\&=\frac{1}{2}\sum_{k=0}^{n-1}(-1)^k\big(\mathrm{F}_{\mathrm{NWA},m,q}(\overline{k}_q\oplus_q 1)+\mathrm{F}_{\mathrm{NWA},m,q}(\overline{k}_q)\big)=\mathrm{RHS}.\end{aligned} \tag{4.215}$$

□

A special case of Φ_q polynomials are the θ_q polynomials of order n, which are obtained by putting $f_n(t) = \frac{g(t)2^n}{(\mathrm{E}_{\frac{1}{q}}(t)+1)^n}$ in (4.104).

Definition 75 A q-analogue of [385, p. 142, (1)]:

$$\frac{2^n}{(\mathrm{E}_{\frac{1}{q}}(t)+1)^n} g(t)\mathrm{E}_q(xt) = \sum_{\nu=0}^{\infty} \frac{t^\nu \theta_{\nu,q}^{(n)}(x)}{\{\nu\}_q!}. \tag{4.216}$$

By (4.115), we obtain a q-analogue of [385], [384, p. 519]:

$$\nabla_{\mathrm{JHC},q}\theta_{\nu,q}^{(n)}(x) = \theta_{\nu,q}^{(n-1)}(x). \tag{4.217}$$

The JHC generalized q-Euler polynomials

We will now define the second q-Euler polynomials, a special case of the θ_q polynomials. There are many similar definitions of these, but we will follow the one in [385, p. 143–147], [138, p. 51], because it is equivalent to the q-Appell polynomials from [15].

Definition 76 The generating function for the second q-Euler polynomials of degree ν and order n, $\mathrm{F}_{\mathrm{JHC},\nu,q}^{(n)}(x)$, is the following q-analogue of [437, p. 102], [385, p. 309], [504, p. 345]:

$$\frac{2^n \mathrm{E}_q(xt)}{(\mathrm{E}_{\frac{1}{q}}(t)+1)^n} = \sum_{\nu=0}^{\infty} \frac{t^\nu}{\{\nu\}_q!} \mathrm{F}_{\mathrm{JHC},\nu,q}^{(n)}(x), \quad |t| < \pi. \tag{4.218}$$

This can be generalized to:

Definition 77 The generating function for the second q-Euler polynomials of degree ν and order n, $\mathrm{F}_{\mathrm{JHC},\nu,q}^{(n)}(x|\omega_1,\ldots,\omega_n)$, is the following q-analogue of [403, p. 143, (78)]:

$$\frac{2^n \mathrm{E}_q(xt)}{\prod_{k=1}^{n}(\mathrm{E}_{\frac{1}{q}}(\omega_k t)+1)} = \sum_{\nu=0}^{\infty} \frac{t^\nu}{\{\nu\}_q!} \mathrm{F}_{\mathrm{JHC},\nu,q}^{(n)}(x|\omega_1,\ldots,\omega_n),$$
$$|t| < \min\big(|\pi/\omega_1|,\ldots,|\pi/\omega_n|\big). \tag{4.219}$$

Corollary 4.3.54

$$\mathrm{E}_q(t\mathrm{F}_{\mathrm{JHC},q}) \doteqdot \frac{2}{\mathrm{E}_{\frac{1}{q}}(t)+1}. \tag{4.220}$$

Obviously, $\mathrm{F}^{(n)}_{\mathrm{JHC},\nu,q}(x|\omega_1,\ldots,\omega_n)$ is symmetric in $\omega_1,\ldots,\omega_n$ and, in particular,

$$\mathrm{F}^{(1)}_{\mathrm{JHC},\nu,q}(x|\omega)=\omega^{\nu}\mathrm{F}_{\mathrm{JHC},\nu,q}\left(\frac{x}{\omega}\right). \tag{4.221}$$

From the relation

$$\underset{\omega_1,\ldots,\omega_n \mathrm{JHC},q}{\nabla^n} \mathrm{F}^{(n)}_{\mathrm{JHC},\nu,q}(x|\omega_1,\ldots,\omega_n)=x^{\nu} \tag{4.222}$$

we obtain

$$\underset{\omega_1,\ldots,\omega_p \mathrm{JHC},q}{\nabla^p} \mathrm{F}^{(n)}_{\mathrm{JHC},\nu,q}(x|\omega_1,\ldots,\omega_n)=\mathrm{F}^{(n-p)}_{\mathrm{JHC},\nu,q}(x|\omega_{p+1},\ldots,\omega_n). \tag{4.223}$$

Theorem 4.3.55 *A q-analogue of* [385, p. 144, (7)], [403, (7), p. 121], [485, p. 378, (28)]:

$$\sum_{k=0}^{\nu}\binom{\nu}{k}_q q^{\binom{k}{2}}\mathrm{F}^{(n)}_{\mathrm{JHC},\nu-k,q}(x)+\mathrm{F}^{(n)}_{\mathrm{JHC},\nu,q}(x)=2\mathrm{F}^{(n-1)}_{\mathrm{JHC},\nu,q}(x). \tag{4.224}$$

With this formula we can compute all second q-Euler polynomials of order n in terms of the polynomials of order $n-1$.

Definition 78 A q-analogue of [402, p. 139], [363, p. 252]. The second generalized q-Euler numbers are given by

$$\mathrm{F}^{(n)}_{\mathrm{JHC},\nu,q}\equiv \mathrm{F}^{(n)}_{\mathrm{JHC},\nu,q}(0). \tag{4.225}$$

Furthermore we put

$$\mathrm{F}_{\mathrm{JHC},k,q}\equiv \mathrm{F}^{(1)}_{\mathrm{JHC},k,q};\qquad \mathrm{F}_{\mathrm{JHC},\nu,q}(x)\equiv \mathrm{F}^{(1)}_{\mathrm{JHC},\nu,q}(x). \tag{4.226}$$

Theorem 4.3.56 *The following operator expression is a q-analogue of* [138, 3.15, p. 51]:

$$\mathrm{F}_{\mathrm{JHC},\nu,q}(x)\equiv \frac{2}{\mathrm{E}_{\frac{1}{q}}(\mathrm{D}_q)+\mathrm{I}}x^{\nu}=\frac{2}{\mathrm{E}(\boxplus_q)+\mathrm{I}}x^{\nu} \doteq (x\oplus_q \mathrm{F}_{\mathrm{JHC},q})^{\nu}. \tag{4.227}$$

The following two recursion formulas are quite useful for the computations of the second q-Euler polynomials.

Theorem 4.3.57 *A q-analogue of* [403, (27), p. 24], [485, p. 378, (29)]:

$$\mathrm{F}_{\mathrm{JHC},\nu,q}(x)+\sum_{k=0}^{\nu}\binom{\nu}{k}_q q^{\binom{\nu-k}{2}}\mathrm{F}_{\mathrm{JHC},k,q}(x)=2x^{\nu}. \tag{4.228}$$

Theorem 4.3.58 *A q-analogue of* [138, 3.16, p. 51], [363, p. 252]:

$$(\mathrm{F}_{\mathrm{JHC},q} \boxplus_q 1)^n + (\mathrm{F}_{\mathrm{JHC},q})^n \doteqdot 2\delta_{0,n}. \tag{4.229}$$

Theorem 4.3.59 *A q-analogue of* [363, p. 261], [98, p. 6, (4.3)], [402, (19), p. 136]:

$$f(x \boxplus_q \mathrm{F}_{\mathrm{JHC},q} \boxplus_q 1) + f(x \oplus_q \mathrm{F}_{\mathrm{JHC},q}) \doteqdot 2f(x). \tag{4.230}$$

We will also state the corresponding equation for $\mathrm{F}^{(n)}_{\mathrm{JHC},\nu,q}$, written in two different forms.

Theorem 4.3.60 *A q-analogue of* [402, (19), p. 150, p. 155], [403, (29), p. 126]:

$$\nabla_{\mathrm{JHC},q} f\big(x \oplus_q \mathrm{F}^{(n)}_{\mathrm{JHC},q}\big) \doteqdot f\big(x \oplus_q \mathrm{F}^{(n-1)}_{\mathrm{JHC},q}\big) \doteqdot \nabla_{\mathrm{JHC},q} f\big(\mathrm{F}^{(n)}_{\mathrm{JHC},q}(x)\big) \doteqdot f\big(\mathrm{F}^{(n-1)}_{\mathrm{JHC},q}(x)\big). \tag{4.231}$$

The following table lists some of the first $\mathrm{F}_{\mathrm{JHC},n,q}$:

$n=0$	$n=1$	$n=2$	$n=3$	$n=4$
1	-2^{-1}	$2^{-2}(1-q)$	$2^{-3}(-1+2q+2q^2-q^3)$	$2^{-4}(1-3q-3q^2+3q^4+3q^5-q^6)$

In contrast to Bernoulli numbers, the even Euler numbers are equal to zero. Compare with the list for the C numbers in Nørlund [403, p. 27].

Theorem 4.3.61

$$\begin{aligned} &\textit{For } \nu \textit{ even}, \ \nu > 0, \quad \mathrm{F}_{\mathrm{NWA},\nu,q} = -\mathrm{F}_{\mathrm{JHC},\nu,q}, \\ &\textit{For } \nu \textit{ uneven}, \quad \mathrm{F}_{\mathrm{NWA},\nu,q} = \mathrm{F}_{\mathrm{JHC},\nu,q}. \end{aligned} \tag{4.232}$$

Proof Assume that $n = 1$. Use the generating function (4.196) with $-t$ and $x = 0$, multiply with $\mathrm{E}_q(t)$ in the denominator and numerator to obtain

$$(-1)^\nu \mathrm{F}_{\mathrm{JHC},\nu,q} = \sum_{k=0}^{\nu} \binom{\nu}{k}_q \mathrm{F}_{\mathrm{NWA},\nu-k,q}. \tag{4.233}$$

Finally compare with the generating function (4.218). □

Theorem 4.3.62 *We have the following operational representation*:

$$\mathrm{F}^{(n)}_{\mathrm{JHC},\nu,q}(\omega_1, \ldots, \omega_n) \doteqdot \left(\bigoplus_{l=1}^{q,n} \omega_l \mathrm{F}_{\mathrm{JHC},q} \right)^\nu. \tag{4.234}$$

Theorem 4.3.63 *The second q-Euler polynomials can be expressed as a finite sum of difference operators on x^n. Almost a q-analogue of* [320, p. 289]:

$$\mathrm{F}_{\mathrm{JHC},n,q}(x)=\sum_{m=0}^{n}\frac{(-1)^m}{2^m}\triangle_{\mathrm{JHC},q}^{m}x^n. \tag{4.235}$$

Theorem 4.3.64 *A generalization of* (4.228):

$$2^{-n}\sum_{k=0}^{n}\binom{n}{k}\mathrm{F}_{\mathrm{JHC},\nu,q}^{(n)}(x\boxplus_q\tilde{k}_q)=x^{\nu}. \tag{4.236}$$

Proof Expand $\nabla_{\mathrm{JHC},q}^{n}\mathrm{F}_{\mathrm{JHC},\nu,q}^{(n)}(x)$. □

Definition 79 A q-analogue of a relation by J. Herschel [277, p. 88]:

$$\sigma_{\mathrm{JHC},m,q}(n)\equiv\sum_{k=0}^{n-1}(-1)^k(\tilde{k}_q)^m. \tag{4.237}$$

Theorem 4.3.65 *A q-analogue of* [138, p. 53], [385, p. 307], [233, p. 136]:

$$\sigma_{\mathrm{JHC},m,q}(n)=\frac{(-1)^{n-1}\mathrm{F}_{\mathrm{JHC},m,q}(\tilde{n}_q)+\mathrm{F}_{\mathrm{JHC},m,q}(\tilde{0}_q)}{2}. \tag{4.238}$$

Proof

$$\begin{aligned}\mathrm{LHS}&=\sum_{k=0}^{n-1}(-1)^k\nabla_{\mathrm{JHC},q}\mathrm{F}_{\mathrm{JHC},m,q}(\tilde{k}_q)\\&=\frac{1}{2}\sum_{k=0}^{n-1}(-1)^k\big(\mathrm{F}_{\mathrm{JHC},m,q}(\tilde{k}_q\boxplus_q 1)+\mathrm{F}_{\mathrm{JHC},m,q}(\tilde{k}_q)\big)=\mathrm{RHS}.\end{aligned} \tag{4.239}$$

□

4.3.6 Several variables; n negative

So far we considered only q-Bernoulli polynomials and q-Euler polynomials of positive order n. As we will see below, it is useful to allow n also to be a negative integer. The following calculations are q-analogues of those by Nørlund [403, p. 133 ff].

Definition 80 As a q-analogue of [403, (50), p. 133], [418, p. 1226, xvi] and [485, p. 378, (19)], we define the first q-Bernoulli polynomials of two variables as

$$\mathrm{B}_{\mathrm{NWA},\nu,q}^{(n+p)}(x \oplus_q y|\omega_1,\ldots,\omega_{n+p}) \equiv \left(\mathrm{B}_{\mathrm{NWA},q}^{(n)}(x|\omega_1,\ldots,\omega_n) \oplus_q \mathrm{B}_{\mathrm{NWA},q}^{(p)}(y|\omega_{n+1},\ldots,\omega_{n+p})\right)^{\nu}, \tag{4.240}$$

where we assume that n and p operate on x and y respectively, and the same for any q-polynomial.

The relation (4.127) together with (4.240) shows that $\mathrm{B}_{\mathrm{NWA},\nu,q}^{(n)}(x|\omega_1,\ldots,\omega_n)$ is a homogeneous function of $x,\omega_1,\ldots,\omega_n$ of degree ν, a q-analogue of [403, p. 134, (55)], i.e.

$$\mathrm{B}_{\mathrm{NWA},\nu,q}^{(n)}(\lambda x|\lambda\omega_1,\ldots,\lambda\omega_n) = \lambda^{\nu}\mathrm{B}_{\mathrm{NWA},\nu,q}^{(n)}(x|\omega_1,\ldots,\omega_n), \quad \lambda \in \mathbb{C}. \tag{4.241}$$

The same holds for q-Euler, Lucas and G-polynomials.

Formula (4.240) can be generalized in at least two ways.

Theorem 4.3.66 *A q-analogue of* [439, p. 704], [403, p. 133]. *If* $\sum_{l=1}^{s} n_l = n$,

$$\mathrm{B}_{\mathrm{NWA},k,q}^{(n)}(x_1 \oplus_q \cdots \oplus_q x_s) = \sum_{m_1+\cdots+m_s=k} \binom{k}{m_1,\ldots,m_s}_q \prod_{j=1}^{s} \mathrm{B}_{\mathrm{NWA},m_j,q}^{(n_j)}(x_j), \tag{4.242}$$

where we assume that n_j operates on x_j. The same holds for any q-polynomial.

Proof In umbral notation we have, as in the classical case:

$$(x_1 \oplus_q \cdots \oplus_q x_s \oplus_q \overline{n}_q\gamma)^k \sim \left(\left(x_1 \oplus_q \overline{n}_{1_q}\gamma'\right) \oplus_q \cdots \oplus_q \left(x_s \oplus_q \overline{n}_{s_q}\gamma''\right)\right)^k, \tag{4.243}$$

where $\gamma',\ldots,\gamma''$ are distinct umbrae, each equivalent to γ. Now use (4.70), (4.74) and (4.130). □

Theorem 4.3.67 *A q-analogue of* [439, p. 704], [403, p. 133]. *If* $\sum_{l=1}^{s} n_l = n$,

$$\mathrm{B}_{\mathrm{NWA},k,q}^{(n)}(x_1 \boxplus_q \cdots \boxplus_q x_s) = \sum_{m_1+\cdots+m_s=k} \binom{k}{m_1,\ldots,m_s}_q \prod_{j=1}^{s} \mathrm{B}_{\mathrm{NWA},m_j,q}^{(n_j)}(x_j) q^{\binom{\vec{i}}{2}}, \tag{4.244}$$

$\vec{i} = (m_2,\ldots,m_n)$. We assume that n_j operates on x_j. The same holds for any q-polynomial.

By (4.119) and (4.195) we obtain

$$\triangle_{\mathrm{NWA},q}^{n}\mathrm{B}_{\mathrm{NWA},\nu,q}^{(n)}(x) = \frac{\{\nu\}_q!}{\{\nu-n\}_q!}x^{\nu-n},$$

$$\nabla_{\mathrm{NWA},q}^{n}\mathrm{F}_{\mathrm{NWA},\nu,q}^{(n)}(x)=x^{\nu},$$

and we have

Definition 81 A q-analogue of [402, p. 177], [403, (66), p. 138]. The first q-Bernoulli polynomials of negative order $-n$ are given by

$$\mathrm{B}_{\mathrm{NWA},\nu,q}^{(-n)}(x|\omega_1,\ldots,\omega_n)\equiv\frac{\{\nu\}_q!}{\{\nu+n\}_q!}\underset{\omega_1,\ldots,\omega_n\mathrm{NWA},q}{\Delta^n}x^{\nu+n} \tag{4.245}$$

and the first q-Euler polynomials of negative order $-n$ by the following q-analogue of [403, (67), p. 138]:

$$\mathrm{F}_{\mathrm{NWA},\nu,q}^{(-n)}(x|\omega_1,\ldots,\omega_n)\equiv\underset{\omega_1,\ldots,\omega_n\mathrm{NWA},q}{\nabla^n}x^{\nu}, \tag{4.246}$$

where $\nu,n\in\mathbb{N}$. This defines the first q-Bernoulli and first q-Euler polynomials of negative order as iterated $\Delta_{\mathrm{NWA},q}$ and $\nabla_{\mathrm{NWA},q}$ operating on positive integer powers of x.

Furthermore,

$$\mathrm{B}_{\mathrm{NWA},\nu,q}^{(-n)}\equiv\mathrm{B}_{\mathrm{NWA},\nu,q}^{(-n)}(0), \tag{4.247}$$

$$\mathrm{F}_{\mathrm{NWA},\nu,q}^{(-n)}\equiv\mathrm{F}_{\mathrm{NWA},\nu,q}^{(-n)}(0). \tag{4.248}$$

A calculation shows that formulas (4.119) and (4.195) hold for negative orders too and we get [12, p. 255, 10.9]

$$\mathrm{B}_{\mathrm{NWA},\nu,q}^{(-n-p)}(x\oplus_q y)\doteq\left(\mathrm{B}_{\mathrm{NWA},q}^{(-n)}(x)\oplus_q\mathrm{B}_{\mathrm{NWA},q}^{(-p)}(y)\right)^{\nu}, \tag{4.249}$$

and the same for first q-Euler polynomials.

A special case is the following q-analogue of [403, p. 139, (71)]:

$$\mathrm{B}_{\mathrm{NWA},\nu,q}^{(-n)}(x\oplus_q y)\doteq\left(\mathrm{B}_{\mathrm{NWA},q}^{(-n)}(x)\oplus_q y\right)^{\nu}, \tag{4.250}$$

and the same for first q-Euler polynomials.

Theorem 4.3.68 *A recurrence formula for the first q-Bernoulli numbers and a recurrence formula for the first q-Euler numbers*:

If $n,p\in\mathbb{Z}$ then

$$\mathrm{B}_{\mathrm{NWA},\nu,q}^{(n+p)}\doteq\left(\mathrm{B}_{\mathrm{NWA},q}^{(n)}\oplus_q\mathrm{B}_{\mathrm{NWA},q}^{(p)}\right)^{\nu}, \tag{4.251}$$

$$\mathrm{F}_{\mathrm{NWA},\nu,q}^{(n+p)}\doteq\left(\mathrm{F}_{\mathrm{NWA},q}^{(n)}\oplus_q\mathrm{F}_{\mathrm{NWA},q}^{(p)}\right)^{\nu}. \tag{4.252}$$

Theorem 4.3.69 *A q-analogue of* [403, p. 140, (72), (73)], [418, p. 1226, xvii]. *The first equation occurred in* [12, p. 255, 10.10].

$$(x \oplus_q y)^\nu \doteqdot \left(\mathrm{B}^{(-n)}_{\mathrm{NWA},q}(x) \oplus_q \mathrm{B}^{(n)}_{\mathrm{NWA},q}(y)\right)^\nu, \tag{4.253}$$

$$(x \oplus_q y)^\nu \doteqdot \left(\mathrm{F}^{(-n)}_{\mathrm{NWA},q}(x) \oplus_q \mathrm{F}^{(n)}_{\mathrm{NWA},q}(y)\right)^\nu. \tag{4.254}$$

Proof Put $p = -n$ in (4.249). □

In particular, for $y = 0$, we obtain a q-analogue of [418, p. 1226, xviii]:

$$x^\nu \doteqdot \left(\mathrm{B}^{(-n)}_{\mathrm{NWA},q} \oplus_q \mathrm{B}^{(n)}_{\mathrm{NWA},q}(x)\right)^\nu, \tag{4.255}$$

$$x^\nu \doteqdot \left(\mathrm{F}^{(-n)}_{\mathrm{NWA},q} \oplus_q \mathrm{F}^{(n)}_{\mathrm{NWA},q}(x)\right)^\nu. \tag{4.256}$$

These recurrence formulas express the first q-Bernoulli and q-Euler polynomials of order n without involving polynomials of negative order.

These can also be expressed in the form

$$x^\nu = \sum_{s=0}^{\nu} \frac{\mathrm{B}^{(-n)}_{\mathrm{NWA},s,q}}{\{s\}_q!} \mathrm{D}_q^s \mathrm{B}^{(n)}_{\mathrm{NWA},\nu,q}(x), \tag{4.257}$$

$$x^\nu = \sum_{s=0}^{\nu} \frac{\mathrm{F}^{(-n)}_{\mathrm{NWA},s,q}}{\{s\}_q!} \mathrm{D}_q^s \mathrm{F}^{(n)}_{\mathrm{NWA},\nu,q}(x). \tag{4.258}$$

We conclude that the first q-Bernoulli and first q-Euler polynomials satisfy linear q-difference equations with constant coefficients.

The following theorem is useful for the computation of q-Bernoulli and q-Euler polynomials of positive order. This is because the polynomials of negative order are of simpler nature and can easily be computed. When the $\mathrm{B}^{(-n)}_{\mathrm{NWA},s,q}$ etc. are known, (4.259) can be used to compute the $\mathrm{B}^{(n)}_{\mathrm{NWA},s,q}$.

Theorem 4.3.70

$$\sum_{s=0}^{\nu} \binom{\nu}{s}_q \mathrm{B}^{(n)}_{\mathrm{NWA},s,q} \mathrm{B}^{(-n)}_{\mathrm{NWA},\nu-s,q} = \delta_{\nu,0}, \tag{4.259}$$

$$\sum_{s=0}^{\nu} \binom{\nu}{s}_q \mathrm{F}^{(n)}_{\mathrm{NWA},s,q} \mathrm{F}^{(-n)}_{\mathrm{NWA},\nu-s,q} = \delta_{\nu,0}. \tag{4.260}$$

Proof Put $x = y = 0$ in (4.253) and (4.254). □

Theorem 4.3.71 *A q-analogue of* [403, p. 142]. *We assume that $f(x)$ is analytic with q-Taylor expansion*

$$f(x)=\sum_{\nu=0}^{\infty}\mathrm{D}_q^{\nu}f(0)\frac{x^{\nu}}{\{\nu\}_q!}. \tag{4.261}$$

Then we can express powers of $\triangle_{\mathrm{NWA},q}$ and $\nabla_{\mathrm{NWA},q}$ operating on $f(x)$ as powers of D_q as follows. These series converge when the absolute value of x is small enough:

$$\underset{\omega_1,\ldots,\omega_n\mathrm{NWA},q}{\triangle^n}f(x)=\sum_{\nu=0}^{\infty}\mathrm{D}_q^{\nu+n}f(0)\frac{\mathrm{B}_{\mathrm{NWA},\nu,q}^{(-n)}(x|\omega_1,\ldots,\omega_n)}{\{\nu\}_q!}, \tag{4.262}$$

$$\underset{\omega_1,\ldots,\omega_n\mathrm{NWA},q}{\nabla^n}f(x)=\sum_{\nu=0}^{\infty}\mathrm{D}_q^{\nu}f(0)\frac{\mathrm{F}_{\mathrm{NWA},\nu,q}^{(-n)}(x|\omega_1,\ldots,\omega_n)}{\{\nu\}_q!}. \tag{4.263}$$

Proof Use (4.119), (4.106) and (4.195), (4.106). □

Now put $f(x)=\mathrm{E}_q(xt)$ to obtain the generating functions of the q-Bernoulli and q-Euler polynomials of negative order:

$$\frac{\prod_{k=1}^{n}(\mathrm{E}_q(\omega_k t)-1)\mathrm{E}_q(xt)}{t^n\prod_{k=1}^{n}\omega_k}=\sum_{\nu=0}^{\infty}\frac{t^{\nu}}{\{\nu\}_q!}\mathrm{B}_{\mathrm{NWA},\nu,q}^{(-n)}(x|\omega_1,\ldots,\omega_n), \tag{4.264}$$

$$\frac{\prod_{k=1}^{n}(\mathrm{E}_q(\omega_k t)+1)\mathrm{E}_q(xt)}{2^n}=\sum_{\nu=0}^{\infty}\frac{t^{\nu}}{\{\nu\}_q!}\mathrm{F}_{\mathrm{NWA},\nu,q}^{(-n)}(x|\omega_1,\ldots,\omega_n). \tag{4.265}$$

The reason for the difference in appearance compared to the original for the following equation is that one of the function arguments is a Ward number.

Theorem 4.3.72 *A q-analogue of* [402, p. 191, (10)], [364, p. 21, (18)]:

$$\mathrm{B}_{\mathrm{NWA},\nu,q}^{(m)}(x\oplus_q\overline{n}_q)=\sum_{k=0}^{\min(\nu,n)}\binom{n}{k}\frac{\{\nu\}_q!}{\{\nu-k\}_q!}\mathrm{B}_{\mathrm{NWA},\nu-k,q}^{(m-k)}(x). \tag{4.266}$$

Proof Use (4.82) and (4.119). □

Example 10 A q-analogue of [385, p. 133, (3)]:

$$(\overline{n}_q)^{\nu}=\sum_{k=0}^{\min(\nu,n)}\binom{n}{k}\frac{\{\nu\}_q!}{\{\nu-k\}_q!}\mathrm{B}_{\mathrm{NWA},\nu-k,q}^{(-k)}. \tag{4.267}$$

Proof Put $m=x=0$ in (4.266). □

4.3.7 q-Euler-Maclaurin expansions

We now present a set of formulas of a new type, which combines the infinite sums of q-Bernoulli or q-Euler polynomials with q-difference operators (Section 4.2), all in the spirit of Nørlund. These formulas have been presented at the conferences OPSFA+ISDE, Munich 2005, Progress of Difference Equations, Salzburg 2008 and Winter Conference on Difference Equations, Homburg (Saar), 2009.

Theorem 4.3.73 *A q-analogue of* [402, (21), p. 163], [418, p. 1220]. *The corresponding formula for $n = 1$ occurred in* [12, p. 254]. *Also compare with* [146, p. 158, (50)].

$$\sum_{k=0}^{\infty} \frac{\mathrm{B}_{\mathrm{NWA},k,q}^{(n)}(x)}{\{k\}_q!} \Delta_{\mathrm{NWA},q}^{n} \mathrm{D}_q^k f(y) = \mathrm{D}_{q,x}^{n} f(x \oplus_q y). \tag{4.268}$$

Proof As in [402, p. 163] replace $f(x)$ by $f(x \oplus_q y)$ in (4.143):

$$f\left(\mathrm{B}_{\mathrm{NWA},q}^{(n)}(x) \oplus_q y \oplus_q 1\right) - f\left(\mathrm{B}_{\mathrm{NWA},q}^{(n)}(x) \oplus_q y\right) = \mathrm{D}_q f\left(\mathrm{B}_{\mathrm{NWA},q}^{(n-1)}(x) \oplus_q y\right). \tag{4.269}$$

Use the umbral formula (4.21) to get

$$\sum_{k=0}^{\infty} \frac{\mathrm{B}_{\mathrm{NWA},k,q}^{(n)}(x)}{\{k\}_q!} \Delta_{\mathrm{NWA},q} \mathrm{D}_q^k f(y) = \sum_{k=0}^{\infty} \frac{\mathrm{B}_{\mathrm{NWA},k,q}^{(n-1)}(x)}{\{k\}_q!} \mathrm{D}_q^{k+1} f(y). \tag{4.270}$$

Apply the operator $\Delta_{\mathrm{NWA},q}^{n-1}$ with respect to y to both sides and use (4.262):

$$\sum_{k=0}^{\infty} \frac{\mathrm{B}_{\mathrm{NWA},k,q}^{(n)}(x)}{\{k\}_q!} \Delta_{\mathrm{NWA},q}^{n} \mathrm{D}_q^k f(y) = \sum_{k=0}^{\infty} \frac{\mathrm{B}_{\mathrm{NWA},k,q}^{(n-1)}(x)}{\{k\}_q!} \sum_{l=0}^{\infty} \mathrm{D}_q^{k+l+n} f(0) \frac{\mathrm{B}_{\mathrm{NWA},l,q}^{(-n+1)}(y)}{\{l\}_q!}. \tag{4.271}$$

Finally, use (4.21), (4.34), (4.253) to rewrite the right-hand side. □

Remark 13 The RHS of (4.268) can also be written as $\mathrm{D}_{q,y}^{n} f(x \oplus_q y)$ or $\mathrm{D}_q^n f(x \oplus_q y)$.

If we put $n = q = 1$ in (4.268), we get an Euler-Maclaurin expansion known from [385, p. 140]. If we also put $y = 0$, we get an expansion of a polynomial in terms of Bernoulli polynomials known from [320, p. 248].

The following formula for $q = 1$, $n = 1$ appeared in Nørlund's work [402, p. 124]. Nørlund had discussed the equation $\Delta f(x) = \mathrm{D}\varphi(x)$ many years in his correspondence with Mittag-Leffler [183].

Corollary 4.3.74 *A q-analogue of Nørlund* [402, p. 164], *Szegő* [493]. *Let $\varphi(x)$ be a polynomial of degree ν. A solution $f(x)$ of the q-difference equation*

$$\underset{\omega_1,\ldots,\omega_n\mathrm{NWA},q}{\triangle^n} f(x) = \mathrm{D}_q^n\varphi(x) \tag{4.272}$$

is given by

$$f(x \oplus_q y) = \sum_{k=0}^{\nu} \frac{\mathrm{B}^{(n)}_{\mathrm{NWA},k,q}(x|\omega_1,\ldots,\omega_n)}{\{k\}_q!}\mathrm{D}_q^k\varphi(y). \tag{4.273}$$

Proof The LHS of (4.273) can be written as $\varphi(\mathrm{B}^{(n)}_{\mathrm{NWA},q}(x|\omega_1,\ldots,\omega_n) \oplus_q y)$, because if we apply $\triangle^n_{\mathrm{NWA},q,x}$ to both sides we get

$$\begin{aligned}\underset{\omega_1,\ldots,\omega_n\mathrm{NWA},q}{\triangle^n} f(x \oplus_q y) &= \mathrm{D}^n_{q,x}\varphi(x \oplus_q y)\\ &= \underset{\omega_1,\ldots,\omega_n\mathrm{NWA},q}{\triangle^n} \varphi\big(\mathrm{B}^{(n)}_{\mathrm{NWA},q}(x|\omega_1,\ldots,\omega_n) \oplus_q y\big).\end{aligned} \tag{4.274}$$

□

Theorem 4.3.75 *A q-analogue of Nørlund* [402, p. 156], [403, p. 127, (31)]. *Compare with* [402, p. 147]. *A special case is found in* [320, p. 307].

$$\sum_{k=0}^{\infty} \frac{\mathrm{F}^{(n)}_{\mathrm{NWA},k,q}(x)}{\{k\}_q!}\nabla^n_{\mathrm{NWA},q}\mathrm{D}_q^k f(y) = f(x \oplus_q y). \tag{4.275}$$

Proof As in [402, p. 155], replace $f(x)$ by $f(x \oplus_q y)$ in (4.209):

$$\frac{1}{2}\big(f\big(\mathrm{F}^{(n)}_{\mathrm{NWA},q}(x) \oplus_q y \oplus_q 1\big) + f\big(\mathrm{F}^{(n)}_{\mathrm{NWA},q}(x) \oplus_q y\big)\big) = f\big(\mathrm{F}^{(n-1)}_{\mathrm{NWA},q}(x) \oplus_q y\big). \tag{4.276}$$

Use the umbral formula (4.21) to get

$$\sum_{k=0}^{\infty} \frac{\mathrm{F}^{(n)}_{\mathrm{NWA},k,q}(x)}{\{k\}_q!}\nabla_{\mathrm{NWA},q}\mathrm{D}_q^k f(y) = \sum_{k=0}^{\infty} \frac{\mathrm{F}^{(n-1)}_{\mathrm{NWA},k,q}(x)}{\{k\}_q!}\mathrm{D}_q^k f(y). \tag{4.277}$$

Apply the operator $\nabla^{n-1}_{\mathrm{NWA},q}$ with respect to y to both sides and use (4.263):

$$\sum_{k=0}^{\infty} \frac{\mathrm{F}^{(n)}_{\mathrm{NWA},k,q}(x)}{\{k\}_q!}\nabla^n_{\mathrm{NWA},q}\mathrm{D}_q^k f(y) = \sum_{k=0}^{\infty} \frac{\mathrm{F}^{(n-1)}_{\mathrm{NWA},k,q}(x)}{\{k\}_q!}\sum_{l=0}^{\infty}\mathrm{D}_q^{k+l} f(0)\frac{\mathrm{F}^{(-n+1)}_{\mathrm{NWA},l,q}(y)}{\{l\}_q!}. \tag{4.278}$$

Finally, use (4.21), (4.34), (4.254) to rewrite the right-hand side. □

If we put $n = q = 1$ in (4.275), we get the Euler-Boole theorem known from [233, p. 128], [385, p. 149].

Corollary 4.3.76 *A q-analogue of Szegő* [493]. *Let $\varphi(x)$ be a polynomial of degree ν. A solution $f(x)$ of the q-difference equation*

$$\underset{\omega_1,\ldots,\omega_n \text{NWA},q}{\nabla^n} f(x) = \varphi(x) \tag{4.279}$$

is given by

$$f(x \oplus_q y) = \sum_{k=0}^{\nu} \frac{\mathrm{F}^{(n)}_{\text{NWA},k,q}(x|\omega_1,\ldots,\omega_n)}{\{k\}_q!} \mathrm{D}_q^k \varphi(y). \tag{4.280}$$

Proof The LHS of (4.280) can be written as $\varphi(\mathrm{F}^{(n)}_{\text{NWA},q}(x|\omega_1,\ldots,\omega_n) \oplus_q y)$, because if we apply $\nabla^n_{\text{NWA},q,x}$ to both sides we get

$$\begin{aligned}\underset{\omega_1,\ldots,\omega_n \text{NWA},q}{\nabla^n} f(x \oplus_q y) &= \varphi(x \oplus_q y)\\ &= \underset{\omega_1,\ldots,\omega_n \text{NWA},q}{\nabla^n} \varphi\big(\mathrm{F}^{(n)}_{\text{NWA},q}(x|\omega_1,\ldots,\omega_n) \oplus_q y\big).\end{aligned} \tag{4.281}$$

□

4.3.8 JHC polynomials of many variables; negative order

The following computations are JHC-q-analogues of Nørlund's work [403, p. 133 ff].

Definition 82 As a q-analogue of [403, (50), p. 133], [418, p. 1226, xvi] and [485, p. 378, (19)], we define the second q-Bernoulli polynomials of two variables as:

$$\begin{aligned}&\mathrm{B}^{(n+p)}_{\text{JHC},\nu,q}(x \oplus_q y|\omega_1,\ldots,\omega_{n+p})\\ &\quad \doteq \big(\mathrm{B}^{(n)}_{\text{JHC},q}(x|\omega_1,\ldots,\omega_n) \oplus_q \mathrm{B}^{(p)}_{\text{JHC},q}(y|\omega_{n+1},\ldots,\omega_{n+p})\big)^{\nu},\end{aligned} \tag{4.282}$$

where we assume that n and p operate on x and y respectively, and the same for any q-polynomial.

The relation (4.168) together with (4.282) show that $\mathrm{B}^{(n)}_{\text{JHC},\nu,q}(x|\omega_1,\ldots,\omega_n)$ is a homogeneous function of $x, \omega_1,\ldots,\omega_n$ of degree ν, a q-analogue of [403, p. 134, (55)], i.e.

$$\mathrm{B}^{(n)}_{\text{JHC},\nu,q}(\lambda x|\lambda\omega_1,\ldots,\lambda\omega_n) = \lambda^{\nu}\mathrm{B}^{(n)}_{\text{JHC},\nu,q}(x|\omega_1,\ldots,\omega_n), \quad \lambda \in \mathbb{C}. \tag{4.283}$$

The same holds for the second q-Euler, Lucas and G-polynomials. The formula (4.282) can be generalized in at least two ways, in the same manner as in the NWA case.

By (4.159) and (4.217), we get

$$\triangle_{\mathrm{JHC},q}^{n}\mathrm{B}_{\mathrm{JHC},\nu,q}^{(n)}(x)=\frac{\{\nu\}_q!}{\{\nu-n\}_q!}x^{\nu-n},$$

$$\nabla_{\mathrm{JHC},q}^{n}\mathrm{F}_{\mathrm{JHC},\nu,q}^{(n)}(x)=x^{\nu},$$

and we have

Definition 83 A q-analogue of [402, p. 177], [403, (66), p. 138]. The second q-Bernoulli polynomials of negative order $-n$ are given by

$$\mathrm{B}_{\mathrm{JHC},\nu,q}^{(-n)}(x|\omega_1,\ldots,\omega_n)\equiv\frac{\{\nu\}_q!}{\{\nu+n\}_q!}\underset{\omega_1,\ldots,\omega_n\,\mathrm{JHC},q}{\triangle^{n}}x^{\nu+n}, \tag{4.284}$$

and the second q-Euler polynomial of negative order $-n$ by the following q-analogue of [403, (67), p. 138]:

$$\mathrm{F}_{\mathrm{JHC},\nu,q}^{(-n)}(x|\omega_1,\ldots,\omega_n)\equiv\underset{\omega_1,\ldots,\omega_n\,\mathrm{JHC},q}{\nabla^{n}}x^{\nu}, \tag{4.285}$$

where $\nu,n\in\mathbb{N}$. This defines second q-Bernoulli and q-Euler polynomials of negative order as iterated $\triangle_{\mathrm{JHC},q}$ and $\nabla_{\mathrm{JHC},q}$ operating on positive integer powers of x.

Furthermore,

$$\mathrm{B}_{\mathrm{JHC},\nu,q}^{(-n)}\equiv\mathrm{B}_{\mathrm{JHC},\nu,q}^{(-n)}(0), \tag{4.286}$$

$$\mathrm{F}_{\mathrm{JHC},\nu,q}^{(-n)}\equiv\mathrm{F}_{\mathrm{JHC},\nu,q}^{(-n)}(0). \tag{4.287}$$

A calculation shows that formulas (4.159) and (4.217) hold for negative orders too and we get

$$\mathrm{B}_{\mathrm{JHC},\nu,q}^{(-n-p)}(x\oplus_q y)\doteq\big(\mathrm{B}_{\mathrm{JHC},q}^{(-n)}(x)\oplus_q\mathrm{B}_{\mathrm{JHC},q}^{(-p)}(y)\big)^{\nu}, \tag{4.288}$$

and the same for second q-Euler polynomials.

A special case is the following q-analogue of [403, p. 139, (71)]:

$$\mathrm{B}_{\mathrm{JHC},\nu,q}^{(-n)}(x\oplus_q y)\doteq\big(\mathrm{B}_{\mathrm{JHC},q}^{(-n)}(x)\oplus_q y\big)^{\nu}, \tag{4.289}$$

and the same for the second q-Euler polynomials.

Theorem 4.3.77 *A recurrence formula for the second q-Bernoulli numbers and a recurrence formula for the second q-Euler numbers.*

If $n, p \in \mathbb{Z}$ *then*

$$\mathrm{B}_{\mathrm{JHC},\nu,q}^{(n+p)} \doteq \left(\mathrm{B}_{\mathrm{JHC},q}^{(n)} \oplus_q \mathrm{B}_{\mathrm{JHC},q}^{(p)}\right)^{\nu}, \tag{4.290}$$

$$\mathrm{F}_{\mathrm{JHC},\nu,q}^{(n+p)} \doteq (\mathrm{F}_{\mathrm{JHC},q}^{(n)} \oplus_q \mathrm{F}_{\mathrm{JHC},q}^{(p)})^{\nu}. \tag{4.291}$$

Theorem 4.3.78 *A q-analogue of* [403, p. 140, (72), (73)], [418, p. 1226, xvii]:

$$(x \oplus_q y)^{\nu} \doteq \left(\mathrm{B}_{\mathrm{JHC},q}^{(-n)}(x) \oplus_q \mathrm{B}_{\mathrm{JHC},q}^{(n)}(y)\right)^{\nu}, \tag{4.292}$$

$$(x \oplus_q y)^{\nu} \doteq \left(\mathrm{F}_{\mathrm{JHC},q}^{(-n)}(x) \oplus_q \mathrm{F}_{\mathrm{JHC},q}^{(n)}(y)\right)^{\nu}. \tag{4.293}$$

Proof Put $p = -n$ in (4.288). □

In particular, for $y = 0$ we get a q-analogue of [418, p. 1226, xviii]:

$$x^{\nu} \doteq \left(\mathrm{B}_{\mathrm{JHC},q}^{(-n)} \oplus_q \mathrm{B}_{\mathrm{JHC},q}^{(n)}(x)\right)^{\nu}, \tag{4.294}$$

$$x^{\nu} \doteq \left(\mathrm{F}_{\mathrm{JHC},q}^{(-n)} \oplus_q \mathrm{F}_{\mathrm{JHC},q}^{(n)}(x)\right)^{\nu}. \tag{4.295}$$

These recurrence formulas express the second q-Bernoulli and q-Euler polynomials of order n without involving polynomials of negative order.

These can also be expressed in the form

$$x^{\nu} = \sum_{s=0}^{\nu} \frac{\mathrm{B}_{\mathrm{JHC},s,q}^{(-n)}}{\{s\}_q!} \mathrm{D}_q^s \mathrm{B}_{\mathrm{JHC},\nu,q}^{(n)}(x), \tag{4.296}$$

$$x^{\nu} = \sum_{s=0}^{\nu} \frac{\mathrm{F}_{\mathrm{JHC},s,q}^{(-n)}}{\{s\}_q!} \mathrm{D}_q^s \mathrm{F}_{\mathrm{JHC},\nu,q}^{(n)}(x). \tag{4.297}$$

We conclude that the second q-Bernoulli and q-Euler polynomials satisfy linear q-difference equations with constant coefficients.

The following theorem is useful for the computation of second q-Bernoulli and q-Euler polynomials of positive order. This is because the polynomials of negative order are of simpler nature and can easily be computed. When the $\mathrm{B}_{\mathrm{JHC},s,q}^{(-n)}$ etc. are known, (4.298) can be used to compute the $\mathrm{B}_{\mathrm{JHC},s,q}^{(n)}$.

Theorem 4.3.79

$$\sum_{s=0}^{\nu} \binom{\nu}{s}_q \mathrm{B}_{\mathrm{JHC},s,q}^{(n)} \mathrm{B}_{\mathrm{JHC},\nu-s,q}^{(-n)} = \delta_{\nu,0}, \tag{4.298}$$

$$\sum_{s=0}^{\nu} \binom{\nu}{s}_q \mathrm{F}_{\mathrm{JHC},s,q}^{(n)} \mathrm{F}_{\mathrm{JHC},\nu-s,q}^{(-n)} = \delta_{\nu,0}. \tag{4.299}$$

Proof Put $x = y = 0$ in (4.292) and (4.293). □

Theorem 4.3.80 *A q-analogue of* [402, p. 182], [403, p. 142]. *Assume that $f(x)$ is analytic with q-Taylor expansion*

$$f(x) = \sum_{\nu=0}^{\infty} \mathrm{D}_q^{\nu} f(0) \frac{x^{\nu}}{\{\nu\}_q!}. \tag{4.300}$$

Then we can express powers of $\Delta_{\mathrm{JHC},q}$ and $\nabla_{\mathrm{JHC},q}$ operating on $f(x)$ as powers of D_q as follows. These series converge when the absolute value of x is small enough:

$$\underset{\omega_1,\ldots,\omega_n \mathrm{JHC},q}{\Delta^n} f(x) = \sum_{\nu=0}^{\infty} \mathrm{D}_q^{\nu+n} f(0) \frac{\mathrm{B}_{\mathrm{JHC},\nu,q}^{(-n)}(x|\omega_1,\ldots,\omega_n)}{\{\nu\}_q!}, \tag{4.301}$$

$$\underset{\omega_1,\ldots,\omega_n \mathrm{JHC},q}{\nabla^n} f(x) = \sum_{\nu=0}^{\infty} \mathrm{D}_q^{\nu} f(0) \frac{\mathrm{F}_{\mathrm{JHC},\nu,q}^{(-n)}(x|\omega_1,\ldots,\omega_n)}{\{\nu\}_q!}. \tag{4.302}$$

Proof Use (4.159), (4.106) and (4.217), (4.106), respectively. □

Now put $f(x) = \mathrm{E}_q(xt)$ to obtain the generating function of the second q-Bernoulli and q-Euler polynomials of negative order:

$$\frac{\prod_{k=1}^{n}(\mathrm{E}_{\frac{1}{q}}(\omega_k t) - 1)\mathrm{E}_q(xt)}{t^n \prod_{k=1}^{n} \omega_k} = \sum_{\nu=0}^{\infty} \frac{t^{\nu}}{\{\nu\}_q!} \mathrm{B}_{\mathrm{JHC},\nu,q}^{(-n)}(x|\omega_1,\ldots,\omega_n), \tag{4.303}$$

$$\frac{\prod_{k=1}^{n}(\mathrm{E}_{\frac{1}{q}}(\omega_k t) + 1)\mathrm{E}_q(xt)}{2^n} = \sum_{\nu=0}^{\infty} \frac{t^{\nu}}{\{\nu\}_q!} \mathrm{F}_{\mathrm{JHC},\nu,q}^{(-n)}(x|\omega_1,\ldots,\omega_n). \tag{4.304}$$

The reason for the difference in appearance compared to the original for the following equation is that one of the function arguments is a Jackson number.

Theorem 4.3.81 *A q-analogue of* [402, p. 191, (10)], [364, p. 21, (18)]:

$$\mathrm{B}_{\mathrm{JHC},\nu,q}^{(m)}(x \oplus_q \tilde{n}_q) = \sum_{k=0}^{\min(\nu,n)} \binom{n}{k} \frac{\{\nu\}_q!}{\{\nu-k\}_q!} \mathrm{B}_{\mathrm{JHC},\nu-k,q}^{(m-k)}(x). \tag{4.305}$$

Proof Use (4.93) and (4.159). □

Example 11 A q-analogue of [385, p. 133, (3)]:

$$(\tilde{n}_q)^{\nu} = \sum_{k=0}^{\min(\nu,n)} \binom{n}{k} \frac{\{\nu\}_q!}{\{\nu-k\}_q!} \mathrm{B}_{\mathrm{JHC},\nu-k,q}^{(-k)}. \tag{4.306}$$

Proof Use $m = x = 0$ in (4.305). □

4.3.9 JHC q-Euler-Maclaurin expansions

Now follows a set of formulas of a different type than before, which combine infinite sums of JHC-q-Bernoulli or q-Euler polynomials times difference operators with q-difference operators, all in the spirit of Nørlund.

Theorem 4.3.82 *A q-analogue of* [402, (21), p. 163], [418, p. 1220]:

$$\sum_{k=0}^{\infty} \frac{\mathrm{B}_{\mathrm{JHC},k,q}^{(n)}(x)}{\{k\}_q!} \triangle_{\mathrm{JHC},q}^{n} \mathrm{D}_q^k f(y) = \mathrm{D}_{q,x}^n f(x \oplus_q y). \tag{4.307}$$

Proof As in [402, p. 163], replace $f(x)$ by $f(x \oplus_q y)$ in (4.185):

$$f\big(\mathrm{B}_{\mathrm{JHC},q}^{(n)}(x) \oplus_q y \boxplus_q 1\big) - f\big(\mathrm{B}_{\mathrm{JHC},q}^{(n)}(x) \oplus_q y\big) = \mathrm{D}_q f\big(\mathrm{B}_{\mathrm{JHC},q}^{(n-1)}(x) \oplus_q y\big). \tag{4.308}$$

Use the umbral formula (4.21) to get

$$\sum_{k=0}^{\infty} \frac{\mathrm{B}_{\mathrm{JHC},k,q}^{(n)}(x)}{\{k\}_q!} \triangle_{\mathrm{JHC},q} \mathrm{D}_q^k f(y) = \sum_{k=0}^{\infty} \frac{\mathrm{B}_{\mathrm{JHC},k,q}^{(n-1)}(x)}{\{k\}_q!} \mathrm{D}_q^{k+1} f(y). \tag{4.309}$$

Apply the operator $\triangle_{\mathrm{JHC},q}^{n-1}$ with respect to y to both sides and use (4.301):

$$\sum_{k=0}^{\infty} \frac{\mathrm{B}_{\mathrm{JHC},k,q}^{(n)}(x)}{\{k\}_q!} \triangle_{\mathrm{JHC},q}^{n} \mathrm{D}_q^k f(y) = \sum_{k=0}^{\infty} \frac{\mathrm{B}_{\mathrm{JHC},k,q}^{(n-1)}(x)}{\{k\}_q!} \sum_{l=0}^{\infty} \mathrm{D}_q^{k+l+n} f(0) \frac{\mathrm{B}_{\mathrm{JHC},l,q}^{(-n+1)}(y)}{\{l\}_q!}. \tag{4.310}$$

Finally use (4.21), (4.34), (4.292) to rewrite the right-hand side. □

If we put $n = q = 1$ in (4.307), we get an Euler-Maclaurin expansion known from [385, p. 140]. If we also put $y = 0$, we get an expansion of a polynomial in terms of Bernoulli polynomials known from [320, p. 248].

Corollary 4.3.83 *A q-analogue of Nørlund* [402, p. 164], *Szegő* [493]. *Let $\varphi(x)$ be a polynomial of degree ν. A solution $f(x)$ of the q-difference equation*

$$\underset{\omega_1,\dots,\omega_n \mathrm{JHC},q}{\triangle^n} f(x) = \mathrm{D}_q^n \varphi(x) \tag{4.311}$$

is given by

$$f(x \oplus_q y) = \sum_{k=0}^{\nu} \frac{\mathrm{B}_{\mathrm{JHC},k,q}^{(n)}(x|\omega_1,\dots,\omega_n)}{\{k\}_q!} \mathrm{D}_q^k \varphi(y). \tag{4.312}$$

Theorem 4.3.84 *A q-analogue of Nørlund* [402, p. 156], [403, p. 127, (31)]. *Compare with* [402, p. 147]. *A special case is found in* [320, p. 307].

$$\sum_{k=0}^{\infty} \frac{\mathrm{F}_{\mathrm{JHC},k,q}^{(n)}(x)}{\{k\}_q!} \nabla_{\mathrm{JHC},q}^{n} \mathrm{D}_q^k f(y) = f(x \oplus_q y). \tag{4.313}$$

Proof As in [402, p. 155], replace $f(x)$ by $f(x \oplus_q y)$ in (4.231):

$$\frac{1}{2}\big(f\big(\mathrm{F}_{\mathrm{JHC},q}^{(n)}(x) \oplus_q y \boxplus_q 1\big) + f\big(\mathrm{F}_{\mathrm{JHC},q}^{(n)}(x) \oplus_q y\big)\big) = f\big(\mathrm{F}_{\mathrm{JHC},q}^{(n-1)}(x) \oplus_q y\big). \tag{4.314}$$

Use the umbral formula (4.21) to get

$$\sum_{k=0}^{\infty} \frac{\mathrm{F}_{\mathrm{JHC},k,q}^{(n)}(x)}{\{k\}_q!} \nabla_{\mathrm{JHC},q} \mathrm{D}_q^k f(y) = \sum_{k=0}^{\infty} \frac{\mathrm{F}_{\mathrm{JHC},k,q}^{(n-1)}(x)}{\{k\}_q!} \mathrm{D}_q^k f(y). \tag{4.315}$$

Apply the operator $\nabla_{\mathrm{JHC},q}^{n-1}$ with respect to y to both sides and use (4.302):

$$\sum_{k=0}^{\infty} \frac{\mathrm{F}_{\mathrm{JHC},k,q}^{(n)}(x)}{\{k\}_q!} \nabla_{\mathrm{JHC},q}^{n} \mathrm{D}_q^k f(y) = \sum_{k=0}^{\infty} \frac{\mathrm{F}_{\mathrm{JHC},k,q}^{(n-1)}(x)}{\{k\}_q!} \sum_{l=0}^{\infty} \mathrm{D}_q^{k+l} f(0) \frac{\mathrm{F}_{\mathrm{JHC},l,q}^{(-n+1)}(y)}{\{l\}_q!}. \tag{4.316}$$

Finally use (4.21), (4.34), (4.293) to rewrite the right-hand side. □

If we put $n = q = 1$ in (4.313), we get the Euler-Boole theorem known from [233, p. 128], [385, p. 149].

Corollary 4.3.85 *A q-analogue of Szegő* [493]. *Let $\varphi(x)$ be a polynomial of degree ν. A solution $f(x)$ of the q-difference equation*

$$\underset{\omega_1,\ldots,\omega_n \mathrm{JHC},q}{\nabla^n} f(x) = \varphi(x) \tag{4.317}$$

is given by

$$f(x \oplus_q y) = \sum_{k=0}^{\nu} \frac{\mathrm{F}_{\mathrm{JHC},k,q}^{(n)}(x|\omega_1,\ldots,\omega_n)}{\{k\}_q!} \mathrm{D}_q^k \varphi(y). \tag{4.318}$$

There are a few formulas similar to the Leibniz theorem. We can express the NWA difference operator in terms of the mean value operator and vice versa.

Theorem 4.3.86

$$\Delta_{\mathrm{NWA},q}^{n}(fg) = 2^n \sum_{i=0}^{n} (-1)^{n-i} \binom{n}{i} \nabla_{\mathrm{NWA},q}^{i} f \big(\nabla_{\mathrm{NWA},q}^{n-i} \mathrm{E}(\oplus_q)^i\big) g. \tag{4.319}$$

Proof Analoguous to Jordan [320, p. 98, (13)]. □

Theorem 4.3.87

$$\nabla^n_{\mathrm{NWA},q}(fg) = \left(-\frac{1}{2}\right)^n \sum_{i=0}^{n} (-2)^i \binom{n}{i} \nabla^i_{\mathrm{NWA},q} f\left(\triangle^{n-i}_{\mathrm{NWA},q} \mathrm{E}(\oplus_q)^i\right) g. \tag{4.320}$$

Proof Analogous to [320, p. 99, (2)]. □

Theorem 4.3.88

$$\nabla^n_{\mathrm{NWA},q}(fg) = \sum_{i=0}^{n} \left(\frac{1}{2}\right)^i \binom{n}{i} \triangle^i_{\mathrm{NWA},q} f\left(\nabla^{n-i}_{\mathrm{NWA},q} \mathrm{E}(\oplus_q)^i\right) g. \tag{4.321}$$

Proof Analogous to [320, p. 99, (3)]. □

Theorem 4.3.89 *A q-analogue of Lagrange* (1772), [320, p. 101], [138, p. 19], [385, p. 37]. *The inverse of the NWA difference operator is given by*

$$\triangle^{-1}_{\mathrm{NWA},q} f(\overline{x}_q)\Big|_0^n = \sum_{k=0}^{n-1} f(\overline{k}_q) \equiv \sum_0^n f(\overline{x}_q)\delta_q(x). \tag{4.322}$$

Proof Use the same idea as Euler, reproduced by F. Schweins [458, p. 9]. □

Theorem 4.3.90 *The inverse of the* ∇ *operator is given by*

$$\nabla^{-1}_{\mathrm{NWA},q}\left(\frac{f(\overline{0}_q) + (-1)^{n-1} f(\overline{n}_q)}{2}\right) = \sum_{k=0}^{n-1} (-1)^k f(\overline{k}_q). \tag{4.323}$$

Theorem 4.3.91 *The analogue of integration by parts is a q-analogue of* [320, p. 105], [138, p. 21], [385, p. 41], [411, p. 19]:

$$\sum_{k=0}^{n-1} f(\overline{k}_q)\triangle_{\mathrm{NWA},q} g(\overline{k}_q) = \left[f(\overline{x}_q)g(\overline{x}_q)\right]_0^n - \sum_{k=0}^{n-1} \mathrm{E}(\oplus_q) g(\overline{k}_q)\triangle_{\mathrm{NWA},q} f(\overline{k}_q). \tag{4.324}$$

4.3.10 Euler symbolic formula

The next, pretty umbral formula, which even for $q = 1$ is rare in the literature, shows how to combine the Bernoulli numbers with the calculus of finite differences. In an upcoming article [182] a similar connection to Cauchy-Vandermonde determinants will be shown.

Theorem 4.3.92 *A q-analogue of Euler's symbolic formula from Glaisher* [229, p. 303], *Lucas* [363, p. 242, (1)]:

$$\sum_{k=0}^{n-1} f(\overline{k}_q) \doteq \int_0^{\overline{n}_q} f(x \oplus_q \mathrm{B}_{\mathrm{NWA},q})\, d_q(x) \equiv \int_0^{\overline{n}_q} f\big(\mathrm{B}_{\mathrm{NWA},q}(x)\big)\, d_q(x)$$
$$\equiv \int_{\mathrm{B}_{\mathrm{NWA},q}}^{\mathrm{B}_{\mathrm{NWA},q} \oplus_q \overline{n}_q} f(x)\, d_q(x). \tag{4.325}$$

Proof Apply $\triangle_{\mathrm{NWA},q}$ to both sides to get

$$f(\overline{x}_q)\big|_0^n \doteq \triangle_{\mathrm{NWA},q} \int_{\overline{0}_q}^{\overline{n}_q} f(x \oplus_q \mathrm{B}_{\mathrm{NWA},q})\, d_q(x). \tag{4.326}$$

Then apply (4.122). □

Corollary 4.3.93 *A q-analogue of Rota, Taylor* [439, p. 701]:

$$\mathrm{J}_{\mathrm{B},\mathrm{N},q} f(x \oplus_q \mathrm{B}_{\mathrm{NWA},q}) \doteq f(x). \tag{4.327}$$

We immediately get a proof of the following formula, which originally ($q = 1$) was found in 1732 by Euler and in 1742 by Maclaurin.

Theorem 4.3.94 *The q-Euler-Maclaurin summation theorem for formal power series. A q-analogue of* [138, p. 54], [229, p. 303], [411, p. 25], [439, p. 706], [320, p. 253]:

$$\sum_{k=0}^{n-1} f(\overline{k}_q) = \int_{\overline{0}_q}^{\overline{n}_q} f(x)\, d_q(x)$$
$$- \frac{1}{\{2\}_q}\big(f(\overline{n}_q) - f(0)\big) + \sum_{k=2}^{\infty} \frac{\mathrm{B}_{\mathrm{NWA},k,q}}{\{k\}_q!}\big(\mathrm{D}_q^{k-1} f(\overline{n}_q) - \mathrm{D}_q^{k-1} f(0)\big). \tag{4.328}$$

Example 12 Put $f(x) = x^m$ in (4.328) to get formula (4.146).

Corollary 4.3.95 *A dual to the q-Euler-Maclaurin summation theorem* (4.328).

The q-integral on the RHS shall be interpreted in the following way: First q-integrate f in the form $f(x)$, then insert the values in the umbral sense according to (4.21):

$$\sum_{k=0}^{n-1} f(\overline{k}_q) \doteq \int_0^{\overline{n}_q} f(\mathrm{B}_{\mathrm{NWA},q})\, d_q(x) + \sum_{k=0}^{\infty} \frac{(\overline{x}_q)^{k+1}}{\{k+1\}_q!} \mathrm{D}_q^k f(\mathrm{B}_{\mathrm{NWA},k,q})\big|_0^n. \tag{4.329}$$

We will now derive analogous results for q-Euler numbers. We start with

Theorem 4.3.96 *A generalization of* (4.214). *Compare with Goldstine* [233, p. 136] ($q = 1$):

$$\sum_{k=0}^{n-1}(-1)^k f(\overline{k}_q) \doteq \frac{(-1)^{x-1}}{2} f\big(\mathrm{F}_{\mathrm{NWA},q}(x)\big)\Big|_{\overline{0}_q}^{\overline{n}_q}. \tag{4.330}$$

Proof Apply $\nabla_{\mathrm{NWA},q}$ to both sides to get by (4.323)

$$\frac{f(\overline{0}_q)+(-1)^{n-1}f(\overline{n}_q)}{2} \doteq \nabla_{\mathrm{NWA},q}\frac{(-1)^{x-1}}{2} f\big(\mathrm{F}_{\mathrm{NWA},q}(x)\big)\Big|_{\overline{0}_q}^{\overline{n}_q}. \tag{4.331}$$

Finally use (4.209) with $n = 1$. □

Theorem 4.3.97 *Almost a q-analogue of Boole's first summation formula for alternate functions* [320, p. 316, (2)]. *A q-analogue of the Euler formula* [229, p. 310], [363, p. 252]:

$$\begin{aligned}\sum_{k=0}^{n-1}(-1)^k f(\overline{k}_q) &= \sum_{k=0}^{\infty}\frac{\mathrm{F}_{\mathrm{NWA},k,q}}{\{k\}_q!}\mathrm{D}_q^k\left(\frac{f(\overline{0}_q)}{2}+(-1)^{n-1}\frac{f(\overline{n}_q)}{2}\right)\\ &\doteq \frac{1}{2}\big[f(\mathrm{F}_{\mathrm{NWA},q})+(-1)^{n-1}f(\mathrm{F}_{\mathrm{NWA},q}\oplus_q \overline{n}_q)\big].\end{aligned} \tag{4.332}$$

Proof Use (4.275). □

There are some further formulas similar to the Leibniz theorem. We can express the JHC difference operator in terms of the mean value operator and vice versa.

Theorem 4.3.98

$$\triangle_{\mathrm{JHC},q}^n(fg) = 2^n\sum_{i=0}^{n}(-1)^{n-i}\binom{n}{i}\nabla_{\mathrm{JHC},q}^i f\big(\nabla_{\mathrm{JHC},q}^{n-i}\mathrm{E}(\boxplus_q)^i\big)g. \tag{4.333}$$

Proof Analoguous to [320, p. 98, (13)]. □

Theorem 4.3.99

$$\nabla_{\mathrm{JHC},q}^n(fg) = \left(-\frac{1}{2}\right)^n\sum_{i=0}^{n}(-2)^i\binom{n}{i}\nabla_{\mathrm{JHC},q}^i f\big(\triangle_{\mathrm{JHC},q}^{n-i}\mathrm{E}(\boxplus_q)^i\big)g. \tag{4.334}$$

Proof Analoguous to [320, p. 99, (2)]. □

Theorem 4.3.100

$$\nabla_{\mathrm{JHC},q}^{n}(fg)=\sum_{i=0}^{n}\left(\frac{1}{2}\right)^{i}\binom{n}{i}\triangle_{\mathrm{JHC},q}^{i}f\left(\nabla_{\mathrm{JHC},q}^{n-i}\mathrm{E}(\boxplus_{q})^{i}\right)g. \tag{4.335}$$

Proof Analoguous to [320, p. 99, (3)]. □

Theorem 4.3.101 *A q-analogue of* [320, p. 101], [138, p. 19], [385, p. 37]. *The inverse of the JHC difference operator is given by*:

$$\triangle_{\mathrm{JHC},q}^{-1}f(\tilde{x}_{q})\big|_{0}^{n}=\sum_{k=0}^{n-1}f(\tilde{k}_{q})\equiv\sum_{0}^{n}f(\tilde{x}_{q})\delta_{q}(x). \tag{4.336}$$

Proof Use the same idea as Euler, reproduced by F. Schweins in [458, p. 9]. □

Theorem 4.3.102 *The inverse of the ∇ operator is given by*:

$$\nabla_{\mathrm{JHC},q}^{-1}\left(\frac{f(\tilde{0}_{q})+(-1)^{n-1}f(\tilde{n}_{q})}{2}\right)=\sum_{k=0}^{n-1}(-1)^{k}f(\tilde{k}_{q}). \tag{4.337}$$

Theorem 4.3.103 *The analogue of integration by parts is a q-analogue of* [320, p. 105], [138, p. 21], [385, p. 41], [411, p. 19]:

$$\sum_{k=0}^{n-1}f(\tilde{k}_{q})\triangle_{\mathrm{JHC},q}g(\tilde{k}_{q})=\left[f(\tilde{x}_{q})g(\tilde{x}_{q})\right]_{0}^{n}-\sum_{k=0}^{n-1}\mathrm{E}(\boxplus_{q})g(\tilde{k}_{q})\triangle_{\mathrm{JHC},q}f(\tilde{k}_{q}). \tag{4.338}$$

Theorem 4.3.104 *A generalization of* (4.238). *Compare with Goldstine* [233, p. 136] ($q=1$):

$$\sum_{k=0}^{n-1}(-1)^{k}f(\tilde{k}_{q})=\frac{(-1)^{x-1}}{2}f\left(\mathrm{F}_{\mathrm{JHC},q}(x)\right)\Big|_{\tilde{0}_{q}}^{\tilde{n}_{q}}. \tag{4.339}$$

Proof Apply $\nabla_{\mathrm{JHC},q}$ to both sides to get, by (4.337),

$$\frac{f(\tilde{0}_{q})+(-1)^{n-1}f(\tilde{n}_{q})}{2}=\nabla_{\mathrm{JHC},q}\frac{(-1)^{x-1}}{2}f\left(\mathrm{F}_{\mathrm{JHC},q}(x)\right)\Big|_{\tilde{0}_{q}}^{\tilde{n}_{q}}. \tag{4.340}$$

Finally, use (4.231) with $n=1$. □

We will continue this part with a theorem involving both q-Bernoulli and q-Euler polynomials.

Theorem 4.3.105 *A q-analogue of Srivastava and Pintér* [485, p. 379]:

$$\begin{aligned}&\mathrm{B}^{(n)}_{\mathrm{NWA},\nu,q}(x \oplus_q y)\\&=\sum_{k=0}^{\nu}\binom{\nu}{k}_q\left(\mathrm{B}^{(n)}_{\mathrm{NWA},k,q}(y)+\frac{\{k\}_q}{2}\mathrm{B}^{(n-1)}_{\mathrm{NWA},k-1,q}(y)\right)\mathrm{F}_{\mathrm{NWA},\nu-k,q}(x).\end{aligned} \tag{4.341}$$

Proof By (4.109) and (4.209), we obtain

$$2\eta^{(n-1)}_{\nu,q}(x)=\eta^{(n)}_{\nu,q}(x)+\sum_{k=0}^{\nu}\binom{\nu}{k}_q\eta^{(n)}_{k,q}(x) \tag{4.342}$$

and

$$x^{\nu}=\frac{1}{2}\left[\mathrm{F}_{\mathrm{NWA},\nu,q}(x)+\sum_{k=0}^{\nu}\binom{\nu}{k}_q\mathrm{F}_{\mathrm{NWA},k,q}(x)\right]. \tag{4.343}$$

Substitution of (4.343) into (4.109) yields

$$\begin{aligned}\mathrm{B}^{(n)}_{\mathrm{NWA},\nu,q}(x \oplus_q y)&=\frac{1}{2}\sum_{k=0}^{\nu}\binom{\nu}{k}_q\mathrm{B}^{(n)}_{\mathrm{NWA},k,q}(y)\Bigg[\mathrm{F}_{\mathrm{NWA},\nu-k,q}(x)\\&\quad+\sum_{j=0}^{\nu-k}\binom{\nu-k}{j}_q\mathrm{F}_{\mathrm{NWA},j,q}(x)\Bigg]\\&=\frac{1}{2}\sum_{k=0}^{\nu}\binom{\nu}{k}_q\mathrm{B}^{(n)}_{\mathrm{NWA},k,q}(y)\mathrm{F}_{\mathrm{NWA},\nu-k,q}(x)\\&\quad+\frac{1}{2}\sum_{j=0}^{\nu}\binom{\nu}{j}_q\mathrm{F}_{\mathrm{NWA},j,q}(x)\sum_{k=0}^{\nu-j}\binom{\nu-j}{k}_q\mathrm{B}^{(n)}_{\mathrm{NWA},k,q}(y)\\&\overset{\text{by (4.109)}}{=}\frac{1}{2}\sum_{k=0}^{\nu}\binom{\nu}{k}_q\left[\mathrm{B}^{(n)}_{\mathrm{NWA},k,q}(y)+\mathrm{B}^{(n)}_{\mathrm{NWA},k,q}(y\oplus_q 1)\right]\\&\quad\times\mathrm{F}_{\mathrm{NWA},\nu-k,q}(x)\overset{\text{by (4.120)}}{=}\text{RHS}.\end{aligned} \tag{4.344}$$

□

4.3.11 Complementary argument theorems

The proofs of the following formulas are carried out using the generating function. Note that we have to change to JHC on the left-hand sides.

Theorem 4.3.106 *A q-analogue of the Raabe-Bernoulli complementary argument theorem* [423, p. 354], [385, p. 128, (1)]:

$$\mathrm{B}_{\mathrm{JHC},\nu,q}(x) = (-1)^{\nu}\mathrm{B}_{\mathrm{NWA},\nu,q}(1 \ominus_q x). \tag{4.345}$$

Theorem 4.3.107 *A q-analogue of the Euler complementary argument theorem by Milne-Thomson* [385, p. 145, (1)]:

$$\mathrm{F}_{\mathrm{JHC},\nu,q}(x) = (-1)^{\nu}\mathrm{F}_{\mathrm{NWA},\nu,q}(1 \ominus_q x). \tag{4.346}$$

4.4 q-Lucas and q-G polynomials

The computations based on q-Bernoulli polynomials and q-Euler polynomials can be continued in the way that is shown in this section. As there are no complementary argument theorems here, we do not need the JHC polynomials. The formulas in this section were predicted by Ward [531]. The following definition is reminiscent of [403, p. 6, (12)]:

Definition 84 The following operators always refer to the variable x:

$$\triangle_{\mathrm{NWA},2,q} \equiv 2\triangle_{\mathrm{NWA},q}\nabla_{\mathrm{NWA},q} \equiv \mathrm{E}(\oplus_q)^{\overline{2}_q} - \mathrm{I}, \tag{4.347}$$

$$\nabla_{\mathrm{NWA},2,q} \equiv \frac{\mathrm{E}(\oplus_q)^{\overline{2}_q} + \mathrm{I}}{2}. \tag{4.348}$$

This can be generalized to:

Definition 85

$$\underset{\omega\,\mathrm{NWA},2,q}{\triangle} \equiv \frac{\mathrm{E}(\oplus_q)^{\overline{2}_q\omega} - \mathrm{I}}{2\omega}, \tag{4.349}$$

$$\underset{\omega\,\mathrm{NWA},2,q}{\nabla} \equiv \frac{\mathrm{E}(\oplus_q)^{\overline{2}_q\omega} + \mathrm{I}}{2}. \tag{4.350}$$

The following formulas apply just as in the previous case:

Theorem 4.4.1

$$\nabla^n_{\mathrm{NWA},q}\triangle^n_{\mathrm{NWA},q} f(x) = 2^{-n}\sum_{k=0}^{n}(-1)^{n-k}\binom{n}{k} f(x \oplus_q \overline{2k}_q), \tag{4.351}$$

$$f(\overline{2n}_q) = \sum_{k=0}^{n}\binom{n}{k}\triangle^k_{\mathrm{NWA},2,q} f(0), \tag{4.352}$$

$$\nabla^n_{\mathrm{NWA},2,q} f(x) = 2^{-n} \sum_{k=0}^{n} \binom{n}{k} f(x \oplus_q \overline{2k_q}). \tag{4.353}$$

In his book Lucas seldom used the generating function technique; instead, he preferred the umbral calculus.

Definition 86 The generating function for the q-Lucas polynomial $\mathrm{L}^{(n)}_{\mathrm{NWA},\nu,q}(x)$ is

$$\frac{(2t)^n}{(\mathrm{E}_q(t\overline{2}_q) - 1)^n} \mathrm{E}_q(xt) = \sum_{\nu=0}^{\infty} \frac{t^\nu \mathrm{L}^{(n)}_{\mathrm{NWA},\nu,q}(x)}{\{\nu\}_q!}, \quad |t| < \pi. \tag{4.354}$$

This can be generalized to:

Definition 87 The generating function for $\mathrm{L}^{(n)}_{\mathrm{NWA},\nu,q}(x|\omega_1, \ldots, \omega_n)$ is

$$\frac{(2t)^n \omega_1 \cdots \omega_n}{\prod_{k=1}^{n} (\mathrm{E}_q(\omega_k t\overline{2}_q) - 1)} \mathrm{E}_q(xt) = \sum_{\nu=0}^{\infty} \frac{t^\nu \mathrm{L}^{(n)}_{\mathrm{NWA},\nu,q}(x|\omega_1, \ldots, \omega_n)}{\{\nu\}_q!},$$
$$|t| < \min\bigl(|\pi/\omega_1|, \ldots, |\pi/\omega_n|\bigr). \tag{4.355}$$

Corollary 4.4.2

$$\mathrm{E}_q(t\mathrm{L}_{\mathrm{NWA},q}) \doteq \frac{2t}{\mathrm{E}_q(t\overline{2}_q) - 1}. \tag{4.356}$$

Obviously, $\mathrm{L}^{(n)}_{\mathrm{NWA},\nu,q}(x|\omega_1, \ldots, \omega_n)$ is symmetric in $\omega_1, \ldots, \omega_n$ and, in particular,

$$\mathrm{L}^{(1)}_{\mathrm{NWA},\nu,q}(x|\omega) = \omega^\nu \mathrm{L}_{\mathrm{NWA},\nu,q}\left(\frac{x}{\omega}\right), \tag{4.357}$$

$$\underset{\omega_1,\ldots,\omega_n \mathrm{NWA},2,q}{\triangle^n} \mathrm{L}^{(n)}_{\mathrm{NWA},\nu,q}(x|\omega_1, \ldots, \omega_n) = \frac{\{\nu\}_q!}{\{\nu - n\}_q!} x^{\nu-n}. \tag{4.358}$$

Theorem 4.4.3 *The successive differences of q-Lucas polynomials can be expressed as q-Lucas polynomials*:

$$\underset{\omega_1,\ldots,\omega_p \mathrm{NWA},2,q}{\triangle^p} \mathrm{L}^{(n)}_{\mathrm{NWA},\nu,q}(x|\omega_1, \ldots, \omega_n) = \frac{\{\nu\}_q!}{\{\nu - p\}_q!} \mathrm{L}^{(n-p)}_{\mathrm{NWA},\nu-p,q}(x|\omega_{p+1}, \ldots, \omega_n), \tag{4.359}$$

$$\mathrm{L}^{(n)}_{\mathrm{NWA},\nu,q}(x|\omega_1, \ldots, \omega_n) \doteq (\mathrm{L}^{(n)}_{\mathrm{NWA},\nu,\omega_1,\ldots,\omega_n,q} \oplus_q x)^\nu. \tag{4.360}$$

The following invertible operator will be useful in this connection.

Definition 88 The operator $\mathrm{S}_{\mathrm{L,N},q}^{n} \in \mathbb{C}(\mathrm{D}_q)$ is given by

$$\mathrm{S}_{\mathrm{L,N},q}^{n} \equiv \frac{(\mathrm{E}_q(\overline{2}_q\mathrm{D}_q) - \mathrm{I})^n}{(2\mathrm{D}_q)^n}. \tag{4.361}$$

This implies

Theorem 4.4.4

$$\nabla_{\mathrm{NWA},q}^{n} \triangle_{\mathrm{NWA},q}^{n} = \mathrm{D}_q^n \mathrm{S}_{\mathrm{L,N},q}^{n}. \tag{4.362}$$

Theorem 4.4.5 *Compare with* [138, p. 43, 3.3]:

$$\sum_{k=0}^{n} (-1)^{n-k} \binom{n}{k} \mathrm{L}_{\mathrm{NWA},\nu,q}^{(n)}(x \oplus_q \overline{2k}_q) = \{\nu - n + 1\}_{n,q} x^{\nu-n} 2^n. \tag{4.363}$$

Proof

$$\begin{aligned}\triangle_{\mathrm{NWA},2,q}^{n} \mathrm{L}_{\mathrm{NWA},\nu,q}^{(n)}(x) &= \mathrm{D}_q^n \mathrm{S}_{\mathrm{L,N},q}^{n} \mathrm{L}_{\mathrm{NWA},\nu,q}^{(n)}(x) 2^n \\ &= \mathrm{D}_q^n x^{\nu} 2^n = \{\nu - n + 1\}_{n,q} x^{\nu-n} 2^n.\end{aligned} \tag{4.364}$$

□

Corollary 4.4.6

$$\sum_{k=0}^{n} (-1)^{n-k} \binom{n}{k} \mathrm{L}_{\mathrm{NWA},\nu,q}^{(n)}(\overline{2k}_q) = \{n\}_q! \delta_{0,\nu-n} 2^n. \tag{4.365}$$

Proof Put $x = 0$ in (4.363). □

By using the generating function we obtain

$$\frac{1}{2} \triangle_{\mathrm{NWA},2,q} \mathrm{L}_{\mathrm{NWA},\nu,q}^{(n)}(x) = \{\nu\}_q \mathrm{L}_{\mathrm{NWA},\nu-1,q}^{(n-1)}(x) = \mathrm{D}_q \mathrm{L}_{\mathrm{NWA},\nu,q}^{(n-1)}(x). \tag{4.366}$$

The following symbolic relations hold:

Theorem 4.4.7

$$\frac{1}{2}\big[\big(\mathrm{L}_{\mathrm{NWA},q}^{(n)} \oplus_q x \oplus_q \overline{2}_q\big)^{\nu} - \big(\mathrm{L}_{\mathrm{NWA},q}^{(n)} \oplus_q x\big)^{\nu}\big] \doteq \{\nu\}_q \big(\mathrm{L}_{\mathrm{NWA},q}^{(n-1)} \oplus_q x\big)^{\nu-1}, \tag{4.367}$$

$$\frac{1}{2}\big[\big(\mathrm{L}_{\mathrm{NWA},q}^{(n)} \oplus_q \overline{2}_q\big)^{\nu} - \mathrm{L}_{\mathrm{NWA},\nu,q}^{(n)}\big] \doteq \{\nu\}_q \mathrm{L}_{\mathrm{NWA},\nu-1,q}^{(n-1)}. \tag{4.368}$$

Theorem 4.4.8

$$\triangle_{\mathrm{NWA},q}\nabla_{\mathrm{NWA},q} f\big(\mathrm{L}_{\mathrm{NWA},\nu,q}^{(n)}(x)\big)$$
$$\equiv \frac{1}{2}\big[f\big(\mathrm{L}_{\mathrm{NWA},\nu,q}^{(n)}(x)\oplus_q \overline{2}_q\big) - f\big(\mathrm{L}_{\mathrm{NWA},\nu,q}^{(n)}(x)\big)\big] \doteq \mathrm{D}_q f\big(\mathrm{L}_{\mathrm{NWA},\nu,q}^{(n-1)}(x)\big). \tag{4.369}$$

Theorem 4.4.9 *The following recurrence holds*:

$$\frac{1}{2}\big[(\mathrm{L}_{\mathrm{NWA},q}\oplus_q \overline{2}_q)^k - \mathrm{L}_{\mathrm{NWA},k,q}\big] \doteq \delta_{1,k}. \tag{4.370}$$

Theorem 4.4.10

$$\frac{1}{2}\sum_{k=1}^{\nu}\binom{\nu}{k}_q \mathrm{L}_{\mathrm{NWA},\nu-k,q}^{(n)}(x)(\overline{2}_q)^k = \{\nu\}_q \mathrm{L}_{\mathrm{NWA},\nu-1,q}^{(n-1)}(x). \tag{4.371}$$

Proof Use Eq. (4.367). □

Theorem 4.4.11 *The q-Lucas polynomials of degree ν and order n can be expressed as*

$$\mathrm{L}_{\mathrm{NWA},\nu,q}^{(n)}(t) = \mathrm{S}_{\mathrm{L,N},q}^{-n} t^{\nu}. \tag{4.372}$$

Proof

$$\mathrm{LHS} = \sum_{k=0}^{\nu}\binom{\nu}{k}_q \mathrm{L}_{\mathrm{NWA},k,q}^{(n)} t^{\nu-k} = \sum_{k=0}^{\infty}\frac{\mathrm{L}_{\mathrm{NWA},k,q}^{(n)}}{\{k\}_q!}\mathrm{D}_q^k t^{\nu} \overset{\text{by (4.354)}}{=} \mathrm{RHS}. \tag{4.373}$$

□

Theorem 4.4.12 *Compare with* [138, 3.15, p. 51], *where the corresponding formula for Euler polynomials was given*:

$$\mathrm{L}_{\mathrm{NWA},\nu,q}(x) \equiv \frac{\{\nu\}_q}{\mathrm{E}_q(\overline{2}_q\mathrm{D}_q) - \mathrm{I}} x^{\nu-1} = \frac{\{\nu\}_q}{\mathrm{E}(\oplus_q)^{\overline{2}_q} - \mathrm{I}} x^{\nu-1} \doteq (x\oplus_q \mathrm{L}_{\mathrm{NWA},q})^{\nu}. \tag{4.374}$$

4.4.1 q-Lucas numbers

The first q-Lucas numbers have the following values:

$$\mathrm{L}_{\mathrm{NWA},0,q} = 1; \qquad \mathrm{L}_{\mathrm{NWA},1,q} = (-3-q)(2+2q)^{-1}, \tag{4.375}$$

$$\mathrm{L}_{\mathrm{NWA},2,q} = \left(1+3q+8q^2+3q^3+q^4\right)\left(4+8q+8q^2+4q^3\right)^{-1}, \tag{4.376}$$

$$\begin{aligned}\mathrm{L}_{\mathrm{NWA},3,q} &= -\left(-1-2q-2q^2-9q^3+9q^4+2q^5+2q^6+q^7\right)\\ &\quad\times\left[8(q+1)^2(q^2+1)\right]^{-1},\end{aligned} \tag{4.377}$$

$$\mathrm{L}_{\mathrm{NWA},4,q} = f(q)\left(16(1+q)^2\{3\}_q\{5\}_q\right)^{-1}, \tag{4.378}$$

where

$$\begin{aligned}f(q) &= 1+3q+q^2-11q^3-18q^4-63q^5-104q^6-130q^7\\ &\quad -104q^8-63q^9-18q^{10}-11q^{11}+q^{12}+3q^{13}+q^{14}.\end{aligned} \tag{4.379}$$

Theorem 4.4.13 *We have the following operational representation*:

$$\mathrm{L}^{(n)}_{\mathrm{NWA},\nu,q}(\omega_1,\ldots,\omega_n) \doteq \left(\bigoplus_{q,l=1}^{n} \omega_l \mathrm{L}_{\mathrm{NWA},q}\right)^{\nu}. \tag{4.380}$$

We will now give a few other equations for q-Lucas polynomials. We start with

Definition 89 A q-analogue of Lucas [363, p. 253], Nielsen [398, p. 401, (7)], Vandiver [517, p. 575]:

$$t_{\mathrm{NWA},m,q}(n) \equiv \sum_{k=0}^{n-1} (\overline{2k+1}_q)^m. \tag{4.381}$$

Theorem 4.4.14 *A q-analogue of Lucas* [363, p. 261]:

$$t_{\mathrm{NWA},m,q}(n) = \frac{\mathrm{L}_{\mathrm{NWA},m+1,q}(\overline{2n+1}_q) - \mathrm{L}_{\mathrm{NWA},m+1,q}(\overline{1}_q)}{\{m+1\}_q}. \tag{4.382}$$

Proof

$$\begin{aligned}\mathrm{LHS} &= \sum_{k=0}^{n-1} \mathrm{L}^{(0)}_{\mathrm{NWA},m,q}(\overline{2k+1}_q)\\ &= \sum_{k=0}^{n-1} \frac{1}{\{m+1\}_q} \triangle_{\mathrm{NWA},2,q} \mathrm{L}_{\mathrm{NWA},m+1,q}(\overline{2k+1}_q) = \mathrm{RHS}.\end{aligned} \tag{4.383}$$

□

Theorem 4.4.15

$$2x^n = \int_x^{x\oplus_q \overline{2}_q} \mathrm{L}_{\mathrm{NWA},n,q}(t)\, d_q(t) = \frac{\mathrm{L}_{\mathrm{NWA},n+1,q}(x\oplus_q \overline{2}_q) - \mathrm{L}_{\mathrm{NWA},n+1,q}(x)}{\{n+1\}_q}. \tag{4.384}$$

Proof q-integrate (4.107) for $n = 1$ and use (4.366). □

This can be rewritten as

$$2x^n = \frac{1}{\{n+1\}_q} \sum_{k=0}^{n} \binom{n+1}{k}_q \mathrm{L}_{\mathrm{NWA},k,q}(x)(\overline{2}_q)^{n+1-k}. \tag{4.385}$$

Definition 90 The q-Lucas operator is given by

$$\mathrm{J}_{\mathrm{L,N},q} f(x) \equiv \int_x^{x \oplus_q \overline{2}_q} f(t)\, d_q(t). \tag{4.386}$$

We have $\mathrm{J}_{\mathrm{L,N},q} \in \mathbb{C}[\![\mathrm{D}_q]\!]$.

Theorem 4.4.16

$$\mathrm{J}_{\mathrm{L,N},q} f(x) = \frac{\Delta_{\mathrm{NWA},2,q}}{2\mathrm{D}_q} f(x). \tag{4.387}$$

Proof By (4.372) and (4.384) we have

$$\int_x^{x \oplus_q \overline{2}_q} \mathrm{L}_{\mathrm{NWA},n,q}(t)\, d_q(t) = \frac{\Delta_{\mathrm{NWA},2,q}}{2\mathrm{D}_q} \mathrm{L}_{\mathrm{NWA},n,q}(x). \tag{4.388}$$

Now any formal power series can be expanded as a sum of $\mathrm{L}_{\mathrm{NWA},n,q}(x)$. □

Theorem 4.4.17

$$f(x) = \sum_{k=0}^{\infty} \int_0^{\overline{2}_q} \mathrm{D}_q^k f(t)\, d_q(t) \frac{\mathrm{L}_{\mathrm{NWA},k,q}(x)}{\{k\}_q!}. \tag{4.389}$$

4.4.2 The q-G polynomials

The following polynomials are introduced under the influence of Nørlund [403], who perhaps would have denoted them by C instead of G.

Definition 91 The generating function for $\mathrm{G}_{\mathrm{NWA},\nu,q}^{(n)}(x)$ is

$$\frac{2^n}{(\mathrm{E}_q(t\overline{2}_q)+1)^n} \mathrm{E}_q(xt) = \sum_{\nu=0}^{\infty} \frac{t^\nu \mathrm{G}_{\mathrm{NWA},\nu,q}^{(n)}(x)}{\{\nu\}_q!}, \quad |t| < \frac{\pi}{2}. \tag{4.390}$$

This implies

$$\nabla_{\mathrm{NWA},2,q} \mathrm{G}_{\mathrm{NWA},\nu,q}^{(n)}(x) = \mathrm{G}_{\mathrm{NWA},\nu,q}^{(n-1)}(x), \tag{4.391}$$

$$\nabla_{\mathrm{NWA},2,q}\mathrm{G}_{\mathrm{NWA},\nu,q}(x)=x^{\nu}. \tag{4.392}$$

This can be generalized to:

Definition 92 The generating function for the q-G polynomials of degree ν and order n, $\mathrm{G}^{(n)}_{\mathrm{NWA},\nu,q}(x|\omega_1,\ldots,\omega_n)$, is given by the following formula, reminding of Nørlund [403, p. 143, (78)]:

$$\frac{2^n\mathrm{E}_q(xt)}{\prod_{k=1}^{n}(\mathrm{E}_q(\omega_k\overline{2}_q)+1)}=\sum_{\nu=0}^{\infty}\frac{t^{\nu}}{\{\nu\}_q!}\mathrm{G}^{(n)}_{\mathrm{NWA},\nu,q}(x|\omega_1,\ldots,\omega_n),$$
$$|t|<\min\bigl(|\pi/2\omega_1|,\ldots,|\pi/2\omega_n|\bigr). \tag{4.393}$$

Corollary 4.4.18

$$\mathrm{E}_q(t\mathrm{G}_{\mathrm{NWA},q})\doteq\frac{2}{\mathrm{E}_q(t\overline{2}_q)+1}. \tag{4.394}$$

Obviously, $\mathrm{G}^{(n)}_{\mathrm{NWA},\nu,q}(x|\omega_1,\ldots,\omega_n)$ is symmetric in $\omega_1,\ldots,\omega_n$ and, in particular,

$$\mathrm{G}^{(1)}_{\mathrm{NWA},\nu,q}(x|\omega)=\omega^{\nu}\mathrm{G}_{\mathrm{NWA},\nu,q}\left(\frac{x}{\omega}\right). \tag{4.395}$$

From

$$\mathop{\nabla^n}_{\omega_1,\ldots,\omega_n\mathrm{NWA},2,q}\mathrm{G}^{(n)}_{\mathrm{NWA},\nu,q}(x|\omega_1,\ldots,\omega_n)=x^{\nu} \tag{4.396}$$

we obtain

$$\mathop{\nabla^p}_{\omega_1,\ldots,\omega_p\mathrm{NWA},2,q}\mathrm{G}^{(n)}_{\mathrm{NWA},\nu,q}(x|\omega_1,\ldots,\omega_n)=\mathrm{G}^{(n-p)}_{\mathrm{NWA},\nu,q}(x|\omega_{p+1},\ldots,\omega_n). \tag{4.397}$$

The following symbolic relations hold.

Theorem 4.4.19 *The first relation below is a q-analogue of* [402, p. 138] ($n=1$), [403, p. 125]. *The second below is a q-analogue of* [402, p. 149 (18)] ($n=1$), [403, p. 124 (22)]:

$$\frac{1}{2}\bigl[(\mathrm{G}^{(n)}_{\mathrm{NWA},q}\oplus_q x\oplus_q\overline{2}_q)^{\nu}+(\mathrm{G}^{(n)}_{\mathrm{NWA},q}\oplus_q x)^{\nu}\bigr]\doteq(\mathrm{G}^{(n-1)}_{\mathrm{NWA},q}\oplus_q x)^{\nu}, \tag{4.398}$$

$$\frac{1}{2}\bigl[(\mathrm{G}^{(n)}_{\mathrm{NWA},q}\oplus_q\overline{2}_q)^{\nu}+\mathrm{G}^{(n)}_{\mathrm{NWA},\nu,q}\bigr]\doteq\mathrm{G}^{(n-1)}_{\mathrm{NWA},\nu,q}. \tag{4.399}$$

Theorem 4.4.20 *A q-analogue of* [402, p. 137] ($n=1$), [403, p. 124]:

$$\nabla_{\mathrm{NWA},2,q}f\bigl(\mathrm{G}^{(n)}_{\mathrm{NWA},q}(x)\bigr)$$
$$\equiv\frac{1}{2}\bigl[f\bigl(\mathrm{G}^{(n)}_{\mathrm{NWA},q}(x)\oplus_q\overline{2}_q\bigr)+f\bigl(\mathrm{G}^{(n)}_{\mathrm{NWA},q}(x)\bigr)\bigr]\doteq f\bigl(\mathrm{G}^{(n-1)}_{\mathrm{NWA},q}(x)\bigr). \tag{4.400}$$

Theorem 4.4.21 *The following recurrence, a q-analogue of* [402, p. 136, (20)], [403, p. 27] *holds*:

$$\left[(\mathrm{G}_{\mathrm{NWA},q}\oplus_q\overline{2}_q)^k+\mathrm{G}_{\mathrm{NWA},k,q}\right]\doteq 2\delta_{0,k}. \tag{4.401}$$

Theorem 4.4.22

$$2^{-n}\sum_{k=0}^{n}\binom{n}{k}\mathrm{G}^{(n)}_{\mathrm{NWA},\nu,q}(x\oplus_q\overline{2k}_q)=x^\nu. \tag{4.402}$$

Proof Use Eqs. (4.391) and (4.353). □

Theorem 4.4.23 *Compare with* [138, 3.15, p. 51], *where the corresponding formula for Euler polynomials was given*:

$$\mathrm{G}_{\mathrm{NWA},\nu,q}(x)\equiv\frac{2}{\mathrm{E}_q(\overline{2}_q\mathrm{D}_q)+\mathrm{I}}x^\nu=\frac{2}{\mathrm{E}(\oplus_q)^{\overline{2}_q}+\mathrm{I}}x^\nu\doteq(x\oplus_q\mathrm{G}_{\mathrm{NWA},q})^\nu. \tag{4.403}$$

The following table lists some of the first $\mathrm{G}_{\mathrm{NWA},n,q}$. These are a q-analogue of the integers in [403, p. 27]:

$n=0$	$n=1$	$n=2$	$n=3$	$n=4$
1	-1	$2^{-1}(-1+q)$	$q(1+q)$	$-2^{-2}(q^3-1)(1+q)^3$

We have the following operational representation:

Theorem 4.4.24

$$\mathrm{G}^{(n)}_{\mathrm{NWA},\nu,q}(\omega_1,\ldots,\omega_n)\doteq\left(\bigoplus_{q,l=1}^{n}\omega_l\mathrm{G}_{\mathrm{NWA},q}\right)^\nu. \tag{4.404}$$

We will now give a few other q-analogues of equations for G polynomials, which belong together with power sums, starting with

Definition 93 A q-analogue of J. Herschel [277, p. 91]:

$$\tau_{\mathrm{NWA},m,q}(n)\equiv\sum_{k=0}^{n-1}(-1)^k(\overline{2k+1}_q)^m. \tag{4.405}$$

Theorem 4.4.25 *A q-analogue of Lucas* [363, p. 237], *Nielsen* [398, p. 401]:

$$\tau_{\mathrm{NWA},m,q}(n)=\frac{(-1)^{n-1}\mathrm{G}_{\mathrm{NWA},m+1,q}(\overline{2n+1}_q)+\mathrm{G}_{\mathrm{NWA},m+1,q}(\overline{1}_q)}{2}. \tag{4.406}$$

Proof

$$\text{LHS} = \sum_{k=0}^{n-1}(-1)^k \nabla_{\text{NWA},2,q} \text{G}_{\text{NWA},m,q}(\overline{2k+1_q})$$
$$= \sum_{k=0}^{n-1} \frac{(-1)^k}{2} \text{G}_{\text{NWA},m,q}(\overline{2k+1_q}) + \text{G}_{\text{NWA},m,q}(\overline{2k+3_q}) = \text{RHS}. \tag{4.407}$$

□

Theorem 4.4.26 *The inverse of the NWA difference operator is given by*

$$\triangle^{-1}_{\text{NWA},2,q} f(\overline{2x+1_q})\Big|_{x=0}^{n} = \sum_{k=0}^{n-1} f(\overline{2k+1_q}). \tag{4.408}$$

The inverse of the $\nabla_{\text{NWA},2,q}$ *operator is given by*

$$\nabla^{-1}_{\text{NWA},2,q}\left(\frac{f(\overline{1}_q) + (-1)^{n-1} f(\overline{2n+1_q})}{2}\right) = \sum_{k=0}^{n-1}(-1)^k f(\overline{2k+1_q}). \tag{4.409}$$

Theorem 4.4.27 *Another integration by parts formula*:

$$\sum_{k=0}^{n-1} f(\overline{2k+1_q}) \triangle_{\text{NWA},q} g(\overline{2k+1_q})$$
$$= \left[f(\overline{2x+1_q}) g(\overline{2x+1_q})\right]_0^n$$
$$- \sum_{k=0}^{n-1} \text{E}(\oplus_q)^{\overline{2}_q} g(\overline{2k+1_q}) \triangle_{\text{NWA},2,q} f(\overline{2k+1_q}). \tag{4.410}$$

Theorem 4.4.28 *Compare with Goldstine* [233, p. 136] ($q = 1$):

$$\sum_{k=0}^{n-1}(-1)^k f(\overline{2k+1_q}) = \frac{(-1)^{x-1}}{2} f\left(\text{G}_{\text{NWA},q}(\overline{2x+1_q})\right)\Big|_0^n. \tag{4.411}$$

Proof Apply $\nabla_{\text{NWA},2,q}$ to both sides to get, by (4.409),

$$\frac{f(\overline{0}_q) + (-1)^{n-1} f(\overline{n}_q)}{2} = \nabla_{\text{NWA},q} \frac{(-1)^{x-1}}{2} f\left(\text{G}_{\text{NWA},q}(\overline{2x+1_q})\right)\Big|_0^n. \tag{4.412}$$

Finally use (4.391) with $n = 1$. □

Example 13 Put $f(x) = x^m$ in (4.411) to get (4.406).

4.4.3 Lucas and G polynomials of negative order

By (4.366) and (4.391) we have

$$\triangle_{\mathrm{NWA},2,q}^{n}\mathrm{L}_{\mathrm{NWA},\nu,q}^{(n)}(x)=\frac{\{\nu\}_q!}{\{\nu-n\}_q!}x^{\nu-n},$$

$$\nabla_{\mathrm{NWA},2,q}^{n}\mathrm{G}_{\mathrm{NWA},\nu,q}^{(n)}(x)=x^{\nu}$$

and we are led to

Definition 94 A q-analogue of the Lucas polynomial of negative order $-n$ is given by

$$\mathrm{L}_{\mathrm{NWA},\nu,q}^{(-n)}(x)\equiv\frac{\{\nu\}_q!}{\{\nu+n\}_q!}\triangle_{\mathrm{NWA},2,q}^{n}x^{\nu+n} \tag{4.413}$$

and the q-G polynomial of negative order $-n$ is given by

$$\mathrm{G}_{\mathrm{NWA},\nu,q}^{(-n)}(x)\equiv\nabla_{\mathrm{NWA},2,q}^{n}x^{\nu}, \tag{4.414}$$

where $\nu,n\in\mathbb{N}$. This defines q-Lucas and q-G polynomials of negative order as iterated $\triangle_{\mathrm{NWA},q}$ and $\nabla_{\mathrm{NWA},q}$ operating on positive integer powers of x. Furthermore,

$$\mathrm{L}_{\mathrm{NWA},\nu,q}^{(-n)}\equiv\mathrm{L}_{\mathrm{NWA},\nu,q}^{(-n)}(0), \tag{4.415}$$

$$\mathrm{G}_{\mathrm{NWA},\nu,q}^{(-n)}\equiv\mathrm{G}_{\mathrm{NWA},\nu,q}^{(-n)}(0). \tag{4.416}$$

A calculation shows that formulas (4.366) and (4.391) hold for negative orders too, and we get

$$\mathrm{L}_{\mathrm{NWA},\nu,q}^{(-n-p)}(x\oplus_q y)\doteq\left(\mathrm{L}_{\mathrm{NWA},q}^{(-n)}(x)\oplus_q\mathrm{L}_{\mathrm{NWA},q}^{(-p)}(y)\right)^{\nu}, \tag{4.417}$$

and the same for q-G polynomials.

A special case is the following:

$$\mathrm{L}_{\mathrm{NWA},\nu,q}^{(-n)}(x\oplus_q y)\doteq\left(\mathrm{L}_{\mathrm{NWA},q}^{(-n)}(x)\oplus_q y\right)^{\nu}, \tag{4.418}$$

and the same for q-G polynomials.

Theorem 4.4.29 *A recurrence formula for the q-Lucas numbers and a recurrence formula for the q-G numbers. If $n,p\in\mathbb{Z}$ then*

$$\mathrm{L}_{\mathrm{NWA},\nu,q}^{(n+p)}\doteq\left(\mathrm{L}_{\mathrm{NWA},q}^{(n)}\oplus_q\mathrm{L}_{\mathrm{NWA},q}^{(p)}\right)^{\nu}, \tag{4.419}$$

$$\mathrm{G}_{\mathrm{NWA},\nu,q}^{(n+p)}\doteq\left(\mathrm{G}_{\mathrm{NWA},q}^{(n)}\oplus_q\mathrm{G}_{\mathrm{NWA},q}^{(p)}\right)^{\nu}. \tag{4.420}$$

Theorem 4.4.30

$$(x \oplus_q y)^\nu \doteq \left(\mathrm{L}_{\mathrm{NWA},q}^{(-n)}(x) \oplus_q \mathrm{L}_{\mathrm{NWA},q}^{(n)}(y)\right)^\nu, \tag{4.421}$$

$$(x \oplus_q y)^\nu \doteq \left(\mathrm{G}_{\mathrm{NWA},q}^{(-n)}(x) \oplus_q \mathrm{G}_{\mathrm{NWA},q}^{(n)}(y)\right)^\nu. \tag{4.422}$$

Proof Put $p = -n$ in (4.417). □

In particular, for $y = 0$

$$x^\nu \doteq \left(\mathrm{L}_{\mathrm{NWA},q}^{(-n)} \oplus_q \mathrm{L}_{\mathrm{NWA},q}^{(n)}(x)\right)^\nu, \tag{4.423}$$

$$x^\nu \doteq \left(\mathrm{G}_{\mathrm{NWA},q}^{(-n)} \oplus_q \mathrm{G}_{\mathrm{NWA},q}^{(n)}(x)\right)^\nu. \tag{4.424}$$

These recurrence formulas express q-Lucas and q-G polynomials of order n without involving polynomials of negative order.

These relations can be restated as

$$x^\nu = \sum_{s=0}^{\nu} \frac{\mathrm{L}_{\mathrm{NWA},s,q}^{(-n)}}{\{s\}_q!} \mathrm{D}_q^s \mathrm{L}_{\mathrm{NWA},\nu,q}^{(n)}(x), \tag{4.425}$$

$$x^\nu = \sum_{s=0}^{\nu} \frac{\mathrm{G}_{\mathrm{NWA},s,q}^{(-n)}}{\{s\}_q!} \mathrm{D}_q^s \mathrm{G}_{\mathrm{NWA},\nu,q}^{(n)}(x). \tag{4.426}$$

We conclude that the q-Lucas and q-G polynomials satisfy linear q-difference equations with constant coefficients. The following corollary is useful for the computation of q-Lucas and q-G polynomials of positive order. This is because the polynomials of negative order are of simpler nature and can easily be computed. When the $\mathrm{L}_{\mathrm{NWa},s,q}^{(-n)}$ etc. are known, (4.427) can be used to compute the $\mathrm{L}_{\mathrm{NWA},s,q}^{(n)}$.

Corollary 4.4.31

$$\sum_{s=0}^{\nu} \binom{\nu}{s}_q \mathrm{L}_{\mathrm{NWA},s,q}^{(n)} \mathrm{L}_{\mathrm{NWA},\nu-s,q}^{(-n)} = \delta_{\nu,0}, \tag{4.427}$$

$$\sum_{s=0}^{\nu} \binom{\nu}{s}_q \mathrm{G}_{\mathrm{NWA},s,q}^{(n)} \mathrm{G}_{\mathrm{NWA},\nu-s,q}^{(-n)} = \delta_{\nu,0}. \tag{4.428}$$

Proof Put $x = y = 0$ in (4.421) and (4.422). □

Theorem 4.4.32 *Assume that $f(x)$ is analytic, with q-Taylor expansion*

$$f(x) = \sum_{\nu=0}^{\infty} \mathrm{D}_q^\nu f(0) \frac{x^\nu}{\{\nu\}_q!}, \tag{4.429}$$

Then we can express powers of $\Delta_{\mathrm{NWA},q}$ and $\nabla_{\mathrm{NWA},q}$ operating on $f(x)$ as powers of D_q as follows, where the series converge when the absolute value of x is small enough:

$$\Delta^n_{\mathrm{NWA},2,q} f(x) = \sum_{\nu=0}^{\infty} \mathrm{D}_q^{\nu+n} f(0) \frac{\mathrm{L}^{(-n)}_{\mathrm{NWA},\nu,q}(x)}{\{\nu\}_q!}, \tag{4.430}$$

$$\nabla^n_{\mathrm{NWA},2,q} f(x) = \sum_{\nu=0}^{\infty} \mathrm{D}_q^{\nu} f(0) \frac{\mathrm{G}^{(-n)}_{\mathrm{NWA},\nu,q}(x)}{\{\nu\}_q!}. \tag{4.431}$$

Proof Use (4.366), (4.106) and (4.391), (4.106), respectively. □

Now put $f(x) = \mathrm{E}_q(xt)$ to get the generating functions for $\mathrm{L}^{(-n)}_{\mathrm{NWA},\nu,q}(x)$ and $\mathrm{G}^{(-n)}_{\mathrm{NWA},\nu,q}(x)$:

$$\big(\mathrm{E}_q(\overline{2}_q t) - 1\big)^n \mathrm{E}_q(xt) = \sum_{\nu=0}^{\infty} \frac{t^{\nu+n}}{\{\nu\}_q!} \mathrm{L}^{(-n)}_{\mathrm{NWA},\nu,q}(x), \tag{4.432}$$

$$\frac{(\mathrm{E}_q(\overline{2}_q t) + 1)^n}{2^n} \mathrm{E}_q(xt) = \sum_{\nu=0}^{\infty} \frac{t^{\nu}}{\{\nu\}_q!} \mathrm{G}^{(-n)}_{\mathrm{NWA},\nu,q}(x). \tag{4.433}$$

The reason for the difference in appearance compared to the original for the following equation is that one of the function arguments is a Ward number.

Theorem 4.4.33

$$\mathrm{L}^{(n)}_{\mathrm{NWA},\nu,q}(x \oplus_q \overline{2n}_q) = \sum_{k=0}^{n} \binom{n}{k} \frac{\{\nu\}_q!}{\{\nu-k\}_q!} \mathrm{L}^{(n-k)}_{\mathrm{NWA},\nu-k,q}(x). \tag{4.434}$$

Proof Use (4.82) and (4.366). □

Theorem 4.4.34

$$\frac{\{\nu\}_q!}{\{\nu-n\}_q!} x^{\nu-n} = \sum_{k=0}^{n} (-1)^{n-k} \binom{n}{k} \mathrm{L}^{(n)}_{\mathrm{NWA},\nu,q}(x \oplus_q \overline{2k}_q). \tag{4.435}$$

Proof Use Eqs. (4.85) and (4.366). □

4.4.4 Expansion formulas

Theorem 4.4.35

$$\frac{1}{2}\sum_{k=0}^{\infty}\frac{\mathrm{L}_{\mathrm{NWA},k,q}^{(n)}(x)}{\{k\}_q!}\triangle_{\mathrm{NWA},2,q}^{n}\mathrm{D}_q^k f(y)=\mathrm{D}_q^n f(x\oplus_q y). \tag{4.436}$$

Proof Replace $f(x)$ by $f(x\oplus_q y)$ in (4.369):

$$\frac{1}{2}\big(f\big(\mathrm{L}_{\mathrm{NWA},q}^{(n)}(x)\oplus_q \overline{2}_q\oplus_q y\big)-f\big(\mathrm{L}_{\mathrm{NWA},q}^{(n)}(x)\oplus_q y\big)\big)\doteq \mathrm{D}_q f\big(\mathrm{L}_{\mathrm{NWA},q}^{(n-1)}(x)\oplus_q y\big). \tag{4.437}$$

Use the umbral formula (4.21) to get

$$\frac{1}{2}\sum_{k=0}^{\infty}\frac{\mathrm{L}_{\mathrm{NWA},k,q}^{(n)}(x)}{\{k\}_q!}\triangle_{\mathrm{NWA},2,q}\mathrm{D}_q^k f(y)=\sum_{k=0}^{\infty}\frac{\mathrm{L}_{\mathrm{NWA},k,q}^{(n-1)}(x)}{\{k\}_q!}\mathrm{D}_q^{k+1} f(y). \tag{4.438}$$

Apply the operator $\triangle_{\mathrm{NWA},2,q}^{n-1}$ with respect to y to both sides and use (4.430):

$$\begin{aligned}&\frac{1}{2}\sum_{k=0}^{\infty}\frac{\mathrm{L}_{\mathrm{NWA},k,q}^{(n)}(x)}{\{k\}_q!}\triangle_{\mathrm{NWA},q}^{n}\mathrm{D}_q^k f(y)\\&\quad=\sum_{k=0}^{\infty}\frac{\mathrm{L}_{\mathrm{NWA},k,q}^{(n-1)}(x)}{\{k\}_q!}\sum_{l=0}^{\infty}\mathrm{D}_q^{k+l+n} f(0)\frac{\mathrm{L}_{\mathrm{NWA},l,q}^{(-n+1)}(y)}{\{l\}_q!}.\end{aligned} \tag{4.439}$$

Finally use (4.421) to rewrite the right-hand side. □

Corollary 4.4.36 *Let $\varphi(x)$ be a polynomial of degree ν. A solution $f(x)$ of the q-difference equation*

$$\frac{1}{2}\,{}_{\omega_1,\ldots,\omega_n}\triangle_{\mathrm{NWA},2,q}^{n} f(x)=\mathrm{D}_q^n\varphi(x) \tag{4.440}$$

is given by

$$f(x\oplus_q y)=\sum_{k=0}^{\nu}\frac{\mathrm{L}_{\mathrm{NWA},k,q}^{(n)}(x|\omega_1,\ldots,\omega_n)}{\{k\}_q!}\mathrm{D}_q^k\varphi(y). \tag{4.441}$$

Proof The LHS of (4.441) can be written as $\varphi(\mathrm{L}^{(n)}_{\mathrm{NWA},q}(x|\omega_1,\ldots,\omega_n)\oplus_q y)$, because if we apply $\frac{1}{2}\triangle^n_{\mathrm{NWA},2,q,x}$ to both sides we get

$$\begin{aligned}\frac{1}{2}\underset{\omega_1,\ldots,\omega_n\mathrm{NWA},2,q}{\triangle^n} f(x\oplus_q y) &= \mathrm{D}^n_{q,x}\varphi(x\oplus_q y)\\ &= \frac{1}{2}\underset{\omega_1,\ldots,\omega_n\mathrm{NWA},2,q}{\triangle^n}\varphi\big(\mathrm{L}^{(n)}_{\mathrm{NWA},q}(x|\omega_1,\ldots,\omega_n)\oplus_q y\big).\end{aligned} \tag{4.442}$$

□

Theorem 4.4.37

$$\sum_{k=0}^{\infty}\frac{\mathrm{G}^{(n)}_{\mathrm{NWA},k,q}(x)}{\{k\}_q!}\nabla^n_{\mathrm{NWA},2,q}\mathrm{D}^k_q f(y) = f(x\oplus_q y). \tag{4.443}$$

Proof Replace $f(x)$ by $f(x\oplus_q y)$ in (4.400):

$$\frac{1}{2}\big(f\big(\mathrm{G}^{(n)}_{\mathrm{NWA},q}(x)\oplus_q \overline{2}_q\oplus_q y\big)+f\big(\mathrm{G}^{(n)}_{\mathrm{NWA},q}(x)\oplus_q y\big)\big)\doteq\big(\mathrm{G}^{(n-1)}_{\mathrm{NWA},q}(x)\oplus_q y\big). \tag{4.444}$$

Use the umbral formula (4.21) to get

$$\sum_{k=0}^{\infty}\frac{\mathrm{G}^{(n)}_{\mathrm{NWA},k,q}(x)}{\{k\}_q!}\nabla_{\mathrm{NWA},2,q}\mathrm{D}^k_q f(y) = \sum_{k=0}^{\infty}\frac{\mathrm{G}^{(n-1)}_{\mathrm{NWA},k,q}(x)}{\{k\}_q!}\mathrm{D}^k_q f(y). \tag{4.445}$$

Apply the operator $\nabla^{n-1}_{\mathrm{NWA},2,q}$ with respect to y to both sides and use (4.431):

$$\begin{aligned}&\sum_{k=0}^{\infty}\frac{\mathrm{G}^{(n)}_{\mathrm{NWA},k,q}(x)}{\{k\}_q!}\nabla^n_{\mathrm{NWA},2,q}\mathrm{D}^k_q f(y)\\ &\quad=\sum_{k=0}^{\infty}\frac{\mathrm{G}^{(n-1)}_{\mathrm{NWA},k,q}(x)}{\{k\}_q!}\sum_{l=0}^{\infty}\mathrm{D}^{k+l}_q f(0)\frac{\mathrm{G}^{(-n+1)}_{\mathrm{NWA},l,q}(y)}{\{l\}_q!}.\end{aligned} \tag{4.446}$$

Finally use (4.422) to rewrite the right-hand side. □

Corollary 4.4.38 *Let $\varphi(x)$ be a polynomial of degree ν. A solution $f(x)$ of the q-difference equation*

$$\underset{\omega_1,\ldots,\omega_n\mathrm{NWA},2,q}{\nabla^n} f(x)=\varphi(x) \tag{4.447}$$

is given by

$$f(x\oplus_q y)=\sum_{k=0}^{\nu}\frac{\mathrm{G}^{(n)}_{\mathrm{NWA},k,q}(x|\omega_1,\ldots,\omega_n)}{\{k\}_q!}\mathrm{D}^k_q\varphi(y). \tag{4.448}$$

Proof The LHS of (4.448) can be written as $\varphi(\mathrm{G}^{(n)}_{\mathrm{NWA},q}(x|\omega_1,\ldots,\omega_n)\oplus_q y)$, because if we apply $\nabla^n_{\mathrm{NWA},2,q,x}$ to both sides we get

$$\begin{aligned}\underset{\omega_1,\ldots,\omega_n\mathrm{NWA},2,q}{\nabla^n} f(x\oplus_q y) &= \varphi(x\oplus_q y)\\ &= \underset{\omega_1,\ldots,\omega_n\mathrm{NWA},2,q}{\nabla^n} \varphi\big(\mathrm{G}^{(n)}_{\mathrm{NWA},q}(x|\omega_1,\ldots,\omega_n)\oplus_q y\big).\end{aligned} \tag{4.449}$$

□

4.5 The semiring of Ward numbers

The concept of a semiring is pretty old. The simplest examples are $\mathbb{N}$ and $\mathbb{Q}_+$. A natural q-analogue is given by

Definition 95 Let $(\mathbb{N}_{\oplus_q},\oplus_q,\odot_q)$ denote the Ward numbers $\overline{k}_q$, $k\geq 0$ together with two binary operations: $\oplus_q$ is the usual Ward q-addition, and the multiplication $\odot_q$ is defined as follows:

$$\overline{n}_q\odot_q\overline{m}_q\sim\overline{nm}_q, \tag{4.450}$$

where $\sim$ denotes the equivalence relation in the alphabet.

Theorem 4.5.1 $(\mathbb{N}_{\oplus_q},\oplus_q,\odot_q)$ *is a commutative semiring.*

Proof The associativity for $(\mathbb{N}_{\oplus_q},\oplus_q)$ results from the corresponding property of the commutative semiring. The zero element is $\overline{0}_q$.

The associativity for $(\mathbb{N}_{\oplus_q},\odot_q)$ is shown as follows: Choose three elements $\overline{n}_q,\overline{m}_q,\overline{k}_q\in(\mathbb{N}_{\oplus_q},\odot_q)$. Then

$$(\overline{n}_q\odot_q\overline{m}_q)\odot_q\overline{k}_q\sim\overline{nm}_q\odot_q\overline{k}_q\sim\overline{nmk}_q, \tag{4.451}$$

$$\overline{n}_q\odot_q(\overline{m}_q\odot_q\overline{k}_q)\sim\overline{n}_q\odot_q\overline{mk}_q\sim\overline{nmk}_q. \tag{4.452}$$

The identity is $\overline{1}_q$. The distributive law is proved as follows:

$$\overline{n}_q\odot_q(\overline{m}_q\oplus_q\overline{k}_q)\sim\overline{n}_q\odot_q\overline{m+k}_q\sim\overline{n(m+k)}_q. \tag{4.453}$$

$$(\overline{n}_q\odot_q\overline{m}_q)\oplus_q(\overline{n}_q\odot_q\overline{k}_q)\sim\overline{nm}_q\oplus_q\overline{nk}_q\sim\overline{n(m+k)}_q. \tag{4.454}$$

Finally, $\overline{0}_q$ annihilates $(\mathbb{N}_{\oplus_q},\odot_q)$. □

Chapter 5
q-Stirling numbers

5.1 Introduction

The aim of this chapter is to describe how different q-difference operators combine with q-Stirling numbers to form various q-formulas. Contrary to the previous chapter, where formal power series were considered, here we focus on functions of q^x, or equivalently functions of the q-binomial coefficients. We will find many q-analogues of Stirling number identities from Jordan [320] and the elementary textbooks by Cigler [138] and Schwatt [456].

James Stirling was born in Scotland and studied in Glasgow and Oxford. In 1717 Stirling went to Venice; probably he had been promised a chair of mathematics there, but for some reason the appointment was never realized. In spite of this, he continued his mathematical research. He also attended the University of Padua, where he got to know Nicolaus Bernoulli (I), who held the chair there.

In the eighteenth and nineteenth centuries many articles about Bernoulli numbers contained Stirling numbers in disguise. A pseudo-Stirling number $\mathrm{S}(n,k)k!$ appears in a work by Euler (1755). Tables for these pseudo-Stirling numbers were published in [277, p. 9] and [250, p. 71].

Before Nielsen coined this name, the most frequent appearance of the so-called second Stirling number was as submultiple of the Euler formula for k-th differences of powers [240]. This formula is

$$\mathrm{S}(n,k) = \frac{1}{k!}\sum_{i=0}^{k}\binom{k}{i}(-1)^{k-i}i^n, \tag{5.1}$$

and is very similar to formula (2.19), which forms the basis of q-calculus. As we shall see, the equivalent expression for the second q-Stirling number has a slightly different character than (2.19).

Formulas for Bernoulli polynomials were written as sums with Stirling numbers as coefficients, like in [545, p. 211] and [446, p. 96].

In the history of mathematics, Stirling numbers often appear in disguised form. Cauchy [112, p. 35] used them in 1833 for calculation of power sums. The following

T. Ernst, *A Comprehensive Treatment of q-Calculus*,
DOI 10.1007/978-3-0348-0431-8_5, © Springer Basel 2012

table shows several different notations for $S(n,k)$:

Grünert [251]	Grünert [252]	Björling [74]	Saalschütz [446]	Worpitzky [545]
A_n^k	C_k^n	C_k^n	α_k^p	α_k^n

Schwatt [456, Chap. 5]	Ettingshausen [525, p. 195]	d'Ocagne [157, p. 354]
$a_{n,k}$	C_k^n	K_n^k

In this chapter we will find many new formulas for q-Stirling numbers. This chapter first appeared as preprint in October–December 2005. In February 2006 Johann Cigler reminded the author about some corrections and that he also had written a tutorial on a similar subject [139]. The q-Stirling numbers of Cigler and those of the author are identical. Whenever an equation appears, which also appeared in [139], it will be mentioned and the page number (November 2006) will be given.

q-Stirling numbers are extremely useful in q-calculus. This however has not been fully acknowledged until now. In a book by Don Knuth [340] it is shown that the q-Stirling number of the second kind gives the running time of the algorithm for a computer program. A related result for Markov processes was obtained by Crippa, Simon, and Trunz in [145].

As Sharma and Chak [463, p. 326] remarked, the operator D_q plays the same role for polynomials in x as the difference operator in Section 5.3,

$$\triangle_{\mathrm{CG},q} \equiv \mathrm{E} \boxminus_q \mathrm{I}, \tag{5.2}$$

does for polynomials in q^x.

We will now discuss briefly the different types of content in Sections 5.2–5.4 from a conceptual point of view.

In Section 5.2 we only consider functions in $\mathbb{C}(q)[x]$. In Sections 5.3 and 5.4 we mainly consider functions $f, g \in \mathbb{C}(q)[q^x]$. The main purpose of Section 5.2 is to introduce the q-Stirling numbers and to begin to study their properties.

In each of Sections 5.2, 5.3, 5.4 for the respective Δ_q operator, we will, when possible, find q-analogues of formulas (2.4)–(2.7). For clarity, we will keep the same order of these equations in each section. As a minimum four of these formulas occur, we will refer to them as the quartet of formulas.

In Section 5.2 and 5.3, we find formulas for the q-Stirling number the second kind in terms of the respective $\triangle$-operator.

In 1951 D.B. Sears wrote an important paper [461], where transformations for basic hypergeometric functions were derived using a slightly different difference operator. In Sears' paper, the variable x was instead the index for a q-shifted factorial. By inversion of the basis two different sets of equations were obtained.

In Chapter 4 we used a formal power series approach to find many formulas for q-Bernoulli polynomials etc. The shift operator was two q-additions (Ward, Jackson) and the quartet of formulas was given with ordinary binomial coefficients.

The $\triangle$-operator in Section 5.3 (Carlitz-Gould) and in Chapter 4 (Jackson) are very similar; in fact

$$\triangle_{\mathrm{J},q}^n f(q^x) = q^{-nx-\binom{n}{2}} \triangle_{\mathrm{CG},q}^n f(q^x). \tag{5.3}$$

The two symbols, sometimes called the difference and sum calculus, correspond respectively to differentiation and integration in the continuous calculus. We will find q-analogues of the inverse operators $\triangle$ and $\sum$ in Section 5.3.

In the footsteps of Faulhaber, Fermat, Jacob Bernoulli and De Moivre, we will find an expression for the Carlitz function

$$\mathrm{S}_{\mathrm{C},m,q}(n) \equiv \sum_{i=0}^{n-1} \{i\}_q^m q^i \tag{5.4}$$

in Sections 5.2 and 5.3.

In Section 5.5 we will prove a q-binomial coefficient identity by using the operator $\triangle_{\mathrm{CG},q}$ with corresponding q-Taylor formula. After simplification, we will show that the operator $\triangle_{\mathrm{J},q}$ with corresponding q-Taylor formula gives the same formula.

5.2 The Hahn-Cigler-Carlitz-Johnson approach

The main purpose of this section is to introduce and study the q-Stirling numbers. This section is a partial continuation of papers by Hahn [260, p. 6, 2.2], Cigler [133, p. 102–104], Carlitz [97] and Johnson [317, p. 217]. The last three papers use the same q-Stirling numbers. We start with three definitions followed by three examples. First we introduce a very important polynomial.

Definition 96 A q-analogue of the polynomial in [138, p. 20]. Cigler [133, p. 102], [139, p. 38] calls this polynomial *Hauptfolge*.

$$(x)_{k,q} \equiv \prod_{m=0}^{k-1} \big(x - \{m\}_q\big). \tag{5.5}$$

The following notation of Cigler [134, p. 107], [139, p. 39, 3.25] will be used:

Definition 97

$$\mathrm{E}_{\mathrm{C},q}^{l} f(x) \equiv f\big(xq^l + \{l\}_q\big). \tag{5.6}$$

Definition 98 [133, p. 102], [134, p. 107] This is a special case of the Hahn operator [260, p. 6, 2.2]:

$$\triangle_{\mathrm{H},q} f(x) \equiv \frac{f(qx+1) - f(x)}{1 + (q-1)x}. \tag{5.7}$$

Example 14 [139, p. 39], [133, p. 102], a q-analogue of [138, p. 20, 2.5]

$$\triangle_{\mathrm{H},q} (x)_{k,q} = \{k\}_q (x)_{k-1,q}. \tag{5.8}$$

Example 15 [134, p. 107]

$$\triangle_{\mathrm{H},q}\mathrm{E}_{\mathrm{C},q} = q\mathrm{E}_{\mathrm{C},q}\triangle_{\mathrm{H},q}. \tag{5.9}$$

Example 16 A q-analogue of [2, p. 237, (27)].

$$\big(\{k\}_q\big)_{l,q} = \{l\}_q!\binom{k}{l}_q q^{\binom{l}{2}}. \tag{5.10}$$

The operator $\triangle^n_{\mathrm{H},q}$ can be expressed by the following two operator equations:

Theorem 5.2.1 *A q-Taylor formula* [133, p. 103], *which is q-analogue of Boole* [79, p. 11], [138, p. 25]:

$$f(x) = \sum_{k=0}^{\infty} \frac{(\triangle^k_{\mathrm{H},q} f)(0)}{\{k\}_q!}(x)_{k,q}. \tag{5.11}$$

Theorem 5.2.2 *Cigler* [139, p. 41, (3.34)], [133, p. 103, (40)]:

$$q^{\binom{n}{2}}\triangle^n_{\mathrm{H},q} = \frac{1}{(1+(q-1)x)^n}\prod_{k=0}^{n-1}\big(\mathrm{E}_{\mathrm{C},q} - q^k\big). \tag{5.12}$$

Theorem 5.2.3 [139, p. 41, (3.33)]:

$$\triangle^n_{\mathrm{H},q}(x)_{k,q} = q^{-\binom{n}{2}}\big(1+(q-1)x\big)^{-n}\sum_{l=0}^{n}(-1)^{n-l}q^{\binom{n-l}{2}}\binom{n}{l}_q\big(xq^l + \{l\}_q\big)_{k,q}. \tag{5.13}$$

Proof Induction on n. □

We will rewrite (5.13) in a slightly different way.

Theorem 5.2.4

$$\triangle^n_{\mathrm{H},q} f(x) = q^{-\binom{n}{2}}\big(1+(q-1)x\big)^{-n}\sum_{l=0}^{n}(-1)^{n-l}q^{\binom{n-l}{2}}\binom{n}{l}_q \mathrm{E}^l_{\mathrm{C},q} f(x). \tag{5.14}$$

Proof Operate on $(x)_{k,q}$. □

This formula can be inverted.

Theorem 5.2.5

$$\mathrm{E}^n_{\mathrm{C},q} f(x) = \sum_{i=0}^{n} q^{\binom{i}{2}}\binom{n}{i}_q\big(1+(q-1)x\big)^i\triangle^i_{\mathrm{H},q} f(x). \tag{5.15}$$

Proof Use the above inversion theorem with $f(n)=\binom{n}{2}$. □

Corollary 5.2.6 *A Leibniz-type formula*:

$$\triangle_{\mathrm{H},q}^{n}(fg)=q^{-\binom{n}{2}}\sum_{i=0}^{n}q^{\binom{i}{2}+\binom{n-i}{2}}\binom{n}{i}_{q}\triangle_{\mathrm{H},q}^{i}f\big(\triangle_{H,q}^{n-i}\mathrm{E}_{\mathrm{C},q}^{i}\big)g. \tag{5.16}$$

Proof Analogous to [320, p. 96 f]. □

We are now ready for the definition of q-Stirling numbers, which will be the subject of the next two sections. The second Stirling numbers, as below but for $q=1$ occurred in Stirling's book [490, p. 8]. However Stirling did not use any symbol for these numbers.

Definition 99 The q-Stirling number of the first kind $\mathrm{s}(n,k)_q$ and the q-Stirling number of the second kind $\mathrm{S}(n,k)_q$ are defined in [251, p. 358], [139, p. 38, (3.13–14)], [133, p. 103], [139, p. 38] and [317, p. 217, 4.11] by:

$$(x)_{n,q}\equiv\sum_{k=0}^{n}\mathrm{s}(n,k)_q x^k, \tag{5.17}$$

$$x^n\equiv\sum_{k=0}^{n}\mathrm{S}(n,k)_q(x)_{k,q}. \tag{5.18}$$

Remark 14 We use the same conventions for Stirling numbers as Cigler [138, p. 34], Jordan [320, p. 142], Gould [240], Vein and Dale [518, p. 306] and Milne [383, p. 90]. Other definitions usually differ in sign, as in [79, p. 114], [244], [107], [235], where all (q-)Stirling numbers are positive.

Remark 15 Schwatt [456, Chap. 5] denotes the $\mathrm{S}(n,k)$ by $a_{n,k}$ without knowing that they are Stirling numbers. The book by Schwatt contains some very interesting calculations with series to which we will come back shortly.

The following recursions follow at once [133, p. 103]. The second one, a q-analogue of Grünert [252, p. 248, p. 256], Björling [74, p. 287] and Schwatt [456, p. 81, (4)], also appeared in [251, p. 360], [507, p. 85, 13.1], [317, p. 213, 3.6], [340, 7215, exc 29], [137, p. 146, (9)]:

$$\mathrm{s}(n+1,k)_q=\mathrm{s}(n,k-1)_q-\{n\}_q\,\mathrm{s}(n,k)_q, \tag{5.19}$$

$$\mathrm{S}(n+1,k)_q=\mathrm{S}(n,k-1)_q+\{k\}_q\,\mathrm{S}(n,k)_q. \tag{5.20}$$

The orthogonality relation is the following q-analogue of Jordan [320, p. 182], [138, p. 35].

Theorem 5.2.7 *The two sets of q-Stirling numbers viewed as matrices are inverses of each other*:

$$\sum_k \mathrm{S}(m,k)_q \mathrm{s}(k,n)_q = \delta_{m,n}. \tag{5.21}$$

The following table lists some of the first $\mathrm{s}(n,k)_q$. Compare with [138, p. 34], [320, p. 144] and [518, p. 306].

	$k=0$	$k=1$	$k=2$	$k=3$	$k=4$
$n=0$	1	0	0	0	0
$n=1$	0	1	0	0	0
$n=2$	0	-1	1	0	0
$n=3$	0	$1+q$	$-(2+q)$	1	0
$n=4$	0	$-\{3\}_q!$	$3+4q+3q^2+q^3$	$-3-2q-q^2$	1

And for $n=5$:

$k=1$	$k=2$	$k=3$	$k=4$	$k=5$
$1+3q+5q^2$ $+6q^3+5q^4$ $+3q^5+q^6$	$-(4+9q+12q^2$ $+12q^3+8q^4$ $+4q^5+q^6)$	$6+9q+9q^2$ $+7q^3+3q^4+q^5$	$-(4+3q$ $+2q^2+q^3)$	1

The following table lists some of the first $\mathrm{S}(n,k)_q$. Compare with [138, p. 35].

	$k=0$	$k=1$	$k=2$	$k=3$	$k=4$
$n=0$	1	0	0	0	0
$n=1$	0	1	0	0	0
$n=2$	0	1	1	0	0
$n=3$	0	1	$2+q$	1	0
$n=4$	0	1	$3+3q+q^2$	$3+2q+q^2$	1

and for $n=5$:

$k=0$	$k=1$	$k=2$	$k=3$	$k=4$	$k=5$
0	1	$4+6q$ $+4q^2+q^3$	$6+8q+7q^2$ $+3q^3+q^4$	$4+3q$ $+2q^2+q^3$	1

There are a few simple rules for checking the values of the q-Stirling numbers of the first kind, as the following q-analogues of [320, p. 145 ff] show. Put $x=1$ in (5.17) to get

$$\sum_{k=1}^{n} \mathrm{s}(n,k)_q = 0, \quad n>1. \tag{5.22}$$

Put $x = -1$ in (5.17) to get

$$\sum_{k=1}^{n} \left|\mathrm{s}(n,k)_q\right| = (-1)^{n-1} \prod_{m=0}^{n-1} \left(1 + \{m\}_q\right), \quad q > 0. \tag{5.23}$$

The q-Stirling numbers of the first kind have a particularly simple expression.

Theorem 5.2.8

$$\mathrm{s}(n,k)_q = (-1)^{k-n} \mathrm{e}_{n-k}\left(1, \{2\}_q, \ldots, \{n\}_q\right), \quad n \in \mathbb{N}, \tag{5.24}$$

where e_k *denotes the elementary symmetric polynomial.*

Proof Use (5.17). □

Corollary 5.2.9 [139, p. 38].

$$\mathrm{s}(n,1)_q = (-1)^{n-1} \{n-1\}_q!. \tag{5.25}$$

Proof Put $k = 1$ in (5.24). □

There is an exact formula by Carlitz for $\mathrm{S}(n,k)_q$, which will be quite useful.

Theorem 5.2.10 [139, p. 41, (3.35)], [133, p. 104], *Carlitz* [97, p. 990, 3.3], *Toscano* [507, p. 86, 13.2] ($c = 1$). *A q-analogue of Jordan* [320, p. 169, (3)], [138, p. 37, 2.31], [505, p. 495], *Schwatt* [456, p. 83, (19)]. *Compare with Grünert* [252, p. 257]:

$$\mathrm{S}(n,k)_q = \left(\{k\}_q! q^{\binom{k}{2}}\right)^{-1} \sum_{i=0}^{k} \binom{k}{i}_q (-1)^i q^{\binom{i}{2}} \{k-i\}_q^n. \tag{5.26}$$

This formula can also be written as the following q-analogue of one by Toscano [505]:

$$\mathrm{S}(n,k)_q = \frac{1}{\{k\}_q!} \Delta_{\mathrm{H},q}^{n} (x)_{k,q}|_{x=0}. \tag{5.27}$$

There are a few simple rules for checking the values of the q-Stirling numbers of the second kind. We start with some q-analogues of relations found in Jordan's book [320, p. 170. f] about finite differences intended both for mathematicians and statisticians.

Theorem 5.2.11

$$(-1)^n = \sum_{k=0}^{n} \mathrm{S}(n,k)_q (-1)^k \prod_{m=0}^{k-1} \left(1 + \{k\}_q\right). \tag{5.28}$$

Proof Put $x=-1$ in (5.18). □

Theorem 5.2.12

$$\mathrm{S}(n+1,n)_q-\mathrm{S}(n,n-1)_q=\{n\}_q. \tag{5.29}$$

Proof Put $k=n$ in (5.20). □

The following formula can be found in [406, p. 66]

$$\mathrm{S}(n+k,n)_q=\mathrm{h}_k\big(1,\{2\}_q,\ldots,\{n\}_q\big),\quad n\in\mathbb{N}, \tag{5.30}$$

where h_k denotes the complete symmetric polynomial.

There are two kinds of generating functions for $\mathrm{S}(n,k)_q$. The first is given by

Theorem 5.2.13 [139, p. 42, (3.36)], *Knuth* [340, 7.2.1.5, *answer* 29], *a q-analogue of Jordan* [320, p. 175, (2)], *Graham* [244, p. 337, 7.47], [507, p. 64, 2.6] *and* [138, p. 37, 2.30].

$$\sum_{n=m}^{\infty}\mathrm{S}(n,m)_q t^n=\frac{t^m}{\prod_{l=1}^{m}(1-t\{l\}_q)},\quad |t|<\frac{1}{m}. \tag{5.31}$$

This can be expressed in two other ways. First, as a q-analogue of [320, p. 193, (1)], which serves as definition of the q-reciprocal factorial:

$$\sum_{n=m}^{\infty}\mathrm{S}(n,m)_q z^{-n-1}=\frac{1}{(z)_{m+1,q}}\equiv(z)_{-(m+1),q},\quad z>m. \tag{5.32}$$

Second, as a q-analogue of [320, p. 193, (2)] and [138, p. 36, 2.29]:

$$\sum_{n=m}^{\infty}\mathrm{S}(n,m)_q(-x)^{-n}=\frac{(-1)^m}{\prod_{l=1}^{m}(x+\{l\}_q)},\quad x>m. \tag{5.33}$$

By the orthogonality relation we obtain a q-analogue of [320, p. 193, (3)]:

$$x^{-k}=\sum_{m=k}^{\infty}\frac{|\mathrm{s}(m,k)_q|}{\prod_{l=1}^{m}(x+\{l\}_q)}. \tag{5.34}$$

The q-Stirling numbers can be used to obtain a number of exact formulas for q-derivatives and q-integrals. For example, we have

Theorem 5.2.14

$$\int_0^1(t)_{n,q}\,d_q(t)=\sum_{k=1}^{n}\frac{\mathrm{s}(n,k)_q}{\{k+1\}_q}. \tag{5.35}$$

Proof q-integrate (5.17). □

Theorem 5.2.15 *A q-analogue of* [320, p. 194, (5)]:

$$\mathrm{D}_q^s \frac{1}{(z)_{m+1,q}} = \sum_{n=m}^{\infty} \mathrm{S}(n,m)_q \{-n-s\}_{s,q} z^{-n-1-s}. \tag{5.36}$$

Proof Use (5.32). □

Theorem 5.2.16 *A q-analogue of* [320, p. 194, (6)]:

$$\int^z \frac{1}{(t)_{m+1,q}} d_q(t) = \sum_{n=m}^{\infty} \frac{\mathrm{S}(n,m)_q}{\{-n\}_q z^n} + k. \tag{5.37}$$

Proof Use (5.32). □

The second generating functions for $\mathrm{S}(n,m)_q$ is

Theorem 5.2.17 [139, p. 42, (3.38)]. *A q-analogue of Comtet* [141, p. 206, (2a)], *Toscano* [507, p. 64, 2.7]:

$$\sum_{n=k}^{\infty} \frac{\mathrm{S}(n,k)_q t^n}{\{n\}_q!} = \left(\{k\}_q! q^{\binom{k}{2}}\right)^{-1} \sum_{i=0}^{k} \binom{k}{i}_q (-1)^i q^{\binom{i}{2}} \mathrm{E}_q\left(t\{k-i\}_q\right). \tag{5.38}$$

Proof

$$\begin{aligned} \mathrm{LHS} &= \sum_{n=k}^{\infty} \frac{t^n}{\{n\}_q!} \left(\{k\}_q! q^{\binom{k}{2}}\right)^{-1} \sum_{i=0}^{k} \binom{k}{i}_q (-1)^i q^{\binom{i}{2}} \{k-i\}_q^n \\ &= \left(\{k\}_q! q^{\binom{k}{2}}\right)^{-1} \sum_{i=0}^{k} \binom{k}{i}_q (-1)^i q^{\binom{i}{2}} \sum_{n=0}^{\infty} \frac{(t\{k-i\}_q)^n}{\{n\}_q!} = \mathrm{RHS}. \end{aligned} \tag{5.39}$$

□

There is a generating function for the q-Stirling numbers of the first kind.

Theorem 5.2.18 *A q-analogue of* [320, p. 185, (1)]:

$$\sum_{k=m}^{n} \mathrm{s}(n,k)_q \binom{k}{m} q^m = q^{n-1} \mathrm{s}(n-1,m)_q + q^n \mathrm{s}(n-1,m-1)_q. \tag{5.40}$$

This formula can be inverted.

Theorem 5.2.19 *A q-analogue of* [320, p. 185]:

$$\mathrm{s}(n,k)_q=\sum_{m=k}^{n} q^{n-m-1}(-1)^{m+k}\binom{m}{k}\big[\mathrm{s}(n-1,m)_q+q\mathrm{s}(n-1,m-1)_q\big]. \quad (5.41)$$

Corollary 5.2.20 *A q-analogue of* [320, p. 186, (4)]:

$$\sum_{k=1}^{n}\mathrm{s}(n,k)_q k=\begin{cases}1, & n=1,\\ q^{n-2}(-1)^n\{n-2\}_q!, & n>1.\end{cases} \quad (5.42)$$

Proof Put $m=1$ in (5.40). □

Corollary 5.2.21 *A q-analogue of* [320, p. 186, (5)]:

$$\sum_{n=2}^{m}\mathrm{S}(m,n)_q q^{n-2}(-1)^n\{n-2\}_q!=m-1. \quad (5.43)$$

Proof Apply $\sum_{n=1}^{m}\mathrm{S}(m,n)_q$ to both sides of (5.42). Then use the orthogonality relation. □

The following three theorems are proved in a similar way as in Jordan's work [320].

Theorem 5.2.22 *A q-analogue of* [320, p. 187, (10)]:

$$\sum_{k=1}^{n}\mathrm{s}(n,k)_q\mathrm{S}(k+1,i)_q=\{n\}_q\binom{0}{n-i}+\binom{0}{n+1-i}. \quad (5.44)$$

Theorem 5.2.23 *A q-analogue of* [320, p. 188, (11)]:

$$\sum_{k=0}^{n}\mathrm{S}(n,k)_q\big[\mathrm{s}(k+1,l)_q+\{k\}_q\mathrm{s}(k,l)_q\big]=\delta_{l,n+1}. \quad (5.45)$$

Theorem 5.2.24 *A q-analogue of* [320, p. 188, (15)]:

$$\sum_{k=1}^{n+1}\mathrm{S}(n+1,k)_q=\sum_{k=1}^{n}\big(1+\{k\}_q\big)\mathrm{S}(n,k)_q,\quad n>0. \quad (5.46)$$

The following theorem on q-Stirling numbers of the first kind is proved by induction.

Theorem 5.2.25 *Oruc* [407, p. 59, 4.12]:

$$\mathrm{s}(n+1,m)_q = \sum_{k=m-1}^{n} (-1)^{k-m-1} q^{n-k} \binom{k}{m-1} \mathrm{s}(n,k)_q, \quad m > 1. \tag{5.47}$$

The following vertical recurrence for q-Stirling numbers of the second kind is also proved by induction.

Theorem 5.2.26 *Oruc* [406], [407, p. 59, 4.13]:

$$\mathrm{S}(n+1,m)_q = \sum_{k=m-1}^{n} q^{k-m+1} \binom{n}{k} \mathrm{S}(k,m-1)_q, \quad m > 1. \tag{5.48}$$

The following operator [307] will be useful. In its earliest form with $q = 1$ it dates back to Euler and Abel [1, B.2, p. 41], who used it in differential equations.

Definition 100 Compare with (10.44):

$$\theta_q \equiv x\mathrm{D}_q. \tag{5.49}$$

Theorem 5.2.27 *Cigler* [139, p. 37, (3.8)]. *A q-analogue of the Grünert operator formula* [252, 247], *see also Gould* [240, p. 455, 4.8], *Stephens* [489, p. 181], *Toscano* [507, p. 64, 2.1], *Schwatt* [456, p. 81, (2)], *Jordan* [320, p. 196, (2)], *Melzak* [377, p. 95]:

$$\theta_q^n = \sum_{k=0}^{n} \mathrm{S}(n,k)_q q^{\binom{k}{2}} x^k \mathrm{D}_q^k. \tag{5.50}$$

Proof Induction on n. □

This leads to the following inversion formula.

Theorem 5.2.28 *Cigler* [139, p. 37, (3.9)]. *A q-analogue of* [503, p. 548], [489, p. 183]:

$$q^{\binom{n}{2}} x^n \mathrm{D}_q^n = \sum_{k=1}^{n} \mathrm{s}(n,k)_q \theta_q^k. \tag{5.51}$$

Proof Use the orthogonality relation for q-Stirling numbers. □

The previous formula can be expressed in another way.

Theorem 5.2.29 *Jackson* [303, p. 305]. *A q-analogue of the* 1844 *Boole formula* [78], *see also Stephens* [489, p. 183], *Chatterjea* [119, p. 24, (2.1)]:

$$q^{\binom{n}{2}} x^n \mathrm{D}_q^n = \prod_{k=0}^{n-1} \left(\theta_q - \{k\}_q\right). \tag{5.52}$$

Proof Use (5.17). □

Example 17 A q-analogue of [320, p. 196]. Let $f(x) = (x \oplus_q 1)^n$ and apply (5.50) to get

$$\sum_{k=0}^{n} \binom{n}{k}_q x^k \{k\}_q^m = \sum_{k=0}^{\min(m,n)} \mathrm{S}(m,k)_q q^{\binom{k}{2}} x^k \{n-k+1\}_{k,q} (x \oplus_q 1)^{n-k}. \tag{5.53}$$

Put $x = 1$ to get

$$\sum_{k=0}^{n} \binom{n}{k}_q \{k\}_q^m = \sum_{k=0}^{\min(m,n)} \mathrm{S}(m,k)_q q^{\binom{k}{2}} \{n-k+1\}_{k,q} (1 \oplus_q 1)^{n-k}. \tag{5.54}$$

If we put $m = 1$ or $m = 2$ in (5.54), we get q-analogues of the mean and variance of the binomial distribution from Melzak [377, p. 96].

Put $x = -1$ in (5.53) to get

$$\sum_{k=0}^{n} \binom{n}{k}_q (-1)^k \{k\}_q^m = \sum_{k=0}^{\min(m,n)} \mathrm{S}(m,k)_q q^{\binom{k}{2}} (-1)^k \{n-k+1\}_{k,q} (1 \ominus_q 1)^{n-k}. \tag{5.55}$$

Example 18 Put $f(x) = (x \boxplus_q 1)^n$ and operate with (5.50) to get

$$\sum_{k=0}^{n} \binom{n}{k}_q q^{\binom{k}{2}} x^{n-k} \{n-k\}_q^m = \sum_{k=0}^{\min(m,n)} \mathrm{S}(m,k)_q q^{\binom{k}{2}} x^k \{n-k+1\}_{k,q} (x \boxplus_q 1)^{n-k}. \tag{5.56}$$

Put $x = 1$ to get

$$\sum_{k=0}^{n} \binom{n}{k}_q q^{\binom{k}{2}} \{n-k\}_q^m = \sum_{k=0}^{\min(m,n)} \mathrm{S}(m,k)_q q^{\binom{k}{2}} \{n-k+1\}_{k,q} (1 \boxplus_q 1)^{n-k}. \tag{5.57}$$

Put $x = -1$ to get (5.26).

We continue with a few equations with operator proofs in the spirit of Gould and Schwatt.

Theorem 5.2.30 *Almost a q-analogue of* [240, p. 455, 4.9]:

$$\sum_{k=0}^{n} \{k\}_q{}^p x^k = \sum_{k=0}^{p} \mathrm{S}(p,k)_q q^{\binom{k}{2}} x^k \mathrm{D}_q^k \left(\frac{x^{n+1}-1}{x-1} \right). \tag{5.58}$$

We obtain the following limit relation:

Theorem 5.2.31

$$\sum_{k=0}^{n} \{k\}_q{}^p = \sum_{k=0}^{p} \mathrm{S}(p,k)_q q^{\binom{k}{2}} \lim_{x\to 1} \mathrm{D}_q^k \left(\frac{x^{n+1}-1}{x-1} \right). \tag{5.59}$$

Theorem 5.2.32 *A q-analogue of* [240, p. 456, 4.10], [456, p. 85, (38)], [239, p. 490]:

$$\sum_{k=0}^{\infty} \{k\}_q{}^p x^k = \sum_{k=0}^{p} \mathrm{S}(p,k)_q q^{\binom{k}{2}} \frac{x^k \{k\}_q!}{(x;q)_{k+1}}, \quad |x| < 1. \tag{5.60}$$

Proof Let $n \to \infty$ in (5.58). □

Example 19 A q-analogue of [240, p. 456, 4.11]:

$$\theta_q^n \mathrm{E}_q(x) = \mathrm{E}_q(x) \sum_{k=0}^{n} \mathrm{S}(n,k)_q q^{\binom{k}{2}} x^k. \tag{5.61}$$

To check this, use (5.50).

The following q-analogue of Bell numbers is the same as Milne's [383, p. 99].

Definition 101 Compare with [133, p. 104] ($x = 1$). The q-Bell number is given by

$$\mathcal{B}_q(n) \equiv \sum_{k=1}^{n} \mathrm{S}(n,k)_q q^{\binom{k}{2}}. \tag{5.62}$$

Theorem 5.2.33 *The q-Dobinski theorem* [383, p. 108, 4.5] *is a q-analogue of* (3.11), [456, p. 84]:

$$\mathcal{B}_q(n) = \mathrm{E}_{\frac{1}{q}}(-1) \sum_{k=0}^{\infty} \frac{\{k\}_q^n}{\{k\}_q!}. \tag{5.63}$$

This can be generalized to

Theorem 5.2.34 *A q-analogue of the Schwatt generalization of the Dobinski theorem* [456, p. 84, (26)]:

$$\sum_{k=0}^{\infty} \frac{\{k\}_q{}^p x^k}{\{k\}_q!} = \sum_{k=0}^{p} \mathrm{S}(p,k)_q q^{\binom{k}{2}} x^k \mathrm{E}_q(x). \tag{5.64}$$

We will now return to the Carlitz q-analogue of [138, p. 13], [517, p. 575], [456, p. 86] for sums of powers. This function was also treated by Kim [333] from a different point of view.

Definition 102 Carlitz [97, p. 994–995]

$$\mathrm{S}_{\mathrm{C},m,q}(n) \equiv \sum_{i=0}^{n-1} \{i\}_q^m q^i, \qquad \mathrm{S}_{\mathrm{C},0,q}(1) \equiv 1. \tag{5.65}$$

We will now follow Cigler's computations ($q = 1$) and finally arrive at a formula which expresses $\mathrm{S}_{\mathrm{C},m,q}(n)$ as a general double sum.

Remark 16 Two completely different approaches to the problem of finding q-analogues of sums of consecutive powers of integers were presented in [532] and [454].

Lemma 5.2.35 *A q-analogue of* [138, p. 20, 2.6]:

$$\sum_{i=0}^{n-1} \big(\{i\}_q\big)_{j,q} q^i = \frac{(\{n\}_q)_{j+1,q}}{\{j+1\}_q}, \quad j < n. \tag{5.66}$$

Proof Induction on n. □

The Carlitz sum can be expressed as a double sum of q-Stirling numbers.

Corollary 5.2.36 *A q-analogue of* [138, p. 35]. *Compare with* [97, p. 994, 6.1]:

$$\mathrm{S}_{\mathrm{C},m,q}(n) = \sum_j \frac{\mathrm{S}(m,j)_q}{\{j+1\}_q} \sum_{l=0}^{j+1} \mathrm{s}(j+1,l)_q \{n\}_q^l. \tag{5.67}$$

Proof

$$\begin{aligned}\text{LHS} &= \sum_{i=1}^{n-1}\sum_{j=0}^{m} \mathrm{S}(m,j)_q\big(\{i\}_q\big)_{j,q} q^i \\ &= \sum_{j=0}^{m} \mathrm{S}(m,j)_q \frac{(\{n\}_q)_{j+1,q}}{\{j+1\}_q} = \text{RHS}.\end{aligned} \tag{5.68}$$

□

Tables of sums of powers have preoccupied mathematicians for centuries. Just as one example, De Moivre [150, 5] tabulated sums of powers up to $m = 10$. De Moivre's l corresponds to our $n - 1$. As an example, we compute a q-analogue for $m = 2$.

Example 20 A q-analogue of Järvheden [314, p. 98] ($m = 2$).

$$\mathrm{S}_{\mathrm{C},2,q}(n) = \frac{(\{n\}_q)_{3,q}}{\{3\}_q} + \frac{(\{n\}_q)_{2,q}}{\{2\}_q}, \quad n \geq 2. \tag{5.69}$$

Proof By (5.66),

$$\sum_{i=0}^{n-1} \{i\}_q\big(\{i\}_q - 1\big)q^i = \frac{(\{n\}_q)_{3,q}}{\{3\}_q}. \tag{5.70}$$

□

In general, $\mathrm{S}_{\mathrm{C},m,q}(n)$ contains the factor $(\{n\}_q)_{2,q}$.

5.3 The Carlitz-Gould approach

Definition 103 Let $\{\theta_i\}_0^\infty$ be an arbitrary sequence.

The Carlitz-Gould q-difference is defined by

$$\begin{aligned}&\triangle^1_{\mathrm{CG},q} f(x) \equiv f(x+1) - f(x), \\ &\triangle^{n+1}_{\mathrm{CG},q} f(x) \equiv \triangle^n_{\mathrm{CG},q} f(x+1) - q^n \triangle^n_{\mathrm{CG},q} f(x).\end{aligned} \tag{5.71}$$

Equivalently,

$$\triangle_{\mathrm{CG},q} \sim (\mathrm{E} \boxminus_q \mathrm{I}). \tag{5.72}$$

If the identity operator, the forward shift operator, or the Carlitz-Gould operator operates on m, we denote it by I_m, E_m, or $\triangle_{\mathrm{CG},m,q}$, respectively.

We get the above definition by putting $y=-1$ in Schendel [450].
Now follow the Carlitz-Gould quartet and two examples.

Theorem 5.3.1 *The following q-Taylor formula* [97, 2.5, p. 988], [236, 7.2, p. 856], [234, 2.11, p. 91] *applies*:

$$f(x)=\sum_{k=0}^{\infty}\frac{\triangle_{\mathrm{CG},q}^{k}f(0)}{\{k\}_q!}\{x-k+1\}_{k,q}. \tag{5.73}$$

Proof Apply $\triangle_{\mathrm{CG},q}^{s}$ to both members and finally put $x=0$. □

Theorem 5.3.2 [235, p. 283, 2.13], *Zeng* [548], *Gould* [234, 2.10, p. 91], *Phillips* [415, p. 46, 1.118], *Milne* [383, p. 91], *a q-analogue of* [138, p. 26]. *Compare with Schendel* [450, p. 82].

$$\triangle_{\mathrm{CG},q}^{n}f(x)=\sum_{k=0}^{n}(-1)^k\binom{n}{k}_q q^{\binom{k}{2}}\mathrm{E}^{n-k}f(x), \tag{5.74}$$

where the shift operator E *is given by*

$$\mathrm{E}^n f(x)\equiv f(x+n). \tag{5.75}$$

Proof Induction. □

This formula can be inverted.

Theorem 5.3.3

$$\mathrm{E}^n f(x)=\sum_{i=0}^{n}\binom{n}{i}_q\triangle_{\mathrm{CG},q}^{i}f(x). \tag{5.76}$$

Proof This is again the general inversion formula. □

Corollary 5.3.4 [415, p. 47, 1.122], *a q-analogue of* [320, p. 97, 10], [138, p. 27, 2.13], [385, p. 35, 2]. *Let us assume that the functions $f(x)$ and $g(x)$ depend on q^x. Then we have*

$$\triangle_{\mathrm{CG},q}^{n}(fg)=\sum_{i=0}^{n}\binom{n}{i}_q\triangle_{\mathrm{CG},q}^{i}f\,\triangle_{\mathrm{CG},q}^{n-i}\mathrm{E}^i g. \tag{5.77}$$

Proof Compare with [320, p. 96 f]:

$$\begin{aligned}\text{LHS} &= \sum_{k=0}^{n}(-1)^k\binom{n}{k}_q q^{\binom{k}{2}}\mathrm{E}^{n-k}f\mathrm{E}^{n-k}g\\ &= \sum_{k=0}^{n}(-1)^k\binom{n}{k}_q q^{\binom{k}{2}}\sum_{i=0}^{n-k}\binom{n-k}{i}_q \triangle_{\mathrm{CG},q}^{i} f\,\mathrm{E}^{n-k}g\\ &= \sum_{i=0}^{n}\binom{n}{i}_q \triangle_{\mathrm{CG},q}^{i} f\sum_{k=0}^{n-i}(-1)^k\binom{n-i}{k}_q q^{\binom{k}{2}}\mathrm{E}^{n-k}g = \text{RHS}. \end{aligned} \tag{5.78}$$

□

Example 21

$$\triangle_{\mathrm{CG},q}^{n}\left(q^{mx}\right) = \begin{cases} q^{mx}(-1)^n q^{\binom{n}{2}}\langle 1+m-n;q\rangle_n, & m \geq n,\\ 0, & m < n. \end{cases} \tag{5.79}$$

Proof Use (5.74). □

Example 22 Compare with [383, p. 92, 1.11]. For $m \leq n$

$$\triangle_{\mathrm{CG},q,x}^{m}\langle x+\gamma;q\rangle_n = \langle n-m+1;q\rangle_m\langle x+m+\gamma;q\rangle_{n-m}q^{m(x+\gamma+m-1)}. \tag{5.80}$$

This is equivalent to the following formula of Gould [234, 2.12, p. 91], a q-analogue of [385, p. 26, 6]:

$$\triangle_{\mathrm{CG},q}^{m}\binom{x}{n}_q = \binom{x}{n-m}_q q^{m(x+m-n)}, \quad m \leq n. \tag{5.81}$$

Proof Induction. □

In the following three theorems Greek letters denote arbitrary sequences.

Theorem 5.3.5 *A q-analogue of the Montmort formula. Jackson* [299, (2), p. 145], [304, (7), p. 146]:

$$\sum_{k=1}^{\infty}x^k\theta_k = \sum_{k=1}^{\infty}\frac{x^k}{(x;q)_k}\triangle_{\mathrm{CG},q}^{k-1}\theta_1. \tag{5.82}$$

Proof

$$\begin{aligned}\mathrm{RS}&=\sum_{k=1}^{\infty}x^k\sum_{l=0}^{\infty}x^l\binom{k+l-1}{l}_q\triangle_{\mathrm{CG},q}^{k-1}\theta_1\\&=\sum_{m=0}^{\infty}x^{m+1}\sum_{k=0}^{m}\binom{m}{k}_q\triangle_{\mathrm{CG},q}^{k}\theta_1=\sum_{m=0}^{\infty}x^{m+1}\sum_{k=0}^{m}\binom{m}{k}_q\prod_{l=0}^{k-1}(E-q^l)\theta_1\\&=\sum_{m=0}^{\infty}x^{m+1}\mathrm{E}^m\theta_1=\mathrm{LS}.\end{aligned}\tag{5.83}$$

□

Theorem 5.3.6 *Jackson* [304, (8), p. 146]:

$$\sum_{k=0}^{\infty}\frac{x^k\theta_k}{\{k\}_q!}=\mathrm{E}_q(x)\sum_{k=0}^{\infty}\frac{x^k\triangle_{\mathrm{CG},q}^{k}}{\{k\}_q!}\theta_0.\tag{5.84}$$

Proof Put $s_q\equiv\sum_{s=0}^{\infty}\frac{x^k\triangle_{\mathrm{CG},q}^{k}}{\{k\}_q!}\theta_0$. Then

$$s_q=\sum_{s=0}^{\infty}x^s\sum_{r=0}^{s}\frac{(-1)^{r+s}q^{\binom{s-r}{2}}}{\{r\}_q!\{s-r\}_q!}\theta_r=\mathrm{E}_{\frac{1}{q}}(-x)\sum_{k=0}^{\infty}\frac{x^k\theta_k}{\{k\}_q!}.\tag{5.85}$$

□

The following formula is a q-analogue of the Eulerian generalization of the Montmortian formula.

Theorem 5.3.7 *Jackson* [304, (9), p. 146]. *Let* $\Phi(x)\equiv\sum_{k=0}^{\infty}x^k\phi_k$. *Then*

$$\sum_{k=0}^{\infty}x^k\theta_k\phi_k=\sum_{k=0}^{\infty}\frac{\mathrm{D}_q^k\Phi(x)\triangle_{\mathrm{CG},q}^{k}}{\{k\}_q!}\theta_0.\tag{5.86}$$

Proof

$$\begin{aligned}\mathrm{LHS}&=\sum_{k=0}^{\infty}x^k\phi_k\sum_{n=0}^{k}\binom{k}{n}_q\triangle_{\mathrm{CG}q}^{n}\theta_0\\&=\sum_{k=0}^{\infty}\frac{x^k}{\{k\}_q!}\triangle_{\mathrm{CG}q}^{k}\theta_0\mathrm{D}_q^k\left(\sum_{l=0}^{\infty}x^l\phi_l\right).\end{aligned}\tag{5.87}$$

□

Theorem 5.3.8 *Milne* [383, p. 93]. *An exact formula for the second q-Stirling number with the CG-operator:*

$$\mathrm{S}(n,k)_q = \left(\{k\}_q! q^{\binom{k}{2}}\right)^{-1} \Delta_{\mathrm{CG},q}^n \{x\}_q^n \big| x=0. \tag{5.88}$$

We will now follow Schwatt [456, Chap. 5] and develop a calculus for the Carlitz function $\mathrm{Sc}_{\mathrm{C},m,q}(n)$ from the previous section. First, a lemma.

Lemma 5.3.9 *A q-analogue of* [456, p. 86, (50)]:

$$\sum_{s=k}^{n-1} \binom{s}{k}_q q^s = \binom{n}{k+1}_q q^k. \tag{5.89}$$

Theorem 5.3.10 *A q-analogue of* [456, p. 86, (51)]:

$$\mathrm{Sc}_{\mathrm{C},m,q}(n) = \sum_{k=0}^{m} (-1)^k q^k \binom{n}{k+1}_q \sum_{a=0}^{k} (-1)^a \binom{k}{a}_q q^{\binom{k-a}{2}} \{a\}_q^m. \tag{5.90}$$

Proof Write the LHS as

$$\theta_q^m \sum_{k=1}^{n-1} (xq)^k \Bigg|_{x=1}, \tag{5.91}$$

and use (5.50), (5.89) and (5.26). □

Theorem 5.3.11 *A q-analogue of* [456, p. 87, (63)]:

$$\mathrm{Sc}_{\mathrm{C},m,q}(n) = \sum_{k=0}^{n} \binom{n}{k}_q \sum_{a=0}^{k} (-1)^a \binom{k}{a}_q q^{\binom{a}{2}} \sum_{i=1}^{k-a-1} \{i\}_q^m q^i. \tag{5.92}$$

Proof Use (5.73) and (5.74). □

Now let $b_{m,k,q}$ denote the coefficient of $q^k \binom{n}{k+1}_q$ in $\mathrm{Sc}_{\mathrm{C},m,q}(n)$. Then, by (5.90), the following recurrence obtains, which is almost a q-analogue of [456, p. 88, (69)]:

$$b_{m,k,q} - \{k\}_q b_{m-1,k,q} = q^{k-1} \{k\}_q b_{m-1,k-1,q}. \tag{5.93}$$

We obtain the following expressions for $\mathrm{Sc}_{\mathrm{C},m,q}(n)$ as linear combinations of q-binomial coefficients.

Theorem 5.3.12 *Almost a q-analogue of Schwatt* [456, p. 88, (69)].

$$\mathrm{Sc}_{\mathrm{C},1,q}(n) = q\binom{n}{2}_q, \tag{5.94}$$

$$\mathrm{S}_{\mathrm{C},2,q}(n) = q\binom{n}{2}_q + q^3(1+q)\binom{n}{3}_q, \tag{5.95}$$

$$\mathrm{S}_{\mathrm{C},3,q}(n) = q\binom{n}{2}_q + q^3\big(1+q+(1+q)^2\big)\binom{n}{3}_q + q^6(1+q)\big(1+q+q^2\big)\binom{n}{4}_q, \tag{5.96}$$

$$\mathrm{S}_{\mathrm{C},4,q}(n) = q\binom{n}{2}_q + q^3(1+q)\big(2+q+(1+q)^2\big)\binom{n}{3}_q + q^6\big(1+q+q^2\big) \times \big(1+q+(1+q)^2+\{3\}_q!\big)\binom{n}{4}_q + q^{10}\{4\}_q!\binom{n}{5}_q. \tag{5.97}$$

By the q-Pascal identity we obtain the following q-analogue of Munch [390, p. 14].

Theorem 5.3.13

$$\mathrm{S}_{\mathrm{C},2,q}(n) = q^3\binom{n}{3}_q + q\binom{n+1}{3}_q, \tag{5.98}$$

$$\mathrm{S}_{\mathrm{C},3,q}(n) = q^6\binom{n}{4}_q + 2q^3(1+q)\binom{n+1}{4}_q + q\binom{n+3}{4}_q, \tag{5.99}$$

$$\mathrm{S}_{\mathrm{C},4,q}(n) = q^{10}\binom{n}{5}_q + q^6\big(3+5q+3q^2\big)\binom{n+1}{5}_q + q^3\big(3+5q+3q^2\big)\binom{n+3}{5}_q + q\binom{n+3}{5}_q. \tag{5.100}$$

We can now introduce the sum operator mentioned in the introduction.

Definition 104 The inverse of the CG difference operator is defined by

$$\triangle_{\mathrm{CG},q}^{-1} f(k)|_0^n \equiv \sum_0^n f(x)\delta_q(x) \equiv \sum_{k=0}^{n-1} f(k). \tag{5.101}$$

Example 23

$$\triangle_{\mathrm{CG},q}^{-1}\big(1-q^l\big)\langle n+1;q\rangle_{l-1}q^n \equiv \sum_{k=0}^{n-1}\big(1-q^l\big)\langle k+1;q\rangle_{l-1}q^k = \langle n;q\rangle_l. \tag{5.102}$$

Corollary 5.3.14

$$\{n\}_q\{n+1\}_q = \{2\}_q \sum_{i=1}^{n} \{i\}_q q^{i-1}, \tag{5.103}$$

$$\sum_{i=1}^{n} \{i\}_q q^{2i} = \frac{\{n\}_{2,q}}{\{2\}_q} - \frac{\{n\}_{3,q}(1-q)}{\{3\}_q}. \tag{5.104}$$

By (5.74) and (5.80), we obtain

Theorem 5.3.15 [400, p. 110].

$$\sum_{n=0}^{m} (-1)^n \binom{m}{n}_q q^{\binom{n}{2}} \langle x+1; q\rangle_{m-n} \langle x-y+1-n+m; q\rangle_n$$
$$= \langle y-m+1; q\rangle_m q^{m(x-y+m)}, \quad x, y \in \mathbb{C}. \tag{5.105}$$

We will return to this equation in the next chapter. Now we say goodbye to the Carlitz-Gould approach and continue with a similar operator which has the advantage of being simultaneously a q-derivative and a difference operator.

5.4 The Jackson q-derivative as difference operator

This section will be about how the Jackson q-derivative can be used as a difference operator operating on the space of all q-shifted factorials. We illustrate the technique with some examples. The similarity with the operator from the previous chapter is striking and will apparently lead to multiple q-equations. However, it turns out that most of these are doublets, as is shown in the example from the last section.

For functions of q^x, the Cigler operator ϵ [133] will be replaced by E in q-Leibniz theorems as below.

Theorem 5.4.1

$$\mathrm{D}^n_{q,q^x} \langle \gamma + x; q\rangle_k = (-1)^n \{k-n+1\}_{n,q} \langle \gamma + x + n; q\rangle_{k-n} q^{\binom{n}{2}+n\gamma}, \quad n \le k. \tag{5.106}$$

Example 24 We apply the operator D^m_{q,q^x} to (5.105). Then

$$\mathrm{LHS} = \sum_{n=0}^{m} (-1)^n \binom{m}{n}_q q^{\binom{n}{2}} \sum_{i=0}^{m} \binom{m}{i}_q \mathrm{D}^i_{q,q^x} \langle x+1; q\rangle_{m-n}$$
$$\times \mathrm{E}^i \mathrm{D}^{m-i}_{q,q^x} \langle x-y+1-n+m; q\rangle_n$$

$$
\begin{aligned}
&= \sum_{n=0}^{m} (-1)^n \binom{m}{n}_q q^{\binom{n}{2}} \\
&\times \binom{m}{m-n}_q (-1)^m \{m-n\}_q! q^{\binom{m-n}{2}+m-n} \{n\}_q! q^{\binom{n}{2}+n(-y+1-n+m)} \\
&= (-1)^m q^{\binom{m}{2}+m} \{m\}_q! \sum_{n=0}^{m} (-1)^n \binom{m}{n}_q q^{\binom{n}{2}-ny} \\
&= q^{m^2-my} \langle y+1-m; q\rangle_m \{m\}_q! = \text{RHS}.
\end{aligned} \tag{5.107}
$$

Now instead rewrite (5.105) in the form

$$
\begin{aligned}
&\sum_{n=0}^{m} \binom{m}{n}_q \langle x+1; q\rangle_{m-n} \langle y-x-m; q\rangle_n q^{n(x+m)+y(m-n)} \\
&\quad = \langle y-m+1; q\rangle_m q^{m(x+m)}, \quad x, y \in \mathbb{C},
\end{aligned} \tag{5.108}
$$

and operate with D^m_{q,q^y} on both sides to obtain

$$
\begin{aligned}
\text{LHS} &= \sum_{n=0}^{m} \binom{m}{n}_q \langle x+1; q\rangle_{m-n} q^{n(x+m)} \sum_{i=0}^{m} \binom{m}{i}_q \mathrm{D}^i_{q,q^y} q^{y(m-n)} \\
&\times \mathrm{E}^i \mathrm{D}^{m-i}_{q,q^y} \langle y-x-m; q\rangle_n \\
&= \sum_{n=0}^{m} \binom{m}{n}_q \langle x+1; q\rangle_{m-n} q^{n(x+m)} \\
&\times \binom{m}{m-n}_q \{m-n\}_q! (-1)^n \{n\}_q! q^{\binom{n}{2}-n(x+m)} \\
&= \{m\}_q! \sum_{n=0}^{m} \binom{m}{n}_q \langle x+1; q\rangle_{m-n} (-1)^n q^{\binom{n}{2}}.
\end{aligned} \tag{5.109}
$$

The RHS is

$$
(-1)^m \{m\}_q! q^{\binom{m}{2}+m(x+1)}. \tag{5.110}
$$

After simplification this last equality is equivalent to a confluent form of the second q-Vandermonde identity.

Inspired by the previous calculation we make the following

Definition 105 The Jackson q-difference is defined by

$$\begin{aligned}\triangle_{\mathrm{J},x,q} f\big(q^{x}\big) &\equiv \triangle_{\mathrm{J},q} f\big(q^{x}\big)\\ &\equiv \big(f\big(q^{x+1}\big)-f\big(q^{x}\big)\big)q^{-x}\equiv -(1-q)\mathrm{D}_{q,q^{x}} f\big(q^{x}\big),\end{aligned} \tag{5.111}$$

$$\triangle_{\mathrm{J},q}^{n+1}=\triangle_{\mathrm{J},q}\triangle_{\mathrm{J},q}^{n}. \tag{5.112}$$

The following equation holds.

Theorem 5.4.2

$$\triangle_{\mathrm{J},q}\left(q^{\binom{k}{2}}\binom{x}{k}_{q}\right)=q^{\binom{k-1}{2}}\binom{x}{k-1}_{q}. \tag{5.113}$$

Proof Use the q-Pascal identity. □

Corollary 5.4.3

$$\triangle_{\mathrm{J},q}^{m}\binom{x}{n}_{q}=\binom{x}{n-m}_{q} q^{-mn+\binom{m+1}{2}}, \quad m\le n. \tag{5.114}$$

Example 25

$$\sum_{n=0}^{\infty} q^{\binom{n}{2}}\binom{x}{n}_{q} t^{n}=\mathrm{e}_{q}\big(-tq^{x}\boxplus_{q} t\big). \tag{5.115}$$

Proof Use the q-binomial theorem. □

We now present the Jackson quartet.

Theorem 5.4.4 *The following q-Taylor formula holds*:

$$f(x)=\sum_{k=0}^{\infty}\binom{x}{k}_{q} q^{\binom{k}{2}}\triangle_{\mathrm{J},q}^{k} f(0). \tag{5.116}$$

Theorem 5.4.5 *A q-analogue of* [138, p. 26]. *Compare with* [450, p. 82]:

$$\triangle_{\mathrm{J},q}^{n} f\big(q^{x}\big)=q^{-nx-\binom{n}{2}}\sum_{k=0}^{n}(-1)^{k}\binom{n}{k}_{q} q^{\binom{k}{2}}\mathrm{E}^{n-k} f\big(q^{x}\big). \tag{5.117}$$

Proof Use the corresponding equation for the q-derivative. □

Relation (5.117) can be inverted.

Theorem 5.4.6

$$\mathrm{E}^n f(q^x) = \sum_{k=0}^{n} q^{xk+\binom{k}{2}} \binom{n}{k}_q \triangle_{\mathrm{J},q}^k f(q^x). \tag{5.118}$$

Corollary 5.4.7 *A q-analogue of* [320, p. 97, 10], [138, p. 27, 2.13], [385, p. 35, 2]:

$$\triangle_{\mathrm{J},q}^n \big(f(q^x)g(q^x)\big) = \sum_{k=0}^{n} \binom{n}{k}_q \triangle_{\mathrm{J},q}^k f(q^x)\big(\triangle_{\mathrm{J},q}^{n-k}\mathrm{E}^k\big)g(q^x). \tag{5.119}$$

Proof Use the Leibniz theorem for the q-derivative. □

Example 26

$$\triangle_{\mathrm{J},q}^n \big(q^{mx}\big) = \begin{cases} q^{x(m-n)}(-1)^n \langle 1+m-n; q\rangle_n, & m \geq n, \\ 0, \quad m < n. \end{cases} \tag{5.120}$$

5.5 Applications

The technique developed above leads to easy proofs of q-binomial coefficient identities. The following example can also be proved by using the q-Vandermonde identity.

Example 27 A q-analogue of the important formula [138, p. 27], [433, p. 15, (9)], [440, p. 65]:

$$\binom{x}{m}_q \binom{x}{n}_q = \sum_{k=0}^{m+n} \mathrm{QE}\big((k-n)(k-m)\big) \binom{k}{n}_q \binom{n}{m+n-k}_q \binom{x}{k}_q. \tag{5.121}$$

Proof We have

$$\begin{aligned} \triangle_{\mathrm{CG},q}^k \binom{x}{m}_q \binom{x}{n}_q &= \sum_{l=0}^{k} \binom{k}{l}_q \mathrm{QE}\big(l(x+l-n) + (k-l)(x+k-m)\big) \\ &\quad \times \binom{x+l}{m-k+l}_q \binom{x}{n-l}_q. \end{aligned} \tag{5.122}$$

Now use the q-Taylor formula (5.73) with $y=0$, $f(x) = \binom{x}{m}_q \binom{x}{n}_q$. □

Remark 17 If we use $\triangle_{\mathrm{J},q}$ instead, we get

$$\begin{aligned}\binom{x}{m}_q\binom{x}{n}_q &= \sum_{k=0}^{m+n}\mathrm{QE}\left(-\binom{m}{2}-\binom{n}{2}+\binom{m+n-k}{2}+\binom{k}{2}\right)\\ &\quad\times\binom{k}{n}_q\binom{n}{m+n-k}_q\binom{x}{k}_q. \qquad (5.123)\end{aligned}$$

This equation is however equivalent to (5.121).

Chapter 6
The first q-functions

We can now use our previous knowledge to start the first investigations of q-functions, a necessary prerequisite for the more technical chapter seven. We start with a short rehearsal of the definitions.

6.1 q-analogue, q-factorial, tilde operator

Definition 106 The q-analogues of a complex number a, a natural number n, the factorial and of the semifactorial [135, p. 43], [303, p. 314] are defined as follows:

$$\{a\}_q \equiv \frac{1-q^a}{1-q}, \quad q \in \mathbb{C}\backslash\{0,1\}, \tag{6.1}$$

$$\{n\}_q \equiv \sum_{k=1}^{n} q^{k-1}, \qquad \{0\}_q = 0, \quad q \in \mathbb{C}\backslash\{0,1\}, \tag{6.2}$$

$$\{n\}_q! \equiv \prod_{k=1}^{n} \{k\}_q, \qquad \{0\}_q! \equiv 1, \quad q \in \mathbb{C}\backslash\{0,1\}, \tag{6.3}$$

$$\{2n-1\}_q!! \equiv \prod_{k=1}^{n} \{2k-1\}_q, \quad q \in \mathbb{C}\backslash\{0,1\}, \tag{6.4}$$

$$\{2n\}_q!! \equiv \prod_{k=1}^{n} \{2k\}_q, \quad q \in \mathbb{C}\backslash\{0,1\}. \tag{6.5}$$

The following equation [235, p. 282] applies:

$$\{-\alpha\}_q = -q^{-\alpha}\{\alpha\}_q. \tag{6.6}$$

We also have

$$\{\alpha\}_{\frac{1}{q}} = q^{-\alpha+1}\{\alpha\}_q. \tag{6.7}$$

T. Ernst, *A Comprehensive Treatment of q-Calculus*,
DOI 10.1007/978-3-0348-0431-8_6, © Springer Basel 2012

Theorem 6.1.1 *The following formulae hold for* $0 < |q| < 1$:

$$\langle a;q\rangle_n = \frac{\langle a;q\rangle_\infty}{\langle a+n;q\rangle_\infty}, \tag{6.8}$$

$$\left(aq^{-n};q\right)_n \equiv \frac{1}{(a;q)_{-n}} = \frac{(aq^{-n};q)_\infty}{(a;q)_\infty}. \tag{6.9}$$

Proof These identities follow immediately from the definition. □

Definition 107 The following two formulas are used as definitions for $\langle a;q\rangle_\alpha$ and $(a;q)_\alpha$, $\alpha \in \mathbb{C}$, compare with [261, p. 342].

$$\langle a;q\rangle_\alpha \equiv \frac{\langle a;q\rangle_\infty}{\langle a+\alpha;q\rangle_\infty}, \quad a \neq -m-\alpha, m=0,1,\ldots, \tag{6.10}$$

$$(a;q)_\alpha \equiv \frac{(a;q)_\infty}{(aq^\alpha;q)_\infty}, \quad a \neq q^{-m-\alpha}, m=0,1,\ldots. \tag{6.11}$$

For negative subscripts, the shifted factorial and the q-shifted factorials are defined by

$$(a)_{-n} \equiv \frac{1}{(a-1)(a-2)\cdots(a-n)} \equiv \frac{1}{(a-n)_n} = \frac{(-1)^n}{(1-a)_n}, \tag{6.12}$$

$$\langle a;q\rangle_{-n} \equiv \frac{1}{\langle a-n;q\rangle_n} = \frac{(-1)^n q^{n(1-a)+\binom{n}{2}}}{\langle 1-a;q\rangle_n}. \tag{6.13}$$

1 N, the unit of measurement for force may be compared with the standard formulas for the q-factorial.

1 N	(6.14)–(6.21), (6.31)–(6.42)

We are now prepared to prove a series of simple basic formulas for computations with q-hypergeometric functions.

Theorem 6.1.2 *The following formulas apply whenever* $q \neq 0$ *and* $q \neq e^{2\pi i t}$, $t \in \mathbb{Q}$:

$$\langle -a+1-n;q\rangle_n = \langle a;q\rangle_n (-1)^n q^{-\binom{n}{2}-na}, \tag{6.14}$$

$$\langle a;q\rangle_{n-k} = \frac{\langle a;q\rangle_n}{\langle -a+1-n;q\rangle_k}(-1)^k q^{\binom{k}{2}+k(1-a-n)}, \tag{6.15}$$

$$\langle a+k;q\rangle_{n-k} = \frac{\langle a;q\rangle_n}{\langle a;q\rangle_k}, \tag{6.16}$$

$$\langle a+n_1;q\rangle_{k_1} = \frac{\langle a;q\rangle_{k_2}\langle a+k_2;q\rangle_{n_2}}{\langle a;q\rangle_{n_1}}, \quad n_1+k_1=n_2+k_2, \tag{6.17}$$

$$\langle a+2k;q\rangle_{n-k} = \frac{\langle a;q\rangle_n \langle a+n;q\rangle_k}{\langle a;q\rangle_{2k}}, \tag{6.18}$$

$$\langle a;q\rangle_m\langle a-n;q\rangle_{2n}=\langle a;q\rangle_n\langle a-n;q\rangle_m\langle a+m-n;q\rangle_n, \tag{6.19}$$

$$\langle -n;q\rangle_k=\frac{\langle 1;q\rangle_n}{\langle 1;q\rangle_{n-k}}(-1)^k q^{\binom{k}{2}-nk}, \tag{6.20}$$

$$\langle a-n;q\rangle_k=\frac{\langle a;q\rangle_k\langle 1-a;q\rangle_n}{\langle -a+1-k;q\rangle_n q^{nk}}, \tag{6.21}$$

$$\frac{\langle a;q\rangle_n\langle 1-a;q\rangle_k}{\langle 1-a-n;q\rangle_k\langle a-k;q\rangle_n}=q^{nk}, \tag{6.22}$$

$$\frac{\langle 1-b;q\rangle_{k+m}\langle a-k;q\rangle_k}{\langle 1-b+k;q\rangle_m\langle 1-a;q\rangle_k\langle b-k;q\rangle_k}=q^{k(a-b)}, \tag{6.23}$$

$$\left\langle a;\frac{1}{q}\right\rangle_n=\langle a;q\rangle_n(-1)^n q^{-\binom{n}{2}-na}. \tag{6.24}$$

Proof Formula (6.14) is proved as follows:

$$\begin{aligned}
\text{LHS}&=\left(1-q^{-a+1-n}\right)\left(1-q^{-a+2-n}\right)\cdots\left(1-q^{-a}\right),\\
\text{RHS}&=\left(1-q^{a}\right)\left(1-q^{a+1}\right)\cdots\left(1-q^{a+n-1}\right)(-1)^n q^{-na-(0+1+2+\cdots+(n-1))}\\
&=\left(1-q^{-a}\right)\left(q-q^{-a}\right)\cdots\left(q^{n-1}-q^{-a}\right)q^{-(0+1+2+\cdots+(n-1))}\\
&=\left(1-q^{-a}\right)\left(1-q^{-a-1}\right)\cdots\left(1-q^{-a-n+1}\right)=\text{LHS}.
\end{aligned}$$

Let us prove (6.15). There are two cases to consider. Case 1: $n>k$.

$$\begin{aligned}
\text{LHS}&=\left(1-q^{a}\right)\left(1-q^{a+1}\right)\cdots\left(1-q^{a+n-k-1}\right),\\
\text{RHS}&=\frac{(1-q^{a})(1-q^{a+1})\cdots(1-q^{a+n-1})(-1)^k q^{k(1-a)+0+1+2+\cdots+(k-1)}}{(1-q^{-a+1-n})(1-q^{-a+2-n})\cdots(1-q^{-a+k-n})q^{nk}}\\
&=\frac{(1-q^{a})(1-q^{a+1})\cdots(1-q^{a+n-1})}{(1-q^{a+n-1})(1-q^{a+n-2})\cdots(1-q^{a+n-k})}=\text{LHS}.
\end{aligned}$$

Case 2: $n<k$.

$$\text{LHS}=\left(\left(1-q^{a+n-k}\right)\left(1-q^{a+n-k+1}\right)\cdots\left(1-q^{a-1}\right)\right)^{-1}=\text{RHS}.$$

The identities (6.17)–(6.18) follow easily from the definition (1.30) and the formula (6.20) is proved as follows:

$$\begin{aligned}
\text{LHS}&=\left(1-q^{-n}\right)\left(1-q^{-n+1}\right)\cdots\left(1-q^{-n+k-1}\right),\\
\text{RHS}&=\frac{(1-q)(1-q^2)\cdots(1-q^n)(-1)^k q^{0+1+2+\cdots+(k-1)}}{(1-q)(1-q^2)\cdots(1-q^{n-k})q^{nk}}\\
&=\frac{(1-q^{n-k+1})(1-q^{n-k+2})\cdots(1-q^n)(-1)^k q^{0+1+2+\cdots+(k-1)}}{q^{nk}}\\
&=\left(q^{-k+1}-q^{-n}\right)\left(q^{-k+2}-q^{-n}\right)\cdots\left(1-q^{-n}\right)q^{0+1+2+\cdots+(k-1)}\\
&=\left(1-q^{k-n-1}\right)\left(1-q^{k-n-2}\right)\cdots\left(1-q^{-n}\right)=\text{LHS}.
\end{aligned}$$

The following computation proves (6.21):

$$\text{LHS} = \big(1-q^{a-n}\big)\big(1-q^{a-n+1}\big)\cdots\big(1-q^{a-n+k-1}\big),$$

$$\text{RHS} = \frac{(1-q^a)(1-q^{a+1})\cdots(1-q^{a+k-1})(1-q^{1-a})(1-q^{2-a})\cdots(1-q^{n-a})}{(1-q^{-a+1-k})(1-q^{-a+2-k})\cdots(1-q^{n-a-k})q^{nk}}$$

$$= \frac{(1-q^a)(1-q^{a+1})\cdots(1-q^{a+k-1})(q^{1-a}-1)(q^{2-a}-1)\cdots(q^{n-a}-1)}{(q^{-a+1}-q^k)(q^{-a+2}-q^k)\cdots(q^{n-a}-q^k)}$$

$$= \frac{(1-q^a)(1-q^{a+1})\cdots(1-q^{a+k-1})(1-q^{a-1})(q-q^{a-1})\cdots(q^{n-1}-q^{a-1})}{(1-q^{a+k-1})(q-q^{a+k-1})\cdots(q^{n-1}-q^{a+k-1})}$$

$$= \frac{(1-q^a)(1-q^{a+1})\cdots(1-q^{a+k-1})(1-q^{a-1})(1-q^{a-2})\cdots(1-q^{a-n})}{(1-q^{a+k-1})(1-q^{a+k-2})\cdots(1-q^{a+k-n})} = \text{LHS}.$$

□

The vector product can be compared with the tilde operator. This fits well, as these two operators have a fundamental significance for the respective disciplines.

$a \times b$	$\widetilde{\langle a; q\rangle_n}$

Definition 108 In the following, $\frac{\mathbb{C}}{\mathbb{Z}}$ will denote the space of complex numbers mod $\frac{2\pi i}{\log q}$. This is isomorphic to the cylinder $\mathbb{R} \times \{e^{2\pi i\theta}, \theta \in \mathbb{R}\}$. The operator

$$\tilde{}: \frac{\mathbb{C}}{\mathbb{Z}} \to \frac{\mathbb{C}}{\mathbb{Z}}$$

is defined by the 2-torsion

$$a \mapsto a + \frac{\pi i}{\log q}. \tag{6.25}$$

Furthermore we define

$$\widetilde{\langle a; q\rangle_n} \equiv \langle \widetilde{a}; q\rangle_n. \tag{6.26}$$

By (6.25), we have

$$\widetilde{\langle a; q\rangle_n} = \prod_{m=0}^{n-1}\big(1+q^{a+m}\big), \tag{6.27}$$

where this time the tilde denotes an involution which changes a minus sign to a plus sign in all the n factors of $\langle a; q\rangle_n$.

The following simple rules follow from (6.25). Clearly the first two equations are applicable to q-exponents. Compare with [499, p. 110].

$$\widetilde{a} \pm b \equiv \widetilde{a \pm b} \left(\text{mod}\, \frac{2\pi i}{\log q}\right), \tag{6.28}$$

$$\widetilde{a} \pm \widetilde{b} \equiv a \pm b \left(\text{mod}\, \frac{2\pi i}{\log q}\right), \tag{6.29}$$

$$q^{\widetilde{a}} = -q^a, \tag{6.30}$$

where the second equation is a consequence of the fact that we work mod $\frac{2\pi i}{\log q}$. We now conclude this section with some more formulas for q-shifted factorials.

Theorem 6.1.3

$$\langle a+1;q\rangle_{2n} = \frac{\langle a;q\rangle_{2n}\langle 1+\frac{a}{2};q\rangle_n \langle \widetilde{1+\frac{a}{2};q}\rangle_n}{\langle \frac{a}{2};q\rangle_n \langle \widetilde{\frac{a}{2};q}\rangle_n}, \tag{6.31}$$

$$\frac{\langle a, \frac{2+a}{2}, \widetilde{\frac{2+a}{2}};q\rangle_n}{\langle \frac{a}{2}, \widetilde{\frac{a}{2}};q\rangle_n} = \frac{\langle a+1;q\rangle_{2n}}{\langle a+n;q\rangle_n}. \tag{6.32}$$

Proof These identities follow from the definition (1.30) and the definition of the tilde operator. □

Remark 18 The first two formulae together form a q-analogue of the very important formula [425, p. 22].

Theorem 6.1.4 *The following formulae, which are well-known, take the following form in the new notation. They hold whenever $q \neq 0$ and $q \neq e^{2\pi i t}$, $t \in \mathbb{Q}$.*

$$\langle a;q^2\rangle_n = \langle a;q\rangle_n \langle \widetilde{a;q}\rangle_n, \tag{6.33}$$

$$\langle a;q\rangle_{2n} = \left\langle \frac{a}{2};q^2\right\rangle_n \left\langle \frac{a+1}{2};q^2\right\rangle_n, \tag{6.34}$$

$$\lim_{b\to-\infty} \langle b;q\rangle_j q^{-bj} = (-1)^j q^{\binom{j}{2}}. \tag{6.35}$$

Theorem 6.1.5 *A q-analogue of the Legendre duplication formula for the Γ function written in q-shifted factorial form* [171, p. 179, (83)]:

$$\frac{\langle n;q\rangle_n}{\langle \widetilde{1-n;q}\rangle_{n-1}\langle \frac{1}{2};q^2\rangle_n} = \mathrm{QE}\left(\binom{n}{2}\right). \tag{6.36}$$

Proof We use induction. The relation holds for $n=2$. Assume that (6.36) holds for n, and show that it also holds for $n+1$. Indeed, taking the ratio of the two expressions, for $n+1$ and n, we get

$$\frac{\langle n+1;q\rangle_{n+1}\langle \widetilde{1-n;q}\rangle_{n-1}\langle \frac{1}{2};q^2\rangle_n q^{\binom{n}{2}}}{\langle n;q\rangle_n \langle \widetilde{-n;q}\rangle_n \langle \frac{1}{2};q^2\rangle_{n+1} q^{\binom{n+1}{2}}} = \frac{(1-q^{2n+1})(1-q^{2n})q^{-n}}{(1-q^n)(1+q^{-n})(1-q^{2n+1})} = 1. \tag{6.37}$$

□

Theorem 6.1.6 [171, p. 170, (42)]:

$$\frac{\langle 1-n;q^2\rangle_{\frac{n}{2}}}{\langle \widetilde{1-n;q}\rangle_{n-1}\langle\frac{1}{2};q^2\rangle_{\frac{n}{2}}}=(-1)^{\frac{n}{2}}\mathrm{QE}\left(-\frac{n^2}{4}\right),\quad n\ even. \tag{6.38}$$

Proof We use induction. The result holds for $n=2$. Assume that (6.38) holds for n. We show that it also holds for $n+2$, i.e., that

$$\frac{\langle 1-n-2;q^2\rangle_{\frac{n}{2}+1}}{\langle \widetilde{1-n-2;q}\rangle_{n+1}\langle\frac{1}{2};q^2\rangle_{\frac{n}{2}+1}}=(-1)^{\frac{n}{2}+1}\mathrm{QE}\left(-\frac{(n+2)^2}{4}\right),\quad n\ even. \tag{6.39}$$

Indeed, taking the ratio of the two expressions we get

$$\begin{aligned}\frac{(6.39)}{(6.38)}&=\frac{\langle 1-n-2;q^2\rangle_{\frac{n}{2}+1}\langle \widetilde{1-n;q}\rangle_{n-1}\langle\frac{1}{2};q^2\rangle_{\frac{n}{2}}}{\langle 1-n;q^2\rangle_{\frac{n}{2}}\langle \widetilde{1-n-2;q}\rangle_{n+1}\langle\frac{1}{2};q^2\rangle_{\frac{n}{2}+1}}\\&=\frac{(1-q^{-2n-2})}{(1-q^{n+1})(1+q^{-n-1})}=-q^{-n-1}.\end{aligned} \tag{6.40}$$

□

The formula (6.38) has a different character than (6.36) and has certain similarities with the formula (8.43) (Γ_q functions with negative integer values). The following theorem is an interesting dual to these two formulas.

Theorem 6.1.7 *A q-analogue of Legendre's duplication formula* (3.35) [220, p. 18]:

$$\Gamma_q(2x)\Gamma_{q^2}\left(\frac{1}{2}\right)=(1+q)^{2x-1}\Gamma_{q^2}(x)\Gamma_{q^2}\left(x+\frac{1}{2}\right). \tag{6.41}$$

Theorem 6.1.8 *Euler* [190, p. 271] *found the following formula in connection with partitions*:

$$\langle\widetilde{1};q\rangle_\infty\left\langle\frac{1}{2};q^2\right\rangle_\infty=1. \tag{6.42}$$

Rauch and Lebowitz [428, p. 109] used this formula in connection with product expansions of theta functions. This formula can also be found in Vivanti [524, p. 170], Hardy and Wright [268, p. 366] and Apostol [37, p. 325].

6.2 The q-derivative

Speed can be compared with the q-derivative.

$\frac{ds}{dt}$	D_q

Variants of the q-derivative were used by Euler and Heine, but a real q-derivative was invented first by Jackson in 1908 [298].

Definition 109 Let φ be a continuous real function. Then we define the q-derivative as follows:

$$(D_q\varphi)(x) \equiv \begin{cases} \frac{\varphi(x)-\varphi(qx)}{(1-q)x}, & \text{if } q \in \mathbb{C}\backslash\{1\},\ x \neq 0, \\ \frac{d\varphi}{dx}(x), & \text{if } q = 1, \\ \frac{d\varphi}{dx}(0), & \text{if } x = 0. \end{cases} \tag{6.43}$$

If we want to indicate the variable which the q-difference operator is applied to, we write $(D_{q,x}\varphi)(x, y)$ for the operator.

The limit as q approaches 1 is the derivative

$$\lim_{q\to 1}(D_q\varphi)(x) = \frac{d\varphi}{dx}, \tag{6.44}$$

if φ is differentiable at x.

Peter Lesky [358] took $q \in \mathbb{R}\backslash\{-1, 0, 1\}$ in (6.43) in his study of real q-operator equations of second order. As the analysis in (6.100) shows, q-analysis with $q =$ root of unity has quite a different character.

Example 28

$$D_q\left(x^\alpha\right) = \frac{x^\alpha - (qx)^\alpha}{(1-q)x} = \frac{x^\alpha(1-q^\alpha)}{x(1-q)} = \{\alpha\}_q x^{\alpha-1}, \quad \alpha \in \mathbb{C}. \tag{6.45}$$

Jackson [302]

$$D_q \log x = \frac{\log q}{q-1}\frac{1}{x}, \quad 0 < q < 1,\ x > 0. \tag{6.46}$$

Theorem 6.2.1 *The formulas for the q-derivative of a sum, a product and a quotient of functions are, respectively* [194]:

$$D_q\big(u(x) + v(x)\big) = D_q u(x) + D_q v(x), \tag{6.47}$$

$$D_q\big(u(x)v(x)\big) = D_q u(x)v(x) + u(qx)D_q v(x), \tag{6.48}$$

$$D_q\left(\frac{u(x)}{v(x)}\right) = \frac{v(x)D_q u(x) - u(x)D_q v(x)}{v(qx)v(x)}, \quad v(qx)v(x) \neq 0. \tag{6.49}$$

Applying the Taylor formula to the right-hand side of (6.43) we obtain the following expression for the q-difference operator:

$$\mathrm{D}_q\big(f(x)\big) = \sum_{k=0}^{\infty} \frac{(q-1)^k}{(k+1)!} x^k f^{(k+1)}(x), \tag{6.50}$$

provided that f is analytic.

The following lemma was proved by Daniel Larsson and Sergei Silvestrov [353].

Lemma 6.2.2 *The chain rule for the q-derivative. Let $g(x)$ be a function with the properties $g(x) \neq 0$ and $g(qx) \neq g(x)$. Then*

$$\mathrm{D}_q(f \circ g)(x) = \Big(\mathrm{D}_{\frac{g(qx)}{g(x)}}(f)\Big)\big(g(x)\big) \times \mathrm{D}_q(g)(x). \tag{6.51}$$

6.3 The q-integral

Thomae was a pupil of Heine, who in the year 1869 [497], [498] introduced the so-called q-integral:

Definition 110

$$\int_0^1 f(t,q)\, d_q(t) \equiv (1-q) \sum_{n=0}^{\infty} f\big(q^n, q\big) q^n, \quad 0 < q < 1. \tag{6.52}$$

In fact, Thomae proved that the Heine transformation for ${}_2\phi_1(a,b;c|q;z)$ was a q-analogue of the Euler beta integral (3.25), which can be expressed as a quotient of q-Gamma functions.

In 1910 Jackson [220], [302], [497] defined the general q-integral:

Definition 111

$$\int_a^b f(t,q)\, d_q(t) \equiv \int_0^b f(t,q)\, d_q(t) - \int_0^a f(t,q)\, d_q(t), \quad a, b \in \mathbb{R}, \tag{6.53}$$

where

$$\int_0^a f(t,q)\, d_q(t) = a(1-q) \sum_{n=0}^{\infty} f\big(aq^n, q\big) q^n, \quad 0 < |q| < 1,\ a \in \mathbb{R}. \tag{6.54}$$

Following Jackson we will put

$$\int_0^{\infty} f(t,q)\, d_q(t) \equiv (1-q) \sum_{n=-\infty}^{\infty} f\big(q^n, q\big) q^n, \quad 0 < |q| < 1, \tag{6.55}$$

provided the sum converges absolutely [344]. Here we allow q to be complex, but only real values correspond in the limit $q \to 1$ to ordinary integrals.

The bilateral q-integral is defined by

$$\int_{-\infty}^{\infty} f(t,q)\,d_q(t) \equiv (1-q)\sum_{n=-\infty}^{\infty}\big[f(q^n,q)+f(-q^n,q)\big]q^n, \quad 0<|q|<1. \tag{6.56}$$

If $f(t,q)$ is continuous on $[0,x]\times(0,1)$, then

$$\begin{aligned} &\mathrm{D}_q\left(\int_0^x f(t,q)\,d_q(t)\right) \\ &= \mathrm{D}_q\left(x(1-q)\sum_{n=0}^{\infty} f(xq^n,q)q^n\right) \\ &= \frac{x(1-q)\sum_{n=0}^{\infty} f(xq^n,q)q^n - x(1-q)\sum_{n=0}^{\infty} f(xq^{n+1},q)q^{n+1}}{x(1-q)} = f(x). \end{aligned} \tag{6.57}$$

The following formulas for q-integration by parts will be useful [194], [302]:

Theorem 6.3.1 *If $u(t,q)$ and $v(t,q)$ are continuous on $[a,b]\times(0,1)$, then*

$$\int_a^b u(t,q)\mathrm{D}_q v(t,q)\,d_q(t) = \big[u(t,q)v(t,q)\big]_a^b - \int_a^b v(qt,q)\mathrm{D}_q u(t,q)\,d_q(t), \tag{6.58}$$

$$\int_a^b u(qt,q)\mathrm{D}_q v(t,q)\,d_q(t) = \big[u(t,q)v(t,q)\big]_a^b - \int_a^b v(t,q)\mathrm{D}_q u(t,q)\,d_q(t). \tag{6.59}$$

Theorem 6.3.2 *If $u(t,q)$ and $v(t,q)$ are continuous on $[0,\infty]\times(0,1)$, $0<q<1$, then*

$$\int_0^{\infty} u(t,q)\mathrm{D}_q v(t,q)\,d_q(t) = \big[u(t,q)v(t,q)\big]_0^{\infty} - \int_0^{\infty} v(qt,q)\mathrm{D}_q u(t,q)\,d_q(t), \tag{6.60}$$

provided that all expressions have a meaning.

Proof This follows from a straightforward computation. We observe that $\lim_{n\to+\infty} q^n = 0$ and $\lim_{n\to-\infty} q^n = +\infty$. □

Theorem 6.3.3 *If $u(t,q)$ and $v(t,q)$ are continuous on $[-\infty,\infty]\times(0,1)$, $0<q<1$, then*

$$\int_{-\infty}^{\infty} u(t,q)\mathrm{D}_q v(t,q)\,d_q(t) = \big[u(t,q)v(t,q)\big]_{-\infty}^{\infty} - \int_{-\infty}^{\infty} v(qt,q)\mathrm{D}_q u(t,q)\,d_q(t), \tag{6.61}$$

provided that all expressions have a meaning.

Proof We observe that $\lim_{n\to+\infty} q^n = 0$ and $\lim_{n\to-\infty} q^n = +\infty$.

It will suffice to prove that

$$\sum_{n=-\infty}^{\infty} \left(u(q^n,q)\left(v(q^n,q)-v(q^{n+1},q)\right)-u(-q^n,q)\left(v(-q^n,q)-v(-q^{n+1},q)\right)\right)$$

$$=\left[u(t,q)v(t,q)\right]_{-\infty}^{\infty}-\sum_{m=-\infty}^{\infty}\left(v(q^{m+1},q)\left(u(q^m,q)-u(q^{m+1},q)\right)\right.$$

$$\left.\times\, v(-q^{m+1},q)\left(u(-q^m,q)-u(-q^{m+1},q)\right)\right). \tag{6.62}$$

All terms with different first function argument in both sums cancel. Also, the terms $u(0,q)v(0,q)$ cancel. From the first sum only the term $\lim_{N\to\infty} u(q^{-N},q)\times v(q^{-N},q)$ survives and so does the corresponding term from the second sum. □

Lesky [358, p. 138] has proved the following similar formula, which he calls partial q-summation:

Theorem 6.3.4 *If $u(x)$ and $v(x)$ are two continuous functions, defined on discrete sets and $x_k = Aq^k$, $k = 0, 1, \dots, N$. Then*

$$\sum_{k=0}^{N} u(qx_k)\mathrm{D}_q v(x_k)(q-1)x_k$$

$$=\left[u(x_k)v(x_k)\right]_0^{N+1}-\sum_{k=0}^{N} v(x_k)\mathrm{D}_q u(x_k)(q-1)x_k. \tag{6.63}$$

Richard Askey has stated that one cannot change variables in a q-integral. However a substitution is possible in the linear case.

Lemma 6.3.5 *Linear substitution in a q-integral*:

$$\int_0^x f(t,q)\,d_q(t) = a\int_0^{\frac{x}{a}} f(at,q)\,d_q(t). \tag{6.64}$$

6.4 Two other tilde operators

Acceleration corresponds to two other (generalized) tilde operators.

$\frac{d^2 s}{dt^2}$	$\langle \widetilde{\overset{m}{T}\, a;q}\rangle_n,\ {}_k\widetilde{\langle a;q\rangle_n}$

It is desirable to generalize the tilde operator to be able to treat a general root of unity. In Eqs. (6.65) to (6.74) we assume that $(m,l)=1$.

Definition 112 The generalized tilde operator

$$\widetilde{\tfrac{m}{l}}: \frac{\mathbb{C}}{\mathbb{Z}} \to \frac{\mathbb{C}}{\mathbb{Z}}$$

is defined by

$$a \mapsto a + \frac{2\pi i m}{l \log q}. \tag{6.65}$$

This means

$$\langle \widetilde{\tfrac{m}{l}} a; q \rangle_n \equiv \prod_{m=0}^{n-1} \left(1 - e^{2\pi i \frac{m}{l}} q^{a+m}\right). \tag{6.66}$$

Next we define

$$\widetilde{{}_{\frac{m}{l}}\langle a; q\rangle_n} \equiv \langle \widetilde{\tfrac{m}{l}} a; q \rangle_n. \tag{6.67}$$

We also need another generalization of the tilde operator:

$$\widetilde{{}_k\langle a; q\rangle_n} \equiv \prod_{m=0}^{n-1} \left(\sum_{i=0}^{k-1} q^{i(a+m)} \right). \tag{6.68}$$

Therefore, we obtain

$$\widetilde{{}_2\langle a; q\rangle_n} = \widetilde{\langle a; q\rangle_n} \quad \text{and} \quad \widetilde{{}_1\langle aq\rangle_n} = \mathrm{I}. \tag{6.69}$$

We also put

$$\langle \widetilde{{}_k a}; q\rangle_n \equiv \widetilde{{}_k\langle a; q\rangle_n}. \tag{6.70}$$

The following simple rules follow from (6.65). Some of them were previously known in other representation in the work of Gasper and Rahman [220].

Theorem 6.4.1

$$\widetilde{\tfrac{m}{l}} a \pm b \equiv \widetilde{\tfrac{m}{l}(a \pm b)} \left(\mathrm{mod}\, \frac{2\pi i}{\log q} \right), \tag{6.71}$$

$$\sum_{k=1}^{n} \widetilde{\tfrac{1}{n} \pm a_k} \equiv \sum_{k=1}^{n} \pm a_k \left(\mathrm{mod}\, \frac{2\pi i}{\log q} \right), \tag{6.72}$$

$$\frac{m}{l} \times \widetilde{a} \equiv \widetilde{\tfrac{m}{2l} \frac{am}{l}} \left(\mathrm{mod}\, \frac{2\pi i}{\log q} \right), \tag{6.73}$$

$$\mathrm{QE}(\widetilde{\tfrac{m}{l}} a) = \mathrm{QE}(a) e^{\frac{2\pi i m}{l}}, \tag{6.74}$$

where the second equation is a consequence of the fact that we work mod $\frac{2\pi i}{\log q}$.

Remark 19 Already in 1847 Heine [270, p. 288, (14)] used the operator $\widetilde{\tfrac{1}{4}a}$ in the formula

$$\varphi_6(z) = z_2\phi_1\left(1, \widetilde{\tfrac{1}{4}1}; \widetilde{\tfrac{1}{4}2}|q^2, -q^2 z\right). \tag{6.75}$$

Heine used the following notation:

$$\varphi_6(z) \equiv z_2\phi_1\left(1, 1+\frac{\pi i}{2\log q}; 2+\frac{\pi i}{2\log q}\middle|q^2, -q^2 z\right). \tag{6.76}$$

The advantage of the elected notation in comparison with Heine's is that we do not need the logarithm in the denominator and can write down the primitive root immediately.

We will use the next formula in Section 8.1.

Theorem 6.4.2

$$\langle\widetilde{a}; q^2\rangle_n = \langle\widetilde{\tfrac{1}{4}a}, \widetilde{\tfrac{3}{4}a}; q\rangle_n, \tag{6.77}$$

$$\langle a; q^p\rangle_n = \prod_{k=0}^{p-1}\langle\widetilde{\tfrac{k}{p}a}; q\rangle_n, \quad \textit{where } p \textit{ is an odd prime}, \tag{6.78}$$

$$\langle a; q^k\rangle_n = \langle a; q\rangle_n \times_k \widetilde{\langle a; q\rangle}_n. \tag{6.79}$$

Remark 20 The formula (6.78) occurred in a special case in [550, p. 1416] and in [272, p. 111].

This leads to the following q-analogue of (1.2).

Definition 113 A q-analogue of a notation by MacRobert [365, p. 135] and Srivastava [475]. This notation was also often used for the Meijer G-function and for the Foxian H-function (Shah, Crelle 288, p. 121–128) ($q = 1$).

This is equivalent to a product of nk q-shifted factorials:

$$\langle\lambda; q\rangle_{kn} \equiv \langle\Delta(q; k; \lambda); q\rangle_n \equiv \prod_{m=0}^{k-1}\left\langle\frac{\lambda+m}{k}; q\right\rangle_n \times_k \left\langle\widetilde{\frac{\lambda+m}{k}}; q\right\rangle_n. \tag{6.80}$$

We also use the notation $\Delta(q; k; \lambda)$ as a parameter in q-hypergeometric functions.

If λ is a vector, we mean the corresponding product of vector elements. If λ is replaced by a sequence of numbers, separated by commas, we mean the corresponding product, as in the case of q-factorials.

The last factor in (6.80) corresponds to k^{nk}.

6.5 The Gaussian q-binomial coefficients and the q-Leibniz theorem

This section is about a function known in special cases to Euler and developed more generally by Gauß. The Gaussian q-binomial coefficients can be used to construct a q-theory equivalent to the theory shaped by the q-shifted factorial. In order to do this, we need a q-operator theory. In this section we treat the q-binomial coefficients. In the next section we briefly summarize this equivalent theory due to Cigler. We start by redefining q-analogues of the binomial coefficients.

Definition 114 The Gaussian q-binomial coefficients [225] are defined by

$$\binom{n}{k}_q \equiv \frac{\langle 1;q\rangle_n}{\langle 1;q\rangle_k \langle 1;q\rangle_{n-k}}, \quad k=0,1,\ldots,n \tag{6.81}$$

and by

$$\binom{\alpha}{\beta}_q \equiv \frac{\langle \beta+1, \alpha-\beta+1;q\rangle_\infty}{\langle 1, \alpha+1;q\rangle_\infty} = \frac{\Gamma_q(\alpha+1)}{\Gamma_q(\beta+1)\Gamma_q(\alpha-\beta+1)} \tag{6.82}$$

for complex α and β if $0<|q|<1$.

Theorem 6.5.1 *Gauß* [222]. *The q-binomial coefficient $\binom{n}{k}_q$ is a polynomial of degree $k(n-k)$ in q with integer coefficients, whose sum equals $\binom{n}{k}$* [68].

We will use the following boundary values for these coefficients:

$$\binom{n}{-m}_q = 0, \quad n=0,\pm1,\pm2,\ldots, \; m=1,2,\ldots, \tag{6.83}$$

$$\binom{n}{n+m}_q = 0, \quad n=0,1,2,\ldots, \; m=1,2,\ldots. \tag{6.84}$$

Theorem 6.5.2 *The following special case is sometimes useful:*

$$\binom{a}{k}_q \equiv \frac{\langle a+1-k;q\rangle_k}{\langle 1;q\rangle_k} = \frac{\langle -a;q\rangle_k(-1)^k q^{-\binom{k}{2}+ka}}{\langle 1;q\rangle_k}. \tag{6.85}$$

It follows that

$$\binom{\widetilde{a}}{k}_q = \frac{\langle \widetilde{-a;q}\rangle_k q^{-\binom{k}{2}+ka}}{\langle 1;q\rangle_k}. \tag{6.86}$$

The following equation for q-inversion applies [235, p. 283]:

$$\binom{x}{k}_{\frac{1}{q}} = q^{k(k-x)}\binom{x}{k}_q. \tag{6.87}$$

It follows that

$$\binom{-n}{k}_q = \frac{\{-n\}_q\{-n-1\}_q\cdots\{-n-k+1\}_q}{\{k\}_q!}, \quad n = 1, 2, \ldots, \tag{6.88}$$

and under suitable conditions equation (1.60) holds for all integers [133].

Example 29 Let $q = p^s$ be a prime power. The Galois field is denoted by $GF(q)$. Let V_n denote a vector space of dimension n over $GF(q)$. The number of subspaces of V_n is

$$G_n = \sum_{k=0}^{n} \binom{n}{k}_q, \tag{6.89}$$

and is called the Galois number [231, p. 77].

Some important relations for the Gauß q-binomial coefficients are collected in [234, p. 91].

The following result will be useful in the proof of the Leibniz q-theorem:

Theorem 6.5.3 *The q-Pascal identity* [225], [220, p. 20]:

$$\binom{\alpha+1}{k}_q = \binom{\alpha}{k}_q q^k + \binom{\alpha}{k-1}_q = \binom{\alpha}{k}_q + \binom{\alpha}{k-1}_q q^{\alpha+1-k}, \tag{6.90}$$

where $0 < |q| < 1$.

Proof The first identity is proved by the following calculation:

$$\text{LHS} = \frac{\langle k+1, \alpha-k+2; q\rangle_\infty}{\langle 1, \alpha+2; q\rangle_\infty},$$

$$\text{RHS} = \frac{\langle k+1, \alpha-k+1; q\rangle_\infty}{\langle 1, \alpha+1; q\rangle_\infty} q^k + \frac{\langle k, \alpha-k+2; q\rangle_\infty}{\langle 1, \alpha+1; q\rangle_\infty}$$

$$\overset{\text{by (6.8)}}{=} \frac{\langle k+1, \alpha-k+2; q\rangle_\infty}{\langle 1, \alpha+2; q\rangle_\infty} \frac{q^k(1-q^{\alpha+1-k}) + 1 - q^k}{1-q^{\alpha+1}} = \text{LHS}.$$

The second identity is proved in a similar way:

$$\text{LHS} = \frac{\langle k+1, \alpha-k+2; q\rangle_\infty}{\langle 1, \alpha+2; q\rangle_\infty},$$

$$\text{RHS} = \frac{\langle k+1, \alpha-k+1; q\rangle_\infty}{\langle 1, \alpha+1; q\rangle_\infty} + \frac{\langle k, \alpha-k+2; q\rangle_\infty}{\langle 1, \alpha+1; q\rangle_\infty} q^{\alpha+1-k}$$

$$\overset{\text{by (6.8)}}{=} \frac{\langle k+1, \alpha-k+2; q\rangle_\infty}{\langle 1, \alpha+2; q\rangle_\infty} \frac{1-q^{\alpha+1-k} + q^{\alpha+1-k}(1-q^k)}{1-q^{\alpha+1}} = \text{LHS}.$$ □

If $\alpha \in \mathbb{N}$, (6.90) holds for all q.

Remark 21 There is a relationship with symmetric polynomials due to Oruc [406, p. 66]:

$$q^{\binom{k}{2}}\binom{n}{k}_q = \mathrm{e}_k\big(1, q, \ldots, q^{n-1}\big), \tag{6.91}$$

where e_k denotes the elementary symmetric polynomial, and

$$\binom{n+k-1}{k}_q = \mathrm{h}_k\big(1, q, \ldots, q^{n-1}\big), \tag{6.92}$$

where h_k denotes the complete symmetric polynomial.

We can now prove formula (2.19), the Rothe-von Grüson-Gauß formula.

Theorem 6.5.4

$$\sum_{n=0}^{m}(-1)^n\binom{m}{n}_q q^{\binom{n}{2}}u^n = (u;q)_m. \tag{6.93}$$

Proof The formula is true for $m = 1$. Assume that the formula is proved for $m-1$. Then it is also true for m, because

$$\begin{aligned}
&\sum_{n=0}^{m}(-1)^n\binom{m}{n}_q q^{\binom{n}{2}}u^n \\
&\overset{\text{by (6.90)}}{=} \sum_{n=0}^{m-1}(-1)^n\binom{m-1}{n}_q q^{\binom{n}{2}}u^n \\
&\quad + \sum_{n=0}^{m-1}(-1)^n\binom{m-1}{n-1}_q q^{\binom{n}{2}+m-n}u^n + (-1)^m q^{\binom{m}{2}}u^m \\
&= (-1)^m q^{\binom{m}{2}}u^m + (u;q)_{m-1} + \sum_{n=0}^{m-2}(-1)^{n+1}\binom{m-1}{n}_q q^{\binom{n}{2}+m-1}u^{n+1} \\
&= (-1)^m q^{\binom{m}{2}}u^m + (u;q)_{m-1}\big(1-uq^{m-1}\big) + (-1)^{m-1}q^{\binom{m-1}{2}+m-1}u^m \\
&= (u;q)_m.
\end{aligned} \tag{6.94}$$

□

Theorem 6.5.5 *The Leibniz q-theorem* [260, 2.5], [387, p. 33] *Let $f(x)$ and $g(x)$ be n times q-differentiable functions. Then $fg(x)$ is also n times q-differentiable and*

$$\mathrm{D}_q^n(fg)(x) = \sum_{k=0}^{n}\binom{n}{k}_q \mathrm{D}_q^k(f)\big(xq^{n-k}\big)\mathrm{D}_q^{n-k}(g)(x). \tag{6.95}$$

Proof For $n = 1$ the formula above becomes (6.48). Assume that the formula is proved for $n = m$. Then it is also true for $n = m + 1$, because

$$\begin{aligned}
\mathrm{D}_q^{m+1}(fg)(x) &= \mathrm{D}_q\bigl(\mathrm{D}_q^m(fg)(x)\bigr)\\
&= \mathrm{D}_q \sum_{k=0}^{m} \binom{m}{k}_q \mathrm{D}_q^k(f)\bigl(xq^{m-k}\bigr)\mathrm{D}_q^{m-k}(g)(x)\\
&\overset{\text{by (6.51)}}{=} \sum_{k=0}^{m} \binom{m}{k}_q \bigl(q^{m-k}\mathrm{D}_q^{k+1}(f)\bigl(xq^{m-k}\bigr)\mathrm{D}_q^{m-k}(g)(x)\\
&\quad + \mathrm{D}_q^k(f)\bigl(xq^{m+1-k}\bigr)\mathrm{D}_q^{m+1-k}(g)(x)\bigr)\\
&= \sum_{k=0}^{m} \binom{m}{k}_q \mathrm{D}_q^k(f)\bigl(xq^{m+1-k}\bigr)\mathrm{D}_q^{m+1-k}(g)(x)\\
&\quad + \sum_{k=1}^{m+1} \binom{m}{k-1}_q q^{m+1-k}\mathrm{D}_q^k(f)\bigl(xq^{m+1-k}\bigr)\mathrm{D}_q^{m+1-k}(g)(x)\\
&= f\bigl(xq^{m+1}\bigr)\mathrm{D}_q^{m+1}(g)(x) + \sum_{k=1}^{m}\left(\binom{m}{k}_q + q^{m+1-k}\binom{m}{k-1}_q\right)\\
&\quad \times \mathrm{D}_q{}^k(f)\bigl(xq^{m+1-k}\bigr)\mathrm{D}_q^{m+1-k}(g)(x) + \mathrm{D}_q^{m+1}(f)(x)g(x)\\
&\overset{\text{by (6.90)}}{=} \sum_{k=0}^{m+1} \binom{m+1}{k}_q \mathrm{D}_q^k(f)\bigl(xq^{m+1-k}\bigr)\mathrm{D}_q^{m+1-k}(g)(x).
\end{aligned} \tag{6.96}$$

□

Remark 22 Roman [436], [438, p. 233] has given a variation of the Leibniz q-theorem, where an additional factor $q^{k(k-n)}$ has occurred. According to (6.87) this corresponds to an inversion of the base in the q-binomial coefficient.

This may be generalized as follows to any product of functions.

Theorem 6.5.6 *Generalized Leibniz q-theorem. Let $\{f_k(x)\}_{k=1}^m$ be n times q-differentiable functions. Then $\prod_{k=1}^m f_k(x)$ is also n times q-differentiable and*

$$\begin{aligned}
\mathrm{D}_q^n \prod_{k=1}^{m} f_k(x) &= \sum_{k_1=0}^{n} \binom{n}{k_1}_q \epsilon^{n-k_1}\mathrm{D}_q^{k_1}(f_1)(x) \sum_{k_2=0}^{n-k_1} \binom{n-k_1}{k_2}_q \epsilon^{n-k_1-k_2}\mathrm{D}_q^{k_2}(f_2)(x)\cdots\\
&\quad \times \sum_{k_{m-1}=0}^{n-k_1-\cdots-k_{m-2}} \binom{n-k_1-\cdots-k_{m-2}}{k_{m-1}}_q\\
&\quad \times \epsilon^{n-k_1-\cdots-k_{m-1}}\mathrm{D}_q^{k_{m-1}}(f_{m-1})(x)\mathrm{D}_q^{n-k_1-\cdots-k_{m-1}}(f_m)(x).
\end{aligned} \tag{6.97}$$

6.5.1 Other formulas

Another interesting formula is the following one (compare with Thomae [499, p. 108], Jackson [298, p. 255]), Hahn [262, p. 260], Ryde [444], Moak [387, p. 33]:

$$\mathrm{D}_q^n(f)(x) = (q-1)^{-n}x^{-n}q^{-\binom{n}{2}}\sum_{k=0}^{n}\binom{n}{k}_q(-1)^k q^{\binom{k}{2}} f(q^{n-k}x). \tag{6.98}$$

Proof For $n=1$ the formula is obvious. Assume that the formula is proved for $n=m$. Then it is also true for $n=m+1$, because

$$\begin{aligned}
&\big(\mathrm{D}_q^{m+1}f\big)(x)\\
&= \mathrm{D}_q\left((q-1)^{-m}x^{-m}q^{-\binom{m}{2}}\sum_{k=0}^{m}\binom{m}{k}_q(-1)^k q^{\binom{k}{2}} f(q^{m-k}x)\right)\\
&\overset{\text{by (6.48)}}{=} (q-1)^{-m}q^{-\binom{m}{2}}\left(-\frac{q^m-1}{q-1}q^{-m}x^{-m-1}\sum_{k=0}^{m}\binom{m}{k}_q(-1)^k q^{\binom{k}{2}} f(q^{m-k}x)\right.\\
&\quad\left.+q^{-m}x^{-m}\sum_{k=0}^{m}-\binom{m}{k}_q(-1)^k\frac{1}{(q-1)x}q^{\binom{k}{2}}\big(f(q^{m-k}x)-f(q^{m+1-k}x)\big)\right)\\
&\overset{\text{by (6.90)}}{=} (q-1)^{-m-1}x^{-m-1}q^{-\binom{m}{2}}\sum_{k=0}^{m+1}\binom{m+1}{k}_q(-1)^k q^{\binom{k}{2}} f(q^{m+1-k}x).
\end{aligned} \tag{6.99}$$

□

Corollary 6.5.7 *If $q^p=1$ and p is prime, then*

$$\mathrm{D}_q^p(f)=0. \tag{6.100}$$

Proof For $p=2$ the corollary is true by inspection. For any other p the terms $f(x)$ and $f(q^p x)$ appear with different sign in (6.98) and cancel each other. All other terms in (6.98) contain the factor $\{p\}_q$, which is zero by hypothesis. □

Remark 23 If $f(x)$ is analytic and $q^n=1$, $n>1$, then $\mathrm{D}_q^n(f)=0$ [274].

Formula (6.98) can be inverted, Hahn [262, p. 260]:

$$f(q^n x) = \sum_{k=0}^{n}(q-1)^k x^k q^{\binom{k}{2}}\binom{n}{k}_q \mathrm{D}_q^k(f)(x). \tag{6.101}$$

The number of known q-binomial coefficient identities is by far less than the number of known binomial coefficient identities. We just give a few examples here.

Jensen [315] and Gould [234, p. 92] found the following q-analogue of Cauchy's formula [320, p. 133]:

$$\binom{\alpha+\beta}{n}_q = \sum_{k=0}^{n} \binom{\alpha}{n-k}_q \binom{\beta}{k}_q q^{k(k+\alpha-n)}. \tag{6.102}$$

This was later generalized by Bender [67]. The next formula is a q-analogue of the combinatorial formula of Riordan [433, p. 6], [433, p. 8, (4)]:

$$\binom{n}{m}_q = \sum_{k=0}^{M} \binom{n-1-k}{m-k}_q q^{m-k}, \quad M = \min(m, n-1). \tag{6.103}$$

The following two formulas for the q-binomial coefficients were proved in [128]:

Theorem 6.5.8 *If q is a primitive $(n+1)$-st root of unity, then*

$$\left|\binom{n}{k}_q\right| = 1, \quad k = 0, 1, \ldots, n. \tag{6.104}$$

Theorem 6.5.9 *If q is a primitive $(2m+2)$-nd root of unity, then*

$$\left|\binom{2m}{m}_q\right| \geq 1. \tag{6.105}$$

Another beautiful identity due to Gauß [222] is stated in [231]:

$$\sum_{k=0}^{2n} (-1)^k \binom{2n}{k}_q = \langle 1; q^2 \rangle_n. \tag{6.106}$$

6.6 Cigler's operational method for q-identities

We still need the Cigler multiplication operator, which was first introduced by Jackson.

Definition 115 Multiplication with x is denoted by $\mathbf{x}$. Multiplication with $1+x\gamma$ in numerator or denominator is denoted by $\mathbf{1}+\mathbf{x}\boldsymbol{\gamma}$.

We introduce a scalar product on $P \equiv \mathbb{C}[x]$ [135, p. 26)] by

$$\langle x^k, x^l \rangle \equiv \{k\}_q! \delta_{k,l}. \tag{6.107}$$

This scalar product can be expressed in a different way:

Let L denote the linear functional

$$Lf(x) = f(0) \tag{6.108}$$

on $\mathbb{C}[x]$. This means

$$\left\langle x^k, x^l \right\rangle = L\mathrm{D}_q^k x^l. \tag{6.109}$$

Obviously $\langle\,\rangle$ is symmetric:

$$\left\langle a(x), b(x) \right\rangle = \left\langle b(x), a(x) \right\rangle, \quad a, b \in P. \tag{6.110}$$

If A is a linear operator, we define the transposed operator A^T by

$$\left\langle Aa(x), b(x) \right\rangle = \left\langle a(x), A^T b(x) \right\rangle. \tag{6.111}$$

The following rules for A^T hold:

$$x^T = \mathrm{D}_q; \qquad \mathrm{D}_q^T = x; \qquad \epsilon^T = \epsilon. \tag{6.112}$$

We present a series of identities for linear operators on the vector space P [133, p. 88].

Example 30

$$\mathrm{D}_q \mathbf{x} - q\mathbf{x}\mathrm{D}_q = I, \tag{6.113}$$

$$\mathrm{D}_q \mathbf{x} - \mathbf{x}\mathrm{D}_q = \epsilon. \tag{6.114}$$

Proof Apply both sides to an arbitrarily polynomial. □

These two equations are equivalent to, respectively,

$$\{n+1\}_q - q\{n\}_q = 1, \tag{6.115}$$

$$\{n+1\}_q - \{n\}_q = q^n. \tag{6.116}$$

Furthermore,

$$\mathrm{D}_q \mathbf{x}^k - q^k \mathbf{x}^k \mathrm{D}_q = \{k\}_q \mathbf{x}^{k-1}, \tag{6.117}$$

$$\mathrm{D}_q \mathbf{x}^k - \mathbf{x}^k \mathrm{D}_q = \{k\}_q \mathbf{x}^{k-1}\epsilon. \tag{6.118}$$

Proof Apply both sides to an arbitrarily polynomial. □

These two equations are equivalent to, respectively,

$$\{n+k\}_q - q^k\{n\}_q = \{k\}_q, \tag{6.119}$$

$$\{n+k\}_q - \{n\}_q = \{k\}_q q^n. \tag{6.120}$$

The following two dual identities, which follow by transposition, are equivalent to the formula (6.90) for the q-binomial coefficients:

$$\mathrm{D}_q^k \mathbf{x} - q^k \mathbf{x}\mathrm{D}_q^k = \{k\}_q \mathrm{D}_q^{k-1}, \tag{6.121}$$

$$\mathrm{D}_q^k \mathbf{x} - \mathbf{x}\mathrm{D}_q^k = \epsilon\{k\}_q \mathrm{D}_q^{k-1}. \tag{6.122}$$

Theorem 6.6.1 *By the formula* (6.48) *we obtain the following operator equation* [135, p. 25)], *where* $'$ *means* D_q:

$$\mathrm{D}_q\mathbf{a}(\mathbf{x}) = \mathbf{a}(\mathbf{qx})\mathrm{D}_q + \big(\mathbf{a}'(\mathbf{x})\big), \tag{6.123}$$

or the dual form

$$\mathrm{D}_q\mathbf{a}(\mathbf{x}) = \mathbf{a}(\mathbf{x})\mathrm{D}_q + \big(\mathbf{a}'(\mathbf{x})\big)\epsilon. \tag{6.124}$$

By the Nalli-Ward q-Taylor formula (4.34) *and transposition of* (6.123) *and* (6.124), *respectively, we obtain* [135, p. 27)]:

$$a(\mathrm{D}_q)\mathbf{x} = \mathbf{x}a(q\mathrm{D}_q) + a'(\mathrm{D}_q), \tag{6.125}$$

$$a(\mathrm{D}_q)\mathbf{x} = \mathbf{x}a(\mathrm{D}_q) + \epsilon a'(\mathrm{D}_q). \tag{6.126}$$

Cigler [135, p. 31, (6)–(9)] proved the following operator formulas.

Theorem 6.6.2

$$\mathrm{E}_q(a\mathrm{D}_q)\mathbf{x}\frac{1}{\mathrm{E}_q(a\mathrm{D}_q)} = \mathbf{x} + a\epsilon. \tag{6.127}$$

Proof We have

$$\frac{1}{\mathrm{E}_q(ax)}\mathrm{D}_q\mathrm{E}_q(ax) = \mathrm{D}_q + a\epsilon. \tag{6.128}$$

A transposition gives the result. □

Theorem 6.6.3

$$\frac{1}{\mathrm{E}_q(a\mathrm{D}_q)}(\mathbf{x}\epsilon^{-1})\mathrm{E}_q(a\mathrm{D}_q) = (\mathbf{x} - a)\epsilon^{-1}. \tag{6.129}$$

Proof We have

$$\begin{aligned} &\mathrm{E}_q(ax)\epsilon^{-1}\mathrm{D}_q\frac{1}{\mathrm{E}_q(ax)}f(x) \\ &= \mathrm{E}_q(ax)\epsilon^{-1}\frac{\mathrm{E}_q(ax)\mathrm{D}_q - a\mathrm{E}_q(ax)}{[\mathrm{E}_q(ax)]^2(1+(q-1)ax)}f(x) \overset{\text{by (6.194)}}{=} \epsilon^{-1}(\mathrm{D}_q - a)f(x). \end{aligned} \tag{6.130}$$

A transposition gives the result. □

Finally, we have [135, p. 39)]

$$\mathrm{D}_{\frac{1}{q}} = \epsilon^{-1}\mathrm{D}_q. \tag{6.131}$$

We will provide a further q-exponential function. For that, we need some new symbols.

Definition 116 The generalized q-factorial is defined by

$$\{n\}_{\alpha,q}! \equiv \prod_{k=1}^{n} \frac{\{k\alpha\}_q}{\{\alpha\}_q}, \tag{6.132}$$

and the generalized q-exponential function by

$$\mathrm{E}_{\alpha,q}(x) \equiv \sum_{k=0}^{\infty} \frac{x^k}{\{k\}_{\alpha,q}!}. \tag{6.133}$$

Theorem 6.6.4

$$\frac{1}{\mathrm{E}_{2,q}(\frac{-x^2}{\{2\}_q})} \mathrm{D}_q \mathrm{E}_{2,q}\left(\frac{-x^2}{\{2\}_q}\right) = \mathrm{D}_q - \mathbf{x}\epsilon. \tag{6.134}$$

Theorem 6.6.5

$$\mathrm{E}_{2,q}\left(\frac{qx^2}{\{2\}_q}\right)(\epsilon^{-1}\mathrm{D}_q)\frac{1}{\mathrm{E}_{2,q}(\frac{qx^2}{\{2\}_q})} = \mathrm{D}_{\frac{1}{q}} - \mathbf{x}\epsilon^{-1} = \epsilon^{-1}\mathrm{D}_q - \mathbf{x}\epsilon^{-1}. \tag{6.135}$$

Proof

$$\begin{aligned}
&\mathrm{E}_{2,q}\left(\frac{qx^2}{\{2\}_q}\right)(\epsilon^{-1}\mathrm{D}_q)\frac{1}{\mathrm{E}_{2,q}(\frac{qx^2}{\{2\}_q})} \\
&= \epsilon^{-1}\mathrm{E}_{2,q}\left(\frac{q^3x^2}{\{2\}_q}\right)\mathrm{D}_q\frac{1}{\mathrm{E}_{2,q}(\frac{qx^2}{\{2\}_q})} \\
&\overset{\text{by (6.123)}}{=} \epsilon^{-1}\left(\mathrm{D}_q\mathrm{E}_{2,q}\left(\frac{qx^2}{\{2\}_q}\right) - qx\mathrm{E}_{2,q}\left(\frac{qx^2}{\{2\}_q}\right)\right)\frac{1}{\mathrm{E}_{2,q}(\frac{qx^2}{\{2\}_q})} \\
&= \epsilon^{-1}(\mathrm{D}_q - q\mathbf{x}) = \epsilon^{-1}\mathrm{D}_q - \mathbf{x}\epsilon^{-1}.
\end{aligned} \tag{6.136}$$

□

Cigler has shown certain orthogonality relations for q-Laguerre and q-Hermite polynomials by his operator method. However, here we treat the orthogonality in a different way by means of q-integrals, see Chapter 9.

6.7 Gould and Carlitz q-binomial coefficient identities

Bateman left an unpublished manuscript of 520 pages on binomial coefficient identities. Then Carlitz and Gould continued the study of this highly interesting subject.

The characteristic stratagem in this study is first to find a binomial coefficient identity and then possibly a q-analogue. We illustrate this with a couple of examples.

Theorem 6.7.1 [237, p. 397]. *Let $g(k)$ be a function $\mathbb{N} \to \mathbb{N}$. Let the functions $F(n)$ and $f(k)$ be connected in the following way:*

$$F(n) = q^{\binom{n}{2}} \sum_{k=0}^{n} (-1)^k \binom{n}{k}_q \binom{g(k)}{n}_q q^{k(k-n)} f(k). \tag{6.137}$$

Then

$$\sum_{n=0}^{\infty} (-1)^n F(n) u^n = \sum_{k=0}^{\infty} \binom{g(k)}{k}_q q^{\binom{k}{2}} u^k f(k) \frac{(u;q)_\infty}{(uq^{g(k)-k};q)_\infty}. \tag{6.138}$$

Proof Combining (6.137) and (6.138) we get

$$\begin{aligned}&\sum_{n=0}^{\infty} (-1)^n u^n q^{\binom{n}{2}} \sum_{k=0}^{n} (-1)^k \binom{n}{k}_q \binom{g(k)}{n}_q q^{k(k-n)} f(k)\\ &\quad = \sum_{k=0}^{\infty} \binom{g(k)}{k}_q q^{\binom{k}{2}} u^k f(k) \frac{1}{(uq^{g(k)-k};q)_{k-g(k)}}.\end{aligned} \tag{6.139}$$

By the q-binomial theorem,

$$\begin{aligned}&\sum_{n=0}^{\infty} (-1)^n u^n q^{\binom{n}{2}} \sum_{k=0}^{n} (-1)^k \binom{n}{k}_q \binom{g(k)}{n}_q q^{k(k-n)} f(k)\\ &\quad = \sum_{k=0}^{\infty} \binom{g(k)}{k}_q q^{\binom{k}{2}} u^k f(k) \sum_{m=0}^{\infty} \frac{\langle k-g(k);q\rangle_m}{\langle 1;q\rangle_m} u^m \mathrm{QE}\big(m\big(g(k)-k\big)\big).\end{aligned} \tag{6.140}$$

Equating the coefficients of $f(k)$ yields

$$\begin{aligned}&\sum_{n=0}^{\infty} (-1)^n u^n q^{\binom{n}{2}} (-1)^k \binom{n}{k}_q \binom{g(k)}{n}_q q^{k(k-n)}\\ &\quad = \binom{g(k)}{k}_q q^{\binom{k}{2}} u^k \sum_{m=0}^{\infty} \frac{\langle k-g(k);q\rangle_m}{\langle 1;q\rangle_m} u^m \mathrm{QE}\big(m\big(g(k)-k\big)\big).\end{aligned} \tag{6.141}$$

We set $n = k + m$ and get formula (6.15) with the following values:

$$a' \to 1, \qquad n' \to g(k) - k, \qquad k' \to n - k,$$

where $'$ refers to (6.15). □

In 1973 Gould and Hsu [241] published an important inverse series relation, which had been communicated by Hsu in a personal letter to Gould already in 1965. In the same year Carlitz found a q-analogue.

Theorem 6.7.2 [106] *q-analogue of the Gould-Hsu inverse relation. Let $\{\alpha_i\}$ and $\{\beta_i\}$ be two sequences of complex numbers and let q be an arbitrary complex number such that*

$$\alpha_i + q^{-k}\beta_i \neq 0, \quad i = 1, 2, 3, \ldots, \; k = 0, 1, 2, \ldots, \tag{6.142}$$

and put

$$\psi(k, n, q) \equiv \prod_{i=1}^{n} \left(\alpha_i + q^{-k}\beta_i\right). \tag{6.143}$$

The system of equations

$$f(n) = \sum_{k=0}^{n} (-1)^k q^{\binom{k}{2}} \binom{n}{k}_q \psi(k, n, q) g(k), \quad n = 0, 1, 2, \ldots, N, \tag{6.144}$$

is equivalent to the system

$$g(n) = \sum_{k=0}^{n} (-1)^k q^{\frac{k(k+1)}{2} - kn} \binom{n}{k}_q \left(\alpha_{k+1} + q^{-k}\beta_{k+1}\right) \frac{f(k)}{\psi(n, k+1, q)},$$
$$n = 0, 1, 2, \ldots, N, \tag{6.145}$$

where $N \in \mathbb{N}$.

Proof We show that Eq. (6.145) implies (6.144). Consider the sum

$$\begin{aligned}
&\sum_{k=0}^{n} (-1)^k q^{\binom{k}{2}} \binom{n}{k}_q \psi(k, n, q) g(k) \\
&\quad = \sum_{k=0}^{n} (-1)^k q^{\binom{k}{2}} \binom{n}{k}_q \psi(k, n, q) \\
&\qquad \times \sum_{j=0}^{k} (-1)^j q^{\frac{j(j+1)}{2} - jk} \binom{k}{j}_q \left(\alpha_{j+1} + q^{-j}\beta_{j+1}\right) \frac{f(j)}{\psi(k, j+1, q)} \\
&\quad = \sum_{j=0}^{n} q^{\binom{j}{2}} \binom{n}{j}_q \left(\alpha_{j+1} + q^{-j}\beta_{j+1}\right) f(j) \\
&\qquad \times \sum_{k=j}^{n} (-1)^{k-j} q^{\binom{k}{2} - jk} \binom{n-j}{k-j}_q \frac{\psi(k, n, q)}{\psi(k, j+1, q)}
\end{aligned}$$

$$= \sum_{j=0}^{n} \binom{n}{j}_q (\alpha_{j+1} + q^{-j}\beta_{j+1}) f(j) \sum_{k=0}^{n-j} (-1)^k q^{\binom{k}{2}} \binom{n-j}{k}_q \frac{\psi(k+j,n,q)}{\psi(k+j,j+1,q)}. \tag{6.146}$$

It suffices to show that

$$\binom{n}{j}_q (\alpha_{j+1} + q^{-j}\beta_{j+1}) \sum_{k=0}^{n-j} (-1)^k q^{\binom{k}{2}} \binom{n-j}{k}_q \frac{\psi(k+j,n,q)}{\psi(k+j,j+1,q)} = \delta_{n,j}, \tag{6.147}$$

or, equivalently,

$$(\alpha_{j+1} + q^{-j}\beta_{j+1}) \sum_{k=0}^{n} (-1)^k q^{\binom{k}{2}} \binom{n}{k}_q \frac{\psi(k+j,n+j,q)}{\psi(k+j,j+1,q)} = \delta_{n,0}. \tag{6.148}$$

For $n = 0$ the LHS of (6.148) is equal to 1, by (6.143). For $n > 0$ we get by (6.143)

$$\frac{\psi(k+j,n+j,q)}{\psi(k+j,j+1,q)} = \prod_{i=j+2}^{n+j} (\alpha_i + q^{-k-j}\beta_i). \tag{6.149}$$

The left-hand side of (6.148) is obviously a polynomial in q^{-k} of degree $\leq n-1$. By (2.19) the left-hand side of (6.148) disappears. □

6.8 q-Exponential and q-trigonometric functions

We have now reached the SI unit 1 candela. The q-exponential function is the light of q-analysis, which allows a large amount of operator formulas, see Section 9.1. The q-Appell polynomials in Chapter 4 as well as the so-called pseudo-q-Appell polynomials, both have the q-exponential function in the generating function. The matrix pseudo-groups [179], which are q-analogues of the most popular Lie groups, are also related with the q-exponential function. The q-Pascal matrix is a solution to the corresponding q-difference equation. The q-Cauchy matrix is constructed by means of the corresponding matrix q-exponential function by NWA-q additions.

1 cd	$\mathrm{E}_q(x)$

The corresponding q-trigonometric functions follow here and in Section 6.11.

Definition 117 If $|q| > 1 \vee 0 < |q| < 1$, $|z| < |1-q|^{-1}$, the q-exponential function $\mathrm{E}_q(z)$ is defined by [293], [194]

$$\mathrm{E}_q(z) \equiv \sum_{k=0}^{\infty} \frac{1}{\{k\}_q!} z^k. \tag{6.150}$$

By the Euler equation (6.188), the meromorphic continuation of $\mathrm{E}_q(z)$ is given by

$$\mathrm{E}_q(z) = \frac{1}{(z(1-q);q)_\infty}. \tag{6.151}$$

Thus the meromorphic function $\frac{1}{(z(1-q);q)_\infty}$, with simple poles at $\frac{q^{-k}}{1-q}$, $k \in \mathbb{N}$ is a good substitute for $\mathrm{E}_q(z)$ in the whole complex plane. We shall however continue to call this function $\mathrm{E}_q(z)$, since it plays an important role in operator theory.

In [125] a similar function was used to prove many beautiful q-identities by so-called parameter augmentation, see Section 6.14. According to Cigler [139, p. 19], the function $\mathrm{E}_q(z)$ can be considered as an interpolation between the geometric series and the exponential function.

The q-difference equation for $\mathrm{E}_q(z)$ is

$$\mathrm{D}_q \mathrm{E}_q(az) = a\mathrm{E}_q(az). \tag{6.152}$$

Definition 118 There is another q-exponential function which is entire when $0 < |q| < 1$ and which converges when $|z| < |1-q|^{-1}$ if $|q| > 1$. To obtain it we must invert the base in (6.150), i.e. $q \to \frac{1}{q}$. This is a common theme in q-calculus:

$$\mathrm{E}_{\frac{1}{q}}(z) \equiv \sum_{k=0}^{\infty} \frac{q^{\binom{k}{2}}}{\{k\}_q!} z^k. \tag{6.153}$$

A solitary factor of $q^{\binom{k}{2}}$ has appeared in the general term of the series. Such terms, whose exponents are quadratic in the index of summation of the series concerned, are often characteristic of q-functions [194]. In this case the term results from the summation of an arithmetic series. In many ways, $\mathrm{E}_q(z)$ and $\mathrm{E}_{\frac{1}{q}}(z)$ are mutually complementary [194]. The q-difference equation corresponding to (6.152) is

$$\mathrm{D}_q \mathrm{E}_{\frac{1}{q}}(az) = a\mathrm{E}_{\frac{1}{q}}(qaz), \tag{6.154}$$

which also reduces to the differential equation of the exponential function when q tends to unity, as expected. Compare formulae (6.188) and (6.189), which give two other q-analogues of the exponential function. We immediately obtain

$$\mathrm{E}_{\frac{1}{q}}(z) = \prod_{n=0}^{\infty} \left(1 + (1-q)zq^n\right), \quad 0 < |q| < 1. \tag{6.155}$$

Theorem 6.8.1 *We have*

$$\lim_{x \to \infty} \mathrm{E}_{\frac{1}{q}}(x) = \infty, \quad 0 < q < 1. \tag{6.156}$$

Proof Put $\mathrm{E}_{n,q}(x) \equiv (-x(1-q);q)_n$. Then $\mathrm{E}_{\frac{1}{q}}(x) = \lim_{n\to\infty} \mathrm{E}_{n,q}(x)$. For q fixed, $0 < q < 1$, choose x so big that

$$\left|1 + q^k x(1-q)\right| > a > 1, \quad k \in \mathbb{N}. \tag{6.157}$$

This means $|\mathrm{E}_{n,q}(x)| > a^n$ and

$$\lim_{n\to\infty} \left|\mathrm{E}_{n,q}(x)\right| > \lim_{n\to\infty} a^n = \infty. \tag{6.158}$$

□

Theorem 6.8.2 *The function* $\mathrm{E}_q(x)$ *oscillates between* $\pm\infty$ *if* $\lim_{x\to\infty}$, $0 < q < 1$.

Proof Consider the function

$$f(x) \equiv \mathrm{E}_{\frac{1}{q}}(-x) = \left(-x(1-q);q\right)_\infty. \tag{6.159}$$

Then $f(x)$ has infinitely many zeros with accumulation point $+\infty$. Our original function $\mathrm{E}_q(x) = \frac{1}{f(x)}$ then goes to $\pm\infty$. □

We now prove two inequalities for the two q-exponential functions. Recently, several research papers on this topic have been published for different q-functions. In particular, there are many inequalities for the q-gamma function, such as those in [19], [20], [48], [248], [259], [291], [290], [334], [368], [369], [386], [462].

Theorem 6.8.3 *An inequality for* $\mathrm{E}_q(-x)$:

$$\mathrm{E}_q(-x) > e^{-x}, \quad 0 < q < 1, \ x > 0. \tag{6.160}$$

Proof Denote

$$P_N(x) \equiv \prod_{k=0}^{N} \frac{1}{1 + x(1-q)q^k}. \tag{6.161}$$

Then we have

$$P_N(x) > \exp\left(-\sum_{k=0}^{N} x(1-q)q^k\right) = \exp\left(-x\left(1-q^{N+1}\right)\right), \tag{6.162}$$

whence

$$\mathrm{E}_q(-x) = \lim_{N\to\infty} P_N(x) > e^{-x}. \tag{6.163}$$

□

Theorem 6.8.4 *An inequality for* $\mathrm{E}_{\frac{1}{q}}(-x)$:

$$\mathrm{E}_{\frac{1}{q}}(-x) < e^{-x}, \quad x \neq 0,\ 0 < q < 1. \tag{6.164}$$

Proof Denote

$$P_N(x) \equiv \prod_{k=0}^{N} \bigl(1 - x(1-q)q^k\bigr). \tag{6.165}$$

Then we have

$$P_N(x) < \exp\left(-\sum_{k=0}^{N} x(1-q)q^k\right) = \exp\bigl(-x\bigl(1-q^{N+1}\bigr)\bigr), \tag{6.166}$$

whence

$$\mathrm{E}_{\frac{1}{q}}(-x) = \lim_{N\to\infty} P_N(x) < e^{-x}. \tag{6.167}$$

□

We have now defined Γ_q and $\mathrm{E}_q(z)$ and it is time to quote a theorem about order and type of these functions:

Theorem 6.8.5 [549]. *For* $0 < q < 1$ *and* $1 \leq \Re(x) \leq 2$, Γ_q *has order* 1 *and type* $< 2\pi$. *Also* $\mathrm{E}_q(z)$ *has order* 1.

Two corresponding q-trigonometric functions are:

Definition 119

$$\mathrm{Sin}_q(x) \equiv \frac{1}{2i}\bigl(\mathrm{E}_q(ix) - \mathrm{E}_q(-ix)\bigr) \tag{6.168}$$

and

$$\mathrm{Cos}_q(x) \equiv \frac{1}{2}\bigl(\mathrm{E}_q(ix) + \mathrm{E}_q(-ix)\bigr). \tag{6.169}$$

We now want to show some graphs for the q-trigonometric functions.

Everywhere we have $q = .9$ and $0 \leq x \leq 8$. The following two graphs show $\mathrm{Sin}_q(x)$ and $\mathrm{Cos}_q(x)$, while the two that follow them are the graphs of $\mathrm{Sin}_{\frac{1}{q}}(x)$ and $\mathrm{Cos}_{\frac{1}{q}}(x)(x)$:

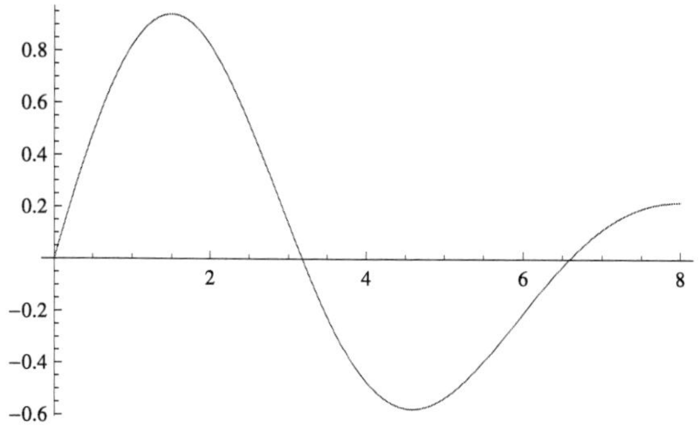
0.8
0.6
0.4
0.2
−0.2
−0.4
−0.6
2
4
6
8

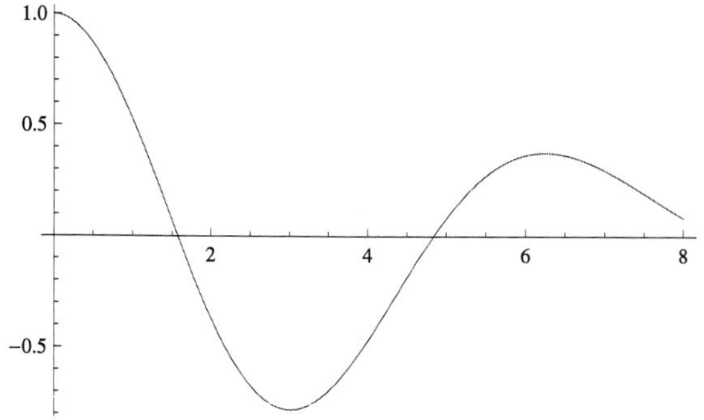
1.0
0.5
−0.5
2
4
6
8

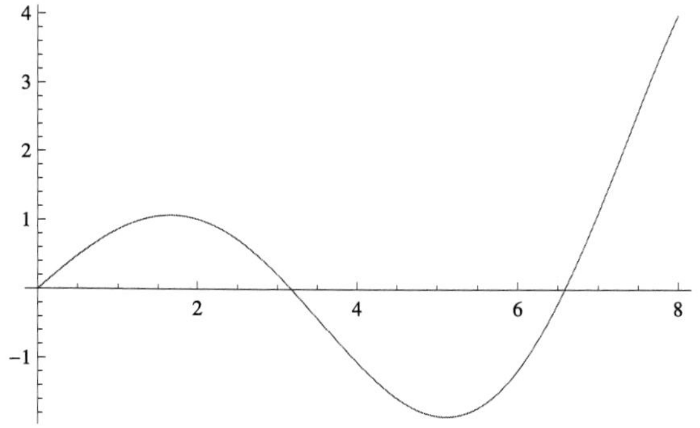
4
3
2
1
−1
2
4
6
8

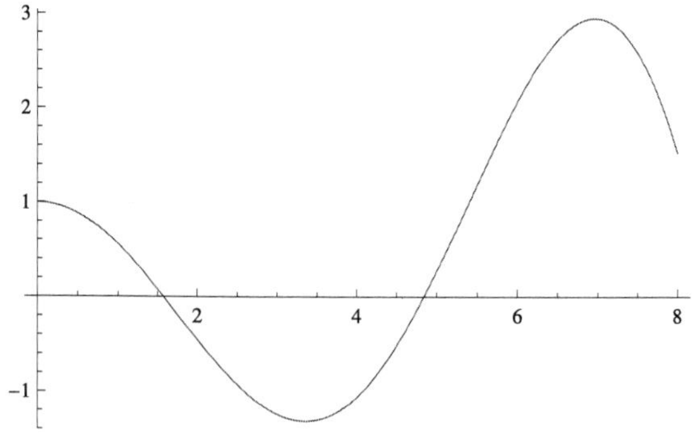
3
2
1
−1
2
4
6
8

The analytic continuations of Sin_q and Cos_q are given by (6.151). The q-differences for the q-trigonometric functions are:

$$\mathrm{D}_q \mathrm{Cos}_q(ax) = -a\,\mathrm{Sin}_q(ax) \tag{6.170}$$

$$\mathrm{D}_q \mathrm{Sin}_q(bx) = b\,\mathrm{Cos}_q(bx). \tag{6.171}$$

The classical q-oscillator

$$\mathrm{D}_q^2 f(x) + \omega^2 f(x) = 0 \tag{6.172}$$

has the solution $f(x) = C_1 \mathrm{Sin}_q(\omega x) + C_2 \mathrm{Cos}_q(\omega x)$.

The following equation is easily proved:

$$\mathrm{Cos}_q(x)\,\mathrm{Cos}_{\frac{1}{q}}(x) + \mathrm{Sin}_q(x)\,\mathrm{Sin}_{\frac{1}{q}}(x) = 1. \tag{6.173}$$

Theorem 6.8.6 *Cigler* [135, p. 30]. *The q-difference equation*

$$\mathrm{D}_q f(x) = axf(x), \tag{6.174}$$

with boundary condition $f(0) = 1$ has the solution

$$f(x) = \mathrm{E}_{2,q}\left(\frac{ax^2}{\{2\}_q}\right). \tag{6.175}$$

Theorem 6.8.7 *Cigler* [135, p. 30, (5)]:

$$\mathrm{E}_{2,q}\left(\frac{x}{\{2\}_q}\right)\mathrm{E}_{2,q}\left(\frac{qx}{\{2\}_q}\right) = \mathrm{E}_q(x). \tag{6.176}$$

Harold Exton has introduced a third variant of q-exponential function:

Definition 120 [194, p. 168]

$$\mathrm{E}_{\mathrm{Ext},q}(x) \equiv \sum_{k=0}^{\infty} \frac{q^{\frac{k(k-1)}{4}}}{\{k\}_q!} x^k. \tag{6.177}$$

This function is invariant under inversion of basis and is entire. No addition theorems and no power series inversion are known. However the following differentiation formula holds:

$$\mathrm{D}_q^n \mathrm{E}_{\mathrm{Ext},q}(x) = \mathrm{QE}\left(\frac{n(n-1)}{4}\right)\mathrm{E}_{\mathrm{Ext},q}\left(xq^{\frac{n}{2}}\right). \tag{6.178}$$

Definition 121 The corresponding q-trigonometric functions are:

$$\mathrm{Sin}_{\mathrm{Ext},q}(x) \equiv \sum_{n=0}^{\infty} (-1)^n \mathrm{QE}\left(\frac{n(2n+1)}{2}\right)\frac{x^{2n+1}}{\{2n+1\}_q!}. \tag{6.179}$$

$$\mathrm{Cos}_{\mathrm{Ext},q}(x) \equiv \sum_{n=0}^{\infty} (-1)^n \mathrm{QE}\left(\frac{n(2n-1)}{2}\right) \frac{x^{2n}}{\{2n\}_q!}, \tag{6.180}$$

where $x \in \mathbb{C}$.

As long as no addition formulas for these q-trigonometric functions exist, these functions are not very interesting from an operational point of view. They have beautiful graphs, exactly resembling the usual trigonometric graphs. Two very similar q-trigonometric functions were given in [209, p. 590].

6.9 The Heine function

We need a larger unit to measure the flux density of an electromagnet.

1 Tesla	$_2\phi_1(a, b; c\|q; z)$

Definition 122 The q-hypergeometric series was developed by Heine in 1846 [269] as a generalization of the hypergeometric series:

$$_2\phi_1(a, b; c|q; z) \equiv \sum_{n=0}^{\infty} \frac{\langle a; q\rangle_n \langle b; q\rangle_n}{\langle 1; q\rangle_n \langle c; q\rangle_n} z^n. \tag{6.181}$$

Theorem 6.9.1 *Pringsheim* [419, p. 924] *Die durch* (6.181) *definierte meromorphe Funktion ist für* $|q| < 1$ *eine in der ganzen Ebene eindeutige und im Endlichen bis auf die einfachen Pole* $z = 1$ *und* $z = (\frac{1}{q})^\nu$ $(\nu \in \mathbb{N})$, *von welchen letzteren bei besonderer Wahl der Parameter auch welche fehlen können, reguläre, also eine gebrochene transzendente Funktion. Der Konvergenzradius ist* 1. *Die hypergeometrische Reihe, die aus der Heineschen Reihe für* $\lim_{q\to 1}$ *entsteht, ist im allgemeinen vieldeutig, und der Punkt* $z = 1$ *ist, abgesehen von ganz speziellen Fällen, Verzweigungspunkt.*

A brief translation is: The meromorphic function (6.181) is uniquely defined in the whole complex plane for $|q| < 1$, with simple poles at $z = 1$ and $z = (\frac{1}{q})^\nu$ $(\nu \in \mathbb{N})$, where the last ones could be missing for certain choices of the parameters. The convergence radius is 1.

The function $_2\phi_1(a, b; c|q; z)$ regarded as a function of the complex variable z, branches only for $z = 0$ and for $z = \infty_{\mathrm{Rie}}$ and the meromorphic continuations of the branches coincide everywhere. It is the differential equation (3.32) that determines the character of the branches of the hypergeometric function. In the same way the meromorphic continuations of $_2\phi_1(a, b; c|q; z)$ are determined by the following q-difference equation:

Lemma 6.9.2 *Heine, Jackson. The series* ${}_2\phi_1(a,b;c|q;z)$ *satisfies the* q*-difference equation*

$$z\big(q^c - zq^{a+b+1}\big)\mathrm{D}_q^2 + \big[\{c\}_q - \big(\{a\}_q q^b + \{b\}_q q^a + q^{a+b}\big)z\big]\mathrm{D}_q - \{a\}_q\{b\}_q \mathrm{I} = 0. \tag{6.182}$$

Proof The q-difference equation can be written

$$-x\{\theta + a\}_q\{\theta + b\}_q + \{\theta\}_q\{\theta + c - 1\}_q = 0, \tag{6.183}$$

where $\{\theta\}_q \equiv x\mathrm{D}_q$. This can be restated as

$$\begin{aligned}
&- z\big(q^a\theta_q + \{a\}_q\big)\big(q^b\theta_q + \{b\}_q\big) + \theta_q\big(q^c\{\theta_q - 1\}_q + \{c\}_q\big)\\
&= -zq^{a+b}\big(qz^2\mathrm{D}_q^2 + z\mathrm{D}_q\big) - z^2\mathrm{D}_q\big(\{a\}_q q^b + \{b\}_q q^a\big) - z\{a\}_q\{b\}_q\\
&\quad + \theta_q\big(q^{c-1}\theta_q - q^{c-1} + \{c\}_q\big)\\
&= -z^3q^{a+b+1}\mathrm{D}_q^2 - z^2q^{a+b}\mathrm{D}_q - z^2\mathrm{D}_q\big(\{a\}_q q^b + \{b\}_q q^a\big) - z\{a\}_q\{b\}_q\\
&\quad + q^c z^2\mathrm{D}_q^2 + z\mathrm{D}_q\{c\}_q = 0.
\end{aligned} \tag{6.184}$$

□

There is also a third form [467, p. 11], which is presented for the generalized series ${}_p\phi_{p-1}(a_1,\ldots,a_p;b_1,\ldots,b_{p-1}|q;z)$. We put $b_p = 1$ and $\mathrm{e}_k =$ elementary symmetric polynomial.

$$\sum_{k=0}^{p}(-1)^k\big(\mathrm{e}_k\big(q^{b_i}\big)q^{-k} - \mathrm{e}_k\big(q^{a_i}\big)x\big)f\big(q^k x\big) = 0. \tag{6.185}$$

The formula (6.185) for $p = 2$ occurred in [186, p. 165, 11.21].

Theorem 6.9.3 [467, p. 57] *If* $c \notin \mathbb{Z}$*, the functions*

$$\begin{aligned}
&u_{1,q}(z) \equiv {}_2\phi_1(a,b;c|q;z),\\
&u_{2,q}(z) \equiv z^{1-c}{}_2\phi_1(a-c+1,b-c+1;2-c|q;z)
\end{aligned} \tag{6.186}$$

form a fundamental system for (6.182) *around* $z = 0$.

Theorem 6.9.4 [467, p. 35] *The functions*

$$\begin{aligned}
&u_{3,q}(z) \equiv z^{-a}{}_3\phi_1\big(a-c+1,a;a-b+1|q;q^{1+c-a-b}z^{-1}\big),\\
&u_{4,q}(z) \equiv z^{-b}{}_2\phi_1\big(b-c+1,b;b-a+1|q;q^{1+c-a-b}z^{-1}\big)
\end{aligned} \tag{6.187}$$

form a fundamental system for (6.182) *around* $z = \infty_{\mathrm{Rie}}$.

Since the point $z = 1$ is not a branch point, in the q-case we do not need the solution (3.45). Instead, we have the solutions (7.52) and (7.53).

Definition 123 Euler [190, Chapter XVI, p. 259] found the following two q-analogues of the exponential function:

$$\begin{aligned} \mathrm{e}_q(z) &\equiv {}_1\phi_0(\infty; -|q; z) \\ &\equiv \sum_{n=0}^{\infty} \frac{z^n}{\langle 1; q\rangle_n} = \frac{1}{(z;q)_\infty}, \quad |z| < 1,\ 0 < |q| < 1, \end{aligned} \tag{6.188}$$

$$\mathrm{e}_{\frac{1}{q}}(z) \equiv {}_0\phi_0(-; -|q; -z) \equiv \sum_{n=0}^{\infty} \frac{q^{\binom{n}{2}}}{\langle 1; q\rangle_n} z^n = (-z; q)_\infty, \quad 0 < |q| < 1, \tag{6.189}$$

where ${}_0\phi_0$ is defined by (7.1).

$\mathrm{e}_{\frac{1}{q}}(z)$ is an entire function, just as the usual exponential function [419, p. 925].

Remark 24 Hardy and Wright [268, p. 366] use two special cases of (6.189) to prove theorems for partitions. The formula (6.188) can also be found in [268, p. 370].

6.10 Oscillations in q-analysis

The oscillation corresponds to the complementary q-exponential function for $x < 0$. Example RC circuit.

Oscillation	$\mathrm{E}_{\frac{1}{q}}(x)$

The simplest approximation method in analysis is the Taylor formula. The domain of convergence can often be clearly defined and there is a formula for the Lagrange remainder. In q-analysis there are four known q-Taylor-series and corresponding formulas with remainder term. In the case of finite differences the corresponding formulas are (2.6) and (4.82). The inverses are (2.5) and (4.85). The last two formulas have the disadvantage that they oscillate, therefore, the result and the domain of convergence are small.

We often use q-exponential functions as weight functions in q-integrals. We have then

$$\lim_{x\to+\infty} \mathrm{E}_q(-x) = 0, \quad 0 < q < 1. \tag{6.190}$$

On the other hand, for $0 < q < 1$, the function $\mathrm{E}_{\frac{1}{q}}(-x)$ oscillates towards zero with decreasing amplitude for positive real values of x.

The two q-additions (10.22), (10.23) for the q-factorial also give oscillations and can therefore only be regarded as formal results.

6.11 The Jackson-Hahn-Cigler q-addition and q-analogues of the trigonometric functions

The scalar product corresponds to the second major operator in q-calculus.

scalar product	q-additions

Jackson and Exton have presented several addition theorems for q-exponential and q-trigonometric functions. We present theses formulas in a more lucid style using two q-additions to be defined shortly. The following polynomial in three variables x, y, q originates from Gauß.

Definition 124 The Jackson-Hahn-Cigler q-addition (JHC), compare [133, p. 91], [261, p. 362], [307, p. 78] is the function

$$(x \boxplus_q y)^n \equiv [x+y]_q^n \equiv \sum_{k=0}^{n} \binom{n}{k}_q q^{\binom{k}{2}} y^k x^{n-k}$$

$$= x^n \left(-\frac{y}{x}; q\right)_n \equiv \mathrm{P}_{n,q}(x, y), \quad n = 0, 1, 2, \ldots, \tag{6.191}$$

$$(x \boxminus_q y)^n \equiv \mathrm{P}_{n,q}(x, -y), \quad n = 0, 1, 2, \ldots. \tag{6.192}$$

Remark 25 The notation $[x+y]_q^n$ is due to Hahn [261, p. 362] and the notation $\mathrm{P}_{n,q}(x, y)$ is due to Cigler [133, p. 91]. We will only use $(x \boxplus_q y)^n$ as it resembles the notation for NWA.

Remark 26 Unlike the Nalli-Ward-Al-Salam q-addition, the Hahn q-addition is neither commutative nor associative, but on the other hand, it can be written as a finite product by (2.19).

The following equation holds:

$$\mathrm{E}_q(x)\mathrm{E}_{\frac{1}{q}}(y) = \mathrm{E}_q(x \boxplus_q y). \tag{6.193}$$

Proof Expand the RHS and use (2.19). □

Cigler [133, p. 91, (13)] has found an interesting dual formula:

$$\mathrm{E}_q(-xq^n)\mathrm{E}_{\frac{1}{q}}(x) = \sum_{k=0}^{n} \binom{n}{k}_q q^{\binom{k}{2}} \big(-x(1-q)\big)^k$$

$$\equiv \mathrm{P}_{n,q}\big(1, -x(1-q)\big), \quad n = 0, 1, 2, \ldots. \tag{6.194}$$

The following two addition theorems [293, p. 32] are proved in the same way as (6.193):

$$\mathrm{Cos}_q(x)\,\mathrm{Cos}_{\frac{1}{q}}(y) - \mathrm{Sin}_q(x)\,\mathrm{Sin}_{\frac{1}{q}}(y) = \mathrm{Cos}_q(x \boxplus_q y), \tag{6.195}$$

$$\mathrm{Sin}_q(x)\,\mathrm{Cos}_{\frac{1}{q}}(y)+\mathrm{Cos}_q(x)\,\mathrm{Sin}_{\frac{1}{q}}(y)=\mathrm{Sin}_q(x\boxplus_q y). \tag{6.196}$$

The following equations hold:

$$\mathrm{D}_q\mathrm{e}_q(x)=\frac{\mathrm{e}_q(x)}{1-q}, \tag{6.197}$$

$$\mathrm{D}_q\mathrm{e}_{\frac{1}{q}}(x)=\frac{\mathrm{e}_{\frac{1}{q}}(qx)}{1-q}. \tag{6.198}$$

Instead Euler proved the equations

$$\triangle^{+}\mathrm{e}_q(x)=\mathrm{e}_q(x), \tag{6.199}$$

$$\triangle^{+}\mathrm{e}_{\frac{1}{q}}(x)=\mathrm{e}_{\frac{1}{q}}(qx), \tag{6.200}$$

where $\triangle^{+}$ is defined by (6.223).

6.11.1 Further q-trigonometric functions

Definition 125 We can now define four other q-analogues of the trigonometric functions [261]:

$$\sin_q(x)\equiv\frac{1}{2i}\big(\mathrm{e}_q(ix)-\mathrm{e}_q(-ix)\big)=\sum_{n=0}^{\infty}(-1)^n\frac{x^{2n+1}}{\langle 1;q\rangle_{2n+1}},\quad |x|<1, \tag{6.201}$$

$$\cos_q(x)\equiv\frac{1}{2}\big(\mathrm{e}_q(ix)+\mathrm{e}_q(-ix)\big)=\sum_{n=0}^{\infty}(-1)^n\frac{x^{2n}}{\langle 1;q\rangle_{2n}},\quad |x|<1, \tag{6.202}$$

$$\sin_{\frac{1}{q}}(x)\equiv\frac{1}{2i}\big(\mathrm{e}_{\frac{1}{q}}(ix)-\mathrm{e}_{\frac{1}{q}}(-ix)\big)=\sum_{n=0}^{\infty}(-1)^n q^{n(2n+1)}\frac{x^{2n+1}}{\langle 1;q\rangle_{2n+1}}, \tag{6.203}$$

$$\cos_{\frac{1}{q}}(x)\equiv\frac{1}{2}\big(\mathrm{e}_{\frac{1}{q}}(ix)+\mathrm{e}_{\frac{1}{q}}(-ix)\big)=\sum_{n=0}^{\infty}(-1)^n q^{n(2n-1)}\frac{x^{2n}}{\langle 1;q\rangle_{2n}}, \tag{6.204}$$

where $x\in\mathbb{C}$ in the last two equations.

These four q-trigonometric functions can also be written in the equivalent forms

$$\sin_q(x)=\frac{x}{1-q}\,{}_4\phi_3\left(\infty,\infty,\infty,\infty;\frac{3}{2},\widetilde{1},\widetilde{\frac{3}{2}}\middle|q;-x^2\right), \tag{6.205}$$

$$\cos_q(x)={}_4\phi_3\left(\infty,\infty,\infty,\infty;\frac{1}{2},\widetilde{1},\widetilde{\frac{1}{2}}\middle|q;-x^2\right), \tag{6.206}$$

$$\sin_{\frac{1}{q}}(x) = \frac{x}{1-q}\,{}_0\phi_3\left(-;\frac{3}{2},\widetilde{1},\widetilde{\frac{3}{2}}\middle|q;-x^2q^3\right), \tag{6.207}$$

$$\cos_{\frac{1}{q}}(x) = {}_0\phi_3\left(-;\frac{1}{2},\widetilde{1},\widetilde{\frac{1}{2}}\middle|q;-x^2q\right). \tag{6.208}$$

The functions $\sin_{\frac{1}{q}}(x)$ and $\cos_{\frac{1}{q}}(x)$ solve the q-difference equation

$$(q-1)^2\mathrm{D}_q^2 f(x) + qf\left(q^2x\right) = 0, \tag{6.209}$$

and the functions $\sin_q(x)$ and $\cos_q(x)$ solve the q-difference equation

$$(q-1)^2\mathrm{D}_q^2 f(x) + f(x) = 0. \tag{6.210}$$

Definition 126 The *q-tangent numbers* $\mathrm{T}_{2n+1}(q)$ are defined by [34, p. 380]:

$$\frac{\sin_q(x)}{\cos_q(x)} \equiv \tan_q(x) = \sum_{n=0}^{\infty}\frac{\mathrm{T}_{2n+1}(q)x^{2n+1}}{\langle 1;q\rangle_{2n+1}}. \tag{6.211}$$

Andrews and Gessel [34, p. 380] have proved that the polynomial $\mathrm{T}_{2n+1}(q)$ is divisible by $\widetilde{\langle 1;q\rangle}_n$.

Definition 127 The *q-Euler numbers* or *q-secant numbers* $\mathcal{S}_{2n}(q)$ are defined by [33, p. 283]:

$$\frac{1}{\cos_q(x)} = \sum_{n=0}^{\infty}\frac{\mathcal{S}_{2n}(q)x^{2n}}{\langle 1;q\rangle_{2n}}. \tag{6.212}$$

The congruence

$$\mathcal{S}_{2n} \equiv 1 \bmod 4 \tag{6.213}$$

was proved by Sylvester [399, p. 260], [33, p. 283].

Andrews and Foata [33, p. 283] have proved that

$$\mathcal{S}_{2n}(q) \equiv q^{2n(n-1)} \bmod (q+1)^2. \tag{6.214}$$

Definition 128 We can now define four q-analogues of the hyperbolic functions [546]. In the first two equations below, $|q|>1$, or $0<|q|<1$ and $|x|<|1-q|^{-1}$.

$$\mathrm{Sinh}_q(x) \equiv \frac{1}{2}\left(\mathrm{E}_q(x) - \mathrm{E}_q(-x)\right) = \sum_{n=0}^{\infty}\frac{x^{2n+1}}{\{2n+1\}_q!}, \tag{6.215}$$

$$\mathrm{Cosh}_q(x) \equiv \frac{1}{2}\left(\mathrm{E}_q(x) + \mathrm{E}_q(-x)\right) = \sum_{n=0}^{\infty}\frac{x^{2n}}{\{2n\}_q!}, \tag{6.216}$$

$$\mathrm{Sinh}_{\frac{1}{q}}(x) \equiv \frac{1}{2}\left(\mathrm{E}_{\frac{1}{q}}(x) - \mathrm{E}_{\frac{1}{q}}(-x)\right) = \sum_{n=0}^{\infty} q^{n(2n+1)} \frac{x^{2n+1}}{\{2n+1\}_q!}, \tag{6.217}$$

$$\mathrm{Cosh}_{\frac{1}{q}}(x) \equiv \frac{1}{2}\left(\mathrm{E}_{\frac{1}{q}}(x) + \mathrm{E}_{\frac{1}{q}}(-x)\right) = \sum_{n=0}^{\infty} q^{n(2n-1)} \frac{x^{2n}}{\{2n\}_q!}, \tag{6.218}$$

where $x \in \mathbb{C}$ in the two last equations.

Theorem 6.11.1 *The following two addition theorems hold*:

$$\mathrm{Cosh}_q(x)\,\mathrm{Cosh}_{\frac{1}{q}}(y) + \mathrm{Sinh}_q(x)\,\mathrm{Sinh}_{\frac{1}{q}}(y) = \mathrm{Cosh}_q(x \boxplus_q y), \tag{6.219}$$

$$\mathrm{Sinh}_q(x)\,\mathrm{Cosh}_{\frac{1}{q}}(y) + \mathrm{Cosh}_q(x)\,\mathrm{Sinh}_{\frac{1}{q}}(y) = \mathrm{Sinh}_q(x \boxplus_q y). \tag{6.220}$$

6.12 The Nalli-Ward-Al-Salam q-addition and some variants of the q-difference operator

In such an interdisciplinary subject as q-calculus, many different definitions have been used and in this section we try to collect some of them. In general, physicists tend to use symmetric operators.

Definition 129 A *symmetric q-difference operator* is defined by

$$\mathrm{D}_{q_1,q^{-1}} f(x) = \frac{f(q_1 x) - f(q^{-1}x)}{(q_1 - q^{-1})x}. \tag{6.221}$$

The relation between $\mathrm{D}_{q_1,q_1^{-1}}$ and D_q is

$$\mathrm{D}_{q_1,q_1^{-1}} = \frac{q_1 - 1}{q_1 - q_1^{-1}} \mathrm{D}_{q_1} + \frac{1 - q_1^{-1}}{q_1 - q_1^{-1}} \mathrm{D}_{q_1^{-1}}. \tag{6.222}$$

Definition 130 Euler used the operator:

$$\Delta^{+}\varphi(x) \equiv \frac{\varphi(x) - \varphi(qx)}{x}. \tag{6.223}$$

The q-Leibniz formula (6.95) also applies for Δ^{+}.

Heine and Thomae used the operator

$$\Delta\varphi(x) \equiv \varphi(qx) - \varphi(x). \tag{6.224}$$

Sometimes Jackson [303] used the operator

$$[\wp]\varphi(x) \equiv \frac{\varphi(x) - \varphi(qx)}{1 - q}, \quad q \in \mathbb{C}\setminus\{1\}. \tag{6.225}$$

Theorem 6.12.1 *Jackson* [303, p. 305] *expressed* (6.98) *as*

$$x^n \mathrm{D}_q^n = q^{-\binom{n}{2}} \prod_{m=0}^{n-1} [\wp - m], \tag{6.226}$$

where the m is counted as the q-analogue of m and $[\wp]$ *is defined by* (6.225).

This formula can be proved by the q-Pascal identity.

In 1994 K. S. Chung, W. S. Chung, S. T. Nam and H. J. Kang [132] rediscovered two q-operations (q-addition and q-subtraction) which lead to new q-binomial formulas and consequently to a new form of the q-derivative.

The NWA proves to be the natural way to work with the quantity on the right side of | in the q-hypergeometric series (6.181). Further examples follow in connection which the q-Appell functions.

Definition 131

$$(a \oplus_q b)^n \equiv \sum_{k=0}^{n} \binom{n}{k}_q a^k b^{n-k}, \quad n = 0, 1, 2, \ldots. \tag{6.227}$$

Remark 27 There is no point in extending the NWA to arbitrary complex exponents since the convergence will be very poor and the associative law does not apply then.

The q-addition is a special case of the so-called Gaussian convolution [232, p. 245] of $\{a_n\}$ and $\{b_n\}$,

$$c_n = \sum_{k=0}^{n} \binom{n}{k}_q a_k b_{n-k}, \quad n = 0, 1, 2, \ldots. \tag{6.228}$$

Theorem 6.12.2 *In* 1936 *Ward* [531, p. 256] *proved the following equations for q-subtraction*:

$$(x \ominus_q y)^{2n+1} = \sum_{k=0}^{n} (-1)^k \binom{2n+1}{k}_q x^k y^k \left(x^{2n+1-2k} - y^{2n+1-2k}\right), \tag{6.229}$$

$$(x \ominus_q y)^{2n} = (-1)^n \binom{2n}{n}_q x^n y^n + \sum_{k=0}^{n-1} (-1)^k \binom{2n}{k}_q x^k y^k \left(x^{2n-2k} + y^{2n-2k}\right). \tag{6.230}$$

Furthermore [531, p. 262]

$$\mathrm{E}_q(x)\mathrm{E}_q(-x) = \sum_{n=0}^{\infty} \frac{(1 \ominus_q 1)^{2n}}{\{2n\}_q!} x^{2n}, \tag{6.231}$$

$$e_q(x)e_q(-x)=\sum_{n=0}^{\infty}\frac{(1\ominus_q 1)^{2n}}{\langle 1;q\rangle_{2n}}x^{2n}. \tag{6.232}$$

We immediately obtain the following rules for the product of two q-exponential functions:

$$\mathrm{E}_q(x)\mathrm{E}_q(y)=\mathrm{E}_q(x\oplus_q y),\qquad e_q(x)e_q(y)=e_q(x\oplus_q y). \tag{6.233}$$

We have

Theorem 6.12.3

$$\mathrm{Cos}_q(\overline{n}_q x)+i\,\mathrm{Sin}_q(\overline{n}_q x)=\big(\mathrm{Cos}_q(x)+i\,\mathrm{Sin}_q(x)\big)^n, \tag{6.234}$$

$$\cos_q(\overline{n}_q x)+i\sin_q(\overline{n}_q x)=\big(\cos_q(x)+i\,\sin_q(x)\big)^n. \tag{6.235}$$

We immediately obtain the following addition theorems for the q-analogues of the trigonometric functions:

$$\mathrm{Cos}_q(x)\,\mathrm{Cos}_q(y)+\mathrm{Sin}_q(x)\,\mathrm{Sin}_q(y)=\mathrm{Cos}_q(x\ominus_q y), \tag{6.236}$$

$$\mathrm{Cos}_q(x)\,\mathrm{Cos}_q(y)-\mathrm{Sin}_q(x)\mathrm{Sin}_q(y)=\mathrm{Cos}_q(x\oplus_q y), \tag{6.237}$$

$$\mathrm{Sin}_q(x)\,\mathrm{Cos}_q(y)+\mathrm{Sin}_q(y)\mathrm{Cos}_q(x)=\mathrm{Sin}_q(x\oplus_q y), \tag{6.238}$$

$$\mathrm{Sin}_q(x)\,\mathrm{Cos}_q(y)-\mathrm{Sin}_q(y)\mathrm{Cos}_q(x)=\mathrm{Sin}_q(x\ominus_q y), \tag{6.239}$$

$$\cos_q(x)\cos_q(y)+\sin_q(x)\sin_q(y)=\cos_q(x\ominus_q y), \tag{6.240}$$

$$\cos_q(x)\cos_q(y)-\sin_q(x)\sin_q(y)=\cos_q(x\oplus_q y), \tag{6.241}$$

$$\sin_q(x)\cos_q(y)+\sin_q(y)\cos_q(x)=\sin_q(x\oplus_q y), \tag{6.242}$$

$$\sin_q(x)\cos_q(y)-\sin_q(y)\cos_q(x)=\sin_q(x\ominus_q y). \tag{6.243}$$

The following definition is equivalent with an inversion of the basis for NWA.

Definition 132 The q-coaddition is defined by [12, p. 240])

$$\big(a\oplus^q b\big)^n\equiv\sum_{k=0}^{n}\binom{n}{k}_q q^{k(k-n)}a^k b^{n-k},\quad n=0,1,2,\ldots. \tag{6.244}$$

The q-cosubtraction is defined by

$$\begin{aligned}&a\ominus^q b\equiv a\oplus^q -b,\\&a\ominus^q b=-\big(b\ominus^q a\big).\end{aligned} \tag{6.245}$$

The following equations for q-cosubtraction hold, compare with [531, p. 256]:

$$\left(x \ominus^q y\right)^{2n+1} = \sum_{k=0}^{n}(-1)^k \binom{2n+1}{k}_q q^{k(k-2n-1)} x^k y^k \left(x^{2n+1-2k} - y^{2n+1-2k}\right), \tag{6.246}$$

$$\begin{aligned}\left(x \ominus^q y\right)^{2n} &= (-1)^n \binom{2n}{n}_q q^{-n^2} x^n y^n \\ &\quad + \sum_{k=0}^{n-1}(-1)^k \binom{2n}{k}_q q^{k(k-2n)} x^k y^k \left(x^{2n-2k} + y^{2n-2k}\right).\end{aligned} \tag{6.247}$$

Furthermore,

$$\mathrm{E}_{\frac{1}{q}}(x)\mathrm{E}_{\frac{1}{q}}(-x) = \sum_{n=0}^{\infty} \frac{(1 \ominus^q 1)^{2n}}{\{2n\}_q!} q^{\binom{2n}{2}} x^{2n}, \tag{6.248}$$

$$\mathrm{e}_{\frac{1}{q}}(x)\mathrm{e}_{\frac{1}{q}}(-x) = \sum_{n=0}^{\infty} \frac{(1 \ominus^q 1)^{2n}}{\langle 1; q\rangle_{2n}} q^{\binom{2n}{2}} x^{2n}. \tag{6.249}$$

We immediately obtain the following rules for the product of two q-exponential functions:

$$\begin{aligned}\mathrm{E}_{\frac{1}{q}}(x)\mathrm{E}_{\frac{1}{q}}(y) &= \mathrm{E}_{\frac{1}{q}}\left(x \oplus^q y\right), \\ \mathrm{e}_{\frac{1}{q}}(x)\mathrm{e}_{\frac{1}{q}}(y) &= \mathrm{e}_{\frac{1}{q}}\left(x \oplus^q y\right).\end{aligned} \tag{6.250}$$

A q-analogue of (3.33) is given by

Theorem 6.12.4

$$(1 \oplus_q x)^n + (1 \ominus_q x)^n = 2{}_4\phi_1\left(-\frac{n}{2}, \frac{1-n}{2}, \infty, \infty; \frac{1}{2}\middle| q^2, x^2 q^{2n-1}\right). \tag{6.251}$$

Proof There are two cases to consider:

1. n even.

$$\mathrm{LHS} = 2\sum_{k=0}^{\frac{n}{2}} \frac{x^{2k}\langle 1; q\rangle_n}{\langle 1; q\rangle_{2k}\langle 1; q\rangle_{n-2k}}$$

$$\overset{\text{by (6.34)}}{=} 2\sum_{k=0}^{\frac{n}{2}} \frac{x^{2k}\langle \frac{1}{2}, 1; q^2\rangle_{\frac{n}{2}}}{\langle \frac{1}{2}, 1; q^2\rangle_k \langle \frac{1}{2}, 1; q^2\rangle_{\frac{n}{2}-k}}$$

$$\overset{\text{by (6.34, 6.20)}}{=} 2\sum_{k=0}^{\frac{n}{2}} \frac{x^{2k}\langle \frac{-n}{2}, \frac{-n+1}{2}; q^2\rangle_k q^{2(-2\binom{k}{2}+\frac{nk}{2}+\frac{(n-1)k}{2})}}{\langle \frac{1}{2}, 1; q^2\rangle_k} = \text{RHS}. \tag{6.252}$$

2. n uneven.

$$\text{LHS} = 2\sum_{k=0}^{\frac{n-1}{2}} \frac{x^{2k}\langle 1; q\rangle_1 \langle 2; q\rangle_{n-1}}{\langle 1; q\rangle_{2k}\langle 1; q\rangle_1\langle 2; q\rangle_{n-1-2k}}$$

$$\overset{\text{by (6.34)}}{=} 2\sum_{k=0}^{\frac{n-1}{2}} \frac{x^{2k}\langle \frac{3}{2}, 1; q^2\rangle_{\frac{n-1}{2}}}{\langle \frac{1}{2}, 1; q^2\rangle_k \langle \frac{3}{2}, 1; q^2\rangle_{\frac{n-1}{2}-k}}$$

$$\overset{\text{by (6.34, 6.20)}}{=} 2\sum_{k=0}^{\frac{n}{2}} \frac{x^{2k}\langle \frac{-n}{2}, \frac{-n+1}{2}; q^2\rangle_k q^{2(-2\binom{k}{2}+\frac{nk}{2}+\frac{(n-1)k}{2})}}{\langle \frac{1}{2}, 1; q^2\rangle_k} = \text{RHS}. \tag{6.253}$$

□

6.13 Weierstraß elliptic functions and sigma functions

In the same way as many beautiful equations in mechanics are connected with the torque (and the angular momentum), the theta functions and the elliptic functions give the most beautiful equations in calculus.

torque	theta functions, elliptic functions

The elliptic functions are included here to explain the relation to Γ_q functions at the end of Section 6.13. Abel derived the infinite double products in the theory of elliptic functions from the multiplication theorem in an elementary way, which does not presuppose any theorems from the general theory of functions. As was pointed out by Ole Jacob Broch (1818–89) in 1864 and Mittag-Leffler in 1876, Abel's proof was not entirely satisfactory. Therefore in 1876 Mittag-Leffler gave another proof of the limiting passage definitely distinct from that of Abel. Mittag-Leffler also wanted to write a book on elliptic functions.

The three Jacobi elliptic functions sn$u$, cn$u$, dn$u$ were denoted λ, μ, ν by Abel. Glaisher formed the other nine Jacobi elliptic functions by taking three reciprocals and six ratios.

The Weierstraß elliptic functions (and sigma functions) are connected with the name of the German mathematician Karl Weierstraß, who first defined them. In order to define the elliptic function sn$u$, we have to start with the definition of the elliptic integral. The formula for the arclength of the ellipse involves an integral of the form

$$E(z) = \int_0^z \frac{\sqrt{(1-(kx)^2)}dx}{\sqrt{1-x^2}}, \tag{6.254}$$

where $0 < k < 1$. Another *elliptic integral* is

$$F(z) = \int_0^z \frac{dx}{\sqrt{(1-x^2)(1-(kx)^2)}}, \tag{6.255}$$

where $0 < k < 1$. $F(z)$ and $E(z)$ are called the Jacobi form for the elliptic integral of the first and second kind. Legendre first worked with these integrals and he was followed by Abel and Jacobi, who in the 1820s, inspired by Gauß, discovered that *inverting* $F(z)$ gave the doubly periodic *elliptic function*:

$$F^{-1}(\omega) = \mathrm{sn}\omega. \tag{6.256}$$

6.13.1 Elliptic functions

Just as the rational functions on the *Riemann sphere* Σ form a field denoted $\mathbb{C}(z)$, the meromorphic functions on the torus $\mathbb{C}/\Omega$ are the doubly-periodic elliptic functions on $\mathbb{C}$, which form a field, denoted $E(\Omega)$, where Ω is a fixed *lattice*. Both $E(\Omega)$ and $\mathrm{E}_1(\Omega)$, the field of even elliptic functions, are extension fields of $\mathbb{C}$ (see Jones and Singerman [319, p. 98]).

Every lattice has a *fundamental region*, e.g. the *Dirichlet domain* $D(\Omega)$ is a fundamental domain for Ω. Just as the rational functions have order equal to the maximum of the degrees in the numerator and in the denominator, the order of an elliptic function f is the sum of the orders of the congruence classes of poles of f. The order is denoted $\mathrm{ord}(f) = N$. By using elementary complex function theory, it is possible to prove the following statements about elliptic functions: An analytic elliptic function must be constant. The sum of the residues of f within a fundamental domain is zero. There are no elliptic functions of order $N = 1$. If f has order $N > 0$, then f takes each value $c \in \Sigma$ exactly N times (see Jones and Singerman [319]).

K. Weierstraß, who lent name to these functions, attended lectures on elliptic functions in Münster (in 1839) given by Cristof Gudermann. Gudermann wrote a standard work on Jacobi elliptic functions in 1844 [256] in which he introduced the standard notation. His former student Weierstraß, who had by then gained fame from his work on Abelian functions, in the mid 1850s modified these and introduced the Weierstraß *sigma function*, $\sigma(z)$, which is z multiplied with the product over all lattice points of those *elementary factors* $\mathcal{E}_2$ which have simple zeros at the lattice points. The sigma function is thus equivalent to the Jacobi theta function, but on a general lattice.

The *Weierstraß zeta function* $\zeta(z)$ is the logarithmic derivative of $\sigma(z)$. The Weierstraß $\wp$ function, which belongs to $\mathrm{E}_1(\Omega)$, is $-\frac{d\zeta}{dz}$. The Weierstraß $\wp$ function satisfies a differential equation, whose coefficients are the *Eisenstein series* for Ω.

Let Ω be a lattice with basis $\{\omega_1, \omega_2\}$ and let

$$\omega_3 = \omega_1 + \omega_2.$$

If P is a fundamental domain with $0, \frac{1}{2}\omega_1, \frac{1}{2}\omega_2, \frac{1}{2}\omega_3$ in its interior, then $\frac{1}{2}\omega_1, \frac{1}{2}\omega_2, \frac{1}{2}\omega_3$ are the zeros of $\frac{d\wp}{dz}$ in P.

Define

$$e_j = \wp\left(\frac{1}{2}\omega_j\right)$$

for $j = 1, 2, 3$. These numbers are mutually distinct.

Although the functions $\zeta(z)$ and $\sigma(z)$ are not elliptic, they enjoy translation properties, which can be proved using the *Legendre relation.*

6.13.2 Connections with the Γ_q function

The following lemmas are just a few examples of Γ_q formulas.

Lemma 6.13.1 *A relation between Γ_q functions with different bases*:

$$\frac{\Gamma_{q^2}(x)}{\Gamma_q(x)} = \frac{\langle\widetilde{1}; q\rangle_\infty}{\langle\widetilde{x}; q\rangle_\infty}(1+q)^{1-x} \equiv (1+q)^{1-x}\langle\widetilde{1}; q\rangle_{x-1}. \tag{6.257}$$

Lemma 6.13.2

$$\frac{\langle\frac{1+a}{2}, 1-b+\frac{a}{2}; q^2\rangle_\infty}{\langle 1+a-b; q^2\rangle_\infty} = \Gamma_q\begin{bmatrix} 1+a-b, 1+\frac{a}{2} \\ 1+a, 1+\frac{a}{2}-b \end{bmatrix} \frac{\langle\widetilde{1+\frac{a}{2}-b}; q\rangle_\infty}{\langle\widetilde{1+\frac{a}{2}}, \widetilde{1+a-b}; q\rangle_\infty}. \tag{6.258}$$

Proof Use the Bailey-Daum theorem, see Section 7.13. □

The Jacobi theta functions are the elliptic analogues of the exponential function and constitute the simplest elements from which the elliptic functions can be constructed. The first Jacobi theta function is related to the Γ_q function via the following q-analogue of the Euler reflection formula:

$$\Gamma(x)\Gamma(1-x) = \frac{\pi}{\sin \pi x}. \tag{6.259}$$

This formula of crucial importance has an interesting history. It was first found by Thomae in 1869 [497, p. 262, (6a)] and 1873 [500, p. 183, (168a)], who used the notation from School (2) in 2.5. Then it was rediscovered by Pia Nalli in 1923 [391, p. 352], who used the notation from School (7) in 2.5. Finally it was given by Atakishiyev 2001 in the notation from the Russian School in 2.2 and School (1), (5) in 2.5. The interesting thing is that both Thomae and Pia Nalli used notations for theta functions, which differ significantly from the modern notation.

The Thomae notation of the School (2) of Section 2.5. reminiscent of the Riemann theta functions and the Pia Nalli notation of School (8) from Section 2.5

may have been influenced by nineteenth century Italian books on elliptic functions. Thomae [497, p. 262] also claimed that his teacher Heine published this equation in [270, p. 310].

Theorem 6.13.3 *A q-analogue of Euler's reflection formula* (6.259) [500, p. 183, (168a)], [50, p. 1326], [391, p. 352]:

$$\Gamma_q(z)\Gamma_q(1-z) = \frac{iq^{\frac{1}{8}}(1-q)(\langle 1;q\rangle_\infty)^3}{q^{\frac{z}{2}}\theta_1(\frac{-iz}{2}\log q, \sqrt{q})}, \tag{6.260}$$

where the first Jacobi theta function is given by

$$\theta_1(z,q) \equiv 2\sum_{n=0}^{\infty}(-1)^n \mathrm{QE}\left(\left(n+\frac{1}{2}\right)^2\right)\sin(2n+1)z. \tag{6.261}$$

This function has period 2π and quasiperiod $\frac{-i}{2}\log q$.

The zeros $z = m\pi$ and $z = \frac{-in}{2}\log q$ of θ_1 correspond to the set of poles $\pm\frac{2m\pi i}{\log q}$ and $-n$ of Γ_q respectively.

The presence of $\sqrt{q}$ in (6.260) is due to the quadratic function values for the product representations of the elliptic functions.

For real z and q in (6.260) the first function value for θ_1 in (6.260) is imaginary. We then have

$$\theta_1(iz,q) = 2i\sum_{n=0}^{\infty}(-1)^n \mathrm{QE}\left(\left(n+\frac{1}{2}\right)^2\right)\sinh(2n+1)z. \tag{6.262}$$

This apparently explains the appearance of the factor i in the numerator of (6.260), $\Gamma_q(z)\Gamma_q(1-z)$ must remain real after the q-deformation.

Remark 28 It is possible that there is some relationship between the Jacobi triple product identity and the q-analogue of Euler's reflection formula, compare with formula (7.56).

C. Zhang [549, p. 62] considered only the range $1 \le \Re(z) \le 2$, and then displayed the quasi-periodicity

$$\Gamma_q\left(x + \frac{2\pi i}{\log q}\right) = (1-q)^{-\frac{2\pi i}{\log q}}\Gamma_q(x). \tag{6.263}$$

Zhang also showed that $\Gamma_q \neq 0$ and there are constants $m, M > 0$, such that [549, p. 62]

$$m \le \left|\Gamma_q(z)\right| \le M.$$

Theorem 6.13.4 [48, p. 128] *q-Bohr-Mollerup theorem (Askey): The Γ_q function is the only function which meets the following three conditions for $x > 0$:*

$$f(x+1) = \{x\}_q f(x), \quad 0 < q < 1, \tag{6.264}$$

$$f(1) = 1, \tag{6.265}$$

$$x \mapsto \log f(x) \quad \textit{is convex}. \tag{6.266}$$

We now want to show some graphs (see Figures 1–6) for the right hand side of (6.260) and make a comparison with the function $\frac{\pi}{\sin \pi x}$.

We start with the interval $[-4.05, -3.9]$. The singularity -4 cuts off the diagram in two parts. The first graph shows $\frac{\pi}{\sin \pi x}$ and the two next (6.260) for $q = .5$ and $q = .2$. Next, consider the interval $[3.9, 4.05]$. The singularity 4 divides the graph into two parts. The first graph shows $\frac{\pi}{\sin \pi x}$ and the two next (6.260) for $q = .5$ and $q = .2$.

Corollary 6.13.5 *A q-analogue of $\Gamma(\frac{1}{2}) = \sqrt{\pi}$ is given by*

$$\Gamma_q\left(\frac{1}{2}\right) = \left(\frac{i(1-q)(\langle 1; q\rangle_\infty)^3}{q^{\frac{1}{8}} \theta_1(\frac{-i}{4} \log q, \sqrt{q})}\right)^{\frac{1}{2}}. \tag{6.267}$$

A similar formula was given by Heine [272, p. 110]:

$$\left[\Omega\left(q^2, -\frac{1}{2}\right)\right]^2 = \frac{K}{\pi} \frac{\sqrt{k}}{\sqrt[4]{q}}. \tag{6.268}$$

The first person to work explicitly with the expression $\Gamma_q(x)\Gamma_q(1-x)$ was Reverend Jackson [302, p. 193], who showed that

$$\Gamma_q(x)\Gamma_q(1-x) = \frac{(\Gamma_q(\frac{1}{2}))^2}{\sigma_q(x)}, \tag{6.269}$$

Fig. 1

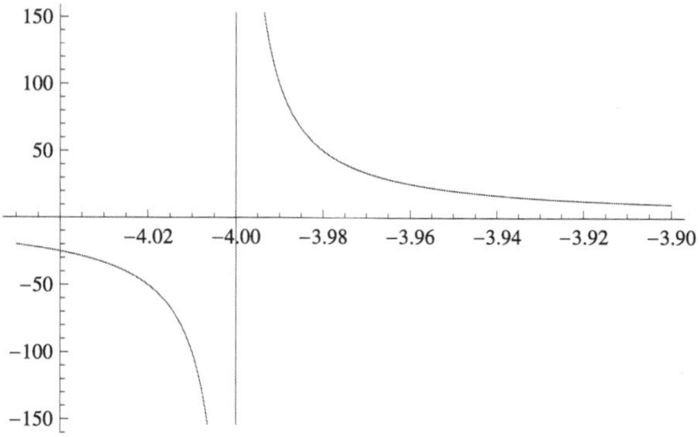

Fig. 2

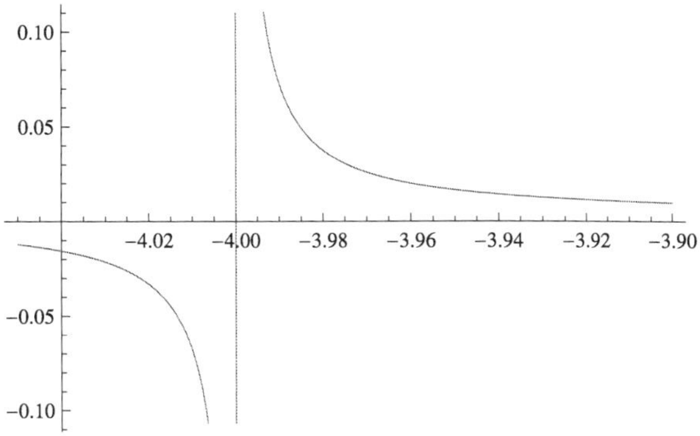

Fig. 3

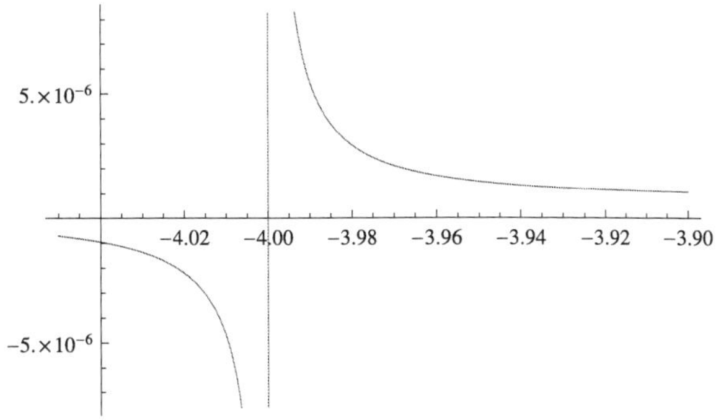

Fig. 4

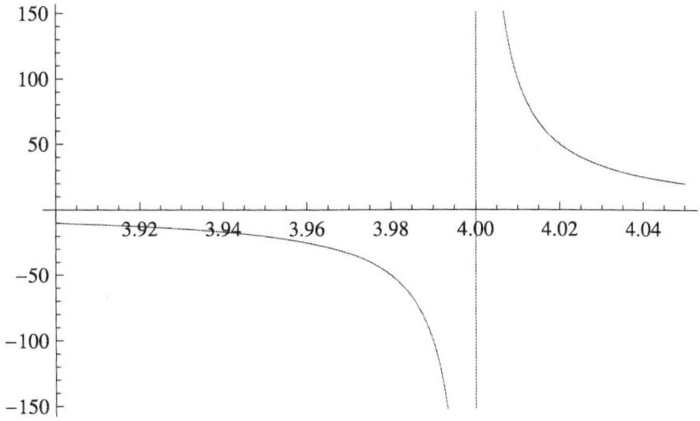

when computing the q-integral $\int_0^\infty \frac{y^{x-1} d_q(y)}{(1\boxplus_q y)^1}$ (this is a special case of formula (8.113)). Here $\sigma_q(x)$ is a certain σ or θ function.

6.14 The Chen-Liu operator or parameter augmentation

In this section x is the variable. All other quantities are called parameters. A function independent of x is called a constant.

Fig. 5

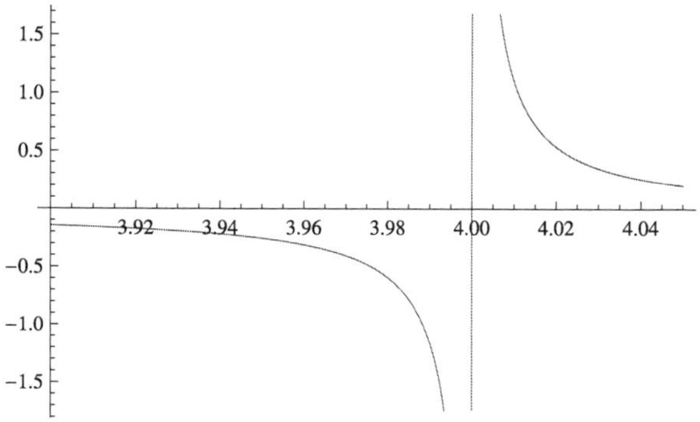

Fig. 6

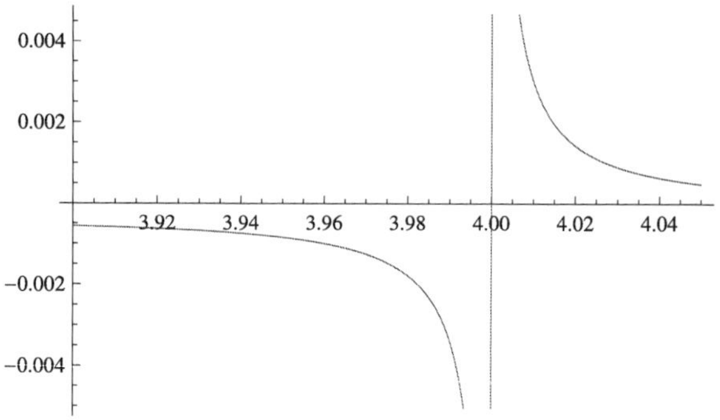

Definition 133

$$\varphi(q) \equiv \epsilon^{-1}\mathrm{D}_q. \tag{6.270}$$

The Chen-Liu operator [125] is defined by

$$E\big(b\varphi(q)\big) \equiv \sum_{n=0}^{\infty} \frac{(b\varphi(q))^n q^{\binom{n}{2}}}{\{n\}_q!}. \tag{6.271}$$

We have $E(b\varphi(q))(\text{constant}) = 1$.

By the q-Leibniz formula, we get [125, p. 114]

Theorem 6.14.1

$$E\big(b\varphi(q)\big)\big\{(xt;q)_\infty\big\} = (xt, bt; q)_\infty, \tag{6.272}$$

$$E\big(b\varphi(q)\big)\big\{(xs, xt;q)_\infty\big\} = \frac{(xs, xt, bs, bt; q)_\infty}{(xbst/q; q)_\infty}. \tag{6.273}$$

In [125] many summation formulas were proved by using this operator, e.g. formula (7.112).

Chapter 7
q-hypergeometric series

This is a modest attempt to present a new notation for q-calculus and, in particular, for q-hypergeometric series, which is compatible with the old notation. Also, a new method that follows from this notation is presented. This notation leads to a new method for computations and classifications of q-special functions. With this notation many formulas of q-calculus become very natural and the q-analogues of many orthogonal polynomials and functions assume a very pleasant form reminding directly of their classical counterparts.

7.1 Definition of q-hypergeometric series

The *stress tensor* (matrix) corresponds to the generalized q-hypergeometric series.

Matrices	${}_p\phi_r((a);(b)\|q;x)$

Definition 134 Generalizing Heine's series (6.181), we define a q-hypergeometric series by (compare with [220, p. 4], [261, p. 345]):

$$
{}_{p+p'}\phi_{r+r'}\left[\begin{matrix}\hat{a}_1,\ldots,\hat{a}_p\\ \hat{b}_1,\ldots,\hat{b}_r\end{matrix}|q;z||\frac{\prod_i f_i(k)}{\prod_j g_j(k)}\right]
\equiv\sum_{k=0}^{\infty}\frac{\langle\hat{a}_1;q\rangle_k\cdots\langle\hat{a}_p;q\rangle_k}{\langle 1,\hat{b}_1;q\rangle_k\cdots\langle\hat{b}_r;q\rangle_k}\left[(-1)^k q^{\binom{k}{2}}\right]^{1+r+r'-p-p'} z^k\frac{\prod_i f_i(k)}{\prod_j g_j(k)}, \tag{7.1}
$$

where

$$
\widehat{a}\equiv a\vee\widetilde{a}\vee\overset{m}{\tau}\,\widetilde{a}\vee_k\widetilde{a}\vee\Delta(q;l;\lambda). \tag{7.2}
$$

In case of $\Delta(q;l;\lambda)$ the index is adjusted accordingly. It is assumed that the denominator contains no zero factors, i.e. $\widehat{b_k}\neq 0$, and also that the $f_i(k)$ and $g_j(k)$ contain p' and r' factors of the form $\langle\widehat{a}(k);q\rangle_k$ or $(s(k);q)_k$ respectively.

The resonance corresponds to the Heine infinity.

T. Ernst, *A Comprehensive Treatment of q-Calculus*,
DOI 10.1007/978-3-0348-0431-8_7, © Springer Basel 2012

Resonance	∞_{H}

In a few cases the parameter $\widehat{a}$ in (7.1) will be the real plus infinity ($0 < |q| < 1$). This corresponds to multiplication by 1. If we want to be formal, we could introduce a symbol ∞_{H}, with the property

$$\langle \infty_{\mathrm{H}}; q\rangle_n = \langle \infty_{\mathrm{H}} + \alpha; q\rangle_n = 1, \quad \alpha \in \mathbb{C},\ 0 < |q| < 1. \tag{7.3}$$

The symbol ∞_{H} corresponds to the parameter 0 in [220, p. 4].

We will denote ∞_{H} by ∞ in the rest of the book.

The following notation is also used:

$$n\infty \equiv \infty, \ldots, \infty, n \text{ times}, \quad n \in \mathbb{N}. \tag{7.4}$$

Furthermore, $b_j \neq 0,\ j = 1, \ldots, r,\ b_j \neq -m,\ j = 1, \ldots, r, m \in \mathbb{N}$,

$$b_j \neq \frac{2m\pi i}{\log q}, \quad j = 1, \ldots, r, m \in \mathbb{N},\ [467].$$

Remark 29 There is a rule analogous to (7.3), but for the Riemann sphere $\hat{\mathbb{C}}$. We find a q-analogue of this in [185].

The use of the extra sign $||$ in (7.1) is best illustrated by the following example, where the vectors $\vec{\lambda}$ and $\vec{\mu}$ have length p:

Example 31 The following notation makes sense:

$${}_{16+4p}\phi_{15+4p}\left[\begin{matrix} \Delta(q; 2; \vec{\lambda}), \Delta(q; 3; \nu + \sigma - 1), 9\infty \\ \Delta(q; 2; \vec{\mu}, \nu, \sigma, \nu + \sigma - 1), \nu, \sigma, \widetilde{1} \end{matrix} |q; -x^2|| \begin{matrix} \langle \widetilde{1 - k}; q\rangle_k \\ - \end{matrix}\right]. \tag{7.5}$$

The $\Delta(q; 3; \nu + \sigma - 1)$ corresponds to $3k$ q-shifted factorials, this explains the 9∞.

The motivation for the extra factor $(-1)q^{\pm\binom{k}{2}}$ is that we need a q-analogue of

$$\begin{aligned} &\lim_{x\to\infty} {}_p\mathrm{F}_r(a_1, \ldots, a_p; b_1, \ldots, b_{r-1}, x; xz) \\ &\quad = {}_p\mathrm{F}_{r-1}(a_1, \ldots, a_p; b_1, \ldots, b_{r-1}; z). \end{aligned} \tag{7.6}$$

This q-analogue is given by

$$\begin{aligned} &\lim_{x\to-\infty} {}_p\phi_r\big(a_1, \ldots, a_p; b_1, \ldots, b_{r-1}, x|q; zq^x\big) \\ &\quad = {}_p\phi_{r-1}(a_1, \ldots, a_p; b_1, \ldots, b_{r-1}|q; z) \quad \big(0 < |q| < 1\big). \end{aligned} \tag{7.7}$$

However, the q-analogue of

$$\begin{aligned} &\lim_{\epsilon\to 0} {}_p\mathrm{F}_r\left(a_1, \ldots, a_{p-1}, \frac{1}{\epsilon}; b_1, \ldots, b_r; \epsilon z\right) \\ &\quad = {}_{p-1}\mathrm{F}_r(a_1, \ldots, a_{p-1}; b_1, \ldots, b_r; z) \end{aligned} \tag{7.8}$$

is given by

$$\lim_{x\to\infty} {}_p\phi_r(a_1,\ldots,a_{p-1},x;b_1,\ldots,b_r|q;z)$$
$$= {}_p\phi_r(a_1,\ldots,a_{p-1},\infty;b_1,\ldots,b_r|q;z) \quad \bigl(0<|q|<1\bigr). \tag{7.9}$$

We have changed the notation for the q-hypergeometric series (7.1) slightly according to the new notation which is introduced in this book. The terms to the left of | in (6.181) and in (7.1) are thought to be exponents. Equalities between these exponents are presented in the form

$$a \equiv b \left(\text{mod} \frac{2\pi i}{\log q}\right). \tag{7.10}$$

The first term to the right of | is the base and the second is a letter in the spirit of Chapter 2. The general equation (7.1), with nonzero parameters to the left of || is used in certain special cases when we need factors $(t;q)_n$ in the q-series. One example is the q-analogue of a bilinear generating formula for Laguerre polynomials.

7.1.1 q-difference equation for ${}_{p+1}\phi_p$

Theorem 7.1.1 *Jackson* [303, p. 311, (15)]. *A q-analogue of* (3.46). *The q-difference equation for ${}_{p+1}\phi_p(a_1,\ldots,a_{p+1};b_1,\ldots,b_p|q;z)$ is*

$$-x\prod_{m=1}^{p+1}\{\theta_q+a_m\}_q+\theta_q\prod_{m=1}^{p}\{\theta_q+b_m-1\}_q=0. \tag{7.11}$$

For $p=1$ this equation can be written in the following form, a q-analogue of [153, p. 775]:

$$\bigl(q^c\theta_q+\{c\}_q\bigr)\mathrm{D}_q-\bigl(q^a\theta_q+\{a\}_q\bigr)\bigl(q^b\theta_q+\{b\}_q\bigr)=0. \tag{7.12}$$

7.2 Balanced and well-poised q-hypergeometric series

In mechanics, one begins with the statics to acquire the force equations. The force equilibrium equations correspond to the balanced hypergeometric series and the balanced Γ functions. The moment equilibrium equations correspond to the well-poised hypergeometric series.

force equilibrium	balanced series
moment equilibrium	well-poised series

We now come to the notions of balanced and well-poised series, which are defined in the same way as for hypergeometric series (Section 3.7.2). The only difference is that we use a congruence sign in equalities. We also introduce the notion pseudo-balanced series.

Definition 135 The series

$$ {}_{r+1}\phi_r(a_1, \ldots, a_{r+1}; b_1, \ldots, b_r | q; z) \tag{7.13}$$

is called *k-balanced* if

$$b_1 + \cdots + b_r \equiv k + a_1 + \cdots + a_{r+1} \left(\operatorname{mod} \frac{2\pi i}{\log q} \right), \tag{7.14}$$

and a 1-balanced series is called balanced (or Saalschützian).

The series

$$ {}_p\phi_r(a_1, \ldots, a_p; b_1, \ldots, b_r | q; z) \tag{7.15}$$

is called pseudo-balanced if

$$b_1 + \cdots + b_r \equiv \widetilde{1} + a_1 + \cdots + a_p \left(\operatorname{mod} \frac{2\pi i}{\log q} \right). \tag{7.16}$$

Extra ∞ in p or r are counted as 0.

Analogous to the hypergeometric case, we shall call the q-hypergeometric series (7.13) *well-poised* if its parameters satisfy the relations

$$1 + a_1 \equiv a_2 + b_1 \equiv a_3 + b_2 \equiv \cdots \equiv a_{r+1} + b_r \left(\operatorname{mod} \frac{2\pi i}{\log q} \right). \tag{7.17}$$

The q-hypergeometric series (7.13) is called *nearly-poised* [56] if its parameters satisfy the relation

$$1 + a_1 \equiv a_{j+1} + b_j \left(\operatorname{mod} \frac{2\pi i}{\log q} \right) \tag{7.18}$$

for all but one value of j in $1 \le j \le r$.

The q-hypergeometric series (7.13) is called *almost poised* [81] if its parameters satisfy the relation

$$\delta_j + a_1 \equiv a_{j+1} + b_j \left(\operatorname{mod} \frac{2\pi i}{\log q} \right), \quad 1 \le j \le r, \tag{7.19}$$

where δ_j is 0, 1 or 2.

If the series (7.13) is well-poised and, in addition,

$$a_2 \equiv 1 + \frac{1}{2} a_1 \left(\operatorname{mod} \frac{2\pi i}{\log q} \right), \qquad a_3 \equiv \widetilde{1 + \frac{1}{2}} a_1 \left(\operatorname{mod} \frac{2\pi i}{\log q} \right), \tag{7.20}$$

then it is called a very-well-poised series.

The series (7.13) is of *type I* if

$$z = q. \tag{7.21}$$

The series (7.13) is of *type II* if

$$z = q^{b_1+\cdots+b_r-a_1-\cdots-a_{r+1}}. \tag{7.22}$$

Remark 30 In [72, p. 534] the extra condition $z = q$ in (7.13) is given, compare with the definition of balanced hypergeometric series.

Example 32 The q-analogue of a ${}_9F_8$ 2-balanced hypergeometric series can be a balanced ${}_{10}\phi_9$ q-hypergeometric series. See (3.63) and Section 7.20.

Let us also introduce an equivalent nomenclature for the Γ_q function. The $j\omega$-method corresponds to the balanced Γ_q functions.

Definition 136 The generalized Γ_q function (1.49) is called balanced if

$$\sum_{k=1}^{p} a_k \equiv \sum_{k=1}^{r} b_k \left(\operatorname{mod} \frac{2\pi i}{\log q}\right), \quad p = r. \tag{7.23}$$

A ratio of infinite q-factorials

$$\frac{\prod_{k=1}^{r} \langle b_k; q\rangle_\infty}{\prod_{k=1}^{p} \langle a_k; q\rangle_\infty} \tag{7.24}$$

is called balanced if

$$\sum_{k=1}^{p} a_k \equiv \sum_{k=1}^{r} b_k \left(\operatorname{mod} \frac{2\pi i}{\log q}\right), \quad p = r. \tag{7.25}$$

It is easy to see that these definitions are equivalent.

Balanced quotients of Γ_q's occur in many equations of q-calculus. To see why this is the case, we need to go back 100 years to the work of Hjalmar Mellin [376]. Mellin was a student of Weierstraß during two periods in Berlin. Mellin discovered that every hypergeometric function can be written as a function of generalized Γ functions. He is now renowned for the Mellin transform. In the same way, every q-hypergeometric function can be written as a function of generalized Γ_q functions [262, p. 258]. In the presence of a balanced quotient of Γ_q's, limits when the parameters tend to $\pm\infty$ usually pose no problem. It would therefore be nice if we could find bounds for quotients of balanced Γ_q's. The first step in this direction was made by Ismail and Muldoon [290, p. 320]. This work was generalized by Alzer [19].

7.3 Advantages of the Heine definition

There are several advantages with this new notation:

1. The theory of hypergeometric series and the theory of q-hypergeometric series will be unified.
2. We work on a logarithmic scale; i.e., we only have to add and subtract exponents in calculations. Compare with the 'index calculus' from [37] (Section 3.1).
3. The conditions for k-balanced hypergeometric series and for k-balanced q-hypergeometric series are the same.
4. The conditions for well-poised and nearly-poised hypergeometric series and for well-poised and nearly-poised q-hypergeometric series are the same. Furthermore, the conditions for almost poised q-hypergeometric series are expressed similarly.
5. The conditions for very-well-poised hypergeometric series and for very-well-poised q-hypergeometric series are similar. In fact, the extra condition for a very-well-poised hypergeometric series is $a_2 = 1 + \frac{1}{2}a_1$ and the extra conditions for a very-well-poised q-hypergeometric series are $a_2 \equiv 1 + \frac{1}{2}a_1$ and $a_3 \equiv \widetilde{1 + \frac{1}{2}a_1}$.
6. We no longer have to distinguish between the notation for integers and non-integers in the q-case.
7. It is easy to translate to the work of Cigler [134] for q-Laguerre-polynomials.

Furthermore, the method is applicable to the mock theta functions.

We summarize the result of the tilde operator for q-hypergeometric series. We content ourselves with an instructive example. The following decomposition of the q-hypergeometric series into even and odd parts is a q-analogue of a relation due to Srivastava, and H. L. Manocha [484, p. 200–201] and an example of a q-analogue of a hypergeometric function with argument $\frac{1}{4}x$:

Theorem 7.3.1 [164, (366), p. 71]

$$
\begin{aligned}
{}_r\phi_s\big((a);(b)|q;z\big) &= {}_{4r}\phi_{4s+3}\left[\begin{matrix}\Delta(q;2;a)\\ \Delta(q;2;b),\frac{1}{2},\widetilde{\tfrac{1}{2}},\widetilde{1}\end{matrix}\Big|q;z^2q^{1+s-r}\right]\\
&\quad+(-1)^{1+s-r}\frac{z}{1-q}\frac{\prod_{j=1}^{r}(1-q^{a_j})}{\prod_{j=1}^{s}(1-q^{b_j})}\\
&\quad\times{}_{4r}\phi_{4s+3}\left[\begin{matrix}\Delta(q;2;a+1)\\ \Delta(q;2;b+1),\frac{3}{2},\widetilde{\tfrac{3}{2}},\widetilde{1}\end{matrix}\Big|q;z^2q^{3(1+s-r)}\right].
\end{aligned}
\tag{7.26}
$$

7.4 q-Binomial theorem

The two Eqs. (6.188) and (6.189) can be generalized to the q-binomial theorem, which was first proved by Cauchy [113] in 1843.

Theorem 7.4.1

$$
{}_1\phi_0(a;-|q;z)\equiv\sum_{n=0}^{\infty}\frac{\langle a;q\rangle_n}{\langle 1;q\rangle_n}z^n=\frac{(zq^a;q)_\infty}{(z;q)_\infty}\equiv\frac{1}{(z;q)_a},\quad |z|<1,\ 0<|q|<1. \tag{7.27}
$$

The proof is due to Heine (1878) [272, p. 97ff].

The proof is a modification of the argument in the book by Gasper and Rahman [220, p. 8].

Proof Put

$$
h_a(z)\equiv\sum_{n=0}^{\infty}\frac{\langle a;q\rangle_n}{\langle 1;q\rangle_n}z^n,\quad |z|<1,\ 0<|q|<1. \tag{7.28}
$$

Compute the difference

$$
\begin{aligned}
h_a(z)-h_{a+1}(z)&=\sum_{n=1}^{\infty}\frac{\langle a;q\rangle_n-\langle a+1;q\rangle_n}{\langle 1;q\rangle_n}z^n\\
&=\sum_{n=1}^{\infty}\frac{\langle a+1;q\rangle_{n-1}}{\langle 1;q\rangle_n}\big(1-q^a-(1-q^{a+n})\big)z^n\\
&=q^a\sum_{n=1}^{\infty}\frac{\langle a+1;q\rangle_{n-1}}{\langle 1;q\rangle_n}(q^n-1)z^n\\
&=-zq^a\sum_{n=1}^{\infty}\frac{\langle a+1;q\rangle_{n-1}}{\langle 1;q\rangle_{n-1}}z^{n-1}=-zq^ah_{a+1}(z).
\end{aligned}
$$

This implies the following recurrence relation for $h_a(z)$:

$$
h_a(z)=\big(1-zq^a\big)h_{a+1}(z). \tag{7.29}
$$

Next compute the difference

$$
\begin{aligned}
h_a(z)-h_a(qz)&=\sum_{n=1}^{\infty}\frac{\langle a;q\rangle_n}{\langle 1;q\rangle_n}\big(z^n-q^{kn}z^n\big)=\sum_{n=1}^{\infty}\frac{\langle a;q\rangle_n}{\langle 1;q\rangle_{n-1}}z^n\\
&=z\big(1-q^a\big)\sum_{n=1}^{\infty}\frac{\langle a+1;q\rangle_{n-1}}{\langle 1;q\rangle_{n-1}}z^{n-1}=z\big(1-q^a\big)h_{a+1}(z).
\end{aligned}
$$

This implies the following equation for $h_a(z)$:

$$
h_a(z)=\frac{1-zq^a}{1-z}h_a(qz). \tag{7.30}
$$

The proof is completed by iterating this n times and by letting n go to infinity. □

Now Eq. (6.188) follows from (7.27) by letting $a \to \infty$ and (6.189) follows from (7.27) by substituting $z \to zq^{-a}$ and letting $a \to -\infty$. This procedure is the q-analogue [186] of the formula

$$\lim_{a\to\infty}\left(1+\frac{z}{a}\right)^a = e^z.$$

7.5 Jacobi's elliptic functions expressed as real and imaginary parts of q-hypergeometric series with exponential argument (Heine)

The interested reader can compare the following equations with the statement of F. Klein from his book about hypergeometric functions: Between the hypergeometric functions and the trigonometric functions there is the same relationship as between the q-hypergeometric functions and the elliptic functions. The following two formulas in the spirit of Heine [272, p. 99] illustrate the use of the tilde operator.

Theorem 7.5.1 *We put*

$$K \equiv \frac{\pi}{2}\sum_{l=0}^{\infty}\left(\frac{(2l-1)!!}{(2l)!!}\right)^2 k^{2l}. \tag{7.31}$$

If $0 < q < 1,\, q = e^{-\pi\frac{K'}{K}}$ *and* $x = \frac{u\pi}{2K}$ *then*

$$\mathrm{sn}u = \frac{2\pi}{KK'}\Im\left(\frac{q^{\frac{1}{2}}e^{ix}}{1-q}{}_2\phi_1\left(1,\frac{1}{2};\frac{3}{2}|q^2;qe^{2ix}\right)\right), \tag{7.32}$$

$$\mathrm{cn}u = \frac{2\pi}{KK'}\Re\left(\frac{q^{\frac{1}{2}}e^{ix}}{1+q}{}_2\phi_1\left(1,\frac{\widetilde{1}}{2};\frac{\widetilde{3}}{2}|q^2;qe^{2ix}\right)\right), \tag{7.33}$$

$$\mathrm{dn}u = \frac{\pi}{2K}\Re(-1+2{}_2\phi_1(1,\widetilde{0};\widetilde{1}|q^2;qe^{2ix})). \tag{7.34}$$

7.6 The Jacobi triple product identity

Our next aim is to prove the famous Gauß-Jacob triple product identity from 1829 with the help of the formulas (6.188) and (6.189). The following formula [310], [523], [37], [220] for the theta function holds:

Theorem 7.6.1

$$\sum_{n=-\infty}^{\infty} q^{n^2}z^n = (q^2,-qz,-qz^{-1};q^2)_\infty, \tag{7.35}$$

where $z \in \mathbb{C}\backslash\{0\},\, 0 < |q| < 1$.

The following proof is due to Andrews [23] and Vilenkin and Klimyk [523]:

Proof If $|z| > |q|$ and $0 < |q| < 1$, then taking into account the fact that $\langle 1+n; q^2\rangle_\infty = 0$ for negative n, we obtain

$$
\begin{aligned}
\left(-qz; q^2\right)_\infty &\overset{\text{by (6.189)}}{=} \sum_{n=0}^{\infty} \frac{(zq)^n q^{n(n-1)}}{\langle 1; q^2\rangle_n} \\
&\overset{\text{by (6.8)}}{=} \frac{1}{\langle 1; q^2\rangle_\infty} \sum_{n=-\infty}^{\infty} z^n q^{n^2} \left\langle 1+n; q^2\right\rangle_\infty \\
&\overset{\text{by (6.189)}}{=} \frac{1}{\langle 1; q^2\rangle_\infty} \sum_{n=-\infty}^{\infty} z^n q^{n^2} \sum_{m=0}^{\infty} \frac{(-1)^m q^{2m+2mn+m(m-1)}}{\langle 1; q^2\rangle_m} \\
&= \frac{1}{\langle 1; q^2\rangle_\infty} \sum_{m=0}^{\infty} \frac{(-1)^m z^{-m} q^m}{\langle 1; q^2\rangle_m} \sum_{n=-\infty}^{\infty} z^{m+n} q^{(m+n)^2} \\
&= \frac{1}{\langle 1; q^2\rangle_\infty} \sum_{m=0}^{\infty} \frac{(-\frac{q}{z})^m}{\langle 1; q^2\rangle_m} \sum_{n=-\infty}^{\infty} z^n q^{n^2} \\
&\overset{\text{by (6.188)}}{=} \frac{1}{\langle 1; q^2\rangle_\infty (-\frac{q}{z}; q^2)_\infty} \sum_{n=-\infty}^{\infty} z^n q^{n^2}.
\end{aligned}
\tag{7.36}
$$

If $|z| > |q|$ and $0 < |q| < 1$, then the series converges absolutely. Transferring $\langle 1; q^2\rangle_\infty (-\frac{q}{z}; q^2)_\infty$ to the left-hand side and continuing analytically in z completes the proof of (7.35). □

7.7 q-contiguity relations

We use abbreviations in accord with Rainville [425, p. 50], which are illustrated by the following:

$$\phi\left(a_1^{\pm}\right) \equiv {}_p\phi_r(a_1 \pm 1, \ldots, a_p; b_1, \ldots, b_r | q; x), \tag{7.37}$$

$$\phi\left(b_1^{\pm}\right) \equiv {}_p\phi_r(a_1, \ldots, a_p; b_1 \pm 1, \ldots, b_r | q; x), \tag{7.38}$$

$$c_n \equiv \frac{\langle a_1, \ldots, a_p; q\rangle_n}{\langle 1, b_1, \ldots, b_r; q\rangle_n} \left[(-1)^n q^{\binom{n}{2}}\right]^{1+r-p}. \tag{7.39}$$

We obtain the following q-contiguity relations:

$$\phi\left(a_k^{+}\right) = \sum_{n=0}^{\infty} \frac{c_n x^n}{\langle 1; q\rangle_n} \frac{1 - q^{a_k+n}}{1 - q^{a_k}}, \tag{7.40}$$

$$\phi(a_k^-) = \sum_{n=0}^{\infty} \frac{c_n x^n}{\langle 1; q\rangle_n} \frac{1 - q^{a_k - 1}}{1 - q^{a_k + a_n - 1}}, \tag{7.41}$$

$$\phi(b_k^+) = \sum_{n=0}^{\infty} \frac{c_n x^n}{\langle 1; q\rangle_n} \frac{1 - q^{b_k}}{1 - q^{b_k + n}}, \tag{7.42}$$

$$\phi(b_k^-) = \sum_{n=0}^{\infty} \frac{c_n x^n}{\langle 1; q\rangle_n} \frac{1 - q^{b_k + n - 1}}{1 - q^{b_k - 1}}. \tag{7.43}$$

A calculation gives the following q-contiguity relations, which are q-analogues of [425, p. 51].

$$(q^{a_k}\theta_q + \{a_k\}_q)\phi = \{a_k\}_q \phi(a_k^+), \quad k = 1, \ldots, p, \tag{7.44}$$

$$(q^{b_k - 1}\theta_q + \{b_k - 1\}_q)\phi = \{b_k - 1\}_q \phi(b_k^-), \quad k = 1, \ldots, r, \tag{7.45}$$

$$q^{-a_1}\{a_1\}_q \phi(a_1^+) - q^{-a_k}\{a_k\}_q \, \phi(a_k^+) = (q^{-a_1}\{a_1\}_q - q^{-a_k}\{a_k\}_q)\phi,$$
$$k = 2, \ldots, p, \tag{7.46}$$

$$q^{-a_1}\{a_1\}_q \phi(a_1^+) - q^{-(b_k - 1)}\{b_k - 1\}_q \phi(b_k^-) = q^{-a_1}\{a_1 - b_k + 1\}_q \phi,$$
$$k = 1, \ldots, r. \tag{7.47}$$

7.8 Heine q-transformations

In his work on continued fractions Heine [271], [272] proved the following transformation formula for ${}_2\phi_1$, which turned out to be a meromorphic continuation of ${}_2\phi_1(a, b; c|q; z)$ outside the domain of convergence:

Theorem 7.8.1 [270, p. 306, (79)]:

$${}_2\phi_1(a, b; c|q; q^z) = \frac{\langle b, a + z; q\rangle_\infty}{\langle c, z; q\rangle_\infty} {}_2\phi_1(c - b, z; a + z|q; q^b), \tag{7.48}$$

where $|q^z| < 1$ *and* $|q^b| < 1$.

Proof

$${}_2\phi_1(a, b; c|q; q^z) = \sum_{n=0}^{\infty} \frac{\langle a; q\rangle_n \langle b; q\rangle_n}{\langle 1; q\rangle_n \langle c; q\rangle_n} q^{zn}$$

$$\overset{\text{by (6.8)}}{=} \frac{\langle b; q\rangle_\infty}{\langle c; q\rangle_\infty} \sum_{n=0}^{\infty} \frac{\langle a; q\rangle_n \langle c + n; q\rangle_\infty}{\langle 1; q\rangle_n \langle b + n; q\rangle_\infty} q^{zn}$$

$$\overset{\text{by (7.27)}}{=} \frac{\langle b;q\rangle_\infty}{\langle c;q\rangle_\infty}\sum_{n=0}^{\infty}\frac{\langle a;q\rangle_n}{\langle 1;q\rangle_n}q^{zn}\sum_{m=0}^{\infty}\frac{\langle c-b;q\rangle_m}{\langle 1;q\rangle_m}q^{m(b+n)}$$

$$= \frac{\langle b;q\rangle_\infty}{\langle c;q\rangle_\infty}\sum_{m=0}^{\infty}\frac{\langle c-b;q\rangle_m}{\langle 1;q\rangle_m}q^{mb}\sum_{n=0}^{\infty}\frac{\langle a;q\rangle_n}{\langle 1;q\rangle_n}q^{n(z+m)}$$

$$\overset{\text{by (7.27)}}{=} \frac{\langle b;q\rangle_\infty}{\langle c;q\rangle_\infty}\sum_{m=0}^{\infty}\frac{\langle c-b;q\rangle_m}{\langle 1;q\rangle_m}q^{mb}\frac{\langle a+z+m;q\rangle_\infty}{\langle z+m;q\rangle_\infty}$$

$$\overset{\text{by (6.8)}}{=} \frac{\langle b;q\rangle_\infty}{\langle c;q\rangle_\infty}\sum_{m=0}^{\infty}\frac{\langle c-b,z;q\rangle_m}{\langle 1,a+z;q\rangle_m}\frac{\langle a+z;q\rangle_\infty}{\langle z;q\rangle_\infty}q^{mb}$$

$$= \frac{\langle b,a+z;q\rangle_\infty}{\langle c,z;q\rangle_\infty}{}_2\phi_1(c-b,z;a+z|q;q^b). \tag{7.49}$$

□

Remark 31 The condition $|q^z|<1$ is equivalent to $\Re[z\log q]<0$.
Heine used this formula to prove equations for theta functions.

The transformation formula (7.48) can be rewritten as a q-integral expression, first found by Thomae 1869 [497], which is the q-analogue of (3.26) and takes the following form in the new notation (we use formal equality because of the q-integral):

Theorem 7.8.2 [303, p. 313], [220, p. 19, (1.11.9)]:

$${}_2\phi_1(a,b;c|q;z) = \frac{\Gamma_q(c)}{\Gamma_q(b)\Gamma_q(c-b)}\int_0^1 t^{b-1}\frac{(qt;q)_{c-b-1}}{(zt;q)_a}\,d_q(t). \tag{7.50}$$

Proof The RHS of (7.50) is equal to

$$\frac{\langle b;q\rangle_\infty}{\langle c;q\rangle_\infty}\sum_{m=0}^{\infty}\frac{\langle c-b;q\rangle_m}{\langle 1;q\rangle_m}q^{mb}\frac{\langle a+z+m;q\rangle_\infty}{\langle z+m;q\rangle_\infty}\overset{\text{by (7.48)}}{=}\text{LHS}. \tag{7.51}$$

Here we used the formulas (1.53), (1.45) and (6.8). □

The following formula was proved by Heine 1878 [272, p. 115], [220, (1.4.3), p. 10].

Theorem 7.8.3 *Another meromorphic continuation of* ${}_2\phi_1(a,b;c|q;z)$, *the RHS corresponds to the solution* $u_{13}(z)$ [275, p. 157]:

$${}_2\phi_1(a,b;c|q;z) = \frac{1}{(z;q)_{a+b-c}}{}_2\phi_1(c-a,c-b;c|q;zq^{a+b-c}). \tag{7.52}$$

Proof Just iterate (7.48) as follows:

$$\begin{aligned} {}_2\phi_1\big(a, b; c|q; q^z\big) &= \frac{\langle b, a+z; q\rangle_\infty}{\langle c, z; q\rangle_\infty} {}_2\phi_1\big(c-b, z; a+z|q; q^b\big) \\ &= \frac{\langle c-b, b+z; q\rangle_\infty}{\langle c, z; q\rangle_\infty} {}_2\phi_1\big(a+b+z-c, b; b+z|q; q^{c-b}\big) \\ &= \frac{\langle a+b+z-c; q\rangle_\infty}{\langle z; q\rangle_\infty} {}_2\phi_1\big(c-a, c-b; c|q; q^{a+b+z-c}\big). \end{aligned}$$ □

Theorem 7.8.4 [220, (1.4.5), p. 10]:

$$ {}_2\phi_1\big(a, b; c|q; q^z\big) = \frac{\langle c-b; q\rangle_b}{\langle z; q\rangle_b} {}_2\phi_1\big(a+z+b-c, b; b+z|q; q^{c-b}\big). \tag{7.53}$$

7.8.1 The q-beta function

Definition 137 The q-beta function is defined by

$$\mathrm{B}_q(x, y) \equiv \frac{\Gamma_q(x)\Gamma_q(y)}{\Gamma_q(x+y)}. \tag{7.54}$$

Theorem 7.8.5 [303, p. 312] *A q-analogue of* (3.25). *Let* $\Re(x) > 0$. *Then the q-Euler beta integral is given by* [220, (1.11.7), p. 19], [194, (2.2.3.9), p. 43]

$$\mathrm{B}_q(x, y) = \int_0^1 t^{x-1} \frac{(tq; q)_\infty}{(tq^y; q)_\infty} d_q(t), \quad y \neq 0, -1, -2, \ldots. \tag{7.55}$$

Proof Put $a = 0$ in (7.50). □

With the help of the q-analogue of Euler's reflection formula (6.260) we get

$$\int_0^1 t^{-y}(tq; q)_{y-1}\, d_q(t) = \frac{iq^{\frac{1}{8}}(1-q)(\langle 1; q\rangle_\infty)^3}{q^{\frac{y}{2}}\theta_1(\frac{-iy}{2}\log q, \sqrt{q})}. \tag{7.56}$$

7.9 Heines q-analogue of the Gauß summation formula

Heines q-analogue of the Gauß summation formula [220, (1.5.1), p. 10] takes the following form in the new notation:

Theorem 7.9.1 [270, p. 307, (80)]:

$$ {}_2\phi_1\big(a, b; c|q; q^{c-a-b}\big) = \frac{\langle c-a, c-b; q\rangle_\infty}{\langle c, c-a-b; q\rangle_\infty}, \quad \big|q^{c-a-b}\big| < 1. \tag{7.57}$$

Proof Put $z = c - a - b$ in (7.48). Assuming that $|q^b| < 1$, $|q^{c-a-b}| < 1$ we obtain

$$
{}_2\phi_1\big(a, b; c|q; q^{c-a-b}\big) = \frac{\langle b, c-b; q\rangle_\infty}{\langle c, c-a-b; q\rangle_\infty} {}_1\phi_0\big(c-a-b; -|q; q^b\big)
$$

$$
\stackrel{\text{by (7.27)}}{=} \frac{\langle c-a, c-b; q\rangle_\infty}{\langle c, c-a-b; q\rangle_\infty}. \tag{7.58}
$$

By meromorphic continuation, we can drop the assumption $|q^b| < 1$ and require only $|q^{c-a-b}| < 1$ in (7.57). □

In the terminating case $a = -n$, (7.57) reduces to [220, p. 11, (1.5.2)]:

$$
{}_2\phi_1\big(-n, b; c|q; q^{c+n-b}\big) \equiv \sum_{k=0}^{n} \frac{\langle b; q\rangle_k}{\langle c; q\rangle_k} (-1)^k \binom{n}{k}_q q^{\binom{k}{2}+k(c-b)}
$$

$$
\stackrel{\text{by (6.8)}}{=} \frac{\langle c-b; q\rangle_n}{\langle c; q\rangle_n}, \quad n = 0, 1, \ldots, \tag{7.59}
$$

which is one q-analogue of (3.29). It is also a generalization of (2.19), which is obtained by putting $c = b$.

Jackson proved a similar formula in 1905:

Theorem 7.9.2 [297, p. 2, (4)], *Dyson* [58, p. 4]:

$$
{}_2\phi_1\big(a, b; c|q; q^{c-a-b-1}\big) = \frac{\langle c-a, c-b; q\rangle_\infty}{\langle c, c-a-b; q\rangle_\infty} \frac{\{c-a-1\}_q + q^{c-1}\{-b\}_q}{\{c-a-b-1\}_q}, \tag{7.60}
$$

provided $|q^{c-a-b}| < 1$.

The following equation is another form of (7.59).

Example 33 A q-analogue of [425, p. 49] [249, (2), p. 98]:

$$
{}_2\phi_1\left(-\frac{n}{2}, \frac{1-n}{2}; \mu + \frac{1}{2}|q^2, q^{2(\mu+n)}\right) = \frac{\langle \widetilde{\mu}, \widetilde{\mu + \frac{n}{2}}; q\rangle_{\frac{n}{2}} \langle \mu; q\rangle_n}{\langle 2\mu; q\rangle_n}, \quad n \text{ even}. \tag{7.61}
$$

Proof

$$
\text{LHS} = \frac{\langle \mu + \frac{n}{2}, \widetilde{\mu + \frac{n}{2}}; q\rangle_{\frac{n}{2}}}{\langle \mu + \frac{1}{2}, \widetilde{\mu + \frac{1}{2}}; q\rangle_{\frac{n}{2}}} = \left\langle \mu + \frac{n}{2}, \widetilde{\mu + \frac{n}{2}}; q\right\rangle_{\frac{n}{2}} \frac{\langle \mu, \widetilde{\mu}; q\rangle_{\frac{n}{2}}}{\langle 2\mu; q\rangle_n} = \text{RHS}.
$$

□

We proceed to prove Jackson's 1910 [304] q-analogue of (3.28) [220, p. 11, (1.5.4)], which takes the following form in the new notation:

Theorem 7.9.3 *Another meromorphic continuation of* ${}_2\phi_1(a,b;c|q;z)$; *this is a* q*-analogue of the solution* $u_7(z)$ [275, p. 156]:

$$ {}_2\phi_1(a,b;c|q;t) \cong \frac{1}{(t;q)_a} {}_2\phi_2\big(a, c-b; c|q; tq^b||-; tq^a\big), \tag{7.62}$$

$|z| < 1$ *and* $|\frac{z}{z-1}| < 1$.

Proof We will prove the following equivalent formula:

$$\begin{aligned} {}_2\phi_1\big(a,b;c|q;q^z\big) &= \frac{\langle a+z;q\rangle_\infty}{\langle z;q\rangle_\infty} \sum_{n=0}^{\infty} \frac{\langle a, c-b; q\rangle_n}{\langle 1, c, a+z; q\rangle_n} (-1)^n q^{n(b+z)+\binom{n}{2}} \\ &\equiv \frac{\langle a+z;q\rangle_\infty}{\langle z;q\rangle_\infty} {}_2\phi_2\big(a, c-b; c, a+z|q; q^{b+z}\big). \end{aligned} \tag{7.63}$$

Indeed,

$$\begin{aligned} {}_2\phi_1\big(a,b;c|q;q^z\big) &\overset{\text{by (7.59)}}{=} \sum_{m=0}^{\infty} \frac{\langle a;q\rangle_m}{\langle 1;q\rangle_m} q^{mz} \\ &\quad \times \sum_{n=0}^{m} \frac{\langle -m, c-b; q\rangle_n}{\langle 1, c; q\rangle_n} q^{(b+m)n} \\ &\overset{\text{by (6.20)}}{=} \sum_{n=0}^{\infty}\sum_{m=n}^{\infty} \frac{\langle a;q\rangle_m \langle c-b;q\rangle_n}{\langle 1,c;q\rangle_n \langle 1;q\rangle_{m-n}} q^{mz+bn+\binom{n}{2}} (-1)^n \\ &= \sum_{n=0}^{\infty}\sum_{m=0}^{\infty} \frac{\langle a;q\rangle_{m+n} \langle c-b;q\rangle_n}{\langle 1,c;q\rangle_n \langle 1;q\rangle_{m}} q^{(m+n)z+bn+\binom{n}{2}} (-1)^n \\ &\overset{\text{by (6.16)}}{=} \sum_{n=0}^{\infty} \frac{\langle a, c-b; q\rangle_n}{\langle 1, c; q\rangle_n} (-1)^n q^{(b+z)n+\binom{n}{2}} \sum_{m=0}^{\infty} \frac{\langle a+n;q\rangle_m}{\langle 1;q\rangle_m} q^{mz} \\ &\overset{\text{by (7.27)}}{=} \sum_{n=0}^{\infty} \frac{\langle a+z+n;q\rangle_\infty}{\langle z;q\rangle_\infty} \frac{\langle a, c-b; q\rangle_n}{\langle 1, c; q\rangle_n} (-1)^n q^{n(b+z)+\binom{n}{2}} \\ &\overset{\text{by (6.8)}}{=} \frac{\langle a+z;q\rangle_\infty}{\langle z;q\rangle_\infty} \sum_{n=0}^{\infty} \frac{\langle a, c-b; q\rangle_n}{\langle 1, c, a+z; q\rangle_n} (-1)^n q^{n(b+z)+\binom{n}{2}}. \end{aligned} \tag{7.64}$$

□

Remark 32 A calculation shows that the domain of convergence for general q in (7.62) is somewhat larger than indicated. The right-hand side converges faster than the left. The right-hand side corresponds to the function $u_7(z)$ of Henrici [275, p. 156].

This leads to the following important

Corollary 7.9.4 [7, p. 105] *q-analogue of* [425, p. 125, (2)], [448, p. 673, 16.7–17]:

$$ {}_2\phi_1\big(a,\infty;c|q;t(1-q)\big) \cong \mathrm{E}_q(t)\,{}_1\phi_1\big(c-a;c|q;t(1-q)q^a\big). \tag{7.65}$$

Proof Let $a \to \infty$ in (7.62). □

If $a=-n$, then the series on the right-hand side of (7.63) can be reversed to yield the Sears (1951) transformation formula [461] [220, (1.5.6), p. 11]. In the new notation this formula reads

$$\begin{aligned} {}_2\phi_1\big(-n,b;c|q;q^z\big) &= \frac{\langle c-b;q\rangle_n}{\langle c;q\rangle_n} q^{(b+z-1)n} \\ &\quad \times {}_3\phi_2(-n,1-z,-c+1-n;b-c+1-n,\infty|q;q), \end{aligned}$$
$$0<|q|<1. \tag{7.66}$$

Proof

$$\begin{aligned} &{}_2\phi_1\big(-n,b;c|q;q^z\big) \\ &\overset{\text{by (7.63)}}{=} \frac{\langle -n+z;q\rangle_\infty}{\langle z;q\rangle_\infty} \sum_{l=0}^{n} \frac{\langle -n,c-b;q\rangle_l}{\langle 1,c,-n+z;q\rangle_l} (-1)^l q^{l(b+z)+\binom{l}{2}} \\ &= \langle -n+z;q\rangle_n \sum_{k=0}^{n} \frac{\langle -n,c-b;q\rangle_{n-k}}{\langle 1,c,-n+z;q\rangle_{n-k}} (-1)^{n-k} q^{(n-k)(b+z)+\binom{n-k}{2}} \\ &\quad \text{by } 4\times(6.15),\ (6.20) \\ &= \frac{\langle c-b,-n;q\rangle_n}{\langle c,1;q\rangle_n}(-1)^n q^{(b+z)n+\binom{n}{2}} \\ &\quad \times {}_3\phi_2(-n,1-z,-c+1-n;b-c+1-n,\infty|q;q) \\ &\overset{\text{by (6.14)}}{=} \frac{\langle c-b;q\rangle_n}{\langle c;q\rangle_n} q^{(b+z-1)n} \\ &\quad \times {}_3\phi_2(-n,1-z,-c+1-n;b-c+1-n,\infty|q;q). \end{aligned}$$
□

7.10 A q-analogue of the Pfaff-Saalschütz summation formula

We proceed to prove Jackson's 1910 [304] q-analogue of (3.30) [220, p. 13, (1.7.2)]. In the new notation this formula takes the following form:

Theorem 7.10.1

$$
{}_3\phi_2(a, b, -n; c, 1+a+b-c-n|q; q) = \frac{\langle c-a, c-b; q\rangle_n}{\langle c, c-a-b; q\rangle_n}, \qquad n = 0, 1, \ldots. \tag{7.67}
$$

Proof By the q-binomial theorem (7.27),

$$
\frac{\langle z+a+b-c; q\rangle_\infty}{\langle z; q\rangle_\infty} = \sum_{n=0}^{\infty} \frac{\langle a+b-c; q\rangle_n}{\langle 1; q\rangle_n} q^{zn}
$$

and Eq. (7.52) gives

$$
\sum_{n=0}^{\infty} \frac{\langle a, b; q\rangle_n}{\langle 1, c; q\rangle_n} q^{zn}
$$

$$
= \sum_{k=0}^{\infty} \frac{\langle a+b-c; q\rangle_k}{\langle 1; q\rangle_k} \sum_{j=0}^{\infty} \frac{\langle c-a, c-b; q\rangle_j}{\langle c, 1; q\rangle_j} q^{j(a+b-c)} q^{z(k+j)}.
$$

Hence, equating the coefficients of q^{zn} gives

$$
\frac{\langle a, b; q\rangle_n}{\langle 1, c; q\rangle_n} = \sum_{j=0}^{n} \frac{\langle a+b-c; q\rangle_{n-j}}{\langle 1; q\rangle_{n-j}} \frac{\langle c-a, c-b; q\rangle_j}{\langle 1, c; q\rangle_j} q^{j(a+b-c)}.
$$

By (6.15), this is equivalent to

$$
\frac{\langle a, b; q\rangle_n}{\langle a+b-c, c; q\rangle_n} = \sum_{j=0}^{n} \frac{\langle c-a, c-b, -n; q\rangle_j}{\langle 1, c, 1-a-b+c-n; q\rangle_j} q^j.
$$

Replacing a, b by $c-a, c-b$, respectively, completes the proof of (7.67). □

Recall that $0 < |q| < 1$. Now let $a \to \infty$ in (7.67) and observe that

$$
\lim_{a\to+\infty} \frac{\langle c-a; q\rangle_n}{\langle c-a-b; q\rangle_n} = q^{bn}.
$$

The formula (1.5.3) in Gasper and Rahman [220, p. 11], takes the following form in the new notation:

$$
{}_2\phi_1(-n, b; c|q; q) = \frac{\langle c-b; q\rangle_n}{\langle c; q\rangle_n} q^{bn}, \qquad n = 0, 1, \ldots, \tag{7.68}
$$

which is another q-analogue of (3.29).

7.11 Sears' $_4\phi_3$ transformation

One of the fundamental transformations in the theory of q-series is the Sears $_4\phi_3$ transformation.

Theorem 7.11.1 [360, p. 7]. *The Sears $_4\phi_3$ transformation can be expressed as*

$$
\begin{aligned}
&{}_4\phi_3(-n, a+b+c+d+n-1, b+x, b-x; a+b, b+c, b+d|q;q)\\
&= \frac{\langle a+c, c+d; q\rangle_n}{\langle a+b, b+d; q\rangle_n} q^{n(b-c)}\\
&\quad \times {}_4\phi_3(-n, a+b+c+d+n-1, c+x, c-x; a+c, b+c, c+d|q;q).
\end{aligned} \tag{7.69}
$$

Proof From the q-Chu-Vandermonde formula we know that

$$
\begin{aligned}
\Delta &= \sum_{k=0}^{n}\sum_{m=0}^{n-k} \frac{\langle d-b, d-c; q\rangle_k}{\langle 1, d; q\rangle_k} \frac{\langle b+c-d; q\rangle_m}{\langle 1; q\rangle_m} \frac{\langle -n; q\rangle_{k+m}}{\langle e; q\rangle_{k+m}} q^{k(b+c+1-d)+m}\\
&= \sum_{k=0}^{n} \frac{\langle -n, d-b, d-c; q\rangle_k}{\langle 1, d, e; q\rangle_k} q^{k(b+c+1-d)} \sum_{m=0}^{n-k} \frac{\langle k-n, b+c-d; q\rangle_m}{\langle 1, e+k; q\rangle_m} q^m\\
&= \sum_{k=0}^{n} \frac{\langle -n, d-b, d-c; q\rangle_k}{\langle 1, d, e; q\rangle_k} q^{k(b+c+1-d)}\\
&\quad \times \frac{\langle e+k-b-c+d; q\rangle_{n-k}}{\langle e+k; q\rangle_{n-k}} q^{n(b+c-d)-k(b+c-d)}\\
&= \sum_{k=0}^{n} \frac{\langle -n, d-b, d-c; q\rangle_k}{\langle 1, d, e; q\rangle_k} q^{k+n(b+c-d)} \frac{\langle e+k-b-c+d; q\rangle_{n-k}}{\langle e+k; q\rangle_{n-k}}\\
&= \frac{\langle e-b-c+d; q\rangle_n}{\langle e; q\rangle_n} q^{n(b+c-d)} \sum_{k=0}^{n} \frac{\langle -n, d-b, d-c; q\rangle_k}{\langle 1, d, d+e-b-c; q\rangle_k} q^k\\
&= \frac{\langle e-b-c+d; q\rangle_n}{\langle e; q\rangle_n} q^{n(b+c-d)} {}_3\phi_2(-n, d-b, d-c; d, d+e-b-c|q;q).
\end{aligned} \tag{7.70}
$$

On the other hand, by the q-Pfaff-Saalschütz formula,

$$
\begin{aligned}
\Delta &= \sum_{j=0}^{n} \frac{\langle -n, b+c-d; q\rangle_j q^j}{\langle 1, e; q\rangle_j} \sum_{k=0}^{j} \frac{\langle -j, d-b, d-c; q\rangle_k}{\langle 1, d, 1-j+d-b-c; q\rangle_k} q^k\\
&= \sum_{j=0}^{n} \frac{\langle -n, b+c-d; q\rangle_j q^j}{\langle 1, e; q\rangle_j} {}_3\phi_2(-j, d-b, d-c; d, d+1-j-b-c|q;q)\\
&= {}_3\phi_2(-n, b, c; d, e|q;q).
\end{aligned} \tag{7.71}
$$

Thus,

$$
{}_3\phi_2(-n,b,c;d,e|q;q) = \frac{\langle e-b-c+d;q\rangle_n}{\langle e;q\rangle_n} q^{n(b+c)} \times {}_3\phi_2(-n,d-b,d-c;d,d+e-b-c|q;q). \quad (7.72)
$$

Replacing (b,c,e,d) by $(b+x,b-x,a+b,b+c)$ gives

$$
{}_3\phi_2(-n,b+x,b-x;a+b,b+c|q;q) = \frac{\langle a+c;q\rangle_n}{\langle a+b;q\rangle_n} q^{n(b-c)} {}_3\phi_2(-n,c+x,c-x;a+c,b+c|q;q), \quad (7.73)
$$

which can be written as

$$
\sum_{j=0}^{n} \frac{\langle -n,b+x,b-x;q\rangle_j}{\langle 1,b+c;q\rangle_j} \left(q^{a+b+j},q^{a+c+n};q\right)_\infty = q^{n(b+c)} \sum_{j=0}^{n} \frac{\langle -n,c+x,c-x;q\rangle_j}{\langle 1,b+c;q\rangle_j} \left(q^{a+c+j},q^{a+b+n};q\right)_\infty. \quad (7.74)
$$

By applying the operator $E(b\varphi(q))$ to both sides and then using the following relations, obtained using (6.273):

$$
E\big(d\varphi(q)\big)\big\{\left(q^{a+b+j},q^{a+c+n};q\right)_\infty\big\} = \frac{(q^{a+b+j},q^{a+c+n},q^{d+b+j},q^{d+c+n};q)_\infty}{(q^{a+b+c+d+n+j-1};q)_\infty},
$$
$$
E\big(d\varphi(q)\big)\big\{\left(q^{a+c+j},q^{a+b+n};q\right)_\infty\big\} = \frac{(q^{a+c+j},q^{a+b+n},q^{d+b+n},q^{d+c+j};q)_\infty}{(q^{a+b+c+d+n+j-1};q)_\infty}, \quad (7.75)
$$

we obtain (7.69). □

Remark 33 Another proof of Sear's transformation is given in [72, p. 558]. Take the Bailey ${}_{10}\phi_9$ transform (7.130), replace b by $a+1-b$ and e by $a+1-e$, let $a \to \infty$. Finally re-lable the parameters $b \to d,\ d \to b,\ f \to a$ to obtain [220, 3.2.1].

7.12 q-analogues of Thomae's transformations

Theorem 7.12.1 [220, p. 62]:

$$
{}_3\phi_2\big(a,b,c;d,e|q;q^{d+e-a-b-c}\big) = \Gamma_q\left[\begin{matrix} e,d+e-a-b-c \\ e-a,d+e-b-c \end{matrix}\right] {}_3\phi_2\big(a,d-b,d-c;d,d+e-b-c|q;q^{e-a}\big). \quad (7.76)
$$

Proof [220, p. 62]. Write (7.76) in the form

$$\begin{aligned}
&{}_4\phi_3(-n,a,b,c;d,e,a+b+c+1-n-d-e|q;q)\\
&\quad=\frac{\langle e-a,d+e-b-c;q\rangle_n}{\langle e,d+e-a-b-c;q\rangle_n}\\
&\qquad\times {}_4\phi_3(-n,a,d-b,d-c;d,d+e-b-c,a+1-n-e|q;q) \qquad (7.77)
\end{aligned}$$

and let $n\to\infty$. □

Theorem 7.12.2

$$\begin{aligned}
&{}_4\phi_3(-n,a,b,c;d,e,a+b+c+1-n-d-e|q;q)\\
&\quad=\frac{\langle b,d+e-a-b,d+e-b-c;q\rangle_n}{\langle d,e,d+e-a-b-c;q\rangle_n}\\
&\qquad\times {}_4\phi_3(-n,d-b,e-b,d+e-a-b-c;\\
&\qquad\qquad d+e-a-b,d+e-b-c,1-n-b|q;q). \qquad (7.78)
\end{aligned}$$

Proof Iterate (7.77). □

Theorem 7.12.3 *Hall* [264].

$$\begin{aligned}
&{}_3\phi_2\big(a,b,c;d,e|q;q^{d+e-a-b-c}\big)\\
&\quad=\Gamma_q\begin{bmatrix} d,e,d+e-a-b-c\\ b,d+e-a-b,d+e-b-c\end{bmatrix}\\
&\qquad\times {}_3\phi_2\big(d-b,e-b,d+e-a-b-c;\\
&\qquad\qquad d+e-a-b,d+e-b-c|q;q^b\big). \qquad (7.79)
\end{aligned}$$

Proof [220, p. 62]. Assume that $\max(q^b,q^{d+e-a-b-c})<1$ and let $n\to\infty$ in (7.78). □

7.13 The Bailey-Daum summation formula

The following q-analogue of Kummer's first formula (3.39) was proved independently by Bailey in 1941 [57] and Daum in 1942 [149], [220, (1.8.1), p. 14]. We put $\beta\equiv\frac{\log(-1)}{\log q}$.

$$\begin{aligned}
&{}_2\phi_1\big(a,b;1+a-b|q;-q^{1-b}\big)\\
&\quad=\Gamma_q\begin{bmatrix} 1+a-b,1+\frac{a}{2},1+\frac{a}{2}+\beta,1-b+\beta\\ 1+a,1+\frac{a}{2}-b,1+\frac{a}{2}-b+\beta,1+\beta\end{bmatrix}. \qquad (7.80)
\end{aligned}$$

Proof

$$
\begin{aligned}
&{}_2\phi_1\big(a,b;1+a-b|q;-q^{1-b}\big)\\
&\overset{\text{by (7.48)}}{=} \frac{\langle a,\widetilde{1};q\rangle_\infty}{\langle 1+a-b,\widetilde{1-b};q\rangle_\infty}\,{}_2\phi_1\big(1-b,\widetilde{1-b};\widetilde{1}|q;q^a\big)\\
&= \frac{\langle a,\widetilde{1};q\rangle_\infty}{\langle 1+a-b,\widetilde{1-b};q\rangle_\infty}\sum_{n=0}^{\infty}\frac{\langle 1-b,\widetilde{1-b};q\rangle_n}{\langle 1,\widetilde{1};q\rangle_n}q^{an}\\
&\overset{\text{by (6.33)}}{=} \frac{\langle a,\widetilde{1};q\rangle_\infty}{\langle 1+a-b,\widetilde{1-b};q\rangle_\infty}\sum_{n=0}^{\infty}\frac{\langle 1-b;q^2\rangle_n}{\langle 1;q^2\rangle_n}q^{an}\\
&\overset{\text{by (7.27)}}{=} \frac{\langle a,\widetilde{1};q\rangle_\infty}{\langle 1+a-b,\widetilde{1-b};q\rangle_\infty}\frac{\langle 1-b+\frac{a}{2};q^2\rangle_\infty}{\langle \frac{a}{2};q^2\rangle_\infty}\\
&\overset{\text{by (6.34)}}{=} \frac{\langle \widetilde{1};q\rangle_\infty\langle \frac{a+1}{2},1-b+\frac{a}{2};q^2\rangle_\infty}{\langle 1+a-b,\widetilde{1-b};q\rangle_\infty}\\
&= \Gamma_q\begin{bmatrix}1+a-b,1+\frac{a}{2}\\1+a,1+\frac{a}{2}-b\end{bmatrix}\frac{\langle \widetilde{1+\frac{a}{2}-b},\widetilde{1};q\rangle_\infty}{\langle \widetilde{1+\frac{a}{2}},\widetilde{1-b};q\rangle_\infty} = \text{RHS}.
\end{aligned}
\tag{7.81}
$$

We get another proof, if we put $c=\widetilde{\frac{a}{2}}$ in the q-analogue of the Dixon-Schafheitlin theorem (7.113). □

Remark 34 By letting $b\to-\infty$ in (7.80) we obtain the following formula from [26, p. 447]:

$$\sum_{n=0}^{\infty}\frac{\langle a;q\rangle_n}{\langle 1;q\rangle_n}q^{\frac{n^2+n}{2}} = \left\langle \frac{a+1}{2};q^2\right\rangle_\infty \langle \widetilde{1};q\rangle_\infty. \tag{7.82}$$

7.14 A general expansion formula

In this section we show an expansion formula of Gasper and Rahman. First a lemma.

Lemma 7.14.1

$$
\begin{aligned}
&(-1)^j q^{-ji} q^{-\binom{j}{2}}\frac{\langle a+j,-n;q\rangle_j\langle -n+j,a+2j;q\rangle_i}{\langle 1,a+j;q\rangle_i}\\
&= \frac{\langle a+i+j,-i-j;q\rangle_j\langle -n;q\rangle_{i+j}q^j}{\langle 1;q\rangle_{i+j}}.
\end{aligned}
\tag{7.83}
$$

Proof This follows from (6.16), (6.17) and (6.20)

$$\langle -n;q\rangle_{i+j}=\langle -n;q\rangle_j\langle -n+j;q\rangle_i, \tag{7.84}$$

$$\langle a+2j;q\rangle_i\langle a+j;q\rangle_j=\langle a+j;q\rangle_i\langle a+i+j;q\rangle_j, \tag{7.85}$$

$$\langle -i-j;q\rangle_j\langle 1;q\rangle_i=\langle 1;q\rangle_{i+j}(-1)^j q^{-j(i+j)}q^{\binom{j}{2}}. \tag{7.86}$$

□

In the q-Saalschützian formula (7.67) replace

$$n,a,b,c \text{ by } k,1+a-b-c,a+k,1+a-b,$$

and use (6.14) twice to get [220, p. 32, (2.2.1)]. In the new notation this formula takes the following form:

$$\begin{aligned}
&{}_3\phi_2(-k,a+k,1+a-b-c;1+a-b,1+a-c|q;q)\\
&=\frac{\langle c,1-b-k;q\rangle_k}{\langle 1+a-b,c-a-k;q\rangle_k}\\
&=\frac{\langle b,c;q\rangle_k}{\langle 1+a-b,1+a-c;q\rangle_k}q^{(1+a-b-c)k},
\end{aligned} \tag{7.87}$$

so that

$$\begin{aligned}
&\sum_{k=0}^{n}\frac{\langle b,c,-n;q\rangle_k}{\langle 1,1+a-b,1+a-c;q\rangle_k}A_k\\
&\overset{\text{by (7.87)}}{=}\sum_{k=0}^{n}\sum_{j=0}^{k}\frac{\langle -k,a+k,1+a-b-c;q\rangle_j\langle -n;q\rangle_k}{\langle 1,1+a-b,1+a-c;q\rangle_j\langle 1;q\rangle_k}q^{j+k(b+c-1-a)}A_k\\
&=\sum_{j=0}^{n}\sum_{i=0}^{n-j}\frac{\langle -i-j,a+i+j,1+a-b-c;q\rangle_j\langle -n;q\rangle_{i+j}}{\langle 1,1+a-b,1+a-c;q\rangle_j\langle 1;q\rangle_{i+j}}\\
&\times q^{j+(i+j)(b+c-1-a)}A_{i+j}\\
&\overset{\text{by (7.83)}}{=}\sum_{j=0}^{n}\frac{\langle 1+a-b-c,a+j,-n;q\rangle_j}{\langle 1,1+a-b,1+a-c;q\rangle_j}(-1)^j q^{-\binom{j}{2}}\\
&\times\sum_{i=0}^{n-j}\frac{\langle j-n,a+2j;q\rangle_i}{\langle 1,a+j;q\rangle_i}q^{-ij+(i+j)(b+c-1-a)}A_{i+j},
\end{aligned} \tag{7.88}$$

where $\{A_k\}$ is an arbitrary sequence. This is equivalent to Bailey's 1949 [58] lemma. Choosing

$$A_k = \frac{\langle a, a_1, \ldots, a_l; q\rangle_k}{\langle b_1, \ldots, b_{l+1}; q\rangle_k} q^{zk} \tag{7.89}$$

and using (6.16) four times, we obtain the following expansion formula [220, p. 33, (2.2.4)]. In the new notation this formula reads

$$\begin{aligned}
&{}_{l+4}\phi_{l+3}\left[\begin{matrix} a, b, c, a_1, \ldots, a_l, -n \\ 1+a-b, 1+a-c, b_1, \ldots, b_{l+1}\end{matrix} \middle| q; q^z\right] \\
&= \sum_{k=0}^{n} \frac{\langle b, c, -n, a, a_1, \ldots, a_l; q\rangle_k}{\langle 1, 1+a-b, 1+a-c, b_1, \ldots, b_{l+1}; q\rangle_k} q^{zk} \\
&= \sum_{j=0}^{n} \frac{\langle -n, a_1, \ldots, a_l, 1+a-b-c; q\rangle_j \langle a; q\rangle_{2j}}{\langle 1, 1+a-b, 1+a-c, b_1, \ldots, b_{l+1}; q\rangle_j} \\
&\times (-1)^j q^{j(b+c-1-a+z)-\binom{j}{2}} \\
&\times {}_{l+2}\phi_{l+1}\left[\begin{matrix} a+2j, a_1+j, \ldots, a_l+j, j-n \\ b_1+j, \ldots, b_{l+1}+j\end{matrix} \middle| q; q^{b+c+z-1-a-j}\right],
\end{aligned} \tag{7.90}$$

which is a q-analogue of Bailey's formula from 1935 [56, 4.3(1)].

Formula (7.90) enables one to reduce a ${}_{l+4}\phi_{l+3}$ series to a sum of ${}_{l+2}\phi_{l+1}$ series [220, p. 33].

Remark 35 Choosing instead

$$A_k = \frac{\langle a, a_1, \ldots, a_l; q\rangle_k}{\langle b_1, \ldots, b_{l+2}; q\rangle_k} q^{zk+\binom{k}{2}}(-1)^k \tag{7.91}$$

and using (6.16) four times, we obtain the following alternate expansion formula. In the new notation this formula reads

$$\begin{aligned}
&{}_{l+4}\phi_{l+4}\left[\begin{matrix} a, b, c, a_1, \ldots, a_l, -n \\ 1+a-b, 1+a-c, b_1, \ldots, b_{l+2}\end{matrix} \middle| q; q^z\right] \\
&= \sum_{k=0}^{n} \frac{\langle b, c, -n, a, a_1, \ldots, a_l; q\rangle_k}{\langle 1, 1+a-b, 1+a-c, b_1, \ldots, b_{l+2}; q\rangle_k} q^{zk+\binom{k}{2}}(-1)^k \\
&= \sum_{j=0}^{n} \frac{\langle -n, a_1, \ldots, a_l, 1+a-b-c; q\rangle_j \langle a; q\rangle_{2j}}{\langle 1, 1+a-b, 1+a-c, b_1, \ldots, b_{l+2}; q\rangle_j} q^{j(b+c-1-a+z)} \\
&\times {}_{l+2}\phi_{l+2}\left[\begin{matrix} a+2j, a_1+j, \ldots, a_l+j, j-n \\ b_1+j, \ldots, b_{l+2}+j\end{matrix} \middle| q; q^{b+c+z-1-a}\right].
\end{aligned} \tag{7.92}$$

7.15 A summation formula for a terminating very-well-poised $_4\phi_3$ series

Setting $b = 1 + \frac{1}{2}a$, $c = \widetilde{1 + \frac{1}{2}a}$, $a_k = b_k$, $k = 1, 2, \ldots, l$ and $b_{l+1} = a + n + 1$, formula (7.90) implies that relation (2.3.1) in Gasper and Rahman [220, p. 33] takes the following form in the new notation:

$$
{}_4\phi_3\left(a, 1+\frac{1}{2}a, \widetilde{1+\frac{1}{2}a}, -n; \frac{1}{2}a, \widetilde{\frac{1}{2}a}, 1+a+n|q, q^z\right)
$$

$$
\overset{\text{by (6.30)}}{=} \sum_{j=0}^{n} \frac{\langle \widetilde{-1}, -n; q\rangle_j \langle a; q\rangle_{2j}}{\langle 1, \frac{1}{2}a, \widetilde{\frac{1}{2}a}, 1+a+n; q\rangle_j} \times q^{(1+z)j} q^{-\binom{j}{2}}
$$

$$
\times {}_2\phi_1\left(a+2j, j-n; 1+a+n+j|q, -q^{z+1-j}\right). \tag{7.93}
$$

If $z = n$, then the $_2\phi_1$-series (7.93) can be summed by means of the Bailey-Daum summation formula (7.80) [220, p. 34, (2.3.2)].

In the new notation this formula reads

$$
{}_2\phi_1\left(a+2j, j-n; 1+a+n+j|q; -q^{n+1-j}\right)
$$

$$
= \frac{\langle \widetilde{1}; q\rangle_\infty \langle \frac{1+a+2j}{2}, 1+n+\frac{a}{2}; q^2\rangle_\infty}{\langle 1+a+j+n, \widetilde{1+n-j}; q\rangle_\infty}. \tag{7.94}
$$

We obtain

$$
{}_4\phi_3\left(a, 1+\frac{1}{2}a, \widetilde{1+\frac{1}{2}a}, -n; \frac{1}{2}a, \widetilde{\frac{1}{2}a}, 1+a+n|q, q^n\right)
$$

by (7.94), (7.93)

$$
= \sum_{j=0}^{n} \frac{\langle \widetilde{-1}, -n; q\rangle_j \langle a; q\rangle_{2j}}{\langle 1, \frac{1}{2}a, \widetilde{\frac{1}{2}a}, 1+a+n; q\rangle_j} q^{(1+n)j} q^{-\binom{j}{2}}
$$

$$
\times \frac{\langle \widetilde{1}; q\rangle_\infty \langle \frac{1+a+2j}{2}, 1+n+\frac{a}{2}; q^2\rangle_\infty}{\langle 1+a+j+n, \widetilde{1+n-j}; q\rangle_\infty}
$$

by (6.33), (6.34)

$$
= \sum_{j=0}^{n} \frac{\langle \widetilde{-1}, -n; q\rangle_j \langle \frac{1+a}{2}; q^2\rangle_j}{\langle 1, 1+a+n; q\rangle_j} q^{(1+n)j} q^{-\binom{j}{2}}
$$

$$
\times \frac{\langle \widetilde{1}; q\rangle_\infty \langle \frac{1+a+2j}{2}, 1+n+\frac{a}{2}; q^2\rangle_\infty}{\langle 1+a+j+n, \widetilde{1+n-j}; q\rangle_\infty}
$$

by $2 \times$ (6.8)

$$= \sum_{j=0}^{n} \frac{\langle \widetilde{-1}, -n; q\rangle_j \langle \frac{1+a}{2}; q^2\rangle_j \langle \widetilde{1}; q\rangle_{n-j}}{\langle 1; q\rangle_j} q^{(1+n)j} q^{-\binom{j}{2}}$$

$$\times \frac{\langle \frac{1+a+2j}{2}, 1+n+\frac{a}{2}; q^2\rangle_\infty}{\langle 1+a+n; q\rangle_\infty}$$

by (6.15), (6.30)

$$= \sum_{j=0}^{n} \frac{\langle \widetilde{-1}, -n; q\rangle_j \langle \frac{1+a}{2}; q^2\rangle_j \langle \widetilde{1}; q\rangle_n}{\langle 1, \widetilde{-n}; q\rangle_j} q^j \frac{\langle \frac{1+a+2j}{2}, 1+n+\frac{a}{2}; q^2\rangle_\infty}{\langle 1+a+n; q\rangle_\infty}$$

$$\overset{\text{by (6.8)}}{=} \sum_{j=0}^{n} \frac{\langle \widetilde{-1}, -n; q\rangle_j \langle \widetilde{1}; q\rangle_n}{\langle 1, \widetilde{-n}; q\rangle_j} q^j \frac{\langle \frac{1+a}{2}, 1+n+\frac{a}{2}; q^2\rangle_\infty}{\langle 1+a+n; q\rangle_\infty}$$

$$\overset{\text{by (7.96)}}{=} \sum_{j=0}^{n} \frac{\langle \widetilde{-1}, -n; q\rangle_j \langle \widetilde{1}; q\rangle_n}{\langle 1, \widetilde{-n}; q\rangle_j} q^j \frac{\langle \frac{1+a}{2}, 1+\frac{a}{2}; q^2\rangle_\infty \langle 1+a; q\rangle_n}{\langle 1+\frac{a}{2}; q^2\rangle_n \langle 1+a; q\rangle_\infty}$$

by (6.33), (6.34)

$$= \frac{\langle \widetilde{1}, 1+a; q\rangle_n}{\langle 1+\frac{1}{2}a, \widetilde{1+\frac{1}{2}a}; q\rangle_n} {}_2\phi_1(-n, \widetilde{-1}; \widetilde{-n}|q; q), \tag{7.95}$$

where, by applying twice (6.8),

$$\frac{\langle 1+n+\frac{a}{2}; q^2\rangle_\infty}{\langle 1+a+n; q\rangle_\infty} = \frac{\langle 1+\frac{a}{2}; q^2\rangle_\infty \langle 1+a; q\rangle_n}{\langle 1+\frac{a}{2}; q^2\rangle_n \langle 1+a; q\rangle_\infty}. \tag{7.96}$$

Both sides of (7.95) are equal to 1 when $n = 0$. By (6.29), (6.30), (7.68) (see [220, p. 34]), the ${}_2\phi_1$ series in the last row of (7.95) has the sum

$$\frac{\langle 1-n; q\rangle_n}{\langle \widetilde{-n}; q\rangle_n} (-q^{-1})^n, \tag{7.97}$$

when $n = 0, 1, \ldots$. Since $\langle 1-n; q\rangle_n = 0$ unless $n = 0$, it follows that the formula [220, p. 34, (2.3.4)]) takes the following form in the new notation:

$${}_4\phi_3\left(a, 1+\frac{1}{2}a, \widetilde{1+\frac{1}{2}a}, -n; \frac{1}{2}a, \widetilde{\frac{1}{2}a}, 1+a+n|q, q^n\right) = \delta_{n,0}, \tag{7.98}$$

where $\delta_{n,0}$ is the Kronecker delta function. This summation formula will be used in the next section to obtain the sum of a ${}_6\phi_5$ series.

7.16 A summation formula for a terminating very-well-poised ${}_6\phi_5$ series

Let us put

$$a_1 = 1+\frac{1}{2}a, \qquad a_2 = \widetilde{1+\frac{1}{2}a}, \qquad b_1 = \frac{1}{2}a, \qquad b_2 = \widetilde{\frac{1}{2}a}, \qquad b_{l+1} = 1+a+n,$$

and $a_k = b_k$, $k = 3, 4, \ldots, l$. Formula (7.90) now implies that the formula (2.4.1) in [220, p. 34] takes the following form in the new notation:

$$
{}_6\phi_5\left[\begin{matrix} a, b, c, 1+\frac{1}{2}a, \widetilde{1+\frac{1}{2}a}, -n \\ 1+a-b, 1+a-c, \frac{1}{2}a, \widetilde{\frac{1}{2}a}, 1+a+n \end{matrix} \middle| q; q^z \right]
$$

$$
= \sum_{j=0}^{n} (-1)^j \frac{\langle 1+a-b-c, 1+\frac{1}{2}a, \widetilde{1+\frac{1}{2}a}, -n; q\rangle_j \langle a; q\rangle_{2j}}{\langle 1, 1+a-b, 1+a-c, \frac{1}{2}a, \widetilde{\frac{1}{2}a}, 1+a+n; q\rangle_j} q^{j(b+c-1-a+z)-\binom{j}{2}}
$$

$$
\times\, {}_4\phi_3\left[\begin{matrix} a+2j, 1+\frac{1}{2}a+j, \widetilde{1+\frac{1}{2}a+j}, j-n \\ \frac{1}{2}a+j, \widetilde{\frac{1}{2}a+j}, 1+a+n+j \end{matrix} \middle| q; q^{b+c+z-1-a-j} \right]. \tag{7.99}
$$

If $z = 1+a+n-b-c$, then we can sum the above ${}_4\phi_3$ series by means of (7.98) and obtain a finite version of (7.112), [220, p. 34, (2.4.2)]. In the new notation this formula reads

$$
{}_6\phi_5\left[\begin{matrix} a, b, c, 1+\frac{1}{2}a, \widetilde{1+\frac{1}{2}a}, -n \\ 1+a-b, 1+a-c, \frac{1}{2}a, \widetilde{\frac{1}{2}a}, 1+a+n \end{matrix} \middle| q; q^{1+a+n-b-c} \right]
$$

$$
= \frac{\langle 1+a-b-c, 1+\frac{1}{2}a, \widetilde{1+\frac{1}{2}a}, -n; q\rangle_n \langle a; q\rangle_{2n}}{\langle 1, 1+a-b, 1+a-c, \frac{1}{2}a, \widetilde{\frac{1}{2}a}, 1+a+n; q\rangle_n} (-1)^n q^{\binom{n+1}{2}}
$$

by (6.16), (6.20) $(k = n)$, by (6.34), (6.33)

$$
= \frac{\langle 1+a, 1+a-b-c; q\rangle_n}{\langle 1+a-b, 1+a-c; q\rangle_n}. \tag{7.100}
$$

7.17 Watson's transformation formula for a terminating very-well-poised ${}_8\phi_7$ series

We will now use (7.100) to prove Watson's 1929 [535] transformation formula for a terminating very-well-poised ${}_8\phi_7$ series [220, p. 35, (2.5.1)]. Denoting

$$
(\alpha) \equiv \left(a, b, c, d, 1+\frac{1}{2}a, \widetilde{1+\frac{1}{2}a}, e, -n\right), \tag{7.101}
$$

$$(\beta) \equiv \left(1+a-b, 1+a-c, 1+a-d, 1+a-e, \frac{1}{2}a, \widetilde{\frac{1}{2}a}, 1+a+n\right), \quad (7.102)$$

$$(\gamma) \equiv \left(a+2j, d+j, e+j, 1+\frac{1}{2}a+j, 1+\widetilde{\frac{1}{2}a}+j, j-n\right) \quad (7.103)$$

and

$$(\delta) \equiv \left(1+a-d+j, 1+a-e+j, \frac{1}{2}a+j, \widetilde{\frac{1}{2}a}+j, 1+a+n+j\right), \quad (7.104)$$

we find that in the new notation this formula reads

Theorem 7.17.1

$$\begin{aligned} &{}_8\phi_7\left[\begin{matrix}(\alpha)\\(\beta)\end{matrix}|q;q^{2a+2+n-b-c-d-e}\right] \\ &= \frac{\langle 1+a, 1+a-d-e;q\rangle_n}{\langle 1+a-d, 1+a-e;q\rangle_n} \\ &\times {}_4\phi_3\left[\begin{matrix}d, e, 1+a-b-c, -n\\ 1+a-b, 1+a-c, d+e-n-a\end{matrix}|q;q\right]. \end{aligned} \quad (7.105)$$

Proof First observe that by (6.14), 2 × (6.16) and (6.18)

$$\begin{aligned} &\langle 1+a+n-d-e-j;q\rangle_j^{-1} \\ &= \langle -a-n+d+e;q\rangle_j^{-1}(-1)^j q^{(d+e-n-a)j+\binom{j}{2}}, \\ &\langle 1+a-d+j;q\rangle_{n-j} = \frac{\langle 1+a-d;q\rangle_n}{\langle 1+a-d;q\rangle_j}, \\ &\langle 1+a-d-e;q\rangle_{n-j} = \frac{\langle 1+a-d-e;q\rangle_n}{\langle 1+a-d-e+n-j;q\rangle_j}, \\ &\langle 1+a+2j;q\rangle_{n-j} = \frac{\langle 1+a;q\rangle_n \langle 1+a+n;q\rangle_j}{\langle 1+a;q\rangle_{2j}}. \end{aligned} \quad (7.106)$$

In (7.90) put

$$\begin{aligned} &l=4, \quad a_1=d, \quad a_2=1+\frac{1}{2}a, \\ &a_3=1+\widetilde{\frac{1}{2}a}, \quad a_4=e, \quad b_1=1+a-d, \\ &b_2=\frac{1}{2}a, \quad b_3=\widetilde{\frac{1}{2}a}, \quad b_4=1+a-e, \\ &b_5=1+a+n, \quad z=2a+2-b-c-d-e+n, \end{aligned} \quad (7.107)$$

and in (7.100) put

$$a = a + 2j, \qquad b = e + j, \qquad c = d + j, \qquad n = n - j. \tag{7.108}$$

We obtain

$$
\begin{aligned}
&{}_8\phi_7\left[\begin{matrix}(\alpha)\\(\beta)\end{matrix}|q;q^{2a+2+n-b-c-d-e}\right]\\
&\overset{\text{by (7.90)}}{=} \sum_{j=0}^{n} \frac{\langle -n, d, e, 1+\frac{1}{2}a; q\rangle_j}{\langle 1, (\beta); q\rangle_j}\\
&\times \left\langle \widetilde{1+\frac{1}{2}a}, 1+a-b-c; q\right\rangle_j \langle a; q\rangle_{2j}(-1)^j q^{j(1+a+n-d-e)-\binom{j}{2}}\\
&\times {}_6\phi_5\left[\begin{matrix}(\gamma)\\(\delta)\end{matrix}|q;q^{1+a+n-j-d-e}\right]\\
&\overset{\text{by (7.100)}}{=} \sum_{j=0}^{n} \frac{\langle -n, d, e, 1+\frac{1}{2}a, \widetilde{1+\frac{1}{2}a}; q\rangle_j}{\langle 1, \beta; q\rangle_j}\\
&\times \langle 1+a-b-c; q\rangle_j \langle a; q\rangle_{2j}(-1)^j q^{j(1+a+n-d-e)-\binom{j}{2}}\\
&\times \frac{\langle 1+a+2j, 1+a-d-e; q\rangle_{n-j}}{\langle a+j+1-e, a+j+1-d; q\rangle_{n-j}}\\
&\quad \text{by (6.33), by (6.34)}\\
&= \sum_{j=0}^{n} \frac{\langle -n, d, e, 1+a-b-c; q\rangle_j \langle 1+a; q\rangle_{2j}(-1)^j q^{j(1+a+n-d-e)-\binom{j}{2}}}{\langle 1, 1+a-b, 1+a-c, 1+a-d, 1+a-e, 1+a+n; q\rangle_j}\\
&\times \frac{\langle 1+a+2j, 1+a-d-e; q\rangle_{n-j}}{\langle a+j+1-e, a+j+1-d; q\rangle_{n-j}}\\
&\overset{\text{by (7.106)}}{=} \frac{\langle 1+a; q\rangle_n}{\langle 1+a-d; q\rangle_n}\\
&\times \frac{\langle 1+a-d-e; q\rangle_n}{\langle 1+a-e; q\rangle_n} {}_4\phi_3\left[\begin{matrix}d, e, 1+a-b-c, -n\\1+a-b, 1+a-c, d+e-n-a\end{matrix}|q;q\right].
\end{aligned}
\tag{7.109}
$$

□

Note that (7.105) is a q-analogue of (3.58).

7.18 Jackson's sum of a terminating very-well-poised balanced ${}_8\phi_7$ series

The following theorem for a terminating very-well-poised balanced ${}_8\phi_7$ series was published by Jackson in 1921 [305], [220, p. 35, (2.6.2)]. In the new notation this theorem reads

Theorem 7.18.1 *Let the ${}_8\phi_7$ series in* (7.105) *be balanced, i.e., the six parameters a, b, c, d, e and n satisfy the relation*

$$2a+n+1 \equiv b+c+d+e \left(\operatorname{mod} \frac{2\pi i}{\log q}\right). \tag{7.110}$$

Then with the same (β) *as in* (7.102),

$$\begin{aligned} &{}_8\phi_7\left[\begin{matrix} a, b, c, d, 1+\frac{1}{2}a, \widetilde{1+\frac{1}{2}a}, e, -n \\ (\beta) \end{matrix} \middle| q; q\right] \\ &= \frac{\langle 1+a, 1+a-b-c, 1+a-b-d, 1+a-c-d; q\rangle_n}{\langle 1+a-b, 1+a-c, 1+a-d, 1+a-b-c-d; q\rangle_n}, \end{aligned} \tag{7.111}$$

when $n = 0, 1, 2, \ldots$.

Proof This follows directly from (7.105), since the ${}_4\phi_3$ series on the right-hand side of (7.105) becomes a balanced ${}_3\phi_2$ series when (7.110) holds and therefore can be summed by the q-Saalschütz's summation formula (7.67) [220, p. 36]). A permutation of the variables completes the proof. □

Remark 36 Note that (7.111) is a q-analogue of (3.50). It was proved by Jackson already in 1905 [27], [466].

Now eliminate e by (7.110) and let $n \to \infty$ to obtain [194], [125, p. 115], [220, p. 36, (2.7.1)], a q-analogue of (3.51):

$$\begin{aligned} &{}_6\phi_5\left[\begin{matrix} a, b, c, d, 1+\frac{1}{2}a, \widetilde{1+\frac{1}{2}a} \\ 1+a-b, 1+a-c, 1+a-d, \frac{1}{2}a, \widetilde{\frac{1}{2}a} \end{matrix} \middle| q; q^{1+a-b-c-d}\right] \\ &= \frac{\langle 1+a, 1+a-b-c, 1+a-b-d, 1+a-c-d; q\rangle_\infty}{\langle 1+a-b, 1+a-c, 1+a-d, 1+a-b-c-d; q\rangle_\infty}. \end{aligned} \tag{7.112}$$

Finally, put $d = \frac{a}{2}$ to obtain the following q-analogue of Dixon's theorem (3.53):

$$\begin{aligned} &{}_4\phi_3\left[\begin{matrix} a, b, c, \widetilde{1+\frac{1}{2}a} \\ 1+a-b, 1+a-c, \widetilde{\frac{1}{2}a} \end{matrix} \middle| q; q^{1+\frac{a}{2}-b-c}\right] \\ &= \frac{\langle 1+a, 1+a-b-c, 1+\frac{a}{2}-b, 1+\frac{a}{2}-c; q\rangle_\infty}{\langle 1+a-b, 1+a-c, 1+\frac{a}{2}, 1+\frac{a}{2}-b-c; q\rangle_\infty}. \end{aligned} \tag{7.113}$$

7.18.1 Three corollaries

We find three q-analogues of the formula (3.52).

Corollary 7.18.2 [130, p. 313, 3.4b], [131, p. 124, 2.3c]:

$$
{}_4\phi_3\left[\begin{matrix} a, b, c, 1+\frac{1}{2}a \\ 1+a-b, 1+a-c, \frac{1}{2}a \end{matrix} | q; -q^{1+\frac{a}{2}-b-c}\right]
$$

$$
= \Gamma_q\left[\begin{matrix} 1+a-b, 1+a-c, 1+\frac{a}{2}+\beta, 1+\frac{a}{2}-b-c+\beta \\ 1+a, 1+a-b-c, 1+\frac{a}{2}-b+\beta, 1+\frac{a}{2}-c+\beta \end{matrix}\right], \tag{7.114}
$$

where $\beta \equiv \frac{\log(-1)}{\log q}$.

Proof Put $d = \widetilde{\frac{a}{2}}$ in (7.112) and use (6.30). □

Corollary 7.18.3

$$
{}_6\phi_4\left[\begin{matrix} a, -n, c, \infty, 1+\frac{1}{2}a, \widetilde{1+\frac{1}{2}a} \\ 1+a+n, 1+a-c, \frac{1}{2}a, \widetilde{\frac{1}{2}a} \end{matrix} | q; q^{n-c}\right]
$$

$$
= \Gamma_q\left[\begin{matrix} 1+a+n, 1+a-c \\ 1+a, 1+a+n-c \end{matrix}\right] q^{-nc}. \tag{7.115}
$$

Proof Let $d \to \infty$ in (7.112). □

Corollary 7.18.4 *A slight improvement of* [130, p. 313, 3.4c], [131, p. 124, 2.3a]:

$$
{}_5\phi_5\left[\begin{matrix} a, b, c, 1+\frac{1}{2}a, \widetilde{1+\frac{1}{2}a} \\ 1+a-b, 1+a-c, \frac{1}{2}a, \widetilde{\frac{1}{2}a}, \infty \end{matrix} | q; q^{1+a-b-c}\right]
$$

$$
= \Gamma_q\left[\begin{matrix} 1+a-b, 1+a-c \\ 1+a, 1+a-b-c \end{matrix}\right]. \tag{7.116}
$$

Proof Let $d \to -\infty$ in (7.112). □

7.19 Watson's proof of the Rogers-Ramanujan identities

Watson [535] used his transformation formula (7.105) to give a simple proof of the famous Rogers-Ramanujan identities [266]:

Theorem 7.19.1

$$\sum_{j=0}^{\infty}\frac{q^{j^2}}{\langle 1;q\rangle_j}=\frac{\langle \frac{3}{5},\frac{3}{5},1;q^5\rangle_\infty}{\langle 1;q\rangle_\infty},\qquad \sum_{j=0}^{\infty}\frac{q^{j(j+1)}}{\langle 1;q\rangle_j}=\frac{\langle \frac{1}{5},\frac{3}{5},1;q^5\rangle_\infty}{\langle 1;q\rangle_\infty},\tag{7.117}$$

where $|q|<1$.

Proof See Gasper and Rahman [220, pp. 36–37] and Slater [466, p. 104]. In the new notation this proof takes the following form: Let $b,c,d,e\to-\infty$ in (7.105) to obtain

$$\lim_{b,c,d,e\to-\infty} {}_8\phi_7\left[\begin{matrix}\alpha\\ \beta\end{matrix}|q;q^{2a+2+n-b-c-d-e}\right]$$

$$\overset{\text{by (6.33)}}{=}\lim_{b,c,d,e\to-\infty}$$

$$\times\sum_{j=0}^{n}\frac{\langle a,b,c,d,e,-n;q\rangle_j\langle \frac{a}{2}+1;q^2\rangle_j q^{j(2a+2+n-b-c-d-e)}}{\langle 1,1+a-b,1+a-c,1+a-d,1+a-e,1+a+n;q\rangle_j\langle \frac{a}{2};q^2\rangle_j}$$

$$\overset{\text{by (6.35)}}{=}\sum_{j=0}^{n}\frac{\langle a,-n;q\rangle_j\langle \frac{a}{2}+1;q^2\rangle_j q^{j(2a+2j+n)}}{\langle 1,1+a+n;q\rangle_j\langle \frac{a}{2};q^2\rangle_j}$$

$$=\sum_{j=0}^{n}\frac{\langle a,-n;q\rangle_j(1-q^{a+2j})q^{j(2a+2j+n)}}{\langle 1,1+a+n;q\rangle_j(1-q^a)}$$

$$=\lim_{b,c,d,e\to-\infty}\frac{\langle 1+a,1+a-d-e;q\rangle_n}{\langle 1+a-d,1+a-e;q\rangle_n}$$

$$\times\sum_{j=0}^{n}\frac{\langle d,e,1+a-b-c,-n;q\rangle_j}{\langle 1,1+a-b,1+a-c,d+e-n-a;q\rangle_j}q^j$$

$$\overset{\text{by (6.35)}}{=}\langle 1+a;q\rangle_n\sum_{j=0}^{n}\frac{\langle -n;q\rangle_j(-1)^j q^{(a+n+\frac{j+1}{2})j}}{\langle 1;q\rangle_j}.\tag{7.118}$$

Thus we have

$$\sum_{j=0}^{n}\frac{\langle a,-n;q\rangle_j(1-q^{a+2j})q^{j(2a+2j+n)}}{\langle 1,1+a+n;q\rangle_j(1-q^a)}$$

$$=\langle 1+a;q\rangle_n\sum_{j=0}^{n}\frac{\langle -n;q\rangle_j(-1)^j q^{(a+n+\frac{j+1}{2})j}}{\langle 1;q\rangle_j}.\tag{7.119}$$

The next step is to let $n\to\infty$. If

$$A_j(n)=\frac{\langle -n;q\rangle_j q^{jn}}{\langle 1+a+n;q\rangle_j},$$

then by (6.35)

$$\lim_{n\to+\infty} A_j(n) = (-1)^j q^{\binom{j}{2}}, \tag{7.120}$$

for any fixed value of j. For all values of n, $|A_j(n)| < L$, a constant. Hence, in the series on the left-hand side of (7.119) the modulus of each term is strictly less than

$$L\left|\frac{\langle a;q\rangle_j(1-q^{a+2j})q^{j(2a+2j)}}{\langle 1;q\rangle_j(1-q^a)}\right| = C\left|q^{2aj+2j^2}\right|,$$

which is the term of a convergent series. If

$$B_j(n) = \langle 1+a;q\rangle_n\langle -n;q\rangle_j q^{nj},$$

then by (6.35)

$$\lim_{n\to+\infty} B_j(n) = (-1)^j q^{\binom{j}{2}}\langle 1+a;q\rangle_\infty, \tag{7.121}$$

for any fixed value of j. For all values of n, $|B_j(n)| < K$, a constant. Hence, in the series on the right-hand side of (7.119) the modulus of each term is strictly less than

$$K\left|\frac{q^{(a+\frac{j+1}{2})j}}{\langle 1;q\rangle_j}\right| = D\left|q^{(a+\frac{j+1}{2})j}\right|,$$

which is the term of a convergent series. By (6.35),

$$1+\sum_{j=1}^{\infty}\frac{\langle 1+a;q\rangle_{j-1}(1-q^{a+2j})}{\langle 1;q\rangle_j}q^{2aj}q^{\frac{j(5j-1)}{2}}(-1)^j = \langle 1+a;q\rangle_\infty\sum_{j=0}^{\infty}\frac{q^{j(a+j)}}{\langle 1;q\rangle_j}. \tag{7.122}$$

Now put $a=0$ in (7.122):

$$\begin{aligned}
&1+\sum_{j=1}^{\infty}\frac{\langle 1;q\rangle_{j-1}(1-q^{2j})}{\langle 1;q\rangle_j}q^{\frac{j(5j-1)}{2}}(-1)^j\\
&=1+\sum_{j=1}^{\infty}(1+q^j)q^{\frac{j(5j-1)}{2}}(-1)^j\\
&=1+\sum_{j=1}^{\infty}(-1)^j\left(q^{\frac{j(5j-1)}{2}}+q^{\frac{j(5j+1)}{2}}\right)=\sum_{j=-\infty}^{\infty}(-1)^j q^{\frac{j}{2}+\frac{5j^3}{2}}\\
&\overset{\text{by (7.35)}}{=}\left\langle\frac{3}{5},\frac{3}{5},1;q^5\right\rangle_\infty = \langle 1;q\rangle_\infty\sum_{j=0}^{\infty}\frac{q^{j^2}}{\langle 1;q\rangle_j}.
\end{aligned} \tag{7.123}$$

This proves the first of relations (7.117).

Next put $a = 1$ in (7.122):

$$1+\sum_{j=1}^{\infty}\frac{\langle 2;q\rangle_{j-1}(1-q^{1+2j})}{\langle 1;q\rangle_j}q^{\frac{j(5j+3)}{2}}(-1)^j=\langle 2;q\rangle_\infty\sum_{j=0}^{\infty}\frac{q^{j(1+j)}}{\langle 1;q\rangle_j}.$$

Multiply by $(1-q)$ we get

$$\sum_{j=0}^{\infty}\frac{q^{j(1+j)}}{\langle 1;q\rangle_j}=\frac{1}{\langle 1;q\rangle_\infty}\left(1-q+\sum_{j=1}^{\infty}(1-q^{1+2j})q^{\frac{j(5j+3)}{2}}(-1)^j\right)$$

$$=\frac{1}{\langle 1;q\rangle_\infty}\sum_{-\infty}^{\infty}q^{\frac{j(5j+3)}{2}}(-1)^j=\frac{\langle \frac{1}{5},\frac{3}{5},1;q^5\rangle_\infty}{\langle 1;q\rangle_\infty}. \tag{7.124}$$

This proves the second of relations (7.117). □

7.20 Bailey's 1929 transformation formula for a terminating, balanced, very-well-poised ${}_{10}\phi_9$

We illustrate the use of the new method with the following important example of Bailey.

Definition 138 For brevity, we shall sometimes replace

$${}_{r+1}\phi_r\left[\begin{matrix} a_1, 1+\frac{1}{2}a_1, \widetilde{1+\frac{1}{2}a_1}, a_4, a_5, \ldots, a_{r+1} \\ \frac{1}{2}a_1, \widetilde{\frac{1}{2}a_1}, 1+a_1-a_4, 1+a_1-a_5, \ldots, 1+a_1-a_{r+1}\end{matrix}|q;z\right]$$

by the more compact notation

$${}_{r+1}W_r(a_1;a_4,a_5,\ldots,a_{r+1}|q;z). \tag{7.125}$$

We also use the notations

$$(\alpha')\equiv\left(a,b,c,d,e,f,1+\frac{1}{2}a,\widetilde{1+\frac{1}{2}a},\lambda+a+n+1-e-f,-n\right), \tag{7.126}$$

$$(\beta')\equiv\Bigg(1+a-b,1+a-c,1+a-d,1+a-e,\frac{1}{2}a,$$
$$\widetilde{\frac{1}{2}a},a+1-f,e+f-n-\lambda,a+n+1\Bigg), \tag{7.127}$$

$$(\gamma')\equiv\Bigg(\lambda,\lambda+b-a,\lambda+c-a,\lambda+d-a,e,f,1+\frac{1}{2}\lambda,$$
$$\widetilde{1+\frac{1}{2}\lambda},\lambda+a+n+1-e-f,-n\Bigg) \tag{7.128}$$

and

$$(\delta') \equiv \Big(1+a-b, 1+a-c, 1+a-d, 1+\lambda-e, \frac{1}{2}\lambda,$$
$$\widetilde{\frac{1}{2}\lambda}, \lambda+1-f, e+f-n-a, \lambda+n+1\Big). \tag{7.129}$$

We find that this formula takes the following form in the new notation:

Theorem 7.20.1 (*Bailey's* 1929 [54] *transformation formula for a terminating balanced, very-well-poised* ${}_{10}\phi_9$ *q-hypergeometric series.*)

$${}_{10}\phi_9\left[\begin{matrix}(\alpha')\\(\beta')\end{matrix}|q;q\right] = \frac{\langle 1+a, 1+a-e-f, 1+\lambda-e, 1+\lambda-f; q\rangle_n}{\langle 1+a-e, 1+a-f, 1+\lambda-e-f, 1+\lambda; q\rangle_n}$$
$$\times\, {}_{10}\phi_9\left[\begin{matrix}(\gamma')\\(\delta')\end{matrix}|q;q\right],$$
$$\textit{where } 2a+1 \equiv \lambda+b+c+d \left(\operatorname{mod} \frac{2\pi i}{\log q}\right). \tag{7.130}$$

Proof See the book by Gasper and Rahman [220, p. 39]. In the new notation the proof takes the following form: By Jackson's sum,

$${}_8\phi_7\left[\begin{matrix}\lambda, \lambda+b-a, \lambda+c-a, \lambda+d-a, a+m, 1+\frac{1}{2}\lambda, \widetilde{1+\frac{1}{2}\lambda}, a+m, -m\\ 1+a-b, 1+a-c, 1+a-d, 1+\lambda-m-a, \frac{1}{2}\lambda, \widetilde{\frac{1}{2}\lambda}, \lambda+1+m\end{matrix}|q;q\right]$$
$$= \frac{\langle b, c, d, \lambda+1; q\rangle_m}{\langle 1+a-b, 1+a-c, 1+a-d, a-\lambda; q\rangle_m}, \tag{7.131}$$

and the left-hand side of (7.130) is equal to

$$\sum_{m=0}^{n} \frac{\langle a, e, f, \lambda+a+1+n-e-f, -n, a-\lambda; q\rangle_m (1-q^{a+2m})q^m}{\langle 1, a+1-e, a+1-f, e+f-n-\lambda, a+n+1, \lambda+1; q\rangle_m (1-q^a)}$$
$$\times \sum_{j=0}^{m} \frac{\langle \lambda, \lambda+b-a, \lambda+c-a, \lambda+d-a, a+m, -m; q\rangle_j (1-q^{\lambda+2j})q^j}{\langle 1, a+1-b, a+1-c, a+1-d, \lambda+1+m, \lambda+1-m-a; q\rangle_j (1-q^\lambda)}$$

by (6.15), (6.16), (6.20)

$$= \sum_{m=0}^{n}\sum_{j=0}^{m} \frac{\langle a; q\rangle_{m+j} \langle e, f, \lambda+a+1+n-e-f, -n; q\rangle_m (1-q^{a+2m})q^m}{\langle 1; q\rangle_{m-j} \langle a+1-e, a+1-f, e+f-n-\lambda, a+n+1; q\rangle_m (1-q^a)}$$
$$\times \frac{\langle a-\lambda; q\rangle_{m-j} \langle \lambda, \lambda+b-a, \lambda+c-a, \lambda+d-a; q\rangle_j (1-q^{\lambda+2j})q^{(a-\lambda)j}}{\langle \lambda+1; q\rangle_{m+j} \langle 1, a+1-b, a+1-c, a+1-d; q\rangle_j (1-q^\lambda)}$$

by $10\times$ (6.16)

$$= \sum_{j=0}^{n} q^{(a+1-\lambda)j} \frac{\langle a+1;q\rangle_{2j}(1-q^{\lambda+2j})}{\langle \lambda+1;q\rangle_{2j}(1-q^{\lambda})}$$

$$\times \frac{\langle \lambda, \lambda+b-a, \lambda+c-a, \lambda+d-a, e, f, \lambda+a+n+1-e-f, -n; q\rangle_j}{\langle 1, a+1-b, a+1-c, a+1-d, a+1-e, a+1-f, e+f-n-\lambda, a+n+1; q\rangle_j}$$

$$\times\, {}_8W_7(a+2j; e+j, f+j, a-\lambda, \lambda+a+n+j+1-e-f, j-n|q;q)$$

by $2\times$ (6.15), $2\times$ (6.16), $2\times$ (6.18)

$$= \frac{\langle 1+a, 1+a-e-f, 1+\lambda-e, 1+\lambda-f; q\rangle_n}{\langle 1+a-e, 1+a-f, 1+\lambda-e-f, 1+\lambda; q\rangle_n} {}_{10}\phi_9 \left[\begin{matrix} (\gamma') \\ (\delta') \end{matrix} | q; q \right]. \tag{7.132}$$

□

Thus the q-analogue of the ${}_9F_8$ 2-balanced hypergeometric series (3.63) is again a balanced ${}_{10}\phi_9$ q-hypergeometric series.

7.21 Watson's q-analogue of the Barnes contour integral

Theorem 7.21.1 *Watsons q-analogue of the Barnes contour integral* [533], [220, p. 104]:

$${}_2\phi_1(a,b;c|q;z) = \frac{\langle a,b;q\rangle_\infty}{\langle 1,c;q\rangle_\infty} \frac{-1}{2\pi i} \int_C \frac{\langle 1+s, c+s;q\rangle_\infty}{\langle a+s, b+s;q\rangle_\infty} \frac{\pi(-z)^s}{\sin \pi s} d_q(s), \tag{7.133}$$

where $0 < q < 1$, $|z| < 1$, $|\arg(-z)| \le \pi - \delta$, $\delta > 0$. The complex curve integral C runs from $-i\infty$ to $i\infty$. We will now describe the integration path more precisely. The factor $\frac{\langle 1+s;q\rangle_\infty}{\sin \pi s}$ gives poles for $s \in \mathbb{N}$. The factor $\frac{1}{\langle a+s,b+s;q\rangle_\infty}$ has simple poles in $s = -a-n$ and in $s = -b-n$. The integration path C runs from $-i\infty$ along the imaginary axis to $i\infty$. To avoid the poles, we make turns to the left of $\mathbb{N}$ and to the right of $s = -a-n$ and $s = -b-n$. For the proof by the residue theorem, we include in the path of integration the semicircle C_R in the first and fourth quadrant, and let $R \to \infty$.

This formula can also be written

$${}_2\phi_1(a,b;c|q;z) = \Gamma_q \left[\begin{matrix} c \\ a,b \end{matrix} \right] \frac{1}{2\pi i} \int_C \Gamma_q \left[\begin{matrix} a+s, b+s \\ c+s, 1+s \end{matrix} \right]$$
$$\times\, \Gamma(-s)\Gamma(1+s)(-z)^s d_q(s). \tag{7.134}$$

Proof Use the Euler reflection formula. □

7.22 Three q-analogues of the Euler integral formula for the function $\Gamma(x)$

The Euler integral formula

$$\Gamma(x)=\int_0^{\infty} t^{x-1}e^{-t}dt, \quad \Re(x)>0 \tag{7.135}$$

is fundamental for the understanding of the Γ function. If we are looking for a q-analogue of such an integral, we have several choices. We can use different q-exponential functions. We can use the q-integral (6.55) with upper limit $+\infty$, or we can use the finite q-integral (1.53) with upper integration limit $\frac{1}{1-q}$.

In [152] the formula

$$\Gamma_q(z)=\int_0^{\frac{1}{1-q}} t^{z-1}\mathrm{E}_{\frac{1}{q}}(-qt)\,d_q(t), \quad \Re(z)>0 \tag{7.136}$$

was given. This q-integral formula is merely a heavily disguised version of the bilateral summation formula (8.103) of Ramanujan.

By a simple standard procedure we obtain

Theorem 7.22.1 *A q-analogue of* [419, p. 689], [397, p. 143]. *The Γ_q function has the following meromorphic continuation to $\mathbb{C}$:*

$$\Gamma_q(z)=\sum_{n=0}^{\infty}\frac{(-1)^n q^{\frac{n^2+n}{2}}}{\{z+n\}_q\{n\}_q!}+\int_1^{\frac{1}{1-q}} t^{z-1}\mathrm{E}_{\frac{1}{q}}(-qt)\,d_q(t). \tag{7.137}$$

Proof

$$\begin{aligned}\Gamma_q(z)&=\int_0^1 t^{z-1}\sum_{n=0}^{\infty}\frac{(-t)^n q^{\frac{n^2+n}{2}}}{\{n\}_q!}d_q(t)+\int_1^{\frac{1}{1-q}} t^{z-1}\mathrm{E}_{\frac{1}{q}}(-qt)\,d_q(t)\\ &=\sum_{n=0}^{\infty}\frac{(-1)^n q^{\frac{n^2+n}{2}}}{\{z+n\}_q\{n\}_q!}+\int_1^{\frac{1}{1-q}} t^{z-1}\mathrm{E}_{\frac{1}{q}}(-qt)\,d_q(t).\end{aligned} \tag{7.138}$$

□

We remark that we integrate to the second zero of $\mathrm{E}_{\frac{1}{q}}(-qt)$. Because of the inequality (6.160) it makes less sense to do a similar computation for $\mathrm{E}_q(-t)$ by using (7.142). For $q=1$, Pringsheim [419, p. 689] writes the formula (7.137) as

$$\Gamma(z)=F(z)+G(z), \tag{7.139}$$

where $F(z)\equiv\sum_{n=0}^{\infty}\frac{(-1)^n}{n!(z+n)}$ and $G(z)$ denotes an entire function.

Theorem 7.22.2

$$\Gamma_q(z) = \int_0^\infty t^{z-1} \mathrm{E}_{\frac{1}{q}}(-qt)\, d_q(t), \quad \Re(z) > 0. \tag{7.140}$$

Proof q-Integration by parts gives

$$\int_0^\infty t^{z-1} \mathrm{E}_{\frac{1}{q}}(-qt)\, d_q(t)$$
$$= \left[-t^{z-1} \mathrm{E}_{\frac{1}{q}}(-t)\right]_0^\infty + \{z-1\}_q \int_0^\infty t^{z-2} \mathrm{E}_{\frac{1}{q}}(-qt)\, d_q(t). \tag{7.141}$$

The first term on the right is zero because of (6.159), the zeros of $\mathrm{E}_{\frac{1}{q}}(-t)$ approach ∞ quicker than any polynomial. We obtain the required recurrence relation: $\Gamma_q(z) = \{z-1\}_q \Gamma_q(z-1)$. □

Theorem 7.22.3 *Compare with Jackson* [302, p. 200, (22)]:

$$\Gamma_q(z) = q^{\binom{z}{2}} \int_0^\infty t^{z-1} \mathrm{E}_q(-t)\, d_q(t), \quad \Re(z) > 0. \tag{7.142}$$

Proof Similar to above. □

Remark 37 Jackson [302, p. 200, (22)] and Exton [194, 2.2.3.3] discuss (7.142), but fail to give the correct formula. Askey [49] comments that (7.142) is a special case of Ramanujans ${}_1\Psi_1$ formula.

7.23 Inequalities for the Γ_q function

T. Kim and Adiga [334, p. 2] have found the following inequalities for the Γ_q function, $0 < q < 1$, $a \geq 1$ and $x \in [0, 1]$:

Theorem 7.23.1

$$\frac{1}{\Gamma_q(1+a)} \leq \frac{(\Gamma_q(1+x))^a}{\Gamma_q(1+ax)} \leq 1. \tag{7.143}$$

This theorem (7.143) is associated with the log-convexity of the Γ_q function. Horst Alzer [19] has found the following inequalities for the Γ_q function.

Theorem 7.23.2 *Let* $0 < q \neq 1$ *and* $s \in (0, 1)$. *The largest numbers* $\alpha(q, s)$ *and* $\beta(q, s)$, *such that*

$$\left(\frac{1-q^{x+\alpha}}{1-q}\right)^{1-s} < \frac{\Gamma_q(x+1)}{\Gamma_q(x+s)} < \left(\frac{1-q^{x+\beta}}{1-q}\right)^{1-s} \tag{7.144}$$

always holds, are given by

$$\alpha(q,s) = \begin{cases} \log[\frac{q^s-q}{(1-s)(1-q)}]\frac{1}{\log q}, & \textit{if}\, 0<q<1, \\ \frac{s}{2}, & \textit{if}\, q>1, \end{cases} \tag{7.145}$$

$$\beta(q,s) = \log\big[1-(1-q)\big(\Gamma_q(s)\big)^{\frac{1}{s-1}}\big]/\log q. \tag{7.146}$$

Remark 38 In 1983 H. Alzer defended his treatise *Die Nullstellen der Hyperbelfunktionen höherer Ordnung* in Bonn. His tutor was Werner Raab.

7.24 Summary of the umbral method

In this book, a new umbral method is introduced. We briefly summarize the content of the umbral method. For convenience, we refer mostly to q-analogues of the hypergeometric function.

1. By the q-additions we can formally deduce q-analogues of the function argument.
2. In the Watson and Jackson formulas for ${}_8\phi_7$ series, we append an extra tilde in the numerator and the denominator to get the q-analogue.

   In Section 10.5 we sometimes append two extra tilde in the numerator and the denominator. In the formulas (10.112) and (10.121) the hypergeometric function argument $4x$ is replaced by x times two tildes in the numerator.

   The tilde operator can be generalized as follows.
3. Srivastava and Karlsson [482] have found many reduction formulas for Kampé de Fériet functions, which almost all include the operator $\triangle(2;\lambda)$, a function of the Pochhammer symbol. In Section 6.4 we introduced its q-analogue $\triangle(q;l;\lambda)$. In an upcoming article we will present a q-analogue of the function argument $27x$.
4. The notation ∞ (for q-factorial) is equivalent to 0 in [220]. This notation simply means a multiplication by $(-1)^k q^{\pm\binom{k}{2}}$.
5. The terms well-poised and balanced series are also introduced in the q-case, but with the important difference that these definitions are always given mod $\frac{2\pi i}{\log q}$. These definitions can in certain cases be extended to balanced Γ_q functions.
6. By using the Cigler ϵ-operator we are able to write down operator formulas as well as q-analogues of partial differential equations, as in Section 11.3.

Chapter 8
Sundry topics

This chapter contains similar aspects as the previous one, it also gives the first application of a general tilde operator.

8.1 Four q-summation formulas of Andrews

As a basis for the following calculations we are going to use four q-formulas of Andrews. The first is

Theorem 8.1.1 *Andrews's q-analogue of Kummer's second summation formula from* [5, 1.8, p. 526]:

$$
{}_2\phi_2\left[\begin{matrix} a, b \\ \frac{1+a+b}{2}, \widetilde{\frac{1+a+b}{2}} \end{matrix} \Big| q; -q\right] = \Gamma_q\left[\begin{matrix} \frac{1+a+b}{2}, \frac{1}{2} \\ \frac{1+b}{2}, \frac{1+a}{2} \end{matrix}\right] \frac{\langle \widetilde{\frac{1+b}{2}}; q\rangle_{\frac{a}{2}}}{\langle \widetilde{\frac{1}{2}}; q\rangle_{\frac{a}{2}}}
$$

$$
\equiv \Gamma_{q^2}\left[\begin{matrix} \frac{1+a+b}{2}, \frac{1}{2} \\ \frac{1+b}{2}, \frac{1+a}{2} \end{matrix}\right]. \tag{8.1}
$$

Proof By [5, 1.8, p. 526]

$$
\begin{aligned}
\text{LHS} &\overset{\text{by (6.258)}}{=} \Gamma_q\left[\begin{matrix} \frac{1+a+b}{2}, 1+\frac{a}{2} \\ 1+a, \frac{1+b}{2} \end{matrix}\right] \frac{\langle \widetilde{\frac{1+b}{2}}, \widetilde{1}; q\rangle_\infty}{\langle \widetilde{\frac{1+a+b}{2}}, \widetilde{1+\frac{a}{2}}; q\rangle_\infty} \\
&\overset{\text{by (6.257)}}{=} \Gamma_q\left[\begin{matrix} \frac{1+a+b}{2} \\ \frac{1+b}{2}, 1+a \end{matrix}\right] \Gamma_{q^2}\left[\begin{matrix} 1+\frac{a}{2} \\ - \end{matrix}\right] (1+q)^{\frac{a}{2}} \frac{\langle \widetilde{\frac{1+b}{2}}; q\rangle_\infty}{\langle \widetilde{\frac{1+a+b}{2}}; q\rangle_\infty} \\
&= \Gamma_q\left[\begin{matrix} \frac{1+a+b}{2} \\ \frac{1+b}{2} \end{matrix}\right] \Gamma_{q^2}\left[\begin{matrix} \frac{1}{2} \\ \frac{1+a}{2} \end{matrix}\right] (1+q)^{-\frac{a}{2}} \frac{\langle \widetilde{\frac{1+b}{2}}; q\rangle_\infty}{\langle \widetilde{\frac{1+a+b}{2}}; q\rangle_\infty} \\
&\overset{\text{by (6.257)}}{=} \text{RHS},
\end{aligned} \tag{8.2}
$$

T. Ernst, *A Comprehensive Treatment of q-Calculus*,
DOI 10.1007/978-3-0348-0431-8_8, © Springer Basel 2012

where we have used the q-analogue of the Legendre duplication formula in the penultimate step. □

If b is a negative integer, (8.1) may be reformulated in the form:

Theorem 8.1.2

$$
\begin{aligned}
&{}_2\phi_2\left[\begin{matrix} a,-N \\ \frac{1+a-N}{2}, \widetilde{\frac{1+a-N}{2}} \end{matrix} | q;-q\right] \\
&\equiv \sum_{k=0}^{N} \binom{N}{k}_q q^{2\binom{k}{2}+k(1-N)} \frac{(-1)^k \langle a;q\rangle_k}{\langle \frac{1+a-N}{2};q^2\rangle_k} \\
&= \begin{cases} \frac{\langle \frac{1}{2};q^2\rangle_{\frac{N}{2}} q^{-\frac{Na}{2}}}{\langle \frac{1-a}{2};q^2\rangle_{\frac{N}{2}}} \equiv \frac{\langle \frac{1-N}{2};q^2\rangle_{\frac{N}{2}}}{\langle \frac{1+a-N}{2};q^2\rangle_{\frac{N}{2}}}, & N \text{ even}, \\ 0, & N \text{ odd}. \end{cases}
\end{aligned}
\tag{8.3}
$$

The following corollary was influenced by Kim, Rathie and Lee [335]. The proof is very similar. This idea to use the contiguity relations goes back to Kummer [348, p. 134–36].

Corollary 8.1.3

$$
\begin{aligned}
{}_2\phi_2\left[\begin{matrix} a,b \\ \frac{2+a+b}{2}, \widetilde{\frac{2+a+b}{2}} \end{matrix} | q;-q\right] &= \frac{q^a(1-q^b)}{q^a-q^b}\Gamma_q\left[\begin{matrix} \frac{2+a+b}{2}, \frac{1}{2} \\ \frac{2+b}{2}, \frac{1+a}{2} \end{matrix}\right] \frac{\langle \widetilde{\frac{2+b}{2}};q\rangle_{\frac{a}{2}}}{\langle \widetilde{\frac{1}{2}};q\rangle_{\frac{a}{2}}} \\
&\quad - \frac{q^b(1-q^a)}{q^a-q^b}\Gamma_q\left[\begin{matrix} \frac{2+a+b}{2}, \frac{1}{2} \\ \frac{1+b}{2}, \frac{2+a}{2} \end{matrix}\right] \frac{\langle \widetilde{\frac{1+b}{2}};q\rangle_{\frac{a+1}{2}}}{\langle \widetilde{\frac{1}{2}};q\rangle_{\frac{a+1}{2}}},
\end{aligned}
\tag{8.4}
$$

$$
\begin{aligned}
&{}_2\phi_2\left[\begin{matrix} a,b \\ \frac{3+a+b}{2}, \widetilde{\frac{3+a+b}{2}} \end{matrix} | q;-q\right] \\
&= \frac{q^a(1-q^b)}{q^a-q^b}\left[\frac{q^a(1-q^{b+1})}{q^a-q^{b+1}}\Gamma_q\left[\begin{matrix} \frac{3+a+b}{2}, \frac{1}{2} \\ \frac{3+b}{2}, \frac{1+a}{2} \end{matrix}\right] \frac{\langle \widetilde{\frac{3+b}{2}};q\rangle_{\frac{a}{2}}}{\langle \widetilde{\frac{1}{2}};q\rangle_{\frac{a}{2}}}\right. \\
&\quad \left. - \frac{q^{b+1}(1-q^a)}{q^a-q^{b+1}}\Gamma_q\left[\begin{matrix} \frac{3+a+b}{2}, \frac{1}{2} \\ \frac{2+b}{2}, \frac{2+a}{2} \end{matrix}\right] \frac{\langle \widetilde{\frac{2+b}{2}};q\rangle_{\frac{a+1}{2}}}{\langle \widetilde{\frac{1}{2}};q\rangle_{\frac{a+1}{2}}}\right] \\
&\quad - \frac{q^b(1-q^a)}{q^a-q^b}\left[\frac{q^{a+1}(1-q^b)}{q^{a+1}-q^b}\Gamma_q\left[\begin{matrix} \frac{3+a+b}{2}, \frac{1}{2} \\ \frac{2+b}{2}, \frac{2+a}{2} \end{matrix}\right] \frac{\langle \widetilde{\frac{2+b}{2}};q\rangle_{\frac{a+1}{2}}}{\langle \widetilde{\frac{1}{2}};q\rangle_{\frac{a+1}{2}}}\right. \\
&\quad \left. - \frac{q^b(1-q^{a+1})}{q^{a+1}-q^b}\Gamma_q\left[\begin{matrix} \frac{3+a+b}{2}, \frac{1}{2} \\ \frac{1+b}{2}, \frac{3+a}{2} \end{matrix}\right] \frac{\langle \widetilde{\frac{1+b}{2}};q\rangle_{\frac{a+3}{2}}}{\langle \widetilde{\frac{1}{2}};q\rangle_{\frac{a+3}{2}}}\right].
\end{aligned}
\tag{8.5}
$$

Theorem 8.1.4 *Andrews's q-analogue of Kummer's third summation formula* [348, p. 134] *of* [25, 1.9, p. 526] *and* [480, (15), p. 9]:

$$
{}_2\phi_2\left[\begin{matrix} a, 1-a \\ c, \widetilde{1} \end{matrix} | q; -q^c\right] = \Gamma_q\left[\begin{matrix} \frac{c}{2}, \frac{1+c}{2} \\ \frac{1+c-a}{2}, \frac{a+c}{2} \end{matrix}\right] \frac{\langle \widetilde{\frac{1+c-a}{2}}; q\rangle_{\frac{a}{2}}}{\langle \widetilde{\frac{c}{2}}; q\rangle_{\frac{a}{2}}}
$$

$$
\equiv \Gamma_{q^2}\left[\begin{matrix} \frac{c}{2}, \frac{1+c}{2} \\ \frac{1+c-a}{2}, \frac{a+c}{2} \end{matrix}\right]. \tag{8.6}
$$

Proof By [5, 1.9, p. 526]

$$
\text{LHS} = \frac{\langle \frac{1+c-a}{2}, \frac{a+c}{2}; q^2\rangle_\infty}{\langle c; q\rangle_\infty} = \frac{\langle \frac{1+c-a}{2}, \widetilde{\frac{1+c-a}{2}}, \frac{a+c}{2}, \widetilde{\frac{a+c}{2}}; q\rangle_\infty}{\langle \frac{1+c}{2}, \widetilde{\frac{1+c}{2}}, \frac{c}{2}, \widetilde{\frac{c}{2}}; q\rangle_\infty} = \text{RHS}. \tag{8.7}
$$

□

Remark 39 Rainville [425, p. 125, (2)] gives a formula which he calls Kummer's first formula. This formula is nothing else than a limit in the Euler-Pfaff-Kummer transformation (7.62).

If a is a negative integer, (8.6) may be recast as

Theorem 8.1.5

$$
{}_2\phi_2\left[\begin{matrix} -N, 1+N \\ c, \widetilde{1} \end{matrix} | q; -q^c\right] \equiv \sum_{k=0}^{N} \binom{N}{k}_q (-1)^k q^{2\binom{k}{2}+k(c-N)} \frac{\langle 1+N; q\rangle_k}{\langle c, \widetilde{1}; q\rangle_k}
$$

$$
= \begin{cases} \frac{\langle \frac{c-N}{2}; q^2\rangle_{\frac{N}{2}}}{\langle \frac{1+c}{2}; q^2\rangle_{\frac{N}{2}}}, & \text{if } N \text{ even}, \\ \frac{\langle \frac{c-N}{2}; q^2\rangle_{\frac{1+N}{2}}}{\langle \frac{c}{2}; q^2\rangle_{\frac{1+N}{2}}}, & \text{if } N \text{ odd}. \end{cases} \tag{8.8}
$$

We have expressed two of Andrew's formulas by the Γ_q function, and also expressed the corresponding finite sums by quotients of q-shifted factorials. We will continue with generalizations of these two formulas.

Theorem 8.1.6 *Compare with* [221, (II 17), p. 355]. *A q-analogue of the Watson-Schafheitlin summation formula* (3.55).

$$
{}_4\phi_3\left[\begin{matrix} \frac{c}{2}, \widetilde{\frac{c}{2}}, a, -N \\ \frac{-N+1+a}{2}, \widetilde{\frac{-N+1+a}{2}}, c \end{matrix} | q; q\right] = \begin{cases} \frac{\langle \frac{1}{2}, \frac{1+c-a}{2}; q^2\rangle_{\frac{N}{2}}}{\langle \frac{1-a}{2}, \frac{1+c}{2}; q^2\rangle_{\frac{N}{2}}}, & \text{if } N \text{ even}, \\ 0, & \text{if } N \text{ odd}. \end{cases} \tag{8.9}
$$

Proof The proof for the second case of (8.9) is relegated to the proof of (8.16). We will use the method in Andrews [28, p. 334]. For convenience, we will denote the LHS of (8.9) by

$$ {}_4\phi_3\left[\begin{matrix} \frac{c}{2}, \frac{\widetilde{c}}{2}, a, -n \\ \frac{-n+1+a}{2}, \frac{\widetilde{-n+1+a}}{2}, c \end{matrix} \middle| q; q\right]. $$

In Watson's transformation formula for a terminating very-well-poised ${}_8\phi_7$ series (7.105), we make the substitution

$$ \begin{aligned} &a \to \widetilde{-n}, \qquad b \to \frac{-n+1-a}{2}, \qquad c \to \frac{\widetilde{-n+1-a}}{2}, \\ &d \to \frac{c}{2}, \qquad e \to \frac{\widetilde{c}}{2}, \end{aligned} \tag{8.10} $$

to get

$$ \begin{aligned} &{}_8\phi_7\left[\begin{matrix} (\alpha') \\ (\beta') \end{matrix} \middle| q; q^{1+a-c}\right] \frac{\langle \widetilde{-n+1-\frac{c}{2}}, -n+1-\frac{c}{2}; q\rangle_n}{\langle \widetilde{-n+1}, -n+1-c; q\rangle_n} \\ &= {}_4\phi_3\left[\begin{matrix} \frac{c}{2}, \frac{\widetilde{c}}{2}, a, -n \\ \frac{-n+1+a}{2}, \frac{\widetilde{-n+1+a}}{2}, c \end{matrix} \middle| q; q\right], \end{aligned} \tag{8.11} $$

where

$$ (\alpha') \equiv \left(\widetilde{-n}, \frac{-n+1-a}{2}, \frac{\widetilde{-n+1-a}}{2}, \frac{c}{2}, \frac{\widetilde{c}}{2}, {}^{\frac{1}{4}}\widetilde{1-\frac{1}{2}n}, {}^{\frac{3}{4}}\widetilde{1-\frac{1}{2}n}, -n\right), \tag{8.12} $$

$$ (\beta') \equiv \left(\frac{-n+1+a}{2}, \frac{\widetilde{-n+1+a}}{2}, 1-n-\frac{c}{2}, \widetilde{1-n}-\frac{c}{2}, {}^{\frac{1}{4}}\frac{\widetilde{-n}}{2}, {}^{\frac{3}{4}}\frac{\widetilde{-n}}{2}, \widetilde{1}\right). \tag{8.13} $$

Now we have

$$ \langle \widetilde{a}; q^2\rangle_n = \langle \widetilde{{}^{\frac{1}{4}}a}, \widetilde{{}^{\frac{3}{4}}a}; q\rangle_n. \tag{8.14} $$

By the q-Dixon formula we have, assuming n even, that

$$ \begin{aligned} &{}_4\phi_3\left[\begin{matrix} \frac{c}{2}, \frac{\widetilde{c}}{2}, a, -n \\ \frac{-n+1+a}{2}, \frac{\widetilde{-n+1+a}}{2}, c \end{matrix} \middle| q; q\right] \\ &= {}_4\phi_3\left[\begin{matrix} \frac{-n+1-a}{2}, \widetilde{1-\frac{n}{2}}, \frac{c}{2}, -n \\ \frac{1-n+a}{2}, 1-n-\frac{c}{2}, \widetilde{-\frac{n}{2}} \end{matrix} \middle| q^2, q^{1+a-c}\right] \frac{\langle \widetilde{-n+1-\frac{c}{2}}, -n+1-\frac{c}{2}; q\rangle_n}{\langle \widetilde{-n+1}, -n+1-c; q\rangle_n} \\ &= \frac{\langle \widetilde{-n+1-\frac{c}{2}}, -n+1-\frac{c}{2}; q\rangle_n}{\langle \widetilde{-n+1}, -n+1-c; q\rangle_n} \frac{\langle 1-n, \frac{1-n+a-c}{2}; q^2\rangle_{\frac{n}{2}}}{\langle -n+1-\frac{c}{2}, \frac{1-n+a}{2}; q^2\rangle_{\frac{n}{2}}} \end{aligned} $$

$$\overset{\text{by (6.38)}}{=} \frac{\langle\frac{1}{2},\frac{1+c-a}{2};q^2\rangle_{\frac{n}{2}}}{\langle\frac{1-a}{2},\frac{1+c}{2};q^2\rangle_{\frac{n}{2}}}. \tag{8.15}$$

□

Equation (8.9) may be recast as

Theorem 8.1.7 *Compare with* [28, p. 334]:

$$\begin{aligned}
&{}_4\phi_3\left[\begin{matrix}\frac{c}{2},\widetilde{\frac{c}{2}},a,-n\\ \frac{-n+1+a}{2},\widetilde{\frac{-n+1+a}{2}},c\end{matrix}|q;q\right]\\
&=\frac{\langle\widetilde{1},1-c;q\rangle_\infty}{\langle\widetilde{1-n},-n+1-c;q\rangle_\infty}\\
&\times\frac{\langle\frac{-2n+3}{2},\frac{1+a}{2},\frac{-n+2-c}{2},\frac{a-n+1-c}{2};q^2\rangle_\infty}{\langle\frac{2-c}{2},\frac{1+a-n}{2},\frac{-n+3}{2},\frac{a+1-c}{2};q^2\rangle_\infty}.
\end{aligned}\tag{8.16}$$

Proof

$$\begin{aligned}
\text{LHS} &= {}_4\phi_3\left[\begin{matrix}\frac{-n+1-a}{2},\widetilde{1-\frac{n}{2}},\frac{c}{2},-n\\ \frac{1-n+a}{2},1-n-\frac{c}{2},\widetilde{-\frac{n}{2}}\end{matrix}|q^2,q^{1+a-c}\right]\frac{\langle\widetilde{-n+1-\frac{c}{2}},-n+1-\frac{c}{2};q\rangle_n}{\langle\widetilde{-n+1},-n+1-c;q\rangle_n}\\
&=\frac{\langle\widetilde{1},1-c;q\rangle_\infty\langle\frac{a+1}{2},-n+1,\frac{-n+2-c}{2},\frac{-n+a+1-c}{2};q^2\rangle_\infty}{\langle\widetilde{-n+1},-n+1-c;q\rangle_\infty\langle\frac{a-n+1}{2},\frac{-n+3}{2},\frac{2-c}{2},\frac{a+1-c}{2};q^2\rangle_\infty}\\
&=\frac{\langle\widetilde{1};q\rangle_\infty\langle\frac{a+1}{2},\frac{-n+1}{2},\frac{1-c}{2},\frac{-n+a+1-c}{2};q^2\rangle_\infty}{\langle\frac{a-n+1}{2},\frac{-n+1-c}{2},\frac{a+1-c}{2};q^2\rangle_\infty}\\
&=\frac{\langle\widetilde{1},1-c;q\rangle_\infty}{\langle\widetilde{1-n},-n+1-c;q\rangle_\infty}\frac{\langle\frac{-2n+3}{2},\frac{1+a}{2},\frac{-n+2-c}{2},\frac{a-n+1-c}{2};q^2\rangle_\infty}{\langle\frac{2-c}{2},\frac{1+a-n}{2},\frac{-n+3}{2},\frac{a+1-c}{2};q^2\rangle_\infty}.
\end{aligned}\tag{8.17}$$

□

Remark 40 The equivalent expression

$$\frac{\langle\frac{a+1}{2},\frac{-n+1}{2},\frac{1-c}{2},\frac{-n+a+1-c}{2};q^2\rangle_\infty}{\langle\frac{1}{2},\frac{a-n+1}{2},\frac{-n+1-c}{2},\frac{a+1-c}{2};q^2\rangle_\infty}\equiv q^{\frac{na}{2}}\Gamma_{q^2}\left[\begin{matrix}\frac{1}{2},\frac{c+1}{2},\frac{1+a-n}{2},\frac{1-a+n+c}{2}\\ \frac{a+1}{2},\frac{-n+1}{2},\frac{1-a+c}{2},\frac{1+n+c}{2}\end{matrix}\right] \tag{8.18}$$

for the RHS of (8.16) was given in [258, 1.4]. The simple proof uses the Euler formula (6.42) [190, p. 271]. This was the beginning of the investigation that led to the above formulas.

Now let $c\to-\infty$ in (8.16) and (8.9) to arrive at (8.1) and (8.3).

Corollary 8.1.8 *A q-analogue of Kummer's second summation formula*:

$$
{}_3\phi_2\left[\begin{matrix} a, -2N, \infty \\ \frac{1+a-2N}{2}, \widetilde{\frac{1+a-2N}{2}} \end{matrix} | q; q\right] = \frac{\langle \frac{1}{2}; q^2\rangle_N}{\langle \frac{1-a}{2}; q^2\rangle_N}. \tag{8.19}
$$

Proof Let $c \to \infty$ in (8.9). □

Corollary 8.1.9 *A first q-analogue of* (3.56).

$$
\begin{aligned}
{}_3\phi_2\left[\begin{matrix} \frac{c}{2}, \widetilde{\frac{c}{2}}, -N \\ c, \infty \end{matrix} | q; q\right] &\equiv \sum_{k=0}^{N} \binom{N}{k}_q q^{\binom{k}{2}+k(1-N)} \frac{(-1)^k \langle \frac{c}{2}; q^2\rangle_k}{\langle c; q\rangle_k} \\
&= \begin{cases} \frac{\langle \frac{1}{2}; q^2\rangle_{\frac{N}{2}} q^{\frac{Nc}{2}}}{\langle \frac{1+c}{2}; q^2\rangle_{\frac{N}{2}}}, & \text{if } N \text{ even}, \\ 0, & \text{if } N \text{ odd}. \end{cases}
\end{aligned} \tag{8.20}
$$

Proof Let $a \to \infty$ in (8.9). □

Corollary 8.1.10 *A second q-analogue of* (3.56). *Compare with* [221, ex. 3.4, p. 101]:

$$
\begin{aligned}
{}_3\phi_1\left[\begin{matrix} \frac{c}{2}, \widetilde{\frac{c}{2}}, -N \\ c \end{matrix} | q; -q^N\right] &\equiv \sum_{k=0}^{N} \binom{N}{k}_q \frac{(-1)^k \langle \frac{c}{2}; q^2\rangle_k}{\langle c; q\rangle_k} \\
&= \begin{cases} \frac{\langle \frac{1}{2}; q^2\rangle_{\frac{N}{2}}}{\langle \frac{1+c}{2}; q^2\rangle_{\frac{N}{2}}}, & \text{if } N \text{ even}, \\ 0, & \text{if } N \text{ odd}. \end{cases}
\end{aligned} \tag{8.21}
$$

Proof Let $a \to -\infty$ in (8.9). □

Theorem 8.1.11 *A balanced summation formula*, [28, p. 333], *compare with* [220, p. 237, II.19].

$$
\begin{aligned}
&{}_4\phi_3\left[\begin{matrix} \frac{c}{2}, \widetilde{\frac{c}{2}}, 1+n, -n \\ c+1-e, \widetilde{1}, e \end{matrix} | q; q\right] \\
&= \frac{\langle \widetilde{e}, e-c; q\rangle_\infty}{\langle \widetilde{e-n}, -n+e-c; q\rangle_\infty} \frac{\langle \frac{e+1+n}{2}, \frac{-n+e+1-c}{2}, -n+e; q^2\rangle_\infty}{\langle \frac{-n+e+1}{2}, \frac{n+1+e-c}{2}, e; q^2\rangle_\infty}.
\end{aligned} \tag{8.22}
$$

Proof We will again use Watson's [535] transformation formula for a terminating very-well-poised ${}_8\phi_7$ series. Now make the substitution

$$\begin{aligned}&a \to \widetilde{-n+e-1}, \qquad b \to -n+e-1, \qquad c \to \widetilde{-n},\\ &d \to \frac{c}{2}, \qquad e \to \frac{\widetilde{c}}{2},\end{aligned} \tag{8.23}$$

to obtain

$$\begin{aligned}&{}_8\phi_7\left[\begin{matrix}(\alpha')\\(\beta')\end{matrix}|q;q^{e+1-a-c}\right]\frac{\langle\widetilde{-n+e}-\frac{c}{2},-n+e-\frac{c}{2};q\rangle_n}{\langle\widetilde{-n+e},-n+e-c;q\rangle_n}\\ &\quad={}_4\phi_3\left[\begin{matrix}\frac{c}{2},\frac{\widetilde{c}}{2},1+n,-n\\ e,\widetilde{1},c+1-e\end{matrix}|q;q\right],\end{aligned} \tag{8.24}$$

where

$$(\alpha')\equiv\left(\widetilde{-n},-n+e-1,\widetilde{-n+e}-1,\frac{c}{2},\frac{\widetilde{c}}{2},{}^{\frac14}\frac{\widetilde{1+e-n}}{2},{}^{\frac34}\frac{\widetilde{1+e-n}}{2},-n\right), \tag{8.25}$$

$$(\beta')\equiv\left(e,\widetilde{e},e-n-\frac{c}{2},\widetilde{e-n}-\frac{c}{2},{}^{\frac14}\frac{\widetilde{-n+e-1}}{2},{}^{\frac34}\frac{\widetilde{-n+e-1}}{2},\widetilde{1}\right). \tag{8.26}$$

We know that

$$\langle\widetilde{a};q^2\rangle_n=\langle\widetilde{{}^{\frac14}a},\widetilde{{}^{\frac34}a};q\rangle_n. \tag{8.27}$$

Hence, we have

$$\begin{aligned}&{}_4\phi_3\left[\begin{matrix}\frac{c}{2},\frac{\widetilde{c}}{2},1+n,-n\\ e,\widetilde{1},c+1-e\end{matrix}|q;q\right]\\ &={}_4\phi_3\left[\begin{matrix}-n-1+e,\widetilde{\frac{1-n+e}{2}},\frac{c}{2},-n\\ e,e-n-\frac{c}{2},\widetilde{\frac{-n+e-1}{2}}\end{matrix}|q^2,q^{e+1+n-c}\right]\\ &\quad\times\frac{\langle\widetilde{-n+e}-\frac{c}{2},-n+e-\frac{c}{2};q\rangle_n}{\langle\widetilde{-n+e},-n+e-c;q\rangle_n}\\ &=\frac{\langle\widetilde{e},e-c;q\rangle_\infty\langle\frac{n+e+1}{2},-n+e,\frac{-n+e+1-c}{2};q^2\rangle_\infty}{\langle\widetilde{-n+e},-n+e-c;q\rangle_\infty\langle e,\frac{-n+e+1}{2},\frac{n+e+1-c}{2};q^2\rangle_\infty}\\ &=\frac{\langle\widetilde{e},e-c;q\rangle_\infty}{\langle\widetilde{e-n},-n+e-c;q\rangle_\infty}\end{aligned}$$

$$\times \frac{\langle \frac{-2n+2e-c}{2}, \frac{e+1+n}{2}, \frac{-n+e+1-c}{2}, \frac{2e-c}{2}, -n+e; q^2\rangle_\infty}{\langle \frac{2e-c}{2}, \frac{-2n+2e-c}{2}, \frac{-n+e+1}{2}, \frac{n+1+e-c}{2}, e; q^2\rangle_\infty} = \text{RHS}. \qquad (8.28)$$

□

Equation (8.22) may be recast as

Theorem 8.1.12 *A q-analogue of the Whipple formula* (3.57):

$$
\begin{aligned}
&{}_4\phi_3\left[\begin{matrix}\frac{c}{2}, \widetilde{\frac{c}{2}}, 1+2N, -2N \\ c+1-e, \widetilde{1}, e\end{matrix} | q; q\right] \\
&\equiv \sum_{k=0}^{2N} \binom{2N}{k}_q q^{\binom{k}{2}+k(1-2N)} \frac{(-1)^k \langle 1+2N; q\rangle_k \langle \frac{c}{2}; q^2\rangle_k}{\langle c+1-e, \widetilde{1}, e; q\rangle_k} \\
&= \frac{(-1)^N \mathrm{QE}(N^2+N+Nc-Ne)\langle \frac{e-2N}{2}, \frac{e+1-c}{2}; q^2\rangle_N}{\langle \frac{e+1}{2}, \frac{2+c-e}{2}; q^2\rangle_N}, \qquad (8.29)
\end{aligned}
$$

$$
\begin{aligned}
&{}_4\phi_3\left[\begin{matrix}\frac{c}{2}, \widetilde{\frac{c}{2}}, 2+2N, -2N-1 \\ c+1-e, \widetilde{1}, e\end{matrix} | q; q\right] \\
&\equiv \sum_{k=0}^{2N+1} \binom{2N+1}{k}_q q^{\binom{k}{2}-2kN} \frac{(-1)^k \langle 2+2N; q\rangle_k \langle \frac{c}{2}; q^2\rangle_k}{\langle c+1-e, \widetilde{1}, e; q\rangle_k} \\
&= \frac{(-1)^{N+1} \mathrm{QE}((N+1)^2+(c-e)(N+1))\langle \frac{e-2N-1}{2}, \frac{e-c}{2}; q^2\rangle_{N+1}}{\langle \frac{e}{2}, \frac{1+c-e}{2}; q^2\rangle_{N+1}}. \qquad (8.30)
\end{aligned}
$$

Now let $c \to -\infty$ in (8.29) and (8.30) to arrive at (8.8).

Corollary 8.1.13 *Another q-analogue of Kummer's third summation formula*:

$$
\begin{aligned}
{}_3\phi_2\left[\begin{matrix}-N, 1+N, \infty \\ c, \widetilde{1}\end{matrix} | q; q\right] &\equiv \sum_{k=0}^{N} \binom{N}{k}_q (-1)^k q^{\binom{k}{2}+k-Nk} \frac{\langle 1+N; q\rangle_k}{\langle c, \widetilde{1}; q\rangle_k} \\
&= \begin{cases} \dfrac{\langle \frac{c-N}{2}; q^2\rangle_{\frac{N}{2}}}{\langle \frac{1+c}{2}; q^2\rangle_{\frac{N}{2}}} q^{\frac{N^3}{2}+\frac{N}{2}}, & \text{if } N \text{ even}, \\ \dfrac{\langle \frac{c-N}{2}; q^2\rangle_{\frac{1+N}{2}}}{\langle \frac{c}{2}; q^2\rangle_{\frac{1+N}{2}}} q^{\binom{N+1}{2}}, & \text{if } N \text{ odd}. \end{cases} \qquad (8.31)
\end{aligned}
$$

Proof Let $c \to \infty$ in (8.29) and (8.30). □

We collect the results of this section in a table, similar to Section 3.7. Recall the table therein:

Dougall (3.50)	–	–
Slater (3.51)	–	–
Dixon-Schafheitlin	Watson-Schafheitlin	Whipple
$c \to -\infty$	$c \to \infty; a \to \infty$	$c \to \infty$
Kummer 1	Kummer 2; (3.56)	Kummer 3

As there is less space, we next write only the numbers of the formulas.

(7.111)	–	–
(7.112)	–	–
(7.113)	(8.9)	(8.29)+(8.30)
$c = \widetilde{\frac{a}{2}}; c \to -\infty; c \to \infty$	$c \to \pm\infty; a \to \pm\infty$	$c \to \infty; c \to -\infty$
(7.80); (8.39); (8.40)	(8.1); (8.19); (8.20); (8.21)	(8.31); (8.8)

The first class contains, from top to bottom, two very-well-poised balanced series, two well-poised series. The second class contains from top to bottom, one balanced series, two pseudo-balanced series (8.1), (8.19). The third class contains on top two balanced series.

8.2 Some quadratic q-hypergeometric transformations

The dynamics therefore corresponds to the formal equalities.

We are going to find q-analogues of the quadratic transformations (3.37) and (3.38). Because the proofs involve the q-binomial theorem, the obtained formulas are only formal. It should however be possible to find some analytic (meromorphic) continuation of the formulas outside the domain of convergence in the spirit of Kaneko [326]. There are some similar approaches to this formal procedure in the literature. In Chen and Liu's paper [125] a parameter augmentation method for a reciprocal of a q-shifted factorial was used to obtain q-summation formulas.

In [165] the equivalent approach (by the q-binomial theorem) of employing the q-exponential function E_q to obtain formulas for q-Laguerre polynomials was used. For a short discussion of generalization to n variables see [169]. Now back to quadratic transformations.

Theorem 8.2.1 *A q-analogue of Kummer's quadratic transformation* (3.37):

$$
\begin{aligned}
&{}_2\phi_1\big(a, b; 1+a-b|q; zq^{\frac{a+1}{2}-b}\big)\\
&\cong \sum_{m=0}^{\infty} \frac{\langle \frac{a}{2}, \frac{a+1}{2}-b, \widetilde{\frac{a}{2}}, \widetilde{\frac{a+1}{2}}; q\rangle_m (-z)^m q^{-\binom{m}{2}}}{\langle 1, 1+a-b; q\rangle_m (zq^{-m}; q)_{a+2m}}\\
&\equiv {}_5\phi_3\left[\begin{matrix} \frac{a}{2}, \frac{a+1}{2}-b, \widetilde{\frac{a}{2}}, \widetilde{\frac{a+1}{2}}, \infty \\ 1+a-b \end{matrix}|q; z||\begin{matrix} - \\ (zq^{-k}; q)_{a+2k} \end{matrix}\right].
\end{aligned} \tag{8.32}
$$

A more practical form is the finite version

$$
\begin{aligned}
&{}_2\phi_1\big(-n, b; 1-n-b|q; zq^{\frac{-n+1}{2}-b}\big)\\
&\cong \sum_{m=0}^{\frac{n}{2}} \frac{\langle \frac{-n}{2}, \frac{-n+1}{2}-b, \widetilde{\frac{-n}{2}}, \widetilde{\frac{-n+1}{2}}; q\rangle_m (-z)^m q^{-\binom{m}{2}}}{\langle 1, 1-n-b; q\rangle_m (zq^{-m}; q)_{-n+2m}}\\
&\equiv {}_5\phi_3\left[\begin{matrix} \frac{-n}{2}, \frac{-n+1}{2}-b, \widetilde{\frac{-n}{2}}, \widetilde{\frac{-n+1}{2}}, \infty \\ 1-n-b \end{matrix} |q; z|| \overline{(zq^{-k}; q)_{-n+2k}}\right], \quad n \text{ even}.
\end{aligned}
\tag{8.33}
$$

Proof By the q-binomial theorem, the RHS can be written as

$$
\sum_{m,k=0}^{\infty} \frac{\langle \frac{a}{2}, \frac{a+1}{2}-b, \widetilde{\frac{a}{2}}, \widetilde{\frac{a+1}{2}}; q\rangle_m \langle a+2m; q\rangle_k (-z)^m z^k q^{-mk-\binom{m}{2}}}{\langle 1, 1+a-b; q\rangle_m \langle 1; q\rangle_k}. \tag{8.34}
$$

The coefficient for z^n is

$$
\begin{aligned}
&\sum_{m=0}^{n} \frac{\langle \frac{a}{2}, \frac{a+1}{2}-b, \widetilde{\frac{a}{2}}, \widetilde{\frac{a+1}{2}}; q\rangle_m \langle a+2m; q\rangle_{n-m} (-1)^m q^{\frac{m^2+m}{2}-nm}}{\langle 1, 1+a-b; q\rangle_m \langle 1; q\rangle_{n-m}}\\
&= \frac{\langle a; q\rangle_n}{\langle 1; q\rangle_n} \sum_{m=0}^{n} \frac{\langle a+n, -n, \frac{a+1}{2}-b; q\rangle_m}{\langle 1+a-b, 1, \frac{a+1}{2}; q\rangle_m} q^m = \frac{\langle 1-b-n, a, \frac{a+1}{2}; q\rangle_n}{\langle 1+a-b, 1, \frac{1-a}{2}-n; q\rangle_n}\\
&= \frac{\langle b, a; q\rangle_n}{\langle 1+a-b, 1; q\rangle_n} q^{n(\frac{1+a}{2}-b)}.
\end{aligned}
\tag{8.35}
$$

□

To find the second q-analogue, we will use the following formula, which follows from [28].

Lemma 8.2.2 *An improved version of Verma* [521, p. 426, 2.2]:

$$
{}_4\phi_3\left[\begin{matrix} a+\frac{1}{2}+l, \widetilde{a+\frac{1}{2}+l}, 2a+n, -n \\ a+\frac{1}{2}, a+\frac{1}{2}, 2a+1+2l \end{matrix} |q; q\right]
= \begin{cases} q^{n(a+\frac{1}{2}+l)} \frac{\langle \frac{1}{2}, -l; q^2\rangle_{\frac{n}{2}}}{\langle a+\frac{1}{2}, 1+a+l; q^2\rangle_{\frac{n}{2}}}, & n \text{ even},\\ 0, & n \text{ odd}. \end{cases}
\tag{8.36}
$$

Theorem 8.2.3 *A q-analogue of the Gaussian formula* (3.38):

$$
{}_2\phi_1\left[\begin{matrix} a, -l \\ a+1+l, \end{matrix} |q^2, y^2\right]
$$

$$
\cong {}_5\phi_3\left[\begin{matrix} a+l+\frac{1}{2}, \widetilde{a+l+\frac{1}{2}}, a, \widetilde{a}, \infty \\ 2a+1+2l \end{matrix} |q; yq^{-a-l-\frac{3}{2}} || \overline{(yq^{-k-a-\frac{1}{2}-l}; q)_{2a+2k}}\right]. \tag{8.37}
$$

Proof

$$
\begin{aligned}
&{}_2\phi_1\left[\begin{matrix} a, -l \\ a+1+l, \end{matrix} |q^2, y^2\right] \\
&= \sum_{n=0}^{\infty} \frac{\langle a, -l; q^2\rangle_n}{\langle 1, a+1+l; q^2\rangle_n} y^{2n} \\
&= \sum_{n=0}^{\infty} \frac{\langle 2a; q\rangle_{2n} \langle \frac{1}{2}, -l; q^2\rangle_n}{\langle 1; q\rangle_{2n} \langle a+\frac{1}{2}, a+1+l; q^2\rangle_n} y^{2n} \\
&\overset{\text{by (8.36)}}{=} \sum_{n=0}^{\infty} \frac{\langle 2a; q\rangle_n y^n}{\langle 1; q\rangle_n} q^{-n(a+\frac{1}{2}+l)} \sum_{k=0}^{n} \frac{\langle -n, 2a+n, a+l+\frac{1}{2}, \widetilde{a+l+\frac{1}{2}}; q\rangle_k q^k}{\langle a+\frac{1}{2}, \widetilde{a+\frac{1}{2}}, 1, 2a+1+2l; q\rangle_k} \\
&= \sum_{k=0}^{\infty} \frac{\langle a+l+\frac{1}{2}, \widetilde{a+l+\frac{1}{2}}; q\rangle_k \langle a; q^2\rangle_k q^k}{\langle 1, 2a+1+2l; q\rangle_k} q^{-k(a+\frac{1}{2}+l)} \\
&\quad \times \sum_{n=0}^{\infty} \frac{\langle -n-k; q\rangle_k y^{n+k} \langle 2a+2k; q\rangle_n}{\langle 1; q\rangle_{n+k}} q^{-n(a+\frac{1}{2}+l)} \\
&= \sum_{k=0}^{\infty} \frac{\langle a+l+\frac{1}{2}, \widetilde{a+l+\frac{1}{2}}; q\rangle_k}{\langle 1, 2a+1+2l; q\rangle_k} \langle a; q^2\rangle_k q^{-k(a+l)-\frac{k^2}{2}} (-y)^k \\
&\quad \times \sum_{n=0}^{\infty} \frac{y^n \langle 2a+2k; q\rangle_n}{\langle 1; q\rangle_n} q^{-n(k+a+\frac{1}{2}+l)} \\
&= \sum_{k=0}^{\infty} \frac{\langle a+l+\frac{1}{2}, \widetilde{a+l+\frac{1}{2}}; q\rangle_k}{\langle 1, 2a+1+2l; q\rangle_k} \langle a; q^2\rangle_k \frac{q^{-k(a+l+1)-\frac{k^2}{2}} (-y)^k}{(yq^{-k-a-\frac{1}{2}-l}; q)_{2a+2k}} = \text{RHS}.
\end{aligned} \tag{8.38}
$$

□

For another q-analogue, see Gasper and Rahman [220, p. 68].

8.3 The Kummer $_2F_1(-1)$ formula and Jacobi's theta function

In (7.80) we used a q-analogue of the Dixon-Schafheitlin theorem to prove a q-analogue of Kummer's first summation formula [348], the Bailey-Daum summation formula.

We can easily find two related formulas.

Theorem 8.3.1 *A second q-analogue of Kummer's first summation formula*:

$$ {}_3\phi_3\left(a, b, \widetilde{1+\frac{a}{2}}; 1+a-b, \frac{\widetilde{a}}{2}, \infty|q; q^{1+\frac{a}{2}-b}\right) = \Gamma_q\left[\begin{matrix}1+a-b, 1+\frac{a}{2}\\ 1+a, 1+\frac{a}{2}-b\end{matrix}\right], \quad (8.39) $$

where $1+a-b \neq 0, -1, -2, \ldots$.

Proof Let $c \to -\infty$ in (7.113). This proof is a q-analogue of Bailey [56, p. 13]. □

Theorem 8.3.2 *A third q-analogue of Kummer's first summation formula. If* $a \equiv -n \vee b \equiv -n \pmod{\frac{2\pi i}{\log q}}$, *then we have*

$$ {}_4\phi_2\left(a, b, \widetilde{1+\frac{a}{2}}, \infty; 1+a-b, \frac{\widetilde{a}}{2}|q; q^{-\frac{a}{2}-b}\right) \cong \Gamma_q\left[\begin{matrix}1+a-b, 1+\frac{a}{2}\\ 1+a, 1+\frac{a}{2}-b\end{matrix}\right] q^{\frac{ab}{2}}, \quad (8.40) $$

where $1+a-b \neq 0, -1, -2, \ldots$.

Proof Let $c \to +\infty$ in (7.113). □

This formula is used in (8.49). In [166] we found the following special case of (7.80):

$$ \begin{aligned} &{}_2\phi_1(-2N, b; 1-2N-b|q; -q^{1-b}) \\ &\equiv \sum_{k=0}^{2N} \binom{2N}{k}_q q^{\binom{k}{2}+k(1-2N-b)} \frac{\langle b; q\rangle_k}{\langle 1-2N-b; q\rangle_k} \\ &\equiv 2\sum_{k=0}^{N-1} \binom{2N}{k}_q (-1)^k \frac{\langle b; q\rangle_k}{\langle 2N+b-k; q\rangle_k} + \binom{2N}{N}_q (-1)^N \\ &\quad \times \frac{\langle b; q\rangle_N}{\langle N+b; q\rangle_N} = \frac{\langle \widetilde{b}; q\rangle_N}{\langle N+b; q\rangle_N} \left\langle \frac{1}{2}; q^2\right\rangle_N; \qquad (8.41) \\ &{}_2\phi_1(-N, b; 1-N-b|q; -q^{1-b}) = 0, \quad N \text{ odd}. \end{aligned} $$

Remark 41 For $b = +\infty$ this is a result of Gauß [222], [231]. The previous equation is a q-analogue of Petkovsek, Wilf and Zeilberger [412, p. 43]. The previous formula was already given by Karlsson and Srivastava [483, p. 250, (47)] without proof.

The following corollary enables us to find a relation for Γ_q functions with negative integer argument.

Corollary 8.3.3

$$\lim_{n\to N}\Gamma_q\begin{bmatrix}1-2N-b,1-n\\1-2n,1-N-b\end{bmatrix}=\frac{\widetilde{\langle 1-N;q\rangle_N}}{\langle b+N;q\rangle_N}\left\langle\frac{1}{2};q^2\right\rangle_N q^{\binom{N}{2}+Nb}. \tag{8.42}$$

Proof Just equate (7.80) and (8.41). □

Corollary 8.3.4

$$\lim_{n\to N}\Gamma_q\begin{bmatrix}1-n\\1-2n\end{bmatrix}=\frac{\widetilde{\langle 1-N;q\rangle_N}}{(1-q)^N}\left\langle\frac{1}{2};q^2\right\rangle_N(-1)^N q^{-N^2}. \tag{8.43}$$

We give another proof of (8.43):

Proof We define

$$H(x,q)\equiv\theta_1\left(\frac{-ix}{2}\log q,\sqrt{q}\right). \tag{8.44}$$

$H(x,q)$ has zeros for $x\in\mathbb{Z}$. We want to compute the limit

$$\lim_{x\to N}\frac{H(2N,q)}{H(N,q)},\qquad N>0. \tag{8.45}$$

With Mathematica we get

$$\frac{\frac{\partial H}{\partial x}(2N,q)}{\frac{\partial H}{\partial x}(N,q)}=(-1)^N q^{-\frac{3N^3}{2}}, \tag{8.46}$$

where $\frac{\partial H}{\partial x}$ denotes the partial derivative with respect to x. Next we put

$$\epsilon\equiv x-N, \tag{8.47}$$

where ϵ denotes an arbitrarily small number. By the Taylor formula we get

$$\begin{aligned}\frac{H(2x,q)}{H(x,q)}&=\frac{H(2N,q)+2\epsilon\frac{\partial H}{\partial x}(2N,q)+2\epsilon^2\frac{\partial H}{\partial x}^2(2N,q)+\cdots}{H(N,q)+\epsilon\frac{\partial H}{\partial x}(N,q)+\frac{1}{2}\epsilon^2\frac{\partial H}{\partial x}^2(N,q)+\cdots}\\&\approx\frac{2\epsilon\frac{\partial H}{\partial x}(2N,q)+A(x,q)\epsilon^3}{\epsilon\frac{\partial H}{\partial x}(N,q)+B(x,q)\epsilon^2}\overset{\text{by (8.46)}}{\approx}2(-1)^N q^{-\frac{3N^3}{2}},\end{aligned} \tag{8.48}$$

where $A(x,q)$ and $B(x,q)$ are bounded functions in the neighbourhood of $(2N,q)$ and (N,q), respectively. □

8.4 Another proof of the q-Dixon formula

We show that the q-analogue of the Bailey proof [56, p. 13], [425, 93] of (7.113) requires another q-analogue of Kummer's first summation formula, namely (8.40):

$$
\begin{aligned}
&\Gamma_q\left[\begin{matrix} a,b,c,1+\frac{a}{2}+\beta \\ 1+a-b,1+a-c,\frac{a}{2}+\beta \end{matrix}\right] {}_4\phi_3\left[\begin{matrix} a,b,c,\widetilde{1+\frac{1}{2}a} \\ 1+a-b,1+a-c,\widetilde{\frac{1}{2}a} \end{matrix}|q;q^{1+\frac{a}{2}-b-c}\right] \\
&=\sum_n \Gamma_q\left[\begin{matrix} a+n,b+n,c+n,1+\frac{a}{2}+n+\beta \\ 1+a-b+n,1+a-c+n,1+n,\frac{a}{2}+n+\beta \end{matrix}\right] q^{n(1+\frac{a}{2}-b-c)} \\
&=\sum_n \Gamma_q\left[\begin{matrix} a+n,b+n,c+n,1+\frac{a}{2}+n+\beta \\ 1+a+2n,1+a-b-c,1+n,\frac{a}{2}+n+\beta \end{matrix}\right] \\
&\quad\times {}_2\phi_1\left[\begin{matrix} b+n,c+n \\ 1+a+2n \end{matrix}|q;q^{1+a-b-c}\right] q^{n(1+\frac{a}{2}-b-c)} \\
&=\sum_{n,m} q^{n(1+\frac{a}{2}-b-c)+m(1+a-b-c)} \\
&\quad\times \Gamma_q\left[\begin{matrix} a+n,b+n+m,c+n+m,1+\frac{a}{2}+n+\beta \\ 1+m,1+a+2n+m,1+a-b-c,1+n,n+\frac{a}{2}+\beta \end{matrix}\right] \\
&=\sum_{p=0}^{\infty}\sum_{n=0}^{p} q^{n(1+\frac{a}{2}-b-c)} q^{(p-n)(1+a-b-c)} \\
&\quad\times \Gamma_q\left[\begin{matrix} a+n,b+p,c+p,1+\frac{a}{2}+n+\beta \\ 1+a+n+p,1+a-b-c,1+p-n,1+n,n+\frac{a}{2}+\beta \end{matrix}\right] \\
&=\sum_{p=0}^{\infty} q^{p(1+a-b-c)} \Gamma_q\left[\begin{matrix} a,b+p,c+p,1+\frac{a}{2}+\beta \\ 1+p,1+a+p,1+a-b-c,\frac{a}{2}+\beta \end{matrix}\right] \\
&\quad\times \sum_{n=0}^{p} \frac{\langle a,-p,\widetilde{1+\frac{a}{2}};q\rangle_n}{\langle 1,1+a+p,\widetilde{\frac{a}{2}};q\rangle_n} q^{n(-\frac{a}{2}+p)-\binom{n}{2}}(-1)^n \\
&=\sum_{p=0}^{\infty} \Gamma_q\left[\begin{matrix} a,b+p,c+p,1+\frac{a}{2},1+\frac{a}{2}+\beta \\ 1+p,1+a-b-c,1+a,1+\frac{a}{2}+p,\frac{a}{2}+\beta \end{matrix}\right] q^{p(1+\frac{a}{2}-b-c)} \\
&=\Gamma_q\left[\begin{matrix} a,b,c,1+\frac{a}{2}+\beta \\ 1+a,1+a-b-c,\frac{a}{2}+\beta \end{matrix}\right] {}_2\phi_1\left[\begin{matrix} b,c \\ 1+\frac{a}{2} \end{matrix}|q;q^{1+\frac{a}{2}-b-c}\right] \\
&=\Gamma_q\left[\begin{matrix} a,b,c,1+\frac{a}{2},1+\frac{a}{2}-b-c,1+\frac{a}{2}+\beta \\ 1+a,1+a-b-c,1+\frac{a}{2}-b,1+\frac{a}{2}-c,\frac{a}{2}+\beta \end{matrix}\right],
\end{aligned}
\tag{8.49}
$$

where $\beta \equiv \frac{\log(-1)}{\log q}$. The proof of (7.113) is finished again.

8.5 A finite version of the q-Dixon formula

The following formula will be useful in conjunction with reduction formulas fur multiple q-series.

Corollary 8.5.1 *Finite version of the q-Dixon formula*

$$\begin{cases} {}_4\phi_3\left[\begin{matrix} -2k, b, c, \widetilde{1-k} \\ 1-2k-b, 1-2k-c, \widetilde{-k}\end{matrix} |q, q^{1-k-b-c}\right] \\ \quad \equiv \sum_{j=0}^{2k} \binom{2k}{j}_q \frac{\langle b, c, \widetilde{1-k}; q\rangle_j (-1)^j \mathrm{QE}(\binom{j}{2} + j(1-3k-b-c))}{\langle 1-2k-b, 1-2k-c, \widetilde{-k}; q\rangle_j} \\ \quad = \frac{\langle 1-2k-b-c, \widetilde{1-k}, \widetilde{b}; q\rangle_k}{\langle 1-2k-c, \widetilde{1-k-b}, b+k; q\rangle_k} \langle \tfrac{1}{2}; q^2\rangle_k; \\ {}_4\phi_3\left[\begin{matrix} -k, b, c, \widetilde{1-\frac{k}{2}} \\ 1-k-b, 1-k-c, \widetilde{-\frac{k}{2}}\end{matrix} |q, q^{1-\frac{k}{2}-b-c}\right] = 0, \quad k \text{ odd.} \end{cases} \tag{8.50}$$

Proof Use the formula (8.42). □

8.6 The Jackson summation formula for a finite, 2-balanced, well-poised ${}_5\phi_4$ series

The following formula needs a proof.

Theorem 8.6.1 *Jacksons theorem* [306] *for a finite,* 2*-balanced, well-poised* ${}_5\phi_4$ *series*:

$$\begin{aligned} &{}_5\phi_4\left[\begin{matrix} -2n, b, c, d, 1-3n-b-c-d \\ 1-2n-b, 1-2n-c, 1-2n-d, b+c+d+n\end{matrix} |q; q^2\right] \\ &= \Gamma_q\left[\begin{matrix} 1+2n, b+c+2n, c+d+2n, b+d+2n, b+n, c+n, d+n, b+c+d+n \\ 1+n, b+c+n, c+d+n, b+d+n, b+2n, c+2n, d+2n, b+c+d+2n\end{matrix}\right]. \end{aligned} \tag{8.51}$$

We only show that this formula is equivalent to the Jacksonian form. We have

$$\begin{aligned} &{}_5\phi_4\left[\begin{matrix} -2n, b, c, d, 1-3n-b-c-d \\ 1-2n-b, 1-2n-c, 1-2n-d, b+c+d+n\end{matrix} |q; q^2\right] \\ &= q^{-n(b+c+d)-\frac{n(3n-1)}{2}}(-1)^n \\ &\quad \times \frac{\langle b, c, d, 1-3n-b-c-d, n+1, b+c+n, c+d+n, b+d+n; q\rangle_n}{\langle b, c, d, 1-3n-b-c-d; q\rangle_{2n}} \\ &= \frac{\langle 1+n, b+c+n, c+d+n, b+d+n; q\rangle_n}{\langle b+n, c+n, d+n, b+c+d+n; q\rangle_n} = \text{RHS}. \end{aligned} \tag{8.52}$$

Both Jackson [306] and Carlitz [105] tried in vain to prove formula (8.51).

8.7 The Jackson finite q-analogue of the Dixon formula

In 1941 Jackson [306] published a finite q-analogue of the Dixon formula, for n even. Carlitz [105] gave the corresponding formula for n odd.

Theorem 8.7.1 [220, p. 237], [306]:

$$ {}_3\phi_2\left[\begin{matrix}-2n, b, c\\ 1-2n-b, 1-2n-c\end{matrix}\middle| q; q^{2-n-b-c}\right] = \frac{\langle b, c; q\rangle_n \langle 1, b+c; q\rangle_{2n}}{\langle 1, b+c; q\rangle_n \langle b, c; q\rangle_{2n}}. \tag{8.53}$$

Theorem 8.7.2 *Carlitz* 1969 [105]:

$$ {}_3\phi_2\left[\begin{matrix}-2n-1, b, c\\ -2n-b, -2n-c\end{matrix}\middle| q; q^{2-n-b-c}\right] = \frac{\langle b, c; q\rangle_{n+1} \langle 1, b+c; q\rangle_{2n+1}}{\langle 1; q\rangle_n \langle b+c; q\rangle_{n+1} \langle b, c; q\rangle_{2n+1}}. \tag{8.54}$$

Corollary 8.7.3

$$ {}_3\phi_1\left(-2n, b, \infty; 1-2n-b|q; q^{1+n-b}\right) = \frac{\langle \widetilde{1}; q\rangle_N}{\langle n+b; q\rangle_n}\left\langle \frac{1}{2}; q^2\right\rangle_n. \tag{8.55}$$

Proof We let $c \to \infty$ in (8.53). □

The last formula can also be written in the form

$$ {}_3\phi_1\left(-2n, b, \infty; 1-2n-b|q; q^{1+n-b}\right) = \Gamma_q\left[\begin{matrix}1-2n-b, 1-n\\ 1-2n, 1-n-b\end{matrix}\right] q^{-\binom{n}{2}-bn}\frac{\langle \widetilde{1}; q\rangle_n}{\langle \widetilde{1-n}; q\rangle_n}. \tag{8.56}$$

8.8 Other examples of q-special functions

Jackson [294] introduced the following functions:

Definition 139 The two Jackson q-Bessel functions are defined by

$$ \mathrm{J}_\alpha^{(1)}(z;q) \equiv \frac{\langle \alpha+1; q\rangle_\infty}{\langle 1; q\rangle_\infty}\left(\frac{z}{2}\right)^\alpha {}_2\phi_1\left(\infty, \infty; \alpha+1|q, -\frac{z^2}{4}\right), \tag{8.57}$$

$$ \mathrm{J}_\alpha^{(2)}(z;q) = \frac{\langle \alpha+1; q\rangle_\infty}{\langle 1; q\rangle_\infty}\left(\frac{z}{2}\right)^\alpha {}_0\phi_1\left(-; \alpha+1|q, -\frac{z^2 q^{\alpha+1}}{4}\right). \tag{8.58}$$

The Hahn-Exton-q-Bessel function is defined by

$$ \mathrm{J}_\alpha^{(3)}(z;q) \equiv \frac{\langle \alpha+1; q\rangle_\infty}{\langle 1; q\rangle_\infty}\left(\frac{z}{2}\right)^\alpha {}_1\phi_1\left(\infty; \alpha+1|q, \left(\frac{z}{2}\right)^2\right). \tag{8.59}$$

Definition 140 The q-Krawtchouk polynomials are defined by

$$K_n(x;a,b;q)=\frac{\langle a,b;q\rangle_n}{\langle -1;q\rangle_n}q^{-n}{}_2\phi_2(-n,x;-a,-b|q,q). \tag{8.60}$$

The q-Hahn polynomials are defined by

$$Q_n(x;a,b,N;q)\equiv {}_3\phi_2(-n,a+b+n+1,-x;a+1,-N|q,q). \tag{8.61}$$

The Askey-Wilson (q-Wilson) polynomials are defined by

$$\begin{aligned}&p_n\big(\cos\theta;q^a,q^b,q^c,q^d\mid q\big)\\&\quad\equiv q^{-an}\langle a+b,a+c,a+d;q\rangle_n\\&\qquad\times {}_4\phi_3(-n,a+b+c+d+n-1,a-iy,a+iy;a+b,a+c,a+d|q,q),\end{aligned} \tag{8.62}$$

where $e^{i\theta}=q^{iy}$.

Definition 141 The Al-Salam-Chihara polynomials are obtained from the Askey-Wilson polynomials by the limit transition $c,d\to\infty$.

The continuous big q-Hermite polynomials are obtained from the Askey-Wilson polynomials by the limit transition $b,c,d\to\infty$.

The continuous q-Hermite polynomials are obtained from the Askey-Wilson polynomials by the limit transition $a,b,c,d\to\infty$.

Definition 142 The continuous q-Jacobi polynomials are defined by

$$\begin{aligned}P_n^{(\alpha,\beta)}(x|q)\equiv\frac{\langle\alpha+1;q\rangle_n}{\langle 1;q\rangle_n}{}_4\phi_3\bigg(&-n,n+\alpha+\beta+1,\frac{2\alpha+1}{4}-ix,\\&\frac{2\alpha+1}{4}+ix;\alpha+1,\frac{\widetilde{\alpha+\beta+1}}{2},\frac{\widetilde{\alpha+\beta+3}}{2}|q,q\bigg).\end{aligned} \tag{8.63}$$

The q-Racah polynomials are defined by

$$\begin{aligned}&p_n(\lambda_j)_q\\&\quad\equiv {}_4\phi_3(-n,a+b+n+1,-j,j+c+d+1;a+1,c+1,b+d+1|q,q).\end{aligned} \tag{8.64}$$

The q-Lommel polynomials are defined by

$$R_{n,q}(x,\alpha)\equiv\langle\alpha;q\rangle_n(2x)^n{}_4\phi_3\bigg(\Delta(q;2;-n);\alpha,-n,1-\alpha-n|q,-\frac{1}{x^2}\bigg), \tag{8.65}$$

$x\neq 0$. This series is 0-balanced.

8.9 q-analogues of two formulas by Brown and Eastham

Brown and Eastham have found two new reformulations of hypergeometric formulas in their recent paper [82].

The first is a restatement of Kummer's first summation formula:

$$
{}_2\mathrm{F}_1(a,b;1+a-b;-1)=2\cos\left(\frac{1}{2}\pi a\right)\Gamma\left[\begin{matrix}-a,1+a-b\\ -\frac{a}{2},1+\frac{a}{2}-b\end{matrix}\right]. \tag{8.66}
$$

The second one can be written as a new form of Dixon's formula.

$$
\begin{aligned}
&{}_3\mathrm{F}_2(a,b,c;1+a-b,1+a-c)\\
&=2\cos\left(\frac{1}{2}\pi a\right)\Gamma\left[\begin{matrix}1+a-c,-a\\ -\frac{a}{2},1+\frac{a}{2}-c\end{matrix}\right]{}_2\mathrm{F}_1\left(\frac{a}{2},c;1+a-b;1\right).
\end{aligned} \tag{8.67}
$$

Equation (8.66) occurred in a slightly different version in [412, p. 43].

It is now a simple matter to deduce the following two formulas.

Theorem 8.9.1 *We put* $\beta\equiv\frac{\log(-1)}{\log q}$. *A* q*-analogue of* (8.66) *is given by*

$$
\begin{aligned}
{}_2\phi_1\big(a,b;1+a-b|q,-q^{1-b}\big)&=\Gamma_q\left[\begin{matrix}1+a-b,-a,1+\frac{a}{2}+\beta,1-b+\beta\\ -\frac{a}{2},1+\frac{a}{2}-b,1+\frac{a}{2}-b+\beta,1+\beta\end{matrix}\right]\\
&\quad\times q^{-\frac{a}{4}}\frac{\theta_1((\frac{ia}{2})\log q,\sqrt{q})}{\theta_1(\frac{ia}{4}\log q,\sqrt{q})}.
\end{aligned} \tag{8.68}
$$

Proof

$$
\mathrm{LHS}=\Gamma_q\left[\begin{matrix}1+a-b,1+\frac{a}{2}\\ 1+a,1+\frac{a}{2}-b\end{matrix}\right]\frac{\langle\widetilde{1+\frac{a}{2}-b},\widetilde{1};q\rangle_\infty}{\langle 1+\frac{a}{2},\widetilde{1-b};q\rangle_\infty}\overset{\text{by (6.260)}}{=}\mathrm{RHS}. \tag{8.69}
$$

□

Theorem 8.9.2 *A* q*-analogue of* (8.67) *is given by*

$$
\begin{aligned}
&{}_4\phi_3\left[\begin{matrix}a,b,c,\widetilde{1+\frac{1}{2}a}\\ 1+a-b,1+a-c,\widetilde{\frac{1}{2}a}\end{matrix}|q,q^{1+\frac{a}{2}-b-c}\right]\\
&=\frac{\theta_1(\frac{ia}{2}\log q,\sqrt{q})}{\theta_1(\frac{ia}{4}\log q,\sqrt{q})}\Gamma_q\left[\begin{matrix}1+a-c,-a\\ -\frac{a}{2},1+\frac{a}{2}-c\end{matrix}\right]q^{-\frac{a}{4}}\\
&\quad\times{}_2\phi_1\left(\frac{a}{2},c;1+a-b|q,q^{1-b-c+\frac{a}{2}}\right).
\end{aligned} \tag{8.70}
$$

Proof

$$\text{LHS} = \frac{\langle 1+a, 1+a-b-c, 1+\frac{a}{2}-b, 1+\frac{a}{2}-c; q\rangle_\infty}{\langle 1+a-b, 1+a-c, 1+\frac{a}{2}, 1+\frac{a}{2}-b-c; q\rangle_\infty}$$

$$= \Gamma_q\left[\begin{matrix} -a, c-\frac{a}{2} \\ -\frac{a}{2}, c-a \end{matrix}\right] {}_2\phi_1\left(\frac{a}{2}, c; 1+a-b|q, q^{1-b-c+\frac{a}{2}}\right)$$

$$\times \frac{\theta_1(\frac{ia}{2}\log q, \sqrt{q})\theta_1(\frac{-i}{2}(c-\frac{a}{2})\log q, \sqrt{q})}{\theta_1(\frac{ia}{4}\log q, \sqrt{q})\theta_1(\frac{-i}{2}(c-a)\log q, \sqrt{q})} = \text{RHS}. \quad (8.71)$$

□

8.10 The q-analogue of Truesdell's function

Clifford Truesdell (1919–2000) was an American physicist with a special fondness for special functions. Influenced by his teacher Bateman, he wrote in 1948 the book *A unified theory of special functions based upon the functional equation* $\frac{\partial}{\partial z}F(z,\alpha) = F(z,\alpha+1)$ [509], where he treated the special functions in a completely new way. This style is a mix of differences and derivation, as the book title says. We know that Toscano treated special functions using difference methods. Although Truesdell knew Italian, he did not cite Toscano, through the methods of these two authors still have some similarities. Truesdell has not published anything more about special functions; instead he became one of the biggest experts on theoretical mechanics of his time. The following function is very similar to the MacRobert E-function.

Definition 143 The generalized q-hypergeometric function (in Γ_q notation) is defined by

$${}_p\Phi_{p-1}\left[\begin{matrix} a_1, \ldots, a_p \\ b_1, \ldots, b_{p-1} \end{matrix} |q, z\right]$$

$$\equiv \sum_{k=0}^{\infty} \Gamma_q\left[\begin{matrix} a_1+k, \ldots, a_p+k \\ b_1+k, \ldots, b_{p-1}+k, 1+k \end{matrix}\right] z^k$$

$$\equiv \Gamma_q\left[\begin{matrix} a_1, \ldots, a_p \\ b_1, \ldots, b_{p-1}, 1 \end{matrix}\right] \sum_{k=0}^{\infty} \frac{\langle a_1, \ldots, a_p; q\rangle_k}{\langle 1, b_1, \ldots, b_{p-1}; q\rangle_k} z^k. \quad (8.72)$$

The q-analogue of Truesdell's function [509, p. 17] is defined by

Definition 144

$$F(z, \alpha, q) \equiv {}_p\Phi_{p-1}\left[\begin{matrix} \alpha+a_1, \ldots, \alpha+a_p \\ \alpha+b_1, \ldots, \alpha+b_{p-1} \end{matrix} |q; z\right]$$

$$\equiv \sum_{k=0}^{\infty} \Gamma_q \begin{bmatrix} \alpha + a_1 + k, \ldots, \alpha + a_p + k \\ \alpha + b_1 + k, \ldots, \alpha + b_{p-1} + k, 1 + k \end{bmatrix} z^k. \tag{8.73}$$

Theorem 8.10.1 *A q-analogue of the Truesdell equation* [509, p. 15]:

$$\mathrm{D}_{q,z} F(z, \alpha, q) = F(z, \alpha + 1, q). \tag{8.74}$$

If $p = 2$ and $a, b \neq -\mathbb{N}$, we can take limits of the summands with singular Γ_q values to obtain a q-analogue of the improved version of [522, p. 118]

$$\begin{aligned} &{}_2\Phi_1 \begin{bmatrix} a, b \\ -N \end{bmatrix} |q; z \end{bmatrix} \\ &= z^{N+1} \frac{\Gamma_q(a + N + 1)\Gamma_q(b + N + 1)}{\{N + 1\}_q!} {}_2\phi_1 \begin{bmatrix} a + N + 1, b + N + 1 \\ N + 2 \end{bmatrix} |q; z \end{bmatrix}. \end{aligned} \tag{8.75}$$

8.11 The Bailey transformation for q-series

The following application of Bailey's transform of Exton [194, p. 81] takes the following form in the new notation.

Definition 145 Let us suppose that

$$(u),\ (e),\ (v),\ (f),\ (d),\ (g),\ (a),\ (h)$$

are vectors with U, E, V, F, D, G, A, H elements satisfying

$$U + D + V = 1 + E + F + G. \tag{8.76}$$

Furthermore, let

$$\begin{aligned} u_r &\equiv \frac{\langle (u); q \rangle_r u^r}{\langle (e); q \rangle_r \langle 1; q \rangle_r}, & v_r &\equiv \frac{\langle (v); q \rangle_r v^r}{\langle (f); q \rangle_r}, \\ \delta_r &\equiv \frac{\langle (d); q \rangle_r x^r}{\langle (g); q \rangle_r}, & \alpha_r &\equiv \frac{\langle (a); q \rangle_r y^r}{\langle (h); q \rangle_r \langle 1; q \rangle_r}. \end{aligned} \tag{8.77}$$

Then

Theorem 8.11.1

$$\begin{aligned}\gamma_n &= \sum_{s=0}^{\infty} \delta_{s+n} u_s v_{s+2n} = \sum_{s=0}^{\infty} \frac{\langle (d); q\rangle_{s+n} x^{s+n} \langle (u); q\rangle_s u^s \langle (v); q\rangle_{s+2n} \varphi^{s+2n}}{\langle (g); q\rangle_{s+n} \langle (e), 1; q\rangle_s \langle (f); q\rangle_{s+2n}} \\ &= \frac{\langle (d); q\rangle_n \langle (v); q\rangle_{2n}}{\langle (g); q\rangle_n \langle (f); q\rangle_{2n}} x^n \varphi^{2n} \sum_{s=0}^{\infty} \frac{\langle (d+n), (u), (v+2n); q\rangle_s (xu\varphi)^s}{\langle (g+n), (e), (f+2n), 1; q\rangle_s} \\ &= \frac{\langle (d); q\rangle_n \langle (v); q\rangle_{2n}}{\langle (g); q\rangle_n \langle (f); q\rangle_{2n}} x^n \varphi^{2n} {}_{U+D+V}\phi_{E+F+G}\left[\begin{matrix}(u), (d+n), (v+2n) \\ (e), (g+n), (f+2n)\end{matrix} \Big| q, xu\varphi\right],\end{aligned} \tag{8.78}$$

$$\begin{aligned}\beta_n &= \sum_{r=0}^{n} \alpha_r u_{n-r} v_{n+r} = \sum_{r=0}^{n} \frac{\langle (a); q\rangle_r y^r \langle (u); q\rangle_{n-r} u^{n-r} \langle (v); q\rangle_{n+r} \varphi^{n+r}}{\langle (h), 1; q\rangle_r \langle (e), 1; q\rangle_{n-r} \langle (f); q\rangle_{n+r}} \\ &= \frac{\langle (u), (v); q\rangle_n}{\langle (e), (f), 1; q\rangle_n} u^n \varphi^n \\ &\quad \times \sum_{r=0}^{n} \frac{\langle (a), (1-n-e), (v+n), -n; q\rangle_r y^r u^{-r} \varphi^r (-1)^{r(U-E-1)}}{\langle (h), 1, (1-n-u), (f+n); q\rangle_r} \\ &\quad \times \mathrm{QE}\left(\binom{r}{2}(U-E-1) + r\left(\sum_{i=1}^{E} e_i - \sum_{i=1}^{U} u_i + (U-E)(1-n) + n\right)\right) \\ &= \frac{\langle (u), (v); q\rangle_n}{\langle (e), (f), 1; q\rangle_n} u^n \varphi^n {}_{A+B+E+V+1}\phi_{C+U+F+H}\left[\begin{matrix}(\theta) \\ (\epsilon)\end{matrix} \Big| q, \frac{y\varphi}{u} \right. \\ &\quad \left. \times (-1)^{C+F+H-B-A-V+1} \mathrm{QE}\left(\sum_{i=1}^{E} e_i - \sum_{i=1}^{U} u_i + (U-E)(1-n) + n\right)\right],\end{aligned} \tag{8.79}$$

where

$$(\theta) \equiv \left(-n, (a), (1-n-e), (v+n), \infty_1, \ldots, \infty_B\right), \tag{8.80}$$

$$(\epsilon) \equiv \left((h), (1-n-u), (f+n), \infty_1, \ldots, \infty_C\right), \tag{8.81}$$

$$B - C = F + H - A - V + 1, \tag{8.82}$$

and $B = 0 \vee C = 0$. *B and C give the number of* ∞ *in* ${}_{A+B+E+V+1}\phi_{C+U+F+H}$ *and they are calculated by* (8.82).

According to Bailey's transform [466, p. 59],

$$\sum_{n=0}^{\infty} \alpha_n \gamma_n = \sum_{n=0}^{\infty} \beta_n \delta_n, \tag{8.83}$$

and we get

$$\begin{aligned}
&\sum_{n=0}^{\infty} \frac{\langle (a), (d); q\rangle_n \langle (v); q\rangle_{2n}}{\langle (h), (g), 1; q\rangle_n \langle (f); q\rangle_{2n}} y^n x^n \varphi^{2n} \\
&\quad \times {}_{U+D+V}\phi_{E+F+G}\left[\begin{matrix}(u), (d+n), (v+2n) \\ (e), (g+n), (f+2n)\end{matrix} \middle| q, xz\varphi\right] \\
&= \sum_{n=0}^{\infty} \frac{\langle (d), (u), (v); q\rangle_n}{\langle (e), (f), (g), 1; q\rangle_n} z^n \varphi^n x^n \\
&\quad \times {}_{A+B+E+V+1}\phi_{C+U+F+H}\left[\begin{matrix}(\theta) \\ (\epsilon)\end{matrix} \middle| q, \frac{y\varphi}{z}(-1)^{C+F+H-B-A-V+1} \right. \\
&\quad \left. \times \operatorname{QE}\left(\sum_{i=1}^{E} e_i - \sum_{i=1}^{U} u_i + (U-E)(1-n) + n\right)\right].
\end{aligned} \tag{8.84}$$

This equation is a q-analogue of [466, p. 60] and [193, p. 23].

We could also choose an arbitrary q-hypergeometric series in the LHS of (8.84), which would result in a similar formula.

8.12 q-Taylor formulas with remainder; the mean value theorem

Our next aim is to prove a new q-Taylor formula, which is a generalization of Jackson's 1909 [301] q-Taylor formula.

This q-Taylor formula is equivalent to the Taylor formula, which can be seen from the following formula.

Theorem 8.12.1 *The following equation relates the n-th q-difference operator at zero to the n-th derivative at zero for an analytic function $f(z)$* [263]:

$$\mathrm{D}_q^n f(0) = \frac{\{n\}_q!}{n!} f^{(n)}(0). \tag{8.85}$$

Proof

$$
\begin{aligned}
\mathrm{D}_q^n f(0) &\overset{\text{by (6.98)}}{=} \lim_{z\to 0} \frac{\sum_{k=0}^{n} \binom{n}{k}_q (-1)^k q^{\binom{k}{2}} f(q^{n-k}z)}{(q-1)^n z^n q^{\binom{n}{2}}} \\
&= \lim_{z\to 0} \frac{\sum_{k=0}^{n} \binom{n}{k}_q (-1)^k q^{\binom{k}{2}} q^{n(n-k)} f^{(n)}(q^{n-k}z)}{(q-1)^n n! q^{\binom{n}{2}}} \\
&= \frac{f^{(n)}(0)(-1)^n(1-q^n)(q-q^n)\cdots(q^{n-1}-q^n)}{(q-1)^n n! q^{0+1+\cdots+(n-1)}} \\
&= \frac{f^{(n)}(0)(q^n-1)(q^{n-1}-1)\cdots(q-1)}{(q-1)^n n!} \\
&= \frac{\{n\}_q!}{n!} f^{(n)}(0).
\end{aligned} \tag{8.86}
$$

In the proof the limit $\lim_{z\to 0}$ in the numerator was zero because of (2.19) and we could use l'Hôpital's rule n times. □

Remark 42 Another proof of this theorem was given in [316, p. 119]. The case when z is complex was proved in [341].

For the proof of (8.90) we need the following lemma:

Lemma 8.12.2

$$
\mathrm{D}_{q,t}\left(-\frac{\mathrm{P}_{m-1,q}(x,-t)}{\{m-1\}_q!}\right) = \frac{\mathrm{P}_{m-2,q}(x,-qt)}{\{m-2\}_q!}, \quad m = 2, 3, \ldots. \tag{8.87}
$$

Proof We use induction. The lemma is true for $m = 2$. Assume that it is true for $m - 1$ and use (6.48) and (6.51) to prove that it is also true for m. So, assume that

$$
\mathrm{D}_{q,t}\left(-\frac{\mathrm{P}_{m-2,q}(x,-t)}{\{m-2\}_q!}\right) = \frac{\mathrm{P}_{m-3,q}(x,-qt)}{\{m-3\}_q!}. \tag{8.88}
$$

Then we obtain

$$
\begin{aligned}
&\mathrm{D}_{q,t}\left(-\frac{\mathrm{P}_{m-1,q}(x,-t)}{\{m-1\}_q!}\right)\\
&=\mathrm{D}_{q,t}\left(-\frac{\mathrm{P}_{m-2,q}(x,-qt)}{\{m-2\}_q!}\frac{x-t}{\{m-1\}_q}\right)\\
&\quad\text{by (6.48), (6.51), (8.88)}\\
&=\frac{1}{\{m-1\}_q}\left(\frac{\mathrm{P}_{m-2,q}(x,-qt)}{\{m-2\}_q!}+\frac{\mathrm{P}_{m-3,q}(x,-q^2t)}{\{m-3\}_q!}\right)\\
&=\frac{\mathrm{P}_{m-3,q}(x,-q^2t)}{\{m-2\}_q!\{m-1\}_q}(x-qt+q(x-qt)\{m-2\}_q)\\
&=\frac{\mathrm{P}_{m-2,q}(x,-qt)}{\{m-2\}_q!\{m-1\}_q}(1+q\{m-2\}_q)\frac{\mathrm{P}_{m-2,q}(x,-qt)}{\{m-2\}_q!}.
\end{aligned}
\tag{8.89}
$$

□

Theorem 8.12.3 *Let $0<|q|<1$ and let the function f be n times q-differentiable in the closed interval $[a,x]$. Then the following generalization of Jackson's formula holds for $n=1,2,\ldots$:*

$$
f(x)=\sum_{k=0}^{n}\frac{\mathrm{P}_{k,q}(x,-a)}{\{k\}_q!}(\mathrm{D}_q^k f)(a)+\frac{1}{\{n\}_q!}\int_a^x \mathrm{P}_{n,q}(x,-qt)(\mathrm{D}_q^{n+1}f)(t)\,d_q(t).
\tag{8.90}
$$

The last term can be written as Lagrange remainder as follows:

$$
\frac{\mathrm{P}_{n,q}(x,-q\xi)}{\{n\}_q!}(\mathrm{D}_q^{n+1}f)(\xi)(x-a),\quad a<\xi<x.
\tag{8.91}
$$

Proof The theorem is true for $n=1$ by inspection. Assume that the theorem is true for $n=m-1$ and use (8.87) to prove that it is also true for $n=m$.

We have

$$
f(x)=\sum_{k=0}^{m-2}\frac{\mathrm{P}_{k,q}(x,-a)}{\{k\}_q!}(\mathrm{D}_q^k f)(a)+\int_{t=a}^{x}\frac{\mathrm{P}_{m-2,q}(x,-qt)}{\{m-2\}_q!}(\mathrm{D}_q^{m-1}f)(t)\,d_q(t)
$$

$$
\begin{aligned}
&\overset{\text{by (8.87)}}{=}\sum_{k=0}^{m-2}\frac{\mathrm{P}_{k,q}(x,-a)}{\{k\}_q!}(\mathrm{D}_q^k f)(a)+\left[-\frac{\mathrm{P}_{m-1,q}(x,-t)}{\{m-1\}_q!}(\mathrm{D}_q^{m-1}f)(t)\right]_{t=a}^{t=x}\\
&\quad+\int_{t=a}^{x}\frac{\mathrm{P}_{m-1,q}(x,-qt)}{\{m-1\}_q!}(\mathrm{D}_q^{m}f)(t)\,d_q(t)
\end{aligned}
$$

$$= \sum_{k=0}^{m-1} \frac{\mathrm{P}_{k,q}(x,-a)}{\{k\}_q!} (\mathrm{D}_q^k f)(a)$$
$$+ \int_{t=a}^{x} \frac{\mathrm{P}_{m-1,q}(x,-qt)}{\{m-1\}_q!} (\mathrm{D}_q^m f)(t)\, d_q(t). \tag{8.92}$$

In the last step, use the mean value theorem (8.95). □

As an example of application of the q-Taylor theorem we obtain

$$\sum_{n=0}^{m} \binom{m}{n}_q P_{n,q}(x,-1) = x^m, \tag{8.93}$$

which is a kind of inverse relation to (2.19).

Rajković, Stanković and Marinković generalized (8.90) to the Lagrange form:

Theorem 8.12.4 [427, p. 176] *There is a unique $q' \in (0,1)$, such that for the function $f(x,q)$ defined on $[b,c] \times (q',1)$, and $x, a \in (b,c)$, a point $\xi \in (b,c)$ can be found between x and a which satisfies*

$$f(x,q) = \sum_{k=0}^{n} \frac{\mathrm{P}_{k,q}(x,a)}{\{k\}_q!} (\mathrm{D}_q^k f)(a,q) + \frac{\mathrm{P}_{n+1,q}(x,a)}{\{n+1\}_q!} (\mathrm{D}_q^{n+1} f)(\xi,q). \tag{8.94}$$

8.12.1 The mean value theorem in q-analysis

Theorem 8.12.5 [427, p. 175]. *The mean value theorem in q-analysis. Let $f(x)$ be a continuous function on $[a,b]$. Then there exists $\widehat{q} \in (0,1)$, such that*

$$\forall q \in (\widehat{q},1), \exists \xi \in (a,b): \quad \mathrm{I}_q(f) = \int_a^b f(t)\, d_q(t) = f(\xi)(b-a). \tag{8.95}$$

8.13 Bilateral series

The bilateral hypergeometric series is defined by

$${}_p\mathrm{H}_p \left[\begin{matrix} a_1, \ldots, a_p \\ b_1, \ldots, b_p \end{matrix} \hat{}\, z \right] \equiv \sum_{n=-\infty}^{\infty} \frac{(a_1, \ldots, a_p)_n}{(b_1, \ldots, b_p)_n} z^n, \quad |z| = 1. \tag{8.96}$$

We define the general bilateral q-hypergeometric series by (compare with [220, p. 125])

$$
\begin{aligned}
{}_p\Psi_r(z) &\equiv {}_p\Psi_r(\widehat{a_1},\ldots,\widehat{a_p};\widehat{b_1},\ldots,\widehat{b_r}|q;z) \equiv {}_p\Psi_r\left[\begin{matrix}\widehat{a_1},\ldots,\widehat{a_p}\\ \widehat{b_1},\ldots,\widehat{b_r}\end{matrix}|q;z\right] \\
&= \sum_{n=-\infty}^{\infty} \frac{\langle\widehat{a_1},\ldots,\widehat{a_p};q\rangle_n}{\langle\widehat{b_1},\ldots,\widehat{b_r};q\rangle_n}\left[(-1)^n q^{\frac{n^2-n}{2}}\right]^{r-p} z^n, \quad \text{where } \widehat{a}\equiv a\vee\widetilde{a}.
\end{aligned}
\tag{8.97}
$$

By (6.13) we obtain

$$
\begin{aligned}
{}_p\Psi_r(z) &= \sum_{n=0}^{\infty} \frac{\langle\widehat{a_1},\ldots,\widehat{a_p};q\rangle_n}{\langle\widehat{b_1},\ldots,\widehat{b_r};q\rangle_n}\left[(-1)^n q^{\frac{n^2-n}{2}}\right]^{r-p} z^n \\
&\quad + \sum_{n=1}^{\infty} \frac{\langle\widehat{1-b_1},\ldots,\widehat{1-b_r};q\rangle_n}{\langle\widehat{1-a_1},\ldots,\widehat{1-a_p};q\rangle_n}\frac{(-1)^{tn}}{z^n}\mathrm{QE}\left(n\left(\sum_{i=1}^{r} b_i - \sum_{i=1}^{p} a_i\right)\right),
\end{aligned}
\tag{8.98}
$$

where t denotes the total number of tildes in ${}_p\Psi_r(z)$.

Compared with bilateral hypergeometric series, bilateral q-series are more interesting since they have greater domains of convergence and have applications in interesting fields such as elliptic functions and number theory. The special case

$$
{}_1\Psi_1\big(a;b|q;q^z\big) = \sum_{n=0}^{\infty}\frac{\langle a;q\rangle_n}{\langle b;q\rangle_n}q^{zn} + \sum_{n=1}^{\infty}\frac{\langle 1-b;q\rangle_n}{\langle 1-a;q\rangle_n}q^{n(b-a-z)}
\tag{8.99}
$$

is convergent for

$$
\Re\big[(b-a)\log q\big] < \Re[z\log q] < 0.
\tag{8.100}
$$

The domain

$$
\pi > \mathrm{Im}(z\log q) > -\pi, \qquad \Re\big[(b-a)\log q\big] < \Re[z\log q] < 0
\tag{8.101}
$$

is a parallelogram in the complex plane, which is the image of the rectangle

$$
\pi > \mathrm{Im}(w) > -\pi, \qquad 0 > \Re(w) > \left|\log q^{b-a}\right|
\tag{8.102}
$$

under the conformal mapping $z = \frac{w}{\log q}$. If q is real this is a rectangle in the right-half plane.

The bilateral summation formula

$$
{}_1\Psi_1\big(a;b|q;q^z\big) = \frac{\langle b-a, a+z, 1, 1-a-z; q\rangle_\infty}{\langle b, b-a-z, z, 1-a; q\rangle_\infty},
\tag{8.103}
$$

an extension of the q-binomial theorem, was first stated by Ramanujan [267]. The following proof is due to Andrews and Askey (1978) [32], see the book by Gasper and Rahman [220, p. 126–27]. A first version of this proof was given by Nathan Fine [202, p. 19].

Proof First regard ${}_1\Psi_1(a;b|q;q^z)$ as a function of b, say $f(b)$. Then $f(b)$ is analytic in the domain (8.100). Now

$$
\begin{aligned}
&{}_1\Psi_1\big(a;b|q;q^z\big) - q^a\,{}_1\Psi_1\big(a;b|q;q^{z+1}\big)\\
&\quad= \sum_{n=-\infty}^{\infty}\left[\frac{\langle a;q\rangle_n}{\langle b;q\rangle_n} - q^{a+n}\frac{\langle a;q\rangle_n}{\langle b;q\rangle_n}\right]q^{zn}\\
&\quad= \sum_{n=-\infty}^{\infty}\frac{\langle a;q\rangle_{n+1}}{\langle b;q\rangle_n}q^{zn} = q^{-z}\big(1-q^{b-1}\big)\sum_{n=-\infty}^{\infty}\frac{\langle a;q\rangle_{n+1}}{\langle b-1;q\rangle_{n+1}}q^{z(n+1)}\\
&\quad= q^{-z}\big(1-q^{b-1}\big)\,{}_1\Psi_1\big(a;b-1|q;q^z\big),
\end{aligned}
\tag{8.104}
$$

whence

$$
f(b+1) - q^{-z}\big(1-q^b\big)f(b) = q^a\,{}_1\Psi_1\big(a;b+1|q;q^{z+1}\big). \tag{8.105}
$$

On the other hand,

$$
\begin{aligned}
q^a\,{}_1\Psi_1\big(a;b+1|q;q^{z+1}\big) &= q^a\sum_{n=-\infty}^{\infty}\frac{\langle a;q\rangle_n}{\langle b+1;q\rangle_n}q^{n(z+1)}\\
&= -q^{a-b}\sum_{n=-\infty}^{\infty}\frac{\langle a;q\rangle_n}{\langle b+1;q\rangle_n}\big(1-q^{b+n}-1\big)q^{nz}\\
&= -q^{a-b}\big(1-q^b\big)f(b) + q^{a-b}f(b+1).
\end{aligned}
\tag{8.106}
$$

By combining (8.105) and (8.104) we obtain the functional equation

$$
f(b+1) - q^{-z}\big(1-q^b\big)f(b) = -q^{a-b}\big(1-q^b\big)f(b) + q^{a-b}f(b+1), \tag{8.107}
$$

that is,

$$
f(b) = \frac{1-q^{b-a}}{(1-q^b)(1-q^{b-a-z})}f(b+1). \tag{8.108}
$$

Now iterate (8.108) to get

$$
f(b) = \frac{\langle b-a;q\rangle_n}{\langle b,b-a-z;q\rangle_n}f(b+n). \tag{8.109}
$$

Now $f(b)$ is analytic for $\Re(b) > \Re(a+z)$ and by letting $n\to+\infty$ we get

$$
f(b) = \frac{\langle b-a;q\rangle_\infty}{\langle b,b-a-z;q\rangle_\infty}f(+\infty). \tag{8.110}
$$

On the other hand, by the q-binomial theorem,

$$f(1) = \frac{\langle a+z;q\rangle_\infty}{\langle z;q\rangle_\infty} \overset{\text{by (8.110)}}{=} \frac{\langle 1-a;q\rangle_\infty}{\langle 1,1-a-z;q\rangle_\infty} f(+\infty). \tag{8.111}$$

Finally, we only have to put this into (8.110) to complete the proof:

$$f(b) = \frac{\langle b-a,a+z,1,1-a-z;q\rangle_\infty}{\langle b,b-a-z,z,1-a;q\rangle_\infty}. \tag{8.112}$$

□

The formula (8.103) can be expressed as a q-integral as follows [48, p. 137, 5.12]:

$$\int_0^\infty t^{a-1} \frac{(tq^{a+b+c};q)_\infty}{(tq^c;q)_\infty} d_q(t) = \Gamma_q \begin{bmatrix} 1-c, c \\ c+a, 1-a-c \end{bmatrix} \mathrm{B}_q(a,b). \tag{8.113}$$

We close this section with some definitions for bilateral series.

Definition 146 A bilateral series ${}_p\Psi_p(z)$ is called *balanced* if

$$b_1 + \cdots + b_p \equiv 2 + a_1 + \cdots + a_p \left(\operatorname{mod} \frac{2\pi i}{\log q}\right), \tag{8.114}$$

well-poised if

$$a_1 + b_1 \equiv a_2 + b_2 = \cdots = a_p + b_p \left(\operatorname{mod} \frac{2\pi i}{\log q}\right), \tag{8.115}$$

and *nearly-poised* if only $p-1$ of these equations hold [26, p. 475].

The following theorem holds [26, p. 475]:

Theorem 8.13.1 *The properties balanced, well-poised and nearly-poised are invariant under shifting the index of summation or reversing the order of summation.*

8.14 Fractional q-integrals

We need the following generalization of $\mathrm{P}_{n,q}(x,a)$ to complex indices:

Definition 147

$$\mathrm{P}_{\alpha,q}(x,a) \equiv x^\alpha \frac{(\frac{a}{x};q)_\infty}{(\frac{a}{x}q^\alpha;q)_\infty}, \quad \frac{a}{x} \neq q^{-m-\alpha}, m = 0,1,\ldots. \tag{8.116}$$

Definition 148 The fractional q-integral is defined in the following way, $\nu \in \mathbb{C}$:

$$\mathrm{D}_q^{-\nu} f(x) \equiv \frac{1}{\Gamma_q(\nu)} \int_0^x \mathrm{P}_{\nu-1,q}(x, qt) f(t)\, d_q(t). \tag{8.117}$$

As a consequence, we have

Theorem 8.14.1

$$\mathrm{D}_q^{-\nu} x^{\mu} = \frac{\Gamma_q(\mu+1)}{\Gamma_q(\mu+\nu+1)} x^{\mu+\nu}. \tag{8.118}$$

With this technique, we can solve fractional q-difference equations.

Chapter 9
q-orthogonal polynomials

9.1 Ciglerian q-Laguerre polynomials

9.1.1 The different Laguerre-philosophies

The French philosophy under Appell and Kampé de Fériet laid the foundations for the so-called Hungarian-French-Italian hypergeometric philosophy. In 1936 Ervin Feldheim (Budapest) wrote together with Paul Erdös on interpolation. The Feldheim family tree, just as the Pringsheim and Heine family trees, consists of famous mathematicians, artists and musicians.

In 1943 Feldheim [199] wrote down the expansion formulas of Burchnall-Chaundy [89, (26), (27), (29)–(31), (38), (39), (42), (43)] from 1940. Some of these formulas were expressed as integral formulas. In the second section Feldheim derived the generating function of Laguerre polynomials and different results for orthogonal polynomials. The paper [199] also included formulas for Laguerre polynomials of two variables. Feldheim summarized the relationship between JLH in [200].

In this chapter we will find four q-analogues of the formula

$$\binom{m+n}{m}\mathrm{L}_{m+n}^{(\alpha)}(x)=\sum_{k=0}^{\min(m,n)}\frac{(-x)^k}{k!}\mathrm{L}_{m-k}^{(\alpha+n+k)}(x)\mathrm{L}_{n-k}^{(\alpha+k)}(x). \tag{9.1}$$

Feldheim [200, p. 134, (43)] has given the rather similar formula

$$\binom{m+n}{m}\mathrm{L}_{m+n}^{(\alpha)}(x)=\sum_{k=0}^{\min(m,n)}(-1)^k\binom{m+n+\alpha}{k}\mathrm{L}_{m-k}^{(\alpha+n)}(x)\mathrm{L}_{n-k}^{(\alpha+m)}(x). \tag{9.2}$$

Feldheim's formulas have attracted the interest of Carlitz. Letterio Toscano did, in fact, continue the work of Feldheim and Schwatt. He did publish 192 articles in the period 1930–1980; like Carlitz, he did not write any book. At the beginning he wrote in French, then in his native Italian and, in the recent years, alternately in French and English. In Messina he took over the legacy of Pia Nalli, but without

T. Ernst, *A Comprehensive Treatment of q-Calculus*,
DOI 10.1007/978-3-0348-0431-8_9, © Springer Basel 2012

her fame. The reason was that he, in contrast to Feldheim, did not master the real Analysis and could therefore usually publish only in journals of secondary importance. But do not forget that special functions are not real Analysis; there is even a theory of orthogonality in q-integrals! Nevertheless, Toscano's contributions to the theory of orthogonal polynomials and hypergeometric functions of many variables were certainly no less interesting than those of Feldheim. Toscano dealt with polynomial difference operators (see Chapter 5) and obtained some deep results. There are also many of Toscano's findings in Chapter 4 (Bernoulli and Euler numbers) and Chapter 5 (Stirling numbers and derivative formulas). In principle Toscano used a difference method similar to that of Sears [460], [461] to prove transformations for special functions. Sears did, however, investigate infinite series.

Some polynomials in the so-called Askey tableau can be expressed in the form of Laguerre polynomials. The same goes for the q-case: some polynomials in the so-called q-Askey-Tableau can be expressed in the form of q-Laguerre polynomials. It is our hope to be able to treat also other q-orthogonal polynomials in other books. The work on q-Hermite polynomials is already finished and the q-Jacobi polynomials, q-Legendre polynomials and Carlitz-Al-Salam polynomials are presented in Sections 9.2 and 9.3.

9.1.2 The q-Laguerre polynomials

The electrical current corresponds to the q-Laguerre polynomial.

1 A	q-Laguerre polynomial

We will use the generating function technique by Rainville [425] to prove recurrences for q-Laguerre polynomials, which are q-analogues of results in [425]. Some of these recurrences were stated already by Moak [387].

We will also find q-analogues of Carlitz's [99] operator expression for Laguerre polynomials. The notation for Cigler's [134] operational calculus will be used when needed. As an application, q-analogues of bilinear generating formulas for Laguerre polynomials of Chatterjea [122, p. 57], [121, p. 88] will be found.

In this book we will be working with two different q-Laguerre polynomials. The polynomial $\mathrm{L}_{n,q,\mathrm{C}}^{(\alpha)}(x)$ was used by Cigler [134]:

Definition 149

$$\begin{aligned}
\mathrm{L}_{n,q,\mathrm{C}}^{(\alpha)}(x) &\equiv \sum_{k=0}^{n} \binom{n+\alpha}{n-k}_q \frac{\{n\}_q!}{\{k\}_q!} q^{k^2+\alpha k}(-1)^k x^k \\
&= \sum_{k=0}^{n} \frac{\langle 1+\alpha; q\rangle_n}{\langle 1+\alpha; q\rangle_k} \frac{\langle -n; q\rangle_k}{\langle 1; q\rangle_k} \frac{q^{\frac{k^2+k}{2}+kn+\alpha k}(1-q)^k x^k}{(1-q)^n} \\
&= \frac{\langle \alpha+1; q\rangle_n}{(1-q)^n} {}_1\phi_1\big(-n; \alpha+1 | q; -x(1-q)q^{n+\alpha+1}\big) \qquad (9.3)
\end{aligned}$$

The polynomial $\mathrm{L}_{n,q,\mathrm{C}}^{(\alpha)}(x)$ arises frequently in operator expressions.

Definition 150 The most common q-Laguerre polynomial $\mathrm{L}_{n,q}^{(\alpha)}(x)$ is defined as follows. Except for the notation, this definition is equivalent to [387], [220] and [492]:

$$\mathrm{L}_{n,q}^{(\alpha)}(x) \equiv \frac{\mathrm{L}_{n,q,\mathrm{C}}^{(\alpha)}(x)}{\{n\}_q!}. \tag{9.4}$$

Definition 151 Cigler [134] has defined a similar q-Laguerre polynomial $l_{n,q,\mathrm{C}}^{(\alpha)}(x)$:

$$\begin{aligned} l_{n,q,\mathrm{C}}^{(\alpha)}(x) &\equiv (-1)^n (1 \boxminus_q \mathrm{D}_q)^{n+\alpha} x^n \\ &= \sum_{k=0}^{n} \binom{n+\alpha}{n-k}_q \frac{\{n\}_q!}{\{k\}_q!} q^{\binom{n-k}{2}} (-1)^k x^k \\ &= \sum_{k=0}^{n} \frac{\langle 1+\alpha; q\rangle_n}{\langle 1+\alpha; q\rangle_k} \frac{\langle -n; q\rangle_k}{\langle 1; q\rangle_k} \frac{q^{\frac{n^2-n}{2}+k}(1-q)^k x^k}{(1-q)^n} \\ &= \frac{\langle 1+\alpha; q\rangle_n}{(1-q)^n} \sum_{k=0}^{n} \binom{n}{k}_q \frac{q^{\binom{n}{2}+k-nk+\binom{k}{2}}(-x)^k(1-q)^k}{\langle 1+\alpha; q\rangle_k} \\ &= \frac{\langle \alpha+1; q\rangle_n}{(1-q)^n} q^{\binom{n}{2}} {}_2\phi_1(-n, \infty; \alpha+1|q; x(1-q)q). \end{aligned} \tag{9.5}$$

Theorem 9.1.1 *The following q-difference equations for q-Laguerre polynomials hold*

Exton [194]:

$$\frac{x}{q}\mathrm{D}_q^2\mathrm{L}_{n,q}^{(\alpha)}\left(\frac{x}{q}\right) + \left(\{\alpha+1\}_q - xq^{1+\alpha}\right)\mathrm{D}_q\mathrm{L}_{n,q}^{(\alpha)}(x) + \{n\}_q q^{1+\alpha}\mathrm{L}_{n,q}^{(\alpha)}(x) = 0; \tag{9.6}$$

Cigler [134, p. 116, (36)]:

$$\left(q^{n+\alpha}x\mathrm{D}_q^2 - x\mathrm{D}_q + q^{n-1}\{\alpha+1\}_q\mathrm{D}_q + \{n\}_q\right) l_{n,q,\mathrm{C}}^{(\alpha)}(x) = 0. \tag{9.7}$$

Remark 43 In [342] the q-Laguerre polynomial is defined as

$$\frac{\langle \alpha+1; q\rangle_n}{\langle 1; q\rangle_n} {}_1\phi_1\left(-n; \alpha+1|q; -xq^{n+\alpha+1}\right). \tag{9.8}$$

In the literature there are many definitions of q-Laguerre polynomials, but most of them are related to each other by some transformation.

9.1.3 Generating functions and recurrences

We shall now discuss the generating functions of Rainville [425].

Consider sequences $\sigma_n(x)$ defined by

$$\mathrm{E}_q(t)\Psi(xt) = \sum_{n=0}^{\infty} \sigma_n(x) t^n. \tag{9.9}$$

Let

$$F = \mathrm{E}_q(t)\Psi(xt). \tag{9.10}$$

Then

$$\mathrm{D}_{q,x} F = t\mathrm{E}_q(t)\mathrm{D}_q \Psi, \tag{9.11}$$

$$\mathrm{D}_{q,t} F = \mathrm{E}_q(t)\Psi + x\big(1-(1-q)t\big)\mathrm{E}_q(t)\mathrm{D}_q\Psi. \tag{9.12}$$

Eliminating Ψ and $\mathrm{D}_q\Psi$ from the above equations we obtain

$$x\big(1-(1-q)t\big)\mathrm{D}_{q,x}F - t\mathrm{D}_{q,t}F = -tF, \tag{9.13}$$

and

$$\sum_{n=0}^{\infty} x\mathrm{D}_q\sigma_n(x)t^n - \sum_{n=1}^{\infty} x(1-q)\mathrm{D}_q\sigma_{n-1}(x)t^n - \sum_{n=0}^{\infty}\{n\}_q\sigma_n(x)t^n$$
$$= -\sum_{n=1}^{\infty}\sigma_{n-1}(x)t^n.$$

By equating the coefficients of t^n we obtain the following recurrence relation:

$$\begin{cases} \mathrm{D}_q\sigma_n(x) = 0, & n=0; \\ x\mathrm{D}_q\sigma_n(x) - x(1-q)\mathrm{D}_q\sigma_{n-1}(x) - \{n\}_q\sigma_n(x) = -\sigma_{n-1}(x), & n \geq 1. \end{cases} \tag{9.14}$$

In particular, by (9.29) we obtain the following recurrence relation for the q-Laguerre polynomials, which is a q-analogue of [425, p. 134]:

$$x\mathrm{D}_q\mathrm{L}_{n,q}^{(\alpha)}(x) - x(1-q)\{\alpha+n\}_q\mathrm{D}_q\mathrm{L}_{n-1,q}^{(\alpha)}(x)$$
$$= \{n\}_q\mathrm{L}_{n,q}^{(\alpha)}(x) - \{\alpha+n\}_q\mathrm{L}_{n-1,q}^{(\alpha)}(x). \tag{9.15}$$

Now let us assume that Ψ has the formal power series expansion

$$\Psi(u) = \sum_{n=0}^{\infty}\gamma_n u^n. \tag{9.16}$$

Then

$$\sum_{n=0}^{\infty} \sigma_n(x)t^n = \sum_{n=0}^{\infty}\sum_{k=0}^{n} \frac{\gamma_k x^k t^n}{\{n-k\}_q!}, \tag{9.17}$$

so that

$$\sigma_n(x) = \sum_{k=0}^{n} \frac{\gamma_k x^k}{\{n-k\}_q!}. \tag{9.18}$$

Now by the q-binomial theorem

$$\begin{aligned}
\sum_{n=0}^{\infty}\{c\}_{n,q}\sigma_n(x)t^n &= \sum_{n=0}^{\infty}\sum_{k=0}^{n} \frac{\{c\}_{n,q}\gamma_k x^k t^n}{\{n-k\}_q!} \\
&= \sum_{n=0}^{\infty}\sum_{k=0}^{\infty} \frac{\{c\}_{n+k,q}\gamma_k x^k t^{n+k}}{\{n\}_q!} \\
&= \sum_{n=0}^{\infty}\sum_{k=0}^{\infty} \frac{\{c+k\}_{n,q}t^n}{\{n\}_q!}\frac{\{c\}_{k,q}\gamma_k (xt)^k}{1} \\
&= \sum_{k=0}^{\infty}\{c\}_{k,q}\gamma_k (xt)^k \frac{(tq^{c+k};q)_\infty}{(t;q)_\infty} = \sum_{k=0}^{\infty}\frac{\{c\}_{k,q}\gamma_k (xt)^k}{(t;q)_{c+k}}.
\end{aligned} \tag{9.19}$$

As a special case we get the following generating function, which is a q-analogue of Feldheim [199, p. 43, (73)] and Rainville [425, p. 135, (13)]:

$$\begin{aligned}
\sum_{n=0}^{\infty}\frac{\{c\}_{n,q}\mathrm{L}_{n,q}^{(\alpha)}(x)t^n}{\{1+\alpha\}_{n,q}} &= \sum_{n=0}^{\infty}\frac{\{c\}_{n,q}\, q^{n^2+\alpha n}(-xt)^n}{\{n\}_q!\{1+\alpha\}_{n,q}(t;q)_{c+n}} \\
&\equiv \frac{1}{(t;q)_c}\,{}_1\phi_2\big(c;1+\alpha|q;-xtq^{1+\alpha}(1-q)||-;tq^c\big).
\end{aligned} \tag{9.20}$$

Consider the important case $c = 1+\alpha$ in (9.20). This is equivalent to Moak [387, p. 29, 4.17], Al-Salam [13, p. 132, 4.2], Feldheim [200, p. 120, 11′]. Call the RHS $F(x,t,q,\alpha)$. By computing the q-difference of $F(x,t,q,\alpha)$ with respect to x we obtain

$$\mathrm{D}_{q,x}F = -tq^{1+\alpha}F(qx,t,q,\alpha+1). \tag{9.21}$$

Equating coefficients of t^n, we obtain the following recurrence relation, which is a q-analogue of [425, p. 203] (also compare with [342, p. 109, 3.21.8] and [355, p. 79]):

$$\mathrm{D}_q\mathrm{L}_{n,q}^{(\alpha)}(x) = -q^{1+\alpha}\mathrm{L}_{n-1,q}^{(1+\alpha)}(xq). \tag{9.22}$$

By computing the q-difference of $F(x,t,q,\alpha)$ with respect to t and equating the coefficients of t^n, we obtain

$$\{n+1\}_q \mathrm{L}_{n+1,q}^{(\alpha)}(x) = \{\alpha+1\}_q \mathrm{L}_{n,q}^{(\alpha+1)}(x) + \frac{\mathrm{L}_{n+1,q}^{(\alpha+1)}(\frac{x}{q}) - \mathrm{L}_{n+1,q}^{(\alpha+1)}(x)}{1-q}. \tag{9.23}$$

Proof

$$\begin{aligned}
\mathrm{D}_{q,t}F &= \sum_{n=0}^{\infty} \frac{q^{n^2+n\alpha}(-x)^n((t;q)_{\alpha+1+n}\{n\}_q t^{n-1} - t^n \mathrm{D}_q(t;q)_{\alpha+1+n})}{(tq;q)_{\alpha+1+n}(t;q)_{\alpha+1+n}\{n\}_q!} \\
&= \sum_{n=0}^{\infty} \frac{q^{n^2+n\alpha}(-x)^n t^n((t;q)_{\alpha+1+n}\{n\}_q t^{-1} + \{\alpha+1+n\}_q(tq;q)_{\alpha+n})}{(tq;q)_{\alpha+1+n}(t;q)_{\alpha+1+n}\{n\}_q!} \\
&= \sum_{n=0}^{\infty} \frac{q^{n^2+n\alpha}(-x)^n t^n(\{n\}_q \frac{1-t}{t} + \{\alpha+1+n\}_q)}{(t;q)_{\alpha+2+n}\{n\}_q!} \\
&= \sum_{n=0}^{\infty} \frac{q^{n^2+n\alpha+n}\{\alpha+1\}_q(-xt)^n}{(t;q)_{\alpha+2+n}\{n\}_q!} + \sum_{n=0}^{\infty} \frac{q^{n^2+n\alpha}\{n\}_q(-x)^n t^{n-1}}{(t;q)_{\alpha+2+n}\{n\}_q!} \\
&= \sum_{n=0}^{\infty} \frac{q^{n^2+n\alpha+n}\{\alpha+1\}_q(-xt)^n}{(t;q)_{\alpha+2+n}\{n\}_q!} + \frac{1}{t(1-q)} \sum_{n=0}^{\infty} \frac{q^{n^2+n\alpha+n}(-xt)^n(\frac{1}{q^n}-1)}{(t;q)_{\alpha+2+n}\{n\}_q!} \\
&= \sum_{n=0}^{\infty} t^n\{\alpha+1\}_q \mathrm{L}_{n,q}^{(\alpha+1)}(x) + \frac{1}{1-q}\sum_{n=0}^{\infty} t^{n-1}\left(\mathrm{L}_{n,q}^{(\alpha+1)}\left(\frac{x}{q}\right) - \mathrm{L}_{n,q}^{(\alpha+1)}(x)\right) \\
&= \sum_{n=0}^{\infty} t^n\{\alpha+1\}_q \mathrm{L}_{n,q}^{(\alpha+1)}(x) + \frac{1}{1-q}\sum_{n=0}^{\infty} t^{n}\left(\mathrm{L}_{n+1,q}^{(\alpha+1)}\left(\frac{x}{q}\right) - \mathrm{L}_{n+1,q}^{(\alpha+1)}(x)\right)
\end{aligned} \tag{9.24}$$

Equating coefficients of t^n we are done. □

The last equation can be expressed as

$$\{n+1\}_q \mathrm{L}_{n+1,q}^{(\alpha)}(x) = \{\alpha+1\}_q \mathrm{L}_{n,q}^{(\alpha+1)}(x) - xq^{2+\alpha}\mathrm{L}_{n,q}^{(\alpha+2)}(x). \tag{9.25}$$

Furthermore, the relation $(1-t)F(x,t,q,\alpha+1) = F(x,tq,q,\alpha)$ yields the following mixed recurrence relation, which was already stated by Moak [387, p. 29, 4.12]:

$$\mathrm{L}_{n,q}^{(\alpha+1)}(x) - \mathrm{L}_{n-1,q}^{(\alpha+1)}(x) = q^n \mathrm{L}_{n,q}^{(\alpha)}(x). \tag{9.26}$$

Theorem 9.1.2 *A generalization of* [387, p. 29, 4.10] *and a q-analogue of* [425, p. 209], [13, p. 131, 3.16], [200, p. 130, 38]:

$$\mathrm{L}_{n,q}^{(\alpha)}(x)=\sum_{k=0}^{n}\frac{\langle\alpha-\beta;q\rangle_k}{\langle 1;q\rangle_k}\mathrm{L}_{n-k,q}^{(\beta)}(x)q^{(\alpha-\beta)(n-k)},\quad \alpha,\ \beta\in\mathbb{C}. \tag{9.27}$$

Proof By the q-binomial theorem we get

$$\begin{aligned}\sum_{n=0}^{\infty}\mathrm{L}_{n,q}^{(\alpha)}(x)t^n&=\sum_{n=0}^{\infty}\frac{q^{n^2+\alpha n}(-xt)^n}{\{n\}_q!(t;q)_{1+\alpha+n}}=\frac{1}{(t;q)_{\alpha-\beta}}\sum_{n=0}^{\infty}\frac{q^{n^2+\beta n}(-xt)^n q^{(\alpha-\beta)n}}{\{n\}_q!(tq^{\alpha-\beta};q)_{1+\beta+n}}\\&=\sum_{k=0}^{\infty}\frac{\langle\alpha-\beta;q\rangle_k}{\langle 1;q\rangle_k}t^k\sum_{l=0}^{\infty}\mathrm{L}_{l,q}^{(\beta)}(x)t^l q^{(\alpha-\beta)l}.\end{aligned}$$

Equating the coefficients of t^n we are done. □

By (9.26) and (9.22), the following important recurrence holds:

$$\mathrm{D}_q\big(\mathrm{L}_{n,q}^{(\alpha)}(x)-\mathrm{L}_{n-1,q}^{(\alpha)}(x)\big)=-q^{n+\alpha}\mathrm{L}_{n-1,q}^{(\alpha)}(xq). \tag{9.28}$$

Theorem 9.1.3 *The following generating function can also be found in Koekoek* [342, p. 109, 3.21.13]. *It is a q-analogue of Feldheim* [199, p. 43, (73)″], [200, p. 121, 12′] *and Rainville* [425, p. 130]:

$$\begin{aligned}\sum_{n=0}^{\infty}\frac{\mathrm{L}_{n,q}^{(\alpha)}(x)t^n}{\{1+\alpha\}_{n,q}}&=\mathrm{E}_q(t)_0\phi_1\big(-;1+\alpha|q;q^{1+\alpha}(1-q)^2(-xt)\big)\\&=\Gamma_q(1+\alpha)(xt)^{-\frac{\alpha}{2}}\mathrm{E}_q(t)\mathrm{J}_\alpha^{(2)}\big(2(1-q)\sqrt{xt};q\big).\end{aligned} \tag{9.29}$$

Proof Let $c\to\infty$ in (9.20). □

Remark 44 Another similar generating function is obtained by letting $t\to tq^{-c}$, $c\to-\infty$ in (9.20). These limits are q-analogues of an idea used by Feldheim [199, p. 43], which is not mentioned by Rainville.

Corollary 9.1.4 *A similar formula is the following q-analogue of Rainville* [425, p. 207, (2)]:

$$x^n=\sum_{k=0}^{n}\Gamma_q\begin{bmatrix}1+\alpha+n\\1+\alpha+k\end{bmatrix}\langle -n;q\rangle_k\mathrm{L}_{k,q}^{(\alpha)}(x)(1-q)^{-k}\mathrm{QE}\bigg(\frac{-n^2-n}{2}-\alpha n+k\bigg). \tag{9.30}$$

This can also be written as the following q-analogue of Feldheim [196, p. 28], [198, p. 217]:

$$\frac{x^n}{\{n\}_q!}=\sum_{k=0}^{n}\binom{\alpha+n}{n-k}_q \mathrm{L}_{k,q}^{(\alpha)}(x)(-1)^k\mathrm{QE}\left(\binom{k}{2}+\frac{-n^2-n}{2}-nk-\alpha n+k\right). \tag{9.31}$$

By (9.3), we obtain

$$\mathrm{L}_{n,q}^{(\alpha)}(x)=\sum_{k=0}^{n}\binom{n+\alpha}{n-k}_q\frac{(-1)^k}{\{k\}_q!}q^{k^2+\alpha k}x^k. \tag{9.32}$$

As was pointed out in [196, p. 28], formulas (9.3) and (9.31) are some kind of inverse relations. Often (9.31) is used to express powers of x in terms of $\mathrm{L}_{k,q}^{(\alpha)}(x)$.

Theorem 9.1.5 *Two generating functions for* $\mathrm{L}_{n,q}^{(\alpha)}(x)$:

$$\begin{aligned}&\sum_{n=0}^{\infty}\frac{\mathrm{L}_{2n,q}^{(\alpha)}(x)(-t^2)^n}{\{1+\alpha\}_{2n,q}}\\&\quad=\mathrm{Cos}_q(t)\,{}_0\phi_7\left(-;\Delta(q;2;1+\alpha),\frac{1}{2},\frac{\widetilde{1}}{2},\widetilde{1}|q;-q^{4+2\alpha}(1-q)^4x^2t^2\right)\\&\qquad+\frac{q^{1+\alpha}(1-q)xt}{1-q^{1+\alpha}}\mathrm{Sin}_q(t)\\&\qquad\times{}_0\phi_7\left(-;\Delta(q;2;2+\alpha),\frac{3}{2},\frac{\widetilde{3}}{2},\widetilde{1}|q;-q^{8+2\alpha}(1-q)^4x^2t^2\right)\quad \textit{and}\\&\sum_{n=0}^{\infty}\frac{\mathrm{L}_{2n+1,q}^{(\alpha)}(x)(-t^2)^n}{\{2+\alpha\}_{2n,q}}\\&\quad=\frac{\{1+\alpha\}_q\,\mathrm{Sin}_q(t)}{t}{}_0\phi_7\left(-;\Delta(q;2;1+\alpha),\frac{1}{2},\frac{\widetilde{1}}{2},\widetilde{1}|q;-q^{4+2\alpha}(1-q)^4x^2t^2\right)\\&\qquad-xq^{1+\alpha}\mathrm{Cos}_q(t)\,{}_0\phi_7\left(-;\Delta(q;2;2+\alpha),\frac{3}{2},\frac{\widetilde{3}}{2},\widetilde{1}|q;-q^{8+2\alpha}(1-q)^4x^2t^2\right).\end{aligned} \tag{9.33}$$

Proof Making use of the decomposition of a series into even and odd parts from Srivastava [484, p. 200, 208], we can rewrite (9.29) in the form

$$\sum_{n=0}^{\infty}\frac{\mathrm{L}_{2n,q}^{(\alpha)}(x)t^{2n}}{\{1+\alpha\}_{2n,q}}+\frac{t}{\{1+\alpha\}_q}\sum_{n=0}^{\infty}\frac{\mathrm{L}_{2n+1,q}^{(\alpha)}(x)t^{2n}}{\{2+\alpha\}_{2n,q}}$$
$$=\mathrm{E}_q(t)\Big[{}_0\phi_7\Big(-;\triangle(q;2;1+\alpha),\frac{1}{2},\frac{\widetilde{1}}{2},\widetilde{1}|q;q^{4+2\alpha}(1-q)^4x^2t^2\Big)$$
$$-\frac{q^{1+\alpha}(1-q)xt}{1-q^{1+\alpha}}{}_0\phi_7\Big(-;\triangle(q;2;2+\alpha),\frac{3}{2},\frac{\widetilde{3}}{2},\widetilde{1}|q;q^{8+2\alpha}(1-q)^4x^2t^2\Big)\Big],\tag{9.34}$$

and replacing t in (9.34) by it, we obtain

$$\sum_{n=0}^{\infty}\frac{\mathrm{L}_{2n,q}^{(\alpha)}(x)(-t^2)^n}{\{1+\alpha\}_{2n,q}}+\frac{it}{\{1+\alpha\}_q}\sum_{n=0}^{\infty}\frac{\mathrm{L}_{2n+1,q}^{(\alpha)}(x)(-t^2)^n}{\{2+\alpha\}_{2n,q}}$$
$$=\big(\mathrm{Cos}_q(t)+i\,\mathrm{Sin}_q(t)\big)$$
$$\times\,{}_0\phi_7\Big(-;\triangle(q;2;1+\alpha),\frac{1}{2},\frac{\widetilde{1}}{2},\widetilde{1}||q;-q^{4+2\alpha}(1-q)^4x^2t^2\Big)$$
$$+\frac{q^{1+\alpha}(1-q)xt}{1-q^{1+\alpha}}\big(\mathrm{Sin}_q(t)-i\,\mathrm{Cos}_q(t)\big)$$
$$\times\,{}_0\phi_7\Big(-;\triangle(q;2;2+\alpha),\frac{3}{2},\frac{\widetilde{3}}{2},\widetilde{1}|q;-q^{8+2\alpha}(1-q)^4x^2t^2\Big).\tag{9.35}$$

Equate the real and imaginary parts of both sides. □

Theorem 9.1.6 *The following generating function is a q-analogue of Feldheim* [199, p. 43, (74′)], [200, p. 120, 11″] *and Carlitz* [100, p. 399]:

$$\sum_{n=0}^{\infty}\mathrm{L}_{n,q}^{(\alpha-n)}(x)t^nq^{\binom{n}{2}-n\alpha}=\frac{\mathrm{E}_{\frac{1}{q}}(-xt)}{(-t;q)_{-\alpha}},\quad |t|<1,\ |x|<1.\tag{9.36}$$

Proof

$$\sum_{n=0}^{\infty}\mathrm{L}_{n,q}^{(\alpha-n)}(x)t^nq^{\binom{n}{2}-n\alpha}$$
$$=\sum_{n=0}^{\infty}t^nq^{\binom{n}{2}-n\alpha}\sum_{k=0}^{n}\frac{\langle 1+\alpha-n;q\rangle_n\langle -n;q\rangle_k}{\langle 1+\alpha-n;q\rangle_k\langle 1;q\rangle_k}\frac{q^{-\binom{k}{2}+k^2+k\alpha}(1-q)^kx^k}{\langle 1;q\rangle_n}$$
$$=\sum_{n=0}^{\infty}\sum_{k=0}^{\infty}t^{n+k}q^{\frac{n^2+2nk+k^2-n-k}{2}-(n+k)\alpha}$$

$$\times \frac{\langle 1+\alpha-n-k;q\rangle_{n+k}\langle -n-k;q\rangle_k}{\langle 1+\alpha-n-k;q\rangle_k\langle 1;q\rangle_k}\frac{q^{-\binom{k}{2}+k^2+k\alpha}(1-q)^k x^k}{\langle 1;q\rangle_{n+k}}$$

$$=\sum_{n=0}^{\infty}\sum_{k=0}^{\infty}t^{n+k}q^{\frac{n^2+k^2-n-k}{2}-n\alpha}\frac{\langle 1+\alpha-n;q\rangle_n}{\langle 1;q\rangle_k\langle 1;q\rangle_n}(-1)^k(1-q)^k x^k$$

$$=\sum_{n=0}^{\infty}t^n(-1)^n\frac{\langle -\alpha;q\rangle_n}{\langle 1;q\rangle_n}\sum_{k=0}^{\infty}(1-q)^k x^k(-1)^k\frac{t^k q^{\binom{k}{2}}}{\langle 1;q\rangle_k}=\frac{\mathrm{E}_{\frac{1}{q}}(-xt)}{(-t;q)_{-\alpha}}. \tag{9.37}$$

□

This means that the function $\mathrm{L}_{n,q}^{(\alpha-n)}(-x)q^{\binom{n}{2}-n\alpha}$ is a pseudo-q-Appell polynomial [184].

We call the function $\mathrm{L}_{n,q}^{(\alpha-n)}(x)$ modified q-Laguerre polynomial.

9.1.4 Product expansions

The theory of commuting ordinary differential operators was first explored in depth by Burchnall and Chaundy [86], [87], [88]. This technique was then used to find differential equations for hypergeometric functions in many papers, e.g. [85]. Some similar q-analogues of these results already exist in the literature and we will prove five q-products expansions starting with

Theorem 9.1.7 *A q-analogue of the Carlitz formula* [99, p. 220]:

$$\mathrm{L}_{n,q,\mathrm{C}}^{(\alpha)}(x)=\prod_{k=3}^{n}\left(q^k x\mathrm{D}_q\epsilon^{-1}-xq^{2k+\alpha-1}+\{\alpha+k\}_q\right)$$
$$\times\left(qx\mathrm{D}_q-xq^{3+\alpha}+\{\alpha+2\}_q\right)\left(x\mathrm{D}_q-xq^{1+\alpha}+\{\alpha+1\}_q\right)1, \tag{9.38}$$

where the number of factors to the right is n.

Proof The theorem is true for $n=0$ and one can check that it is true for $n=1,2$. Assume that (9.38) is true for $n-1$, $n\geq 3$. Then it suffices to prove that

$$\sum_{k=0}^{n}\frac{\langle 1+\alpha;q\rangle_n}{\langle 1+\alpha;q\rangle_k}\frac{\langle -n;q\rangle_k}{\langle 1;q\rangle_k}\frac{q^{-\binom{k}{2}+k^2+kn+\alpha k}(1-q)^k x^k}{(1-q)^n}=$$
$$=\left(q^n x\mathrm{D}_q\epsilon^{-1}-xq^{2n+\alpha-1}+\{\alpha+n\}_q\right)$$
$$\times\sum_{k=0}^{n-1}\frac{\langle 1+\alpha;q\rangle_{n-1}}{\langle 1+\alpha;q\rangle_k}\frac{\langle 1-n;q\rangle_k}{\langle 1;q\rangle_k}\frac{q^{\frac{k^2-k}{2}+kn+\alpha k}(1-q)^k x^k}{(1-q)^{n-1}}. \tag{9.39}$$

A calculation shows that the right-hand side is equal to

$$\sum_{k=0}^{n-1} \frac{\langle 1+\alpha;q\rangle_n}{\langle 1+\alpha;q\rangle_k} \frac{\langle 1-n;q\rangle_k}{\langle 1;q\rangle_k} \frac{q^{\frac{k^2-k}{2}+kn+\alpha k}(1-q)^k x^k}{(1-q)^n}$$

$$-\sum_{k=0}^{n-1} \frac{\langle 1+\alpha;q\rangle_{n-1}}{\langle 1+\alpha;q\rangle_k} \frac{\langle 1-n;q\rangle_k}{\langle 1;q\rangle_k} \frac{q^{\frac{k^2-k}{2}+kn+\alpha k} q^{2n+\alpha-1}(1-q)^k x^{k+1}}{(1-q)^{n-1}}$$

$$+q^n \sum_{k=1}^{n-1} \frac{\langle 1+\alpha;q\rangle_{n-1}}{\langle 1+\alpha;q\rangle_k} \frac{\langle 1-n;q\rangle_k}{\langle 1;q\rangle_{k-1}} \frac{q^{\frac{k^2-k}{2}+kn+\alpha k} q^{-k}(1-q)^k x^k}{(1-q)^n}. \tag{9.40}$$

Finally, it suffices to prove that

$$\frac{1-q^{n+\alpha}}{1-q^{k+\alpha}} \frac{1-q^{-n}}{1-q^k} = \frac{q^{-k}(1-q^{n+\alpha})}{1-q^{\alpha+k}} \frac{1-q^{k-n}}{1-q^k} + \frac{q^{n-2k}(1-q^{k-n})}{1-q^{\alpha+k}} - q^{-2k+n}, \tag{9.41}$$

which is easily checked. □

The following theorem is proved in a similar way.

Theorem 9.1.8 *A q-analogue of* [424, p. 374, (2)]:

$$\mathrm{L}_{n,q,\mathrm{C}}^{(\alpha)}(x) = \mathrm{E}_{\frac{1}{q}}(x) \prod_{k=1}^{n} \left(q^{k+\alpha} x \mathrm{D}_q + \{\alpha+k\}_q\right) \mathrm{E}_q(-x). \tag{9.42}$$

Proof The theorem is true for $n=0$. Assume that it is true for $n-1$. Then it suffices to prove that

$$\sum_{k=0}^{n} \frac{\langle 1+\alpha;q\rangle_n}{\langle 1+\alpha;q\rangle_k} \frac{\langle -n;q\rangle_k}{\langle 1;q\rangle_k} \frac{q^{-\binom{k}{2}+k^2+kn+\alpha k}(1-q)^k x^k}{(1-q)^n}$$

$$= \mathrm{E}_{\frac{1}{q}}(x)\left(q^{n+\alpha} x \mathrm{D}_q + \{\alpha+n\}_q\right)$$

$$= \sum_{k=0}^{n-1} \frac{\langle 1+\alpha;q\rangle_{n-1}}{\langle 1+\alpha;q\rangle_k} \frac{\langle -n+1;q\rangle_k}{\langle 1;q\rangle_k} \frac{q^{-\binom{k}{2}+k^2+k(n-1)+\alpha k}(1-q)^k x^k}{(1-q)^{n-1}} \mathrm{E}_q(-x). \tag{9.43}$$

A calculation shows that the right-hand side is equal to

$$\mathrm{E}_{\frac{1}{q}}(x)\Bigg[\{\alpha+n\}_q\sum_{k=0}^{n-1}\frac{\langle 1+\alpha;q\rangle_{n-1}}{\langle 1+\alpha;q\rangle_k}\frac{\langle -n+1;q\rangle_k}{\langle 1;q\rangle_k}\frac{q^{-\binom{k}{2}+k^2+k(n-1)+\alpha k}(1-q)^k x^k}{(1-q)^{n-1}}$$
$$+q^{n+\alpha}x$$
$$\times\Bigg[\sum_{k=1}^{n-1}\frac{\langle 1+\alpha;q\rangle_{n-1}}{\langle 1+\alpha;q\rangle_k}\frac{\langle -n+1;q\rangle_k}{\langle 1;q\rangle_k}\frac{q^{-\binom{k}{2}+k^2+k(n-1)+\alpha k}(1-q)^k(1-q^k)x^{k-1}}{(1-q)^n}$$
$$-\sum_{k=0}^{n-1}\frac{\langle 1+\alpha;q\rangle_{n-1}}{\langle 1+\alpha;q\rangle_k}\frac{\langle -n+1;q\rangle_k}{\langle 1;q\rangle_k}\frac{q^{-\binom{k}{2}+k^2+kn+\alpha k}(1-q)^k x^k}{(1-q)^{n-1}}\Bigg]\Bigg]\mathrm{E}_q(-x). \tag{9.44}$$

It suffices to prove that

$$\begin{aligned}&\frac{1-q^{n+\alpha}}{\langle 1+\alpha;q\rangle_k}\frac{\langle -n;q\rangle_k}{\langle 1;q\rangle_k}\frac{q^{\frac{k^2}{2}+\frac{k}{2}+kn+\alpha k}(1-q)^k}{(1-q)^n}\\ &\quad=\frac{1-q^{n+\alpha}}{\langle 1+\alpha;q\rangle_k}\frac{\langle 1-n;q\rangle_k}{\langle 1;q\rangle_k}\frac{q^{\frac{k^2}{2}-\frac{k}{2}+kn+\alpha k}(1-q)^k}{(1-q)^n}\\ &\qquad+\frac{q^{n+\alpha}}{\langle 1+\alpha;q\rangle_k}\frac{\langle 1-n;q\rangle_k}{\langle 1;q\rangle_k}\frac{q^{\frac{k^2}{2}-\frac{k}{2}+kn+\alpha k}(1-q^k)(1-q)^k}{(1-q)^n}\\ &\qquad-\frac{q^{n+\alpha}}{\langle 1+\alpha;q\rangle_{k-1}}\frac{\langle 1-n;q\rangle_{k-1}}{\langle 1;q\rangle_{k-1}}\frac{q^{\frac{k^2}{2}-\frac{k}{2}+kn-n+\alpha k-\alpha}(1-q)^k}{(1-q)^n},\end{aligned} \tag{9.45}$$

which implies that

$$\frac{1-q^{n+\alpha}}{1-q^{k+\alpha}}\frac{1-q^{-n}}{1-q^k}=\frac{q^{-k}(1-q^{n+\alpha})}{1-q^{\alpha+k}}\frac{1-q^{k-n}}{1-q^k}+\frac{q^{n+\alpha-k}(1-q^{k-n})}{1-q^{\alpha+k}}-q^{-k}, \tag{9.46}$$

which is easily checked. □

The following theorem is a q-analogue of Chatterjea [118, p. 286 ($k=1$)].

Theorem 9.1.9

$$\mathrm{L}_{n,q,\mathrm{C}}^{(\alpha)}(x)=x^{-\alpha}\mathrm{E}_{\frac{1}{q}}(x)(\mathrm{D}_q)^n\big(x^{\alpha+n}\mathrm{E}_q(-x)\big). \tag{9.47}$$

Proof Just use Leibniz' rule for the n-th q-difference of a product of functions. □

Theorem 9.1.10 *A q-analogue of Chak* [116], *see also Chatterjea* [122]:

$$\mathrm{L}_{n,q,\mathrm{C}}^{(\alpha)}(x) = x^{-\alpha-n-1}\mathrm{E}_{\frac{1}{q}}(x)\big(x^2\mathrm{D}_q\big)^n x^{\alpha+1}\mathrm{E}_q(-x). \tag{9.48}$$

Proof The theorem is true for $n = 0$. Assume that (9.48) is true for $n-1$. Then it suffices to prove that

$$\begin{aligned}
&\sum_{k=0}^{n} \frac{\langle 1+\alpha; q\rangle_n}{\langle 1+\alpha; q\rangle_k} \frac{\langle -n; q\rangle_k}{\langle 1; q\rangle_k} \frac{q^{\frac{k^2+k}{2}+kn+\alpha k}(1-q)^k x^k}{(1-q)^n} \\
&= x^{-\alpha-n-1}\mathrm{E}_{\frac{1}{q}}(x)x^2\mathrm{D}_q \sum_{k=0}^{n-1} \frac{\langle 1+\alpha; q\rangle_{n-1}}{\langle 1+\alpha; q\rangle_k} \frac{\langle -n+1; q\rangle_k}{\langle 1; q\rangle_k} \\
&\quad\times \frac{q^{\frac{k^2+k}{2}+k(n-1)+\alpha k}(1-q)^k x^{k+\alpha+n}}{(1-q)^{n-1}} \mathrm{E}_q(-x).
\end{aligned} \tag{9.49}$$

A calculation shows that the right-hand side is equal to

$$\begin{aligned}
&\sum_{k=0}^{n-1} \frac{\langle 1+\alpha; q\rangle_{n-1}}{\langle 1+\alpha; q\rangle_k} \frac{\langle -n+1; q\rangle_k}{\langle 1; q\rangle_k} \\
&\quad\times \frac{q^{\frac{k^2+k}{2}+k(n-1)+\alpha k}(1-q)^k x^k}{(1-q)^{n-1}} \big(\{k+\alpha+n\}_q\big(1+(1-q)x\big)-x\big) \\
&= \frac{\langle 1+\alpha; q\rangle_n}{(1-q)^n} - \frac{\langle -n+1; q\rangle_{n-1}}{\langle 1; q\rangle_{n-1}} q^{\frac{n^2-n}{2}+n^2+\alpha n} x^n \\
&\quad+ \sum_{k=1}^{n-1} \frac{\langle 1+\alpha; q\rangle_{n-1}}{\langle 1+\alpha; q\rangle_k} \frac{\langle -n+1; q\rangle_k}{\langle 1; q\rangle_k} \frac{q^{\frac{k^2+k}{2}+kn-k+\alpha k} x^k}{(1-q)^{n-k}} \big(1-q^{k+\alpha+n}\big) \\
&\quad- \sum_{k=1}^{n-1} \frac{\langle 1+\alpha; q\rangle_{n-1}}{\langle 1+\alpha; q\rangle_{k-1}} \frac{\langle -n+1; q\rangle_{k-1}}{\langle 1; q\rangle_{k-1}} \frac{q^{\frac{k^2-k}{2}+kn+\alpha k} x^k}{(1-q)^{n-k}} = \mathrm{LHS}.
\end{aligned} \tag{9.50}$$

□

The following theorem is a q-analogue of Chatterjea [120] and a generalization of (9.48):

Theorem 9.1.11

$$\mathrm{L}_{n,q,\mathrm{C}}^{(\alpha)}(x) = x^{-\alpha-n-k}\mathrm{E}_{\frac{1}{q}}(x)\big(\{1-k\}_q x + q^{1-k}x^2\mathrm{D}_q\big)^n x^{\alpha+k}\mathrm{E}_q(-x). \tag{9.51}$$

9.1.5 Bilinear generating functions

We will now prove a couple of bilinear generating formulae for Laguerre polynomials. There is much more to be proved as can be seen from the corresponding hypergeometric identities in [484, p. 133–135, 245, (18)], [313, p. 427, 430–431]. With the help of (9.48) we can prove a q-analogue of a bilinear generating formula for modified Laguerre polynomials of Chatterjea [122, p. 57]:

Theorem 9.1.12

$$\sum_{n=0}^{\infty}\{n\}_q!\,\mathrm{L}_{n,q}^{(\alpha-n)}(x)\mathrm{L}_{n,q}^{(\beta-n)}(y)t^n$$
$$=\mathrm{E}_{\frac{1}{q}}(x)\mathrm{E}_{\frac{1}{q}}(y)\sum_{r,s=0}^{\infty}\frac{(-1)^{r+s}x^r y^s}{\{r\}_q!\{s\}_q!}{}_3\phi_0\left(\infty,-r-\alpha,-s-\beta;-|q;\frac{tq^{r+\alpha+s+\beta}}{1-q}\right). \tag{9.52}$$

Proof

$$\text{LHS}=\sum_{n=0}^{\infty}\frac{x^{-\alpha-1}}{\{n\}_q!}\mathrm{E}_{\frac{1}{q}}(x)\big(x^2\mathrm{D}_{q,x}\big)^n x^{\alpha-n+1}\mathrm{E}_q(-x)y^{-\beta-1}$$
$$\times\,\mathrm{E}_{\frac{1}{q}}(y)\big(y^2\mathrm{D}_{q,y}\big)^n y^{\beta-n+1}\mathrm{E}_q(-y)t^n$$
$$=\mathrm{E}_{\frac{1}{q}}(x)\mathrm{E}_{\frac{1}{q}}(y)x^{-\alpha-1}y^{-\beta-1}$$
$$\times\sum_{n=0}^{\infty}\frac{t^n}{\{n\}_q!}(x\theta_1)^n(y\theta_2)^n x^{\alpha-n+1}y^{\beta-n+1}\mathrm{E}_q(-x)\mathrm{E}_q(-y)$$
$$=\mathrm{E}_{\frac{1}{q}}(x)\mathrm{E}_{\frac{1}{q}}(y)x^{-\alpha-1}y^{-\beta-1}\sum_{n=0}^{\infty}\frac{t^n}{\{n\}_q!}(x\theta_1)^n(y\theta_2)^n\sum_{r=0}^{\infty}\frac{(-1)^r}{\{r\}_q!}x^{\alpha+r-n+1}$$
$$\times\sum_{s=0}^{\infty}\frac{(-1)^s}{\{s\}_q!}y^{\beta+s-n+1}=\mathrm{E}_{\frac{1}{q}}(x)\mathrm{E}_{\frac{1}{q}}(y)x^{-\alpha-1}y^{-\beta-1}$$
$$\times\sum_{n=0}^{\infty}\frac{t^n}{\{n\}_q!}\sum_{r=0}^{\infty}\frac{(-1)^r}{\{r\}_q!}\{r+\alpha-n+1\}_{n,q}x^{\alpha+r+1}$$
$$\times\sum_{s=0}^{\infty}\frac{(-1)^s}{\{s\}_q!}\{s+\beta-n+1\}_{n,q}y^{\beta+s+1}$$
$$=\mathrm{E}_{\frac{1}{q}}(x)\mathrm{E}_{\frac{1}{q}}(y)\sum_{r,s=0}^{\infty}\frac{(-1)^{r+s}x^r y^s}{\{r\}_q!\{s\}_q!}$$

$$\times \sum_{n=0}^{\infty} \frac{\langle -r-\alpha, -s-\beta; q\rangle_n}{\langle 1; q\rangle_n (1-q)^n} q^{-2\binom{n}{2}+n(\alpha+r+\beta+s)} t^n = \text{RHS}. \tag{9.53}$$

□

By the same method, we can find a q-analogue of a bilinear generating formula for Laguerre polynomials of Chatterjea [121, p. 88]:

Theorem 9.1.13

$$\begin{aligned}
&\sum_{n=0}^{\infty} \frac{\langle 1, \gamma; q\rangle_n (xyt)^n}{\langle \alpha+1, \beta+1; q\rangle_n} \mathrm{L}_{n,q}^{(\alpha)}(x) \mathrm{L}_{n,q}^{(\beta)}(y) \\
&\quad = \mathrm{E}_{\frac{1}{q}}(x) \mathrm{E}_{\frac{1}{q}}(y) \\
&\qquad \times \sum_{r,s=0}^{\infty} \frac{(-1)^{r+s} x^r y^s}{\{r\}_q! \{s\}_q!} {}_3\phi_2(\gamma, \alpha+r+1, \beta+s+1; \alpha+1, \beta+1 | q; xyt).
\end{aligned} \tag{9.54}$$

Proof

$$\begin{aligned}
\text{LHS} &= \sum_{n=0}^{\infty} \frac{\langle 1, \gamma; q\rangle_n}{\langle \alpha+1, \beta+1; q\rangle_n} \frac{x^{-\alpha-1}}{(\{n\}_q!)^2} \mathrm{E}_{\frac{1}{q}}(x) \big(x^2 \mathrm{D}_{q,x}\big)^n x^{\alpha+1} \mathrm{E}_q(-x) \\
&\quad \times y^{-\beta-1} \mathrm{E}_{\frac{1}{q}}(y) \big(y^2 \mathrm{D}_{q,y}\big)^n y^{\beta+1} \mathrm{E}_q(-y) t^n \\
&= \mathrm{E}_{\frac{1}{q}}(x) \mathrm{E}_{\frac{1}{q}}(y) x^{-\alpha-1} y^{-\beta-1} \sum_{n=0}^{\infty} \frac{\langle 1, \gamma; q\rangle_n t^n}{\langle \alpha+1, \beta+1; q\rangle_n (\{n\}_q!)^2} (x\theta_1)^n \\
&\quad \times \sum_{r=0}^{\infty} \frac{(-1)^r}{\{r\}_q!} x^{\alpha+r+1} (y\theta_2)^n \sum_{s=0}^{\infty} \frac{(-1)^s}{\{s\}_q!} y^{\beta+s+1} \\
&= \mathrm{E}_{\frac{1}{q}}(x) \mathrm{E}_{\frac{1}{q}}(y) \sum_{n=0}^{\infty} \frac{\langle 1, \gamma; q\rangle_n t^n}{\langle \alpha+1, \beta+1; q\rangle_n (\{n\}_q!)^2} \\
&\quad \times \sum_{r,s=0}^{\infty} \frac{(-1)^{r+s} x^{n+r} y^{n+s}}{\{r\}_q! \{s\}_q!} \{r+\alpha+1\}_{n,q} \{s+\beta+1\}_{n,q} \\
&= \mathrm{E}_{\frac{1}{q}}(x) \mathrm{E}_{\frac{1}{q}}(y) \sum_{r,s=0}^{\infty} \frac{(-1)^{r+s} x^r y^s}{\{r\}_q! \{s\}_q!} \sum_{n=0}^{\infty} \frac{\langle \gamma, \alpha+r+1, \beta+s+1; q\rangle_n}{\langle 1, \alpha+1, \beta+1; q\rangle_n} (xyt)^n \\
&= \text{RHS}.
\end{aligned} \tag{9.55}$$

□

Theorem 9.1.14 *Compare with Chatterjea* [121, 2.3, p. 89].

$$\begin{aligned}
&\sum_{n=0}^{\infty} \frac{\langle 1;q\rangle_n t^n}{\langle \alpha+1;q\rangle_n} \mathrm{L}_{n,q}^{(\alpha)}(x)\mathrm{L}_{n,q}^{(\beta)}(y) \\
&= \mathrm{E}_{\frac{1}{q}}(y)\frac{1}{(t;q)_{\beta+1}} \sum_{s=0}^{\infty} \frac{(-y)^s}{\{s\}_q!(tq^{\beta+1};q)_s} \\
&\quad \times {}_1\phi_2(\beta+s+1;\alpha+1|q;-xt(1-q)q^{1+\alpha}||-;\left(tq^{\beta+s+1};q\right)_k.
\end{aligned} \tag{9.56}$$

Proof Put $\gamma = \beta + 1$ in (9.54). Then

$$\begin{aligned}
&\sum_{n=0}^{\infty} \frac{\langle 1;q\rangle_n t^n}{\langle \alpha+1;q\rangle_n} \mathrm{L}_{n,q}^{(\alpha)}(x)\mathrm{L}_{n,q}^{(\beta)}(y) \\
&= \mathrm{E}_{\frac{1}{q}}(x)\mathrm{E}_{\frac{1}{q}}(y) \sum_{r,s=0}^{\infty} \frac{(-1)^{r+s}x^r y^s}{\{r\}_q!\{s\}_q!} {}_2\phi_1(\alpha+r+1,\beta+s+1;\alpha+1|q;t) \\
&= \mathrm{E}_{\frac{1}{q}}(x)\mathrm{E}_{\frac{1}{q}}(y) \sum_{r,s=0}^{\infty} \frac{(-1)^{r+s}x^r y^s}{\{r\}_q!\{s\}_q!} \frac{1}{(t;q)_{\beta+s+1}} \\
&\quad \times {}_2\phi_2(\beta+s+1,-r;\alpha+1|q;tq^{\alpha+r+1}||-;\left(tq^{\beta+s+1};q\right)_k \\
&= \mathrm{E}_{\frac{1}{q}}(x)\mathrm{E}_{\frac{1}{q}}(y) \frac{1}{(t;q)_{\beta+1}} \sum_{s=0}^{\infty} \frac{(-y)^s}{\{s\}_q!(tq^{\beta+1};q)_s} \sum_{r=0}^{\infty} \frac{(-x)^r}{\{r\}_q!} \\
&\quad \times {}_2\phi_2(\beta+s+1,-r;\alpha+1|q;tq^{\alpha+r+1}||-;\left(tq^{\beta+s+1};q\right)_k \\
&= \mathrm{E}_{\frac{1}{q}}(x)\mathrm{E}_{\frac{1}{q}}(y) \frac{1}{(t;q)_{\beta+1}} \sum_{s=0}^{\infty} \frac{(-y)^s}{\{s\}_q!(tq^{\beta+1};q)_s} \sum_{r=0}^{\infty} \frac{(-x)^r}{\{r\}_q!} \\
&\quad \times \sum_{k=0}^{r} \frac{\langle \beta+s+1,-r;q\rangle_k}{\langle 1,\alpha+1;q\rangle_k} \frac{(-t)^k q^{\binom{k}{2}+k(\alpha+r+1)}}{(tq^{\beta+s+1};q)_k} \\
&= \mathrm{E}_{\frac{1}{q}}(y) \frac{1}{(t;q)_{\beta+1}} \sum_{s=0}^{\infty} \frac{(-y)^s}{\{s\}_q!(tq^{\beta+1};q)_s} \\
&\quad \times \sum_{k=0}^{\infty} \frac{\langle \beta+s+1;q\rangle_k(-1)^k}{\langle 1,\alpha+1;q\rangle_k} \frac{(xt)^k(1-q)^k q^{k^2+k\alpha}}{(tq^{\beta+s+1};q)_k} = \text{RHS}.
\end{aligned} \tag{9.57}$$

□

Theorem 9.1.15 *A q-analogue of the Hille-Hardy formula:*

$$\sum_{n=0}^{\infty}\frac{\langle 1;q\rangle_n t^n}{\langle\alpha+1;q\rangle_n}\mathrm{L}_{n,q}^{(\alpha)}(x)\mathrm{L}_{n,q}^{(\alpha)}(y)$$
$$=\frac{\mathrm{E}_{\frac{1}{q}}(x)\mathrm{E}_{\frac{1}{q}}(y)}{(t;q)_{\alpha+1}}\sum_{s,r,k=0}^{\infty}\frac{(-y)^s}{\{s\}_q!}\frac{(-x)^r}{\{r\}_q!}\frac{(1-q)^{2k}(xyt)^k q^{\alpha k+k^2}}{\langle 1,\alpha+1;q\rangle_k (tq^{\alpha+1};q)_{r+2k+s}}. \tag{9.58}$$

Proof Put $\beta=\alpha$ and $\gamma=\alpha+1$ in (9.54). Then

$$\sum_{n=0}^{\infty}\frac{\langle 1;q\rangle_n t^n}{\langle\alpha+1;q\rangle_n}\mathrm{L}_{n,q}^{(\alpha)}(x)\mathrm{L}_{n,q}^{(\alpha)}(y)$$
$$=\mathrm{E}_{\frac{1}{q}}(x)\mathrm{E}_{\frac{1}{q}}(y)\sum_{r,s=0}^{\infty}\frac{(-1)^{r+s}x^r y^s}{\{r\}_q!\{s\}_q!}{}_2\phi_1(\alpha+r+1,\alpha+s+1;\alpha+1,|q;t)$$
$$=\mathrm{E}_{\frac{1}{q}}(x)\mathrm{E}_{\frac{1}{q}}(y)\sum_{r,s=0}^{\infty}\frac{(-1)^{r+s}x^r y^s}{\{r\}_q!\{s\}_q!}\frac{}{(t;q)_{\alpha+r+s+1}}$$
$$\times\, {}_2\phi_1\big(-r,-s;\alpha+1,|q;tq^{\alpha+r+s+1}\big)$$
$$=\frac{\mathrm{E}_{\frac{1}{q}}(x)\mathrm{E}_{\frac{1}{q}}(y)}{(t;q)_{\alpha+1}}\sum_{r=0}^{\infty}\frac{(-x)^r}{\{r\}_q!(tq^{\alpha+1};q)_r}$$
$$\times\sum_{s=0}^{\infty}\frac{(-y)^s}{\{s\}_q!(tq^{\alpha+1+r};q)_s}{}_2\phi_1\big(-r,-s;\alpha+1,|q;tq^{\alpha+r+s+1}\big)$$
$$=\frac{\mathrm{E}_{\frac{1}{q}}(x)\mathrm{E}_{\frac{1}{q}}(y)}{(t;q)_{\alpha+1}}\sum_{r=0}^{\infty}\frac{(-x)^r}{\{r\}_q!(tq^{\alpha+1};q)_r}\sum_{s,k=0}^{\infty}\frac{(-y)^{s+k}}{\{s+k\}_q!(tq^{\alpha+1+r};q)_{s+k}}$$
$$\times\frac{\langle -s-k,-r;q\rangle_k}{\langle 1,\alpha+1;q\rangle_k}t^k q^{(\alpha+r+s+1)k+k^2}$$
$$=\frac{\mathrm{E}_{\frac{1}{q}}(x)\mathrm{E}_{\frac{1}{q}}(y)}{(t;q)_{\alpha+1}}\sum_{r=0}^{\infty}\frac{(-x)^r}{\{r\}_q!(tq^{\alpha+1};q)_r}$$
$$\times\sum_{s,k=0}^{\infty}\frac{(-y)^s(yt)^k(1-q)^{s+k}\langle -r;q\rangle_k q^{(\alpha+r)k+\frac{k^2}{2}+\frac{k}{2}}}{\langle 1;q\rangle_s\langle 1,\alpha+1;q\rangle_k(tq^{\alpha+1+r};q)_{s+k}}$$
$$=\frac{\mathrm{E}_{\frac{1}{q}}(x)\mathrm{E}_{\frac{1}{q}}(y)}{(t;q)_{\alpha+1}}\sum_{s,r,k=0}^{\infty}\frac{(-y)^s}{\{s\}_q!}\frac{(-x)^r}{\{r\}_q!}\frac{(1-q)^{2k}(xyt)^k q^{\alpha k+k^2}}{\langle 1,\alpha+1;q\rangle_k(tq^{\alpha+1};q)_{r+2k+s}}. \tag{9.59}$$

□

9.1.6 Al-Salam operator expressions

Operational formulas were often used with big success in the theory of classical orthogonal polynomials and Bessel functions [246]. The results obtained herewith are of a certain theoretical interest.

The classical JLH polynomials have many things in common, as Feldheim [200] explained beautifully. If we have an equation for the Jacobi polynomials, we automatically obtain a corresponding formula for Laguerre and Hermite polynomials.

The following formulas are useful.

Theorem 9.1.16

$$\mathrm{D}_q^l \frac{1}{(x;q)_\alpha} = \frac{\{\alpha\}_{l,q}}{(x;q)_{\alpha+l}}, \tag{9.60}$$

$$\mathrm{D}_q^r \mathrm{L}_{n,q}^{(\alpha)}(x) = (-1)^r q^{\alpha r+\binom{r+1}{2}+\binom{r}{2}} \mathrm{L}_{n-r,q}^{(\alpha+r)}(xq^r), \tag{9.61}$$

$$\left(\mathrm{D}_q\epsilon^{-1}\right)^n \mathrm{L}_{m,q}^{(\alpha)}(x) = (-1)^n \mathrm{L}_{m-n,q}^{(\alpha+n)}(x)\mathrm{QE}\left(\binom{n}{2}+n\alpha\right), \tag{9.62}$$

$$\mathrm{D}_q^k \frac{x}{1-xq^{1+\alpha}} = \frac{q^{(1+\alpha)(k-1)}\{k\}_q!}{(xq^{1+\alpha};q)_{k+1}}, \quad k>0. \tag{9.63}$$

Definition 152 Let H_q be the set of functions of the form

$$F(x) \equiv \sum_{k=0}^{\infty} x^{\alpha_k} \frac{\gamma_k}{(x;q)_{\beta_k}}, \quad \alpha_k \in \mathbb{N},\ \beta_k \in \mathbb{R},\ \gamma_k \in \mathbb{C}. \tag{9.64}$$

Definition 153 We will use the following two operators on H_q as the basis for our calculations; the special case $\alpha = 0$, $q = 1$ was treated in [13, 1.1]. A related operator was used in [16, p. 4, (2.1)].

$$\theta_{q,\alpha} \equiv x\left(\{1+\alpha\}_q I + q^{1+\alpha} x\mathrm{D}_{q,x}\right), \tag{9.65}$$

$$\Phi_{q,\alpha} \equiv y\left(\{1+\alpha\}_q I + q^{1+\alpha} y\mathrm{D}_{q,y}\right). \tag{9.66}$$

By using the above expression we obtain [13, 1.2], [16, p. 4, (2.2)]

$$\theta_{q,\alpha}^n\left(x^\beta\right) = x^{\beta+n}\{1+\alpha+\beta\}_{n,q}. \tag{9.67}$$

Theorem 9.1.17 [13, 2.1], [16, p. 4, (2.5)]:

$$\theta_{q,\alpha}^n = x^n \prod_{j=1}^{n}\left(\{j+\alpha\}_q I + q^{j+\alpha} x\mathrm{D}_q\right) \equiv \frac{x^n}{(1-q)^n}\left(\epsilon q^{1+\alpha};q\right)_n$$

$$= \frac{x^n}{(1-q)^n}\sum_{k=0}^{n}(-1)^k \binom{n}{k}_q q^{\binom{k}{2}+k(1+\alpha)}\epsilon^k. \tag{9.68}$$

Proof Induction. □

Theorem 9.1.18

$$\theta_{q,\alpha}^{n}\frac{1}{(x;q)_\beta}=\frac{x^n}{(1-q)^n(x;q)_{\beta+n}}\sum_{k=0}^{n}(-1)^k\binom{n}{k}_q q^{\binom{k}{2}+k(1+\alpha)}$$
$$\times\sum_{l=0}^{k}(-1)^l\binom{k}{l}_q q^{\binom{l}{2}}x^l\sum_{m=0}^{n-k}(-1)^m\binom{n-k}{m}_q q^{\binom{m}{2}+m(k+\beta)}x^m. \tag{9.69}$$

Proof

$$\begin{aligned}\text{LHS}&=\frac{x^n}{(1-q)^n}\sum_{k=0}^{n}(-1)^k\binom{n}{k}_q q^{\binom{k}{2}+k(1+\alpha)}\frac{1}{(xq^k;q)_\beta}\\&=\frac{x^n}{(1-q)^n}\sum_{k=0}^{n}(-1)^k\binom{n}{k}_q q^{\binom{k}{2}+k(1+\alpha)}\frac{(x;q)_k(xq^{k+\beta};q)_{n-k}}{(x;q)_{\beta+n}}=\text{RHS}.\end{aligned} \tag{9.70}$$

□

The following special case holds:

Theorem 9.1.19

$$\theta_{q,\alpha}^{n}\frac{1}{(x;q)_\alpha}=\frac{x^n}{(x;q)_{\alpha+n}}\mathrm{F}_n(x), \tag{9.71}$$

where

$$\mathrm{F}_n(x)\equiv\sum_{k=0}^{n}a_{k,n}x^k, \tag{9.72}$$

with coefficients

$$a_{0,n}=\{1+\alpha\}_{n,q},\qquad a_{1,n}=-q^\alpha\{n\}_q\{2+\alpha\}_{n,q},\qquad a_{n,n}=(-1)^nq^{\binom{n}{2}+n\alpha}\{n\}_q!. \tag{9.73}$$

Proof Assume that the theorem is true for $n-1$. Then

$$\begin{aligned}\theta_{q,\alpha}^{n}\frac{1}{(x;q)_\alpha}=&\frac{x^n}{(1-q)(x;q)_{\alpha+n-1}}\sum_{k=0}^{n-1}a_{k,n-1}x^k\\&-\frac{x^nq^{\alpha+n}}{(1-q)(xq;q)_{\alpha+n-1}}\sum_{k=0}^{n-1}a_{k,n-1}x^kq^k.\end{aligned} \tag{9.74}$$

We obtain by induction

$$\begin{aligned} a_{1,n} &= -\frac{q^{\alpha+n-1}}{1-q}\{1+\alpha\}_{n-1,q} - \frac{q^{\alpha}(1-q^{n-1})}{(1-q)^2}\{2+\alpha\}_{n-2,q} \\ &\quad + \frac{q^{\alpha+n}}{1-q}\{1+\alpha\}_{n-1,q} + \frac{q^{2\alpha+n+1}(1-q^{n-1})}{(1-q)^2}\{2+\alpha\}_{n-2,q} \\ &= -\frac{\{2+\alpha\}_{n-2,q}q^{\alpha}}{(1-q)^2}\left(1-q^{\alpha+n}-q^n+q^{\alpha+2n}\right) = -q^{\alpha}\{n\}_q\{2+\alpha\}_{n,q}. \end{aligned} \tag{9.75}$$

The formulas for $a_{0,n}$ and $a_{n,n}$ are proved in a similar way. □

If $F(x)$ and $f(x)$ are formal power series we obtain the following rule [13, 2.2], [16, p. 4, (2.6)]:

$$F(\theta_{q,0})x^{\alpha} f(x) = x^{\alpha} F(\theta_{q,\alpha}) f(x). \tag{9.76}$$

Theorem 9.1.20 *A q-analogue of Cauchy* [111, p. 161]:

$$\mathrm{D}_q^m\left(\mathrm{E}_q(\alpha x) f(x)\right) = \mathrm{E}_q(\alpha x)(\mathrm{D}_q \oplus_q \alpha\epsilon)^m f(x). \tag{9.77}$$

Proof Use the q-Leibniz formula. □

Theorem 9.1.21 *N. A. Al-Salam* [16, p. 4, (2.3)]. *The Leibniz formula is the following q-analogue of W. A. Al-Salam* [13, 2.4]:

$$\theta_{q,\alpha}^n\left(x^{1+\alpha} f(x)g(x)\right) = x^{1+\alpha}\sum_{j=0}^{n}\binom{n}{j}_q q^{(1+\alpha)(n-j)}\left(\theta_{q,\alpha}^j \epsilon^{n-j} g(x)\right)\left(\theta_{q,\alpha}^{n-j} f(x)\right). \tag{9.78}$$

Proof Put $f(x) = x^{-n-1-2\alpha}$, $g(x) = x^{n+\beta+\alpha}$. Then it suffices to prove that

$$\begin{aligned} & x^{n+\beta}\sum_{k=0}^{n}(-1)^k\binom{n}{k}_q \langle -n-\alpha; q\rangle_{n-k}\langle -\beta-2\alpha-n-k; q\rangle_k \\ &\quad \times \mathrm{QE}\left(\binom{k}{2} + (1+\alpha)(n-k) + (n-k)(n+\beta+\alpha) + k(1+n+\beta+2\alpha)\right) \\ &= x^{n+\beta}\langle 1+\beta+\alpha; q\rangle_m q^{m(x-j+m)}. \end{aligned} \tag{9.79}$$

However, this follows from a simple change of variables in the following results of Carlitz 1948 [97, p. 988], see also [164]:

$$\begin{aligned} & \sum_{n=0}^{m}(-1)^n\binom{m}{n}_q q^{\binom{n}{2}}\langle x+1; q\rangle_{m-n}\langle x-j+1-n+m; q\rangle_n \\ &\quad = \langle j-m+1; q\rangle_m q^{m(x-j+m)}, \quad x \in \mathbb{C},\ j < m. \end{aligned} \tag{9.80}$$

□

The previous theorem implies the q-analogue of the corrected version of [13, 2.5]:

$$\begin{aligned}
&\mathrm{E}_q(t\theta_{q,\alpha})\big(x^{1+\alpha}f(x)g(x)\big)\\
&=x^{1+\alpha}\sum_{j,l=0}^{\infty}\frac{t^j q^{j(1+\alpha)}}{\{j\}_q!}\theta_{q,\alpha}^j\big(g(x)\big)\frac{t^l}{\{l\}_q!}\theta_{q,\alpha}^l\epsilon^j\big(f(x)\big)\\
&=x^{1+\alpha}\mathrm{E}_q\big(tq^{1+\alpha}\theta_{q,\alpha,g}\epsilon_f\big)g(x)\mathrm{E}_q\big(tq^{1+\alpha}\theta_{q,\alpha,f}\big)f(x). \qquad (9.81)
\end{aligned}$$

Proof

$$\begin{aligned}
\text{LHS}&=\mathrm{E}_q(t\theta_{q,\alpha})\big(x^{1+\alpha}f(x)g(x)\big)=\sum_{k=0}^{\infty}\frac{t^k}{\{k\}_q!}\theta_{q,\alpha}^k\big(x^{1+\alpha}f(x)g(x)\big)\\
&=\sum_{k=0}^{\infty}\frac{t^k}{\{k\}_q!}x^{1+\alpha}\sum_{j=0}^{k}\frac{\{k\}_q!}{\{j\}_q!\{k-j\}_q!}q^{j(1+\alpha)}\big(\theta_{q,\alpha}^j g(x)\big)\big(\theta_{q,\alpha}^{k-j}\epsilon^j f(x)\big)\\
&=x^{1+\alpha}\sum_{j=0}^{\infty}\frac{t^j q^{j(1+\alpha)}}{\{j\}_q!}\theta_{q,\alpha}^j g(x)\sum_{l=0}^{\infty}\frac{t^l}{\{l\}_q!}\theta_{q,\alpha}^l\epsilon^j f(x)=\text{RHS}. \qquad (9.82)
\end{aligned}$$

□

Definition 154 Compare with [332, p. 87]. If $v(x)$ is the power series $\sum_{j=0}^{\infty}a_j x^j$, we define

$$v_q\left(\frac{x}{[1-t]}\right)\equiv\sum_{j=0}^{\infty}\frac{a_j x^j}{(t;q)_j}. \qquad (9.83)$$

We obtain

Lemma 9.1.22

$$\mathrm{E}_q(t\theta_{q,\alpha})\big(v(x)\big)\cong\frac{1}{(tx;q)_{1+\alpha}}v_q\left(\frac{x}{[1-txq^{\alpha+1}]}\right). \qquad (9.84)$$

Proof

$$\begin{aligned}
\text{LHS}&=\sum_{k=0}^{\infty}\frac{t^k}{\{k\}_q!}\theta_{q,\alpha}^k\sum_{l=0}^{\infty}a_l x^l=\sum_{k=0}^{\infty}t^k\sum_{l=0}^{\infty}\frac{a_l x^{l+k}\langle 1+\alpha+l;q\rangle_k}{\langle 1;q\rangle_k}\\
&=\sum_{l=0}^{\infty}a_l x^l\sum_{k=0}^{\infty}\frac{x^k t^k\langle 1+\alpha+l;q\rangle_k}{\langle 1;q\rangle_k}\\
&\cong\sum_{l=0}^{\infty}a_l x^l\frac{1}{(tx;q)_{l+1+\alpha}}=\frac{1}{(tx;q)_{1-tx}}\sum_{l=0}^{\infty}a_l x^l\frac{1}{(txq;q)_{l+\alpha}}=\text{RHS}.
\end{aligned}$$

□

We also have the following q-analogue of the corrected version of [13, 2.6]:

$$\mathrm{E}_q(t\theta_{q,\alpha})x^\beta \cong \frac{x^\beta}{(tx;q)_{\beta+1+\alpha}}. \tag{9.85}$$

Proof

$$\begin{aligned}\mathrm{LHS} = \mathrm{E}_q(t\theta_{q,\alpha})x^\beta &= \sum_{k=0}^{\infty} \frac{t^k}{\{k\}_q!}\theta_{q,\alpha}^k x^\beta \\ &= \sum_{k=0}^{\infty} \frac{t^k}{\langle 1;q\rangle_k} x^{\beta+k}\langle \beta+1+\alpha;q\rangle_k \cong \frac{x^\beta}{\langle tx;q\rangle_{\beta+1+\alpha}} = \mathrm{RHS}\end{aligned} \tag{9.86}$$

□

Using Lemma 9.1.22, we get

Theorem 9.1.23 *A q-analogue of* [13, 2.7]:

$$\mathrm{E}_q(t\theta_{q,\alpha})x^\beta v(x) = \frac{x^\beta}{(tx;q)_{\beta+\alpha+1}} v_q\left(\frac{x}{[1-xtq^{\beta+\alpha+1}]}\right). \tag{9.87}$$

Proof

$$\begin{aligned}\mathrm{LHS} &= \mathrm{E}_q(t\theta_{q,\alpha})x^\beta v(x) \\ &= x^{1+\alpha}\sum_{j=0}^{\infty}\frac{(tq^{1+\alpha})^j}{\{j\}_q!}\theta_{q,\alpha}^j v \sum_{l=0}^{\infty}\frac{t^l}{\{l\}_q!}\theta_{q,\alpha}^l q^{j(\beta-\alpha-1)}x^{\beta-\alpha-1} \\ &= x^{1+\alpha}\sum_{j=0}^{\infty}\frac{(tq^\beta)^j}{\{j\}_q!}\theta_{q,\alpha}^j(v)\mathrm{E}_q(t\theta_{q,\alpha})x^{\beta-\alpha-1} \\ &= x^{1+\alpha}\mathrm{E}_q\left(tq^\beta\theta_{q,\alpha}\right)v\mathrm{E}_q(t\theta_{q,\alpha})x^{\beta-\alpha-1} \\ &= x^{1+\alpha}\frac{1}{(txq^\beta;q)_{1+\alpha}} v_q\left(\frac{x}{[1-txq^{\beta+\alpha+1}]}\right)\frac{x^{\beta-\alpha-1}}{(tx;q)_\beta} \\ &= \frac{x^\beta}{(tx;q)_{\beta+\alpha+1}} v_q\left(\frac{x}{[1-txq^{\beta+\alpha+1}]}\right) = \mathrm{RHS}.\end{aligned} \tag{9.88}$$

□

Theorem 9.1.24 *A q-analogue of* [13, 2.8]:

$${}_p\phi_r\left((a);(b)|q;t\theta_{q,\alpha}\right)x^\gamma = x^\gamma\,{}_{p+1}\phi_{r+1}\left((a),\gamma+\alpha+1;(b),\infty|q;\frac{tx}{1-q}\right). \tag{9.89}$$

Proof

$$\text{LHS} = \sum_k \frac{\langle (a); q\rangle_k}{\langle 1, (b); q\rangle_k} q^{\binom{k}{2}(1+r-p)(-1)^{1+r-p}} (-1)^{k(1+r-p)} t^k x^{\gamma+k} \frac{\langle 1+\alpha+\gamma; q\rangle_k}{(1-q)^k}. \tag{9.90}$$

□

Corollary 9.1.25 [13, 2.9]

$$\begin{aligned} &{}_p\phi_r\big((a); (b)|q; t\theta_{q,\alpha}\big)x^\gamma \mathrm{E}_q(-x) \\ &= x^\gamma \sum_{l=0}^{\infty} \frac{(-x)^l}{\{l\}_q!} {}_{p+1}\phi_{r+1}\left((a), \gamma+\alpha+l+1; (b), \infty|q; \frac{tx}{1-q}\right). \end{aligned} \tag{9.91}$$

Proof

$$\begin{aligned} \text{LHS} &= {}_p\phi_r \sum_n \frac{x^{\gamma+n}(-1)^n}{\{n\}_q!} \\ &= \sum_n \frac{x^{\gamma+n}(-1)^n}{\{n\}_q!} {}_{p+1}\phi_{r+1}\left((a), \gamma+n+\alpha+1; (b), \infty|q, \frac{tx}{1-q}\right) = \text{RHS}. \end{aligned} \tag{9.92}$$

□

There is an inverse operator [13, 2.11]:

$$\theta_{q,\alpha}^{-n}\left(x^{-\beta}\right) = x^{-\beta-n} \frac{(-1)^n q^{n(\beta-\alpha)+\binom{n}{2}}}{\{-\alpha+\beta\}_{n,q}}. \tag{9.93}$$

Proof

$$\begin{aligned} \text{LHS} = \theta_{q,\alpha}^{-n}\left(x^{-\beta}\right) &= x^{-\beta-n} \frac{\langle 1+\alpha-\beta; q\rangle_{-n}}{(1-q)^{-n}} \\ &= x^{-\beta-n}(1-q)^n \frac{(-1)^n q^{n(\beta-\alpha)+\binom{n}{2}}}{\langle \beta-\alpha; q\rangle_n} = \text{RHS}. \end{aligned} \tag{9.94}$$

□

Theorem 9.1.26 *A q-analogue of the corrected version of* [13, 2.12]:

$$\begin{aligned} &\left(\frac{\Phi_{q,\alpha}}{\theta_{q,\alpha}}\right)^k \left(\frac{y^\delta}{x^{\beta+1}}\right) \\ &= \frac{y^{\delta+k}}{x^{\beta+k+1}} \cdot \frac{\langle 1+\alpha+\delta; q\rangle_k}{\langle 1+\beta-\alpha; q\rangle_k} (-1)^k \mathrm{QE}\left(k(1+\beta-\alpha)+\binom{n}{2}\right), \end{aligned} \tag{9.95}$$

where $\beta-\alpha \neq 0,-1,-2,\dots$.

Corollary 9.1.27 *A q-analogue of the corrected version of* [13, 2.13]:

$$
\begin{aligned}
&{}_p\phi_r\left((a);(b)|q;t\frac{\Phi_{q,\alpha}}{\theta_{q,\alpha}}\right)\frac{y^\delta}{x^{\beta+1}}\\
&\quad=\frac{y^\delta}{x^{\beta+1}}\,{}_{p+1}\phi_{r+2}\left((a),\alpha+\delta+1;(b),-\alpha+\beta+1,\infty|q;\frac{ty}{x}q^{1+\beta-\alpha}\right).
\end{aligned}\tag{9.96}
$$

Proof

$$
\begin{aligned}
&{}_p\phi_r\left((a);(b)|q,t\frac{\Phi_{q,\alpha}}{\theta_{q,\alpha}}\right)\left(\frac{y^\delta}{x^{\beta+1}}\right)\\
&\quad=\sum_{k=0}^{\infty}\frac{\langle(a);q\rangle_k}{\langle 1,(b);q\rangle_k}t^k\frac{y^\delta}{x^{\beta+1}}(-1)^{1-r-p}q^{\binom{k}{2}(1+r-p)}\left(\frac{y}{x}\right)^k\\
&\qquad\times\frac{\langle 1+\alpha+\delta;q\rangle_k}{\langle 1+\beta-\alpha;q\rangle_k}(-1)^k\,\mathrm{QE}\left(k(1+\beta-\alpha)+\binom{k}{2}\right)\\
&\quad=\frac{y^\delta}{x^{\beta+1}}\,{}_{p+1}\phi_{r+2}\left((a),\alpha+\delta+1;(b),-\alpha+\beta+1,\infty|q,\frac{ty}{x}q^{1+\beta-\alpha}\right).
\end{aligned}\tag{9.97}
$$

□

Theorem 9.1.28 *A q-analogue of the corrected version of* [13, 2.14]:

$$
\begin{aligned}
&{}_1\phi_0\left(c;-|q;t\frac{\Phi_{q,\alpha}}{\theta_{q,\alpha}}\right)\frac{y^\delta}{x^{\beta+1}}\\
&\quad=\frac{y^\delta}{x^{\beta+1}}\,{}_2\phi_2\left(c,\alpha+\delta+1;-\alpha+\beta+1,\infty|q;\frac{ty}{x}q^{1+\beta-\alpha}\right).
\end{aligned}\tag{9.98}
$$

Theorem 9.1.29 *A q-analogue of the corrected version of* [13, 2.16]:

$$
{}_2\phi_0\left(\infty,1-\alpha+\beta|q,t\frac{\phi_{q,\alpha}}{\theta_{q,\alpha}}\right)\left(\frac{y^\delta}{x^{\beta+1}}\right)=\frac{y^\delta}{x^{\beta+1}}\cdot\frac{1}{(\frac{ty}{x}q^{1+\beta-\alpha};q)_{1+\alpha+\delta}}.\tag{9.99}
$$

By (9.99) we obtain

Corollary 9.1.30 *A q-analogue of* [13, 2.17]. *If* $F(x)\in\mathbb{C}[[x]]$, *then*

$$
\begin{aligned}
&{}_2\phi_0\left(\infty,1-\alpha+\beta|q,t\frac{\phi_{q,\alpha}}{\theta_{q,\alpha}}\right)x^{-\beta-1}F(y)\\
&\quad=\frac{x^{-\beta-1}}{(\frac{ty}{x}q^{1+\beta-\alpha})_{1+\alpha}}\mathrm{F}_q\left(\frac{y}{[1-\frac{ty}{x}q^{2+\beta}]}\right).
\end{aligned}\tag{9.100}
$$

Theorem 9.1.31 *Compare with* [13, 2.18]:

$$\mathrm{E}_q\left(-\frac{t}{\theta_{q,\alpha}}\right)x^{-\beta-1} = x^{-\beta-1}{}_1\phi_1\left(\infty; -\alpha+\beta+1|q; \frac{-(1-q)^2 t}{x}q^{1+\beta-\alpha}\right). \tag{9.101}$$

Proof

$$\mathrm{LHS} = \sum_{n=0}^{\infty} \frac{(-t)^n}{\{n\}_q!} x^{-\beta-1-n} \frac{(1-q)^n(-1)^n q^{n(\beta+1-\alpha)+\binom{n}{2}}}{\langle \beta-\alpha+1; q\rangle_n} = \mathrm{RHS}. \tag{9.102}$$

□

The next formula somehow combines the methods of Sections 9.1.6 and 9.1.7.

Theorem 9.1.32 *The q-Gould-Hopper formula* [331, p. 77, 2.4], [13, 3.4]:

$$\begin{aligned}\frac{1}{x^n}\theta_{q,\alpha}^n &= \prod_{k=1}^{n}\left(q^{k+\alpha}\mathbf{x}\mathrm{D}_q + \{\alpha+k\}_q I\right)\\ &= \sum_{k=0}^{n}\binom{n+\alpha}{n-k}_q \frac{\langle 1; q\rangle_n}{\langle 1; q\rangle_k} q^{k(k+\alpha)} x^k \mathrm{D}_q^k (1-q)^{k-n}.\end{aligned} \tag{9.103}$$

As was pointed out in [16, p. 4], the operator $\theta_{q,\alpha}$ is particularly useful in dealing with q-Laguerre polynomials.

Theorem 9.1.33 *A q-analogue of the corrected version of* [13, 3.9, 3.12, 3.13]:

$$\theta_{q,\alpha}^n \mathrm{E}_q(-x) = x^n \mathrm{E}_q(-x)\mathrm{L}_{n,q,\mathrm{C}}^{(\alpha)}(x). \tag{9.104}$$

9.1.7 The q-Laguerre Rodriguez operator

Definition 155 A q-analogue of Chatterjea [119, p. 245]. The q-Laguerre Rodriguez operator $\Omega_{n,q}^{(\alpha)} : \mathrm{H}_q \to \mathrm{H}_q$ is defined by

$$\Omega_{n,q}^{(\alpha)} f(x) \equiv x^{-\alpha}\mathrm{E}_{\frac{1}{q}}(x)\mathrm{D}_q^n\left(x^{\alpha+n}\mathrm{E}_q(-x)f(x)\right), \quad f(x) \in \mathrm{H}_q. \tag{9.105}$$

We immediately obtain

Theorem 9.1.34 *A q-analogue of* [119, p. 245].

$$\Omega_{n,q}^{(\alpha)} = \sum_{k=0}^{n}\binom{n+\alpha}{n-k}_q \frac{\{n\}_q!}{\{k\}_q!} q^{k^2+\alpha k} x^k (\mathrm{D}_q \ominus_q \epsilon)^k. \tag{9.106}$$

Proof Use the q-Leibniz theorem. □

We will now give a q-analogue of the Carlitz [99, p. 219] operator expression for Laguerre polynomials, and an extension of Khan's q-analogue of that paper [331, p. 79]. It turns out that we obtain an equivalence class of six objects for each element in Carlitz' paper.

Theorem 9.1.35 *A q-analogue of Carlitz* [99, p. 219]. *All the products begin with $k=n$ and end with $k=1$:*

$$\begin{aligned}\Omega_{n,q}^{(\alpha)} &\cong \prod_{k=n}^{1}\big(q^{k+\alpha}\big(I+(1-q)\mathbf{x}\big)\mathbf{x}\mathrm{D}_q+\{\alpha+k\}_q I-q^{k+\alpha}\mathbf{x}\big)\\ &\cong \prod_{k=n}^{1}\big(q^{k+\alpha}\mathbf{x}\mathrm{D}_q+\{\alpha+k\}_q I-q^{k+\alpha}\mathbf{x}\epsilon\big)\\ &\cong \prod_{k=n}^{1}\big(\mathbf{x}\mathrm{D}_q+\{\alpha+k\}_q\epsilon-q^{k+\alpha}\mathbf{x}\epsilon\big)\\ &\cong \prod_{k=n}^{1}\big(q^{k+\alpha}\big(I+(1-q)\mathbf{x}\big)\mathbf{x}\mathrm{D}_q+\{\alpha+k\}_q\big(I+(1-q)\mathbf{x}\big)-\mathbf{x}\big)\\ &\cong \prod_{k=n}^{1}\big(\big(I+(1-q)\mathbf{x}\big)\mathbf{x}\mathrm{D}_q+\{\alpha+k\}_q\big(I+(1-q)\mathbf{x}\big)-\mathbf{x}\big)\\ &\cong \prod_{k=n}^{1}\big(\mathbf{x}\mathrm{D}_q+\{\alpha+k\}_q\big(I+(1-q)\mathbf{x}\big)-\mathbf{x}\epsilon\big).\end{aligned}\tag{9.107}$$

Proof We will use (9.105). We only prove the first identity. The five others are proved in a similar way by permutation of the three functions involved in the q-differentiation. We will use Cigler's relation (6.194) in the computations.

$$\begin{aligned}\Omega_{n+1,q}^{(\alpha)} f(x) &= x^{-\alpha}\mathrm{E}_{\frac{1}{q}}(x)\mathrm{D}_q^n\big[\big(1+(1-q)x\big)q^{n+1+\alpha}\mathrm{E}_q(-x)x^{n+1+\alpha}\mathrm{D}_q\\ &\quad+\{\alpha+n+1\}_q x^{n+\alpha}\mathrm{E}_q(-x)-(xq)^{n+1+\alpha}\mathrm{E}_q(-x)\big]f(x)\\ &=\Omega_{n,q}^{(\alpha)}\big[\{\alpha+n+1\}_q-xq^{n+1+\alpha}+\big(1+(1-q)x\big)q^{n+1+\alpha}x\mathrm{D}_q\big]f(x).\end{aligned}\tag{9.108}$$

□

Remark 45 This was the first occasion where multiple q-analogues occurred because of the q-Leibniz theorem. We had three functions and got $\binom{3}{2}$ q-analogues.

Theorem 9.1.36 *A first q-analogue of Carlitz* [99, (4), p. 219], *Chatterjea* [119, p. 246]:

$$\Omega_{n,q}^{(\alpha)} f(x) = \{n\}_q! \sum_{k=0}^{n} \frac{x^k}{\{k\}_q!} \mathrm{L}_{n-k,q}^{(\alpha+k)}(x) \epsilon^{n-k} \mathrm{D}_q^k f(x). \tag{9.109}$$

Proof

$$\begin{aligned}\Omega_{n,q}^{(\alpha)} f(x) &= x^{-\alpha} \mathrm{E}_{\frac{1}{q}}(x) \sum_{k=0}^{n} \binom{n}{k}_q \mathrm{D}_q^{n-k}\big(x^{\alpha+n} \mathrm{E}_q(-x)\big) \epsilon^{n-k} \mathrm{D}_q^k f(x) \\ &= x^{-\alpha} \mathrm{E}_{\frac{1}{q}}(x) \sum_{k=0}^{n} \binom{n}{k}_q x^{\alpha+k} \mathrm{E}_q(-x) \mathrm{L}_{n-k,q,c}^{(\alpha+k)}(x) \epsilon^{n-k} \mathrm{D}_q^k f(x) = \mathrm{RHS}.\end{aligned} \tag{9.110}$$

□

Ohm's law corresponds to the q-Laguerre Rodriguez formula (9.111).

Theorem 9.1.37 *A special case of* (9.109) *and a q-analogue of* [99, (6), p. 220], *see also Khan* [331, p. 79]:

$$\mathrm{L}_{n,q,\mathrm{C}}^{(\alpha)}(x) = \Omega_{n,q}^{(\alpha)} 1. \tag{9.111}$$

Theorem 9.1.38 *The first q-analogue of Carlitz* [99, (7), p. 221], *Das* [147]:

$$\binom{m+n}{m}_q \mathrm{L}_{m+n,q}^{(\alpha)}(x) = \sum_{k=0}^{\min(m,n)} \frac{(-x)^k}{\{k\}_q!} \mathrm{L}_{m-k,q}^{(\alpha+n+k)}(x) q^{\alpha k + \binom{k+1}{2} + \binom{k}{2}} \mathrm{L}_{n-k,q}^{(\alpha+k)}(xq^m). \tag{9.112}$$

Proof As in [99, (7), p. 221]. □

Theorem 9.1.39 *A second q-analogue of* [99, (4), p. 219], *Chatterjea* [119, p. 246]:

$$\Omega_{n,q}^{(\alpha)} f(x) = \{n\}_q! \sum_{k=0}^{n} \frac{x^k}{\{k\}_q!} q^{k(\alpha+k)} \mathrm{P}_{k,q}\big(1, -(1-q)x\big) \mathrm{L}_{n-k,q}^{(\alpha+k)}\big(xq^k\big) \mathrm{D}_q^k f(x). \tag{9.113}$$

Proof We use (6.194) and get

$$\begin{aligned}\Omega_{n,q}^{(\alpha)} f(x) &= x^{-\alpha} \mathrm{E}_{\frac{1}{q}}(x) \sum_{k=0}^{n} \binom{n}{k}_q \epsilon^k \mathrm{D}_q^{n-k}\big(x^{\alpha+n} \mathrm{E}_q(-x)\big) \mathrm{D}_q^k f(x) \\ &= x^{-\alpha} \mathrm{E}_{\frac{1}{q}}(x) \sum_{k=0}^{n} \binom{n}{k}_q \epsilon^k \big[x^{\alpha+k} \mathrm{E}_q(-x) \mathrm{L}_{n-k,q,c}^{(\alpha+k)}(x)\big] \mathrm{D}_q^k f(x)\end{aligned}$$

$$= x^{-\alpha}\mathrm{E}_{\frac{1}{q}}(x)\sum_{k=0}^{n}\binom{n}{k}_q (xq^k)^{\alpha+k}\mathrm{E}_q(-xq^k)\mathrm{L}_{n-k,q,c}^{(\alpha+k)}(xq^k)\mathrm{D}_q^k f(x)$$

$$= \text{RHS}. \tag{9.114}$$

□

The following formula is the second q-analogue of Carlitz [99, (7), p. 221]:

$$\binom{m+n}{m}_q \mathrm{L}_{m+n,q}^{(\alpha)}(x) = \sum_{k=0}^{\min(m,n)} \frac{(-x)^k}{\{k\}_q!}\mathrm{P}_{k,q}(1,-(1-q)x) \times \mathrm{L}_{m-k,q}^{(\alpha+n+k)}(xq^k)q^{k(2\alpha+n+k)+\binom{k+1}{2}+\binom{k}{2}}\mathrm{L}_{n-k,q}^{(\alpha+k)}(xq^k). \tag{9.115}$$

For the next theorem we need the following definition:

Definition 156 The generalized (not commutative) NWA q-addition is the function

$$(x \oplus_{q,t} y)^n \equiv \sum_{k=0}^{n}\binom{n}{k}_q x^{n-k}y^k q^{t(nk-\binom{k}{2})}, \quad n=0,1,2,\ldots. \tag{9.116}$$

An interesting consequence of (9.112) is the following q-analogue of the generating function of Carlitz [99, (10), p. 222]:

Theorem 9.1.40

$$\sum_{n=0}^{\infty}\binom{m+n}{m}_q \mathrm{L}_{m+n,q}^{(\alpha-n)}(x)t^n q^{\binom{n}{2}-n\alpha} = \mathrm{L}_{m,q}^{(\alpha)}(x \oplus_{q,-1} xtq^{-\alpha})\frac{\mathrm{E}_{\frac{1}{q}}(-xtq^m)}{(-t;q)_{-\alpha}}. \tag{9.117}$$

Proof

$$\text{LHS} = \sum_{n=0}^{\infty}\sum_{k=0}^{\min(m,n)}\frac{(-x)^k}{\{k\}_q!}\mathrm{L}_{m-k,q}^{(\alpha+k)}(x) \times \mathrm{QE}\left((\alpha-n)k+\binom{k+1}{2}+\binom{k}{2}\right)\mathrm{L}_{n-k,q}^{(\alpha-n+k)}(xq^m)t^n q^{\binom{n}{2}-n\alpha}$$

$$= \sum_{n=0}^{\infty}\sum_{k=0}^{m}\frac{(-x)^k}{\{k\}_q!}\mathrm{L}_{m-k,q}^{(\alpha+k)}(x)q^{\frac{k^2-k}{2}}\mathrm{L}_{n,q}^{(\alpha-n)}(xq^m)t^{n+k}q^{\binom{n}{2}-n\alpha} = \text{RHS}, \tag{9.118}$$

where in the last step we have used [165, 5.29, p. 28], (10.34) and (9.62). □

9.1.8 q-orthogonality

We now come to the first q-orthogonality relation. All orthogonality proofs use q-integration by parts, a method equivalent to recurrences. The orthogonality for q-Laguerre polynomials has a weight function, namely, x^α times a q-exponential function with negative function value. This q-exponential function can also be written as inverse q-factorial. We can use the definition of $\mathrm{E}_q(-x)$ for $x < \frac{1}{1-q}$. For larger x we use the inverse q-factorial formula.

To make the full proof of the following theorem, we need a formula for a given q-integral.

Lemma 9.1.41 *Compare Jackson* [302, p. 200, (22)]. *The moments of order n for the q-Laguerre weight function are given by*

$$\int_0^\infty x^{\alpha+n}\mathrm{E}_q(-x)\,d_q(x) = \mathrm{QE}\left(-\binom{n+\alpha+1}{2}\right)\Gamma_q(n+\alpha+1). \tag{9.119}$$

Proof Induction. □

Kirchhoff's law corresponds to q-orthogonality (9.120).

Since the Stieltjes moment problem for q-Laguerre polynomials is indeterminate, there are many orthogonality relations. One of them is the following.

Theorem 9.1.42 *A q-analogue of Tricomi* [508, p. 214, (1.6)]. *Let* $\Re(\alpha) > -1$. *Then*

$$\begin{aligned}&\{n\}_q!\int_0^\infty \mathrm{L}_{n,q}^{(\alpha)}(x)\mathrm{L}_{m,q}^{(\alpha)}(x)x^\alpha\mathrm{E}_q(-x)\,d_q(x)\\&\quad=\delta(m,n)\mathrm{QE}\left(\binom{n}{2}+n\alpha-\binom{n+\alpha+1}{2}\right)\Gamma_q(n+\alpha+1).\end{aligned} \tag{9.120}$$

Proof We use (6.59), with $v = \mathrm{D}_q^{n-1}(x^{\alpha+n}\mathrm{E}_q(-x))$ and $u = \epsilon^{-1}\mathrm{L}_{m,q}^{(\alpha)}(x)$. In the penultimate step we use the q-Leibniz formula for the calculation of D_q^{n-l}.

q-integration by parts implies, for $n \geq m$,

$$\begin{aligned}&\int_0^\infty \mathrm{L}_{n,q}^{(\alpha)}(x)\mathrm{L}_{m,q}^{(\alpha)}(x)x^\alpha\{n\}_q!\mathrm{E}_q(-x)\,d_q(x)\\&\quad=\int_0^\infty \mathrm{L}_{m,q}^{(\alpha)}(x)\mathrm{D}_q^n\big(x^{\alpha+n}\mathrm{E}_q(-x)\big)\,d_q(x)\\&\quad=\big[\epsilon^{-1}\mathrm{L}_{m,q}^{(\alpha)}(x)\mathrm{D}_q^{n-1}\big(x^{\alpha+n}\mathrm{E}_q(-x)\big)\big]_0^\infty\\&\qquad-\int_0^\infty \big(\mathrm{D}_q\epsilon^{-1}\big)\big(\mathrm{L}_{m,q}^{(\alpha}(x)\big)\mathrm{D}_q^{n-1}\big(x^{\alpha+n}\mathrm{E}_q(-x)\big)\,d_q(x)\end{aligned}$$

$$
\begin{aligned}
&= \cdots \\
&= \sum_{l=1}^{n} (-1)^{l+1} \Big[\big(\epsilon^{-1} \mathrm{D}_q\big)^{l-1} \big(\epsilon^{-1} \mathrm{L}_{m,q}^{(\alpha)}(x)\big) \mathrm{D}_q^{n-l} \big(x^{\alpha+n} \mathrm{E}_q(-x)\big) \Big]_0^{\infty} \\
&\quad + (-1)^n \int_0^{\infty} \big(\mathrm{D}_q \epsilon^{-1}\big)^n \big[\mathrm{L}_{m,q}^{(\alpha)}(x)\big] x^{\alpha+n} \mathrm{E}_q(-x)\, d_q(x) \\
&= \sum_{l=1}^{n} \Bigg[(-1)^{l+1} \sum_{k=0}^{n-l} \binom{n-l}{k}_q \{\alpha+1+k+l\}_{n-l-k,q} x^{\alpha+k+l} (-1)^k \\
&\quad \times q^{k(\alpha+k+l)} \mathrm{E}_q(-x) \big(\epsilon^{-1} \mathrm{D}_q\big)^{l-1} \epsilon^{-1} \mathrm{L}_{m,q}^{(\alpha)}(x) \Bigg]_0^{\infty} \\
&\quad + (-1)^n \int_0^{\infty} \big(\mathrm{D}_q \epsilon^{-1}\big)^n \big[\mathrm{L}_{m,q}^{(\alpha)}(x)\big] x^{\alpha+n} \mathrm{E}_q(-x)\, d_q(x) = \mathrm{RS}. \qquad (9.121)
\end{aligned}
$$

□

Remark 46 Incidentally, the integral (9.119) is discussed in [49], [194, 2.2.3.3] and [309, p. 6].

9.2 q-Jacobi polynomials

9.2.1 Definition and the Rodriguez formula

We now come to the unit 1 mol.

We are in a border area between physics and chemistry. Physicists did not really think that 1 mole would fit in the SI system.

We use the following symbols:

p	Pressure
V	Volume
n	Chemical amount
R	Universal gas constant
T	Temperature in Kelvin
N_A	Avogadro constant
F	Faraday constant
e	Elementary charge

1 mol	q-Jacobi-polynomial

We now come to the definition of q-Jacobi polynomials. In the literature there is a very similar so-called little q-Jacobi polynomial [31]. We will however use the original definition, because it leads to a nice Rodriguez formula with corresponding orthogonality. In the limit $q \to 1$ we get the original Jacobi polynomials [311, p. 192], [38], [36, p. 162], [325, p. 467], [197, p. 242, (1)], [143, p. 76–77].

Definition 157

$$\begin{aligned}\mathrm{P}_{n,q}^{(\alpha,\beta)}(x) &\equiv \frac{\langle 1+\alpha;q\rangle_n}{\langle 1;q\rangle_n}{}_2\phi_1\big(-n,\beta+n;1+\alpha|q;xq^{\alpha+1-\beta}\big)\\ &\equiv \frac{\langle 1+\alpha;q\rangle_n}{\langle 1;q\rangle_n}\sum_{k=0}^{n}\binom{n}{k}_q\frac{\langle \beta+n;q\rangle_k}{\langle 1+\alpha;q\rangle_k}(-x)^k q^{\binom{k}{2}+(\alpha+1-\beta-n)k}. \end{aligned}\tag{9.122}$$

Theorem 9.2.1 *We have the following limit*:

$$\lim_{\beta\to-\infty}\mathrm{P}_{n,q}^{(\alpha,\beta)}\big(-x(1-q)\big)=\mathrm{L}_{n,q}^{(\alpha)}(x).\tag{9.123}$$

The law

$$F=eN_A\tag{9.124}$$

corresponds to the Rodriguez formula (9.125).

The following formula is a q-analogue of Jacobi [311, p. 192, (7)], Feldheim [197, p. 242], Toscano [506, p. 220, (1.1)], Appell and Kampé de Fériet [41, p. 99], Courant and Hilbert [143, p. 77]:

Theorem 9.2.2 *Rodriguez formula for q-Jacobi polynomials*, $x\in(0,|q^{\beta-\alpha-1}|)$:

$$\mathrm{P}_{n,q}^{(\alpha,\beta)}(x)=\frac{x^{-\alpha}}{\{n\}_q!(xq^{\alpha+1-\beta};q)_{\beta-\alpha-1}}\mathrm{D}_q^n\left(\frac{x^{\alpha+n}}{(x;q)_{\alpha+1-\beta-n}}\right).\tag{9.125}$$

Proof The q-Leibniz formula gives

$$\begin{aligned}\text{RHS} &= \frac{x^{-\alpha}}{\{n\}_q!(xq^{-\beta+\alpha+1};q)_{\beta-\alpha-1}}\\ &\quad\times\sum_{k=0}^{n}\frac{\langle 1;q\rangle_n\{1+\alpha-\beta-n\}_{k,q}\{1+\alpha+k\}_{n-k,q}x^{\alpha+k}q^{k\alpha+k^2}}{\langle 1;q\rangle_k\langle 1;q\rangle_{n-k}(x;q)_{-\beta+\alpha+k+1-n}}\\ &=\sum_{k=0}^{n}\frac{x^k\langle 1+\alpha-\beta-n;q\rangle_k\langle 1+\alpha;q\rangle_n q^{k\alpha+k^2}}{\langle 1+\alpha;q\rangle_k(xq^{-\beta+\alpha+1};q)_{k-n}\langle 1;q\rangle_k\langle 1;q\rangle_{n-k}}\\ &=\sum_{k=0}^{n}\frac{x^k\langle 1+\alpha-\beta-n,-n;q\rangle_k\langle 1+\alpha;q\rangle_n q^{k\alpha+k^2-\binom{k}{2}+nk}(-1)^k}{\langle 1,1+\alpha;q\rangle_k(xq^{-\beta+\alpha+1};q)_{k-n}\langle 1;q\rangle_n}\\ &=\frac{\langle 1+\alpha;q\rangle_n}{\langle 1;q\rangle_n(xq^{-\beta+\alpha+1};q)_{-n}}\\ &\quad\times{}_2\phi_2\big(-n,-n-\beta+\alpha+1;\alpha+1|q;xq^{n+\alpha+1}||-;\big(xq^{-n-\beta+\alpha+1};q\big)_k\big).\end{aligned}\tag{9.126}$$

The interval for x is chosen to make certain infinite products converge, compare [443, p. 300]. ☐

Corollary 9.2.3

$$\mathrm{P}_{n,q}^{(\alpha,\beta)}\big(xq^{\gamma}\big)=\frac{x^{-\alpha}}{\{n\}_q!(xq^{\alpha+\gamma+1-\beta};q)_{\beta-\alpha-1}}\mathrm{D}_q^n\bigg(\frac{x^{\alpha+n}}{(xq^{\gamma};q)_{\alpha+1-\beta-n}}\bigg). \tag{9.127}$$

Proof As above. □

Corollary 9.2.4 *A function* $F(x)\in \mathrm{H}_q$,

$$F(x)=\sum_{k=0}^{\infty}\frac{\alpha_k x^{\beta_k}}{(x;q)_{\gamma_k}},$$

has n*-th* q*-difference given by*

$$\mathrm{D}_q^n F(x)=\sum_{k=0}^{\infty}\alpha_k \mathrm{P}_{n,q}^{(\beta_k-n,\beta_k+1-2n-\gamma_k)}(x)\{n\}_q!\frac{x^{\beta_k-n}}{(x;q)_{\gamma_k+n}}, \tag{9.128}$$

$x\in(0,|q^{-n-\gamma_k}|)$, $\forall k$.

9.2.2 The q-Jacobi Rodriguez operator

In this section, primarily formal identities for q-Jacobi polynomials and q-Jacobi operators are proved. But from time to time a formula for q-Laguerre polynomials comes in between.

Definition 158 The q-Jacobi Rodriguez operator $\Omega_{n,q}^{(\alpha,\beta)}:\mathrm{H}_q\to\mathrm{H}_q$ is a q-analogue of Singh [464, p. 238]:

$$\Omega_{n,q}^{(\alpha,\beta)}f(x)\equiv\frac{x^{-\alpha}}{\{n\}_q!(xq^{\alpha+1-\beta};q)_{\beta-\alpha-1}}\mathrm{D}_q^n\bigg(\frac{x^{\alpha+n}}{(x;q)_{\alpha+1-\beta-n}}f(x)\bigg), \tag{9.129}$$

$f(x)\in\mathrm{H}_q$, $x\in(0,|q^{\beta-\alpha-1}|)$;

$$\Omega_{0,q}^{(\alpha,\beta)}\equiv I. \tag{9.130}$$

Theorem 9.2.5 *Almost a* q*-analogue of Singh* [464, 2.2, p. 238]:

$$\Omega_{n,q}^{(\alpha,\beta-n+1)}f(x)\cong\prod_{k=2}^{n}\bigg(\frac{1-xq^{\alpha-\beta+1}}{\{k\}_q}\bigg)\Omega_{1,q}^{(\alpha,\beta)}\prod_{k=2}^{n}\Theta_{k,q}^{(\alpha,\beta)}f(x),\quad n\ge 1, \tag{9.131}$$

where $\Theta_{k,q}^{(\alpha,\beta)}$ *is given by one of the following six equivalent expressions.* $\Theta_{k,q}^{(\alpha,\beta)}$ *is a bilinear function of* D_q *and* ϵ *with coefficients in the quotient field of* $\mathbb{C}[\mathbf{x}]$.

$$\begin{aligned}
\Theta_{k,q}^{(\alpha,\beta)} &\cong \frac{q^{k+\alpha}(1-\mathbf{x})}{1-\mathbf{x}q^{2-k+\alpha-\beta}}\mathbf{x}\mathrm{D}_q+\{\alpha+k\}_q I+\frac{q^{k+\alpha}\{2-k+\alpha-\beta\}_q}{1-\mathbf{x}q^{2-k+\alpha-\beta}}\mathbf{x}\\
&\cong \{\alpha+k\}_q I+\frac{q^{k+\alpha}\{2-k+\alpha-\beta\}_q}{1-\mathbf{x}q^{2-k+\alpha-\beta}}\mathbf{x}\epsilon+q^{k+\alpha}\mathbf{x}\mathrm{D}_q\\
&\cong \mathbf{x}\mathrm{D}_q+\{\alpha+k\}_q\epsilon+\frac{q^{k+\alpha}\{2-k+\alpha-\beta\}_q}{1-\mathbf{x}q^{2-k+\alpha-\beta}}\mathbf{x}\epsilon\\
&\cong \frac{\{2-k+\alpha-\beta\}_q}{1-\mathbf{x}q^{2-k+\alpha-\beta}}\mathbf{x}+\frac{q^{k+\alpha}(1-\mathbf{x})}{1-\mathbf{x}q^{2-k+\alpha-\beta}}\mathbf{x}\mathrm{D}_q+\frac{\{\alpha+k\}_q(1-\mathbf{x})}{1-\mathbf{x}q^{2-k+\alpha-\beta}}\\
&\cong \frac{\{2-k+\alpha-\beta\}_q}{1-\mathbf{x}q^{2-k+\alpha-\beta}}\mathbf{x}+\frac{(1-\mathbf{x})}{1-\mathbf{x}q^{2-k+\alpha-\beta}}\mathbf{x}\mathrm{D}_q+\frac{\{\alpha+k\}_q(1-\mathbf{x})}{1-\mathbf{x}q^{2-k+\alpha-\beta}}\epsilon\\
&\cong \mathbf{x}\mathrm{D}_q+\frac{\{\alpha+k\}_q(1-\mathbf{x})}{1-\mathbf{x}q^{2-k+\alpha-\beta}}\epsilon+\frac{\{1+\alpha-\beta\}_q}{1-\mathbf{x}q^{1+\alpha-\beta}}\mathbf{x}\epsilon.
\end{aligned}\tag{9.132}$$

Proof We only prove the first identity for $\Theta_{k,q}^{(\alpha,\beta)}$. The five others are proved in a similar way by permutation of the three functions involved in the q-differentiation.

$$\begin{aligned}
\Omega_{n+1,q}^{(\alpha,\beta-1)}f(x) &\cong \frac{x^{-\alpha}}{\{n+1\}_q!(xq^{\alpha+2-\beta};q)_{\beta-\alpha-2}}\\
&\quad\times \mathrm{D}_q^n\left[\left[\frac{\{\alpha+n+1\}_q x^{\alpha+n}}{(x;q)_{1+\alpha-\beta-n}}+\frac{(xq)^{\alpha+n+1}\mathrm{D}_q}{(xq;q)_{1+\alpha-\beta-n}}\right.\right.\\
&\quad\left.\left.+\frac{(xq)^{\alpha+n+1}\{\alpha+1-\beta-n\}_q}{(x;q)_{2+\alpha-\beta-n}}\right]f(x)\right]\\
&\cong \frac{1-xq^{\alpha+1-\beta}}{\{n+1\}_q}\Omega_{n,q}^{(\alpha,\beta)}\left[\left[\frac{(1-x)q^{n+1+\alpha}}{1-xq^{1+\alpha-\beta-n}}x\mathrm{D}_q+\{\alpha+n+1\}_q\right.\right.\\
&\quad\left.\left.+\frac{xq^{n+1+\alpha}\{1+\alpha-\beta-n\}_q}{1-xq^{1+\alpha-\beta-n}}\right]f(x)\right].
\end{aligned}\tag{9.133}$$

The assertion now follows by induction. □

The following generalization of (9.109) is a first q-analogue of Singh [464, 2.3, p. 239], with the difference that in the present paper Jacobi's original polynomial definition is used.

Theorem 9.2.6

$$\Omega_{n,q}^{(\alpha,\beta)} f(x) \cong \sum_{k=0}^{n} \frac{x^k}{\{k\}_q!} \big(xq^{\alpha+1-k-\beta};q\big)_k \mathrm{P}_{n-k,q}^{(\alpha+k,\beta+2k)}(x)\epsilon^{n-k}\mathrm{D}_q^k f(x). \quad (9.134)$$

Proof

$$\begin{aligned}\Omega_{n,q}^{(\alpha,\beta)} f(x) &\cong \frac{x^{-\alpha}}{\{n\}_q!(xq^{\alpha+1-\beta};q)_{\beta-\alpha-1}} \\ &\quad\times \sum_{k=0}^{n} \binom{n}{k}_q \mathrm{D}_q^{n-k}\left[\frac{x^{\alpha+n}}{(x;q)_{-\beta+\alpha+1-n}}\right]\epsilon^{n-k}\mathrm{D}_q^k f(x) \\ &\cong \frac{1}{\{n\}_q!(xq^{\alpha+1-\beta};q)_{\beta-\alpha-1}} \sum_{k=0}^{n} \binom{n}{k}_q x^k \mathrm{P}_{n-k,q}^{(\alpha+k,\beta+2k)}(x) \\ &\quad\times \{n-k\}_q!\big(xq^{\alpha-k+1-\beta};q\big)_{\beta-\alpha+k-1}\epsilon^{n-k}\mathrm{D}_q^k f(x) \cong \mathrm{RHS}. \quad (9.135)\end{aligned}$$

□

Lemma 9.2.7

$$\begin{aligned}\Omega_{n,q}^{(\alpha,\beta)} \frac{x}{1-xq^{1+\alpha-\beta-n}} &\cong \mathrm{P}_{n,q}^{(\alpha,\beta)}(x)\frac{xq^n}{1-xq^{1+\alpha-\beta}} \\ &\quad+ \sum_{k=1}^{n} x^k \mathrm{P}_{n-k,q}^{(\alpha+k,\beta+2k)}(x)\frac{q^{(\alpha+1-\beta-n)(k-1)}}{1-xq^{1+\alpha-\beta}}. \quad (9.136)\end{aligned}$$

Proof Use (9.134) and (9.63). □

Theorem 9.2.8

$$\begin{aligned}\mathrm{P}_{n+1,q}^{(\alpha,\beta-1)}(x) &\cong \frac{1-xq^{\alpha+1-\beta}}{\{n+1\}_q}\{\alpha+n+1\}_q \mathrm{P}_{n,q}^{(\alpha,\beta)}(x) + \frac{q^{\alpha+n+1}\{\alpha+1-\beta-n\}_q}{\{n+1\}_q} \\ &\quad\times \left[\sum_{k=1}^{n} x^k \mathrm{P}_{n-k,q}^{(\alpha+k,\beta+2k)}(x)q^{(\alpha+1-\beta-n)(k-1)} + \mathrm{P}_{n,q}^{(\alpha,\beta)}(x)xq^n\right].\end{aligned}$$
(9.137)

Proof Apply (9.133) to 1 and use (9.136). □

Corollary 9.2.9

$$\begin{aligned}\mathrm{L}_{n+1,q}^{(\alpha)} &\cong \frac{1}{\{n+1\}_q}\big[\{\alpha+n+1\}_q \mathrm{L}_{n,q}^{(\alpha)}(x) \\ &\quad- q^{n+\alpha+1}\big[x\mathrm{L}_{n-1,q}^{(\alpha+1)}(x) + q^n x \mathrm{L}_{n,q}^{(\alpha)}(x)\big]\big]. \quad (9.138)\end{aligned}$$

Proof

$$\begin{aligned}
\text{LHS} &\cong \lim_{\beta\to-\infty} \frac{1+x(1-q)q^{\alpha+1-\beta}}{\{n+1\}_q}\Bigg[\{\alpha+n+1\}_q \mathrm{P}_{n,q}^{(\alpha,\beta)}\big(-x(1-q)\big) \\
&\quad + q^{\alpha+n+1}\{\alpha+1-\beta-n\}_q\Bigg[\sum_{k=1}^{n}\big(-x(1-q)\big)^k \mathrm{P}_{n-k,q}^{(\alpha+k,\beta+2k)}\big(-x(1-q)\big) \\
&\quad \times \frac{q^{(\alpha+1-\beta-n)(k-1)}}{1+x(1-q)q^{1+\alpha-\beta}} + \mathrm{P}_{n,q}^{(\alpha,\beta)}\big(-x(1-q)\big)\frac{-x(1-q)q^n}{1+x(1-q)q^{1+\alpha-\beta}}\Bigg]\Bigg] \\
&\cong \text{RHS}.
\end{aligned} \tag{9.139}$$

□

The following generalization of (9.113) is the second q-analogue of Singh [464, 2.3, p. 239]:

Theorem 9.2.10

$$\Omega_{n,q}^{(\alpha,\beta)} f(x) \cong \sum_{k=0}^{n} \frac{x^k}{\{k\}_q!} q^{k(\alpha+k)}(x;q)_k \mathrm{P}_{n-k,q}^{(\alpha+k,\beta+2k)}\big(xq^k\big)\mathrm{D}_q^k f(x). \tag{9.140}$$

Proof

$$\begin{aligned}
\Omega_{n,q}^{(\alpha,\beta)} f(x) &\cong \frac{x^{-\alpha}}{\{n\}_q!(xq^{\alpha+1-\beta};q)_{\beta-\alpha-1}} \\
&\quad \times \sum_{k=0}^{n}\binom{n}{k}_q \epsilon^k\left[\mathrm{D}_q^{n-k}\left[\frac{x^{\alpha+n}}{(x;q)_{-\beta+\alpha+1-n}}\right]\right]\mathrm{D}_q^k f(x) \cong \text{RHS}.
\end{aligned} \tag{9.141}$$

□

Lemma 9.2.11

$$\begin{aligned}
&\Omega_{n,q}^{(\alpha,\beta)} \frac{x}{1-xq^{1+\alpha-\beta-n}} \\
&\quad \cong \mathrm{P}_{n,q}^{(\alpha,\beta)}(x)\frac{x}{1-xq^{1+\alpha-\beta-n}} \\
&\qquad + \sum_{k=1}^{n} x^k q^{k(\alpha+k)} \mathrm{P}_{n-k,q}^{(\alpha+k,\beta+2k)}\big(xq^k\big)\frac{q^{(\alpha+1-\beta-n)(k-1)}(x;q)_k}{(xq^{\alpha+1-\beta-n});q)_{k+1}}.
\end{aligned} \tag{9.142}$$

Proof Use (9.140) and (9.63). □

Theorem 9.2.12

$$\begin{aligned}
&\frac{\{n+1\}_q}{1-xq^{\alpha+1-\beta}}\mathrm{P}_{n+1,q}^{(\alpha,\beta-1)}(x)\\
&\quad\cong\{\alpha+n+1\}_q\mathrm{P}_{n,q}^{(\alpha,\beta)}(x)\\
&\qquad+q^{\alpha+n+1}\{\alpha+1-\beta-n\}_q\Bigg(\mathrm{P}_{n,q}^{(\alpha,\beta)}(x)\frac{x}{1-xq^{\alpha+1-\beta-n}}\\
&\qquad+\sum_{k=1}^{n}x^k q^{k(\alpha+k)}(x;q)_k\mathrm{P}_{n-k,q}^{(\alpha+k,\beta+2k)}\big(xq^k\big)\frac{q^{(\alpha+1-\beta-n)(k-1)}}{(xq^{\alpha+1-\beta-n};q)_{k+1}}(x;q)_k\Bigg).
\end{aligned}\tag{9.143}$$

Proof Apply (9.133) on 1 and use (9.142). □

Theorem 9.2.13 *A variation of the Rodriguez formula:*

$$\mathrm{P}_{n,q}^{(\alpha,\beta)}(x)=\frac{x^{-\alpha-n-1}}{\{n\}_q!(xq^{\alpha+1-\beta};q)_{\beta-\alpha-1}}\big(x^2\mathrm{D}_q\big)^n\left(\frac{x^{\alpha+1}}{(x;q)_{\alpha+1-\beta-n}}\right).\tag{9.144}$$

Proof This follows from a q-analogue of Toscano [506, p. 220]. □

The limit to q-Laguerre polynomials for the above equation leads to (9.48).

The second of the following equations shows that the operator $\mathrm{D}_q\epsilon^{-1}$ keeps the same function argument, while D_q does not. This will be important in future applications.

$$\begin{aligned}
\mathrm{D}_q^m\mathrm{P}_{n,q}^{(\alpha,\beta)}(x)&=\mathrm{P}_{n-m,q}^{(\alpha+m,\beta+m)}\big(xq^m\big)\frac{(-1)^m\langle\beta+n;q\rangle_m}{(1-q)^m}\\
&\quad\times\mathrm{QE}\left(\binom{m}{2}+m(\alpha+1-\beta-n)\right),
\end{aligned}\tag{9.145}$$

$$\begin{aligned}
\big(\mathrm{D}_q\epsilon^{-1}\big)^m\mathrm{P}_{n,q}^{(\alpha,\beta)}(x)&=\mathrm{P}_{n-m,q}^{(\alpha+m,\beta+m)}(x)\frac{(-1)^m\langle\beta+n;q\rangle_m}{(1-q)^m}\\
&\quad\times\mathrm{QE}\big(m(\alpha-\beta-n)\big).
\end{aligned}\tag{9.146}$$

By the Ward q-Taylor formula (4.34), the first Jackson q-Taylor formula (4.35), and (10.34) we obtain the following three q-analogues of Manocha and Sharma [367, (8), p. 460]:

Theorem 9.2.14

$$\mathrm{P}_{n,q}^{(\alpha,\beta)}(x \oplus_q y) = \sum_{k=0}^{n} \frac{y^k}{\{k\}_q!} \mathrm{P}_{n-k,q}^{(\alpha+k,\beta+k)}\big(xq^k\big) \frac{(-1)^k \langle \beta+n; q\rangle_k}{(1-q)^k} \times \mathrm{QE}\left(\binom{k}{2} + k(\alpha+1-\beta-n)\right), \tag{9.147}$$

$$\mathrm{P}_{n,q}^{(\alpha,\beta)}(x) = \sum_{k=0}^{n} \frac{(x \boxminus_q y)^k}{\{k\}_q!} \mathrm{P}_{n-k,q}^{(\alpha+k,\beta+k)}\big(yq^k\big) \frac{(-1)^k \langle \beta+n; q\rangle_k}{(1-q)^k} \times \mathrm{QE}\left(\binom{k}{2} + k(\alpha+1-\beta-n)\right), \tag{9.148}$$

$$\mathrm{P}_{n,q}^{(\alpha,\beta)}(x \oplus_{q,-1} y) = \sum_{k=0}^{n} \frac{(-y)^k}{\{k\}_q!} \mathrm{P}_{n-k,q}^{(\alpha+k,\beta+k)}(x) \frac{\langle \beta+n; q\rangle_k}{(1-q)^k} q^{k(\alpha-\beta-n)}. \tag{9.149}$$

Theorem 9.2.15 *The first q-analogue of Manocha, Sharma* [367, (6), p. 459] *and Feldheim* [200, p. 134]:

$$\binom{m+n}{m}_q \big(xq^{\alpha+1+n-\beta}; q\big)_{\beta-n-\alpha-1} \mathrm{P}_{m+n,q}^{(\alpha,\beta-n)}(x) \cong \sum_{k=0}^{\min(m,n)} \frac{\mathrm{P}_{n-k,q}^{(\alpha+k,\beta+2k-n)}(x) x^k}{\{k\}_q! (x;q)_{\alpha+1+n-\beta-k}} \mathrm{P}_{m-k,q}^{(\alpha+n+k,\beta+n+k)}\big(xq^n\big)(-1)^k \frac{\langle \beta+n+m; q\rangle_k}{(1-q)^k} \times \mathrm{QE}\left(\binom{k}{2} + k(\alpha+1-\beta-m)\right). \tag{9.150}$$

Proof

$$\begin{aligned}
\mathrm{P}_{m+n,q}^{(\alpha,\beta-n)}(x) &\cong \frac{x^{-\alpha}}{\{m+n\}_q! (xq^{\alpha+1-\beta+n}; q)_{\beta-\alpha-1-n}} \mathrm{D}_q^{m+n}\left(\frac{x^{\alpha+m+n}}{(x;q)_{\alpha+1-\beta-m}}\right) \\
&\cong \frac{x^{-\alpha}\{m\}_q!}{\{m+n\}_q! (xq^{\alpha+1-\beta+n}; q)_{\beta-\alpha-1-n}} \\
&\quad \times \mathrm{D}_q^n\Big[x^{\alpha+n} \mathrm{P}_{m,q}^{(\alpha+n,\beta+n)}(x) \big(xq^{\alpha+1-\beta}; q\big)_{\beta-\alpha-1}\Big] \\
&\cong \frac{x^{-\alpha}\{m\}_q!}{\{m+n\}_q! (xq^{\alpha+1-\beta+n}; q)_{\beta-\alpha-1-n}} \sum_{k=0}^{n} \binom{n}{k}_q \\
&\quad \times \frac{\mathrm{P}_{n-k,q}^{(\alpha+k,\beta+2k-n)}(x) x^{\alpha+k} \{n-k\}_q!}{(x;q)_{\alpha+1-\beta-k+n}} \epsilon^{n-k} \mathrm{D}_q^k \mathrm{P}_{m,q}^{(\alpha+n,\beta+n)}(x)
\end{aligned}$$

$$\stackrel{\text{by (9.145)}}{\cong} \frac{x^{-\alpha}\{m\}_q!\{n\}_q!}{\{m+n\}_q!(xq^{\alpha+1-\beta+n};q)_{\beta-\alpha-1-n}}$$

$$\times \sum_{k=0}^{\min(m,n)} \frac{\mathrm{P}_{n-k,q}^{(\alpha+k,\beta+2k-n)}(x)x^{\alpha+k}}{\{k\}_q!(x;q)_{\alpha+1-\beta-k+n}} \mathrm{P}_{m-k,q}^{(\alpha+n+k,\beta+n+k)}(xq^n)(-1)^k$$

$$\times \frac{\langle\beta+n+m;q\rangle_k}{(1-q)^k}\mathrm{QE}\left(\binom{k}{2}+k(\alpha+1-\beta-m)\right). \quad (9.151)$$

□

Theorem 9.2.16 *The second q-analogue of Manocha, Sharma* [367, (6), p. 459] *and Feldheim* [200, p. 134]:

$$\binom{m+n}{m}_q (xq^{\alpha+1+n-\beta};q)_{\beta-n-\alpha-1}\mathrm{P}_{m+n,q}^{(\alpha,\beta-n)}(x)$$

$$\cong \sum_{k=0}^{\min(m,n)} \frac{\mathrm{P}_{n-k,q}^{(\alpha+k,\beta+2k-n)}(xq^k)x^k}{\{k\}_q!(xq^k;q)_{\alpha+1+n-\beta-k}} \mathrm{P}_{m-k,q}^{(\alpha+n+k,\beta+n+k)}(xq^k)(-1)^k$$

$$\times \frac{\langle\beta+n+m;q\rangle_k}{(1-q)^k}\mathrm{QE}\left(\binom{k}{2}+k(\alpha+1-\beta-m)+k+\alpha\right). \quad (9.152)$$

In the limit we obtain the following two additional q-analogues of [99, (7), p. 221]:

$$\binom{m+n}{m}_q \mathrm{L}_{m+n,q}^{(\alpha)}(x) = \sum_{k=0}^{\min(m,n)} \frac{(-x)^k}{\{k\}_q!}\mathrm{L}_{m-k,q}^{(\alpha+n+k)}(xq^n)$$

$$\times \mathrm{QE}\left(k(n+\alpha+1)+2\binom{k}{2}\right)\mathrm{L}_{n-k,q}^{(\alpha+k)}(x), \quad (9.153)$$

$$\binom{m+n}{m}_q \mathrm{L}_{m+n,q}^{(\alpha)}(x) \cong \sum_{k=0}^{\min(m,n)} \frac{(-x)^k}{\{k\}_q!}\mathrm{L}_{m-k,q}^{(\alpha+n+k)}(xq^k)\mathrm{P}_{k,q}(1,-(1-q)x)$$

$$\times \mathrm{QE}\left(k(n+\alpha+1)+2\binom{k}{2}+k+\alpha\right)\mathrm{L}_{n-k,q}^{(\alpha+k)}(xq^k). \quad (9.154)$$

The following two formulas are q-analogues of Manocha and Sharma [367, (9), p. 460]:

Theorem 9.2.17

$$\mathrm{P}_{n,q}^{(\alpha+\gamma,\beta+\delta)}(x) \cong \sum_{k=0}^{n} q^{(n-k)(\gamma-k)}\mathrm{P}_{n-k,q}^{(\alpha+k,\beta+k)}(x)\mathrm{P}_{k,q}^{(\gamma-k,1+\delta-k)}(xq^{\alpha+1-\beta}). \quad (9.155)$$

Proof

$$\begin{aligned}
\text{LHS} &\cong \frac{x^{-\alpha-\gamma}}{\{n\}_q!(xq^{\alpha+1+\gamma-\beta-\delta};q)_{\beta+\delta-\alpha-\gamma-1}} \mathrm{D}_q^n\left(\frac{x^{\alpha+\gamma+n}}{(x;q)_{\alpha+\gamma+1-\beta-\delta-n}}\right) \\
&\cong \frac{x^{-\alpha-\gamma}}{\{n\}_q!(xq^{\alpha+1+\gamma-\beta-\delta};q)_{\beta+\delta-\alpha-\gamma-1}} \sum_{k=0}^{n} \binom{n}{k}_q \mathrm{D}_q^{n-k}\left(\frac{x^{\alpha+n}}{(x;q)_{\alpha+1-\beta-n+k}}\right) \\
&\quad \times \epsilon^{n-k}\mathrm{D}_q^k\left(\frac{x^{\gamma}}{(xq^{\alpha+1-\beta-n+k};q)_{\gamma-\delta-k}}\right) \\
&\overset{\text{by (9.127)}}{\cong} \frac{x^{-\alpha-\gamma}}{(xq^{\alpha+1+\gamma-\beta-\delta};q)_{\beta+\delta-\alpha-\gamma-1}} \\
&\quad \times \sum_{k=0}^{n} x^{\alpha+k}\left(xq^{\alpha+1-\beta};q\right)_{\beta-\alpha-1} \mathrm{P}_{n-k,q}^{(\alpha+k,\beta+k)}(x) \\
&\quad \times \epsilon^{n-k}\left[x^{\gamma-k}\left(xq^{\gamma+\alpha-\beta-\delta+1-n+k};q\right)_{\delta-\gamma} \mathrm{P}_{k,q}^{(\gamma-k,1+\delta-k)}\left(xq^{\alpha+1-\beta-n+k}\right)\right] \\
&\cong \text{RHS}.
\end{aligned} \tag{9.156}$$

□

Theorem 9.2.18

$$\begin{aligned}
\mathrm{P}_{n,q}^{(\alpha+\gamma,\beta+\delta)}(x) &\cong \sum_{k=0}^{n} q^{k(\alpha+k)} \mathrm{P}_{n-k,q}^{(\alpha+k,\beta+k)}\left(xq^k\right) \mathrm{P}_{k,q}^{(\gamma-k,1+\delta-k)}\left(xq^{\alpha+1-\beta-n+k}\right) \\
&\quad \times \frac{(x;q)_k}{(xq^{\alpha+1-\beta+\gamma-\delta};q)_{-n+k}(xq^{\alpha+1-\beta-n+k};q)_n}.
\end{aligned} \tag{9.157}$$

Proof A slight modification of the previous proof:

$$\begin{aligned}
\text{LHS} &\cong \frac{x^{-\alpha-\gamma}}{\{n\}_q!(xq^{\alpha+1+\gamma-\beta-\delta};q)_{\beta+\delta-\alpha-\gamma-1}} \\
&\quad \times \sum_{k=0}^{n} \binom{n}{k}_q \epsilon^k \mathrm{D}_q^{n-k}\left(\frac{x^{\alpha+n}}{(x;q)_{\alpha+1-\beta-n+k}}\right) \mathrm{D}_q^k\left(\frac{x^{\gamma}}{(xq^{\alpha+1-\beta-n+k};q)_{\gamma-\delta-k}}\right) \\
&\overset{\text{by (9.127)}}{\cong} \frac{x^{-\alpha-\gamma}}{(xq^{\alpha+1+\gamma-\beta-\delta};q)_{\beta+\delta-\alpha-\gamma-1}} \\
&\quad \times \sum_{k=0}^{n} \epsilon^k\left[x^{\alpha+k}\left(xq^{\alpha+1-\beta};q\right)_{\beta-\alpha-1} \mathrm{P}_{n-k,q}^{(\alpha+k,\beta+k)}(x)\right] \\
&\quad \times x^{\gamma-k}\left(xq^{\gamma+\alpha-\beta-\delta+1-n+k};q\right)_{\delta-\gamma} \mathrm{P}_{k,q}^{(\gamma-k,1+\delta-k)}\left(xq^{\alpha+1-\beta-n+k}\right) \cong \text{RHS}.
\end{aligned} \tag{9.158}$$

□

9.2.3 More generating functions and recurrences

The connection between generating functions and recurrence relations is well-known. We will give an example of how the operator technique can be used to obtain a new generating function for the q-Laguerre polynomials, compare with W. A. Al-Salam [13, p. 132].

Theorem 9.2.19

$$\sum_{n=0}^{\infty} \mathrm{L}_{n,q}^{(\alpha)}(x)\mathrm{E}_q(-x)(xt)^n = \frac{1}{(tx;q)_{\alpha+1}}\mathrm{E}_q\left(\frac{-x}{[1-xtq^{\alpha+1}]}\right). \tag{9.159}$$

Proof Apply $\mathrm{E}_q(t\theta_{q,\alpha})$ to $\mathrm{E}_q(-x)$ in two different ways and use (9.87) and (9.104). □

Theorem 9.2.20 *Three q-analogues of the recurrence relation due to Lebedev* [355, p. 78, 4.18.1], *which are all different from* [387, p. 26, 3.2]:

$$\begin{aligned}
&\{n+1\}_q\mathrm{L}_{n+1,q}^{(\alpha)}(x) + xq^{1+\alpha+n}\mathrm{L}_{n,q}^{(\alpha)}(qx) - \{n\}_q(1+q)\mathrm{L}_{n,q}^{(\alpha)}(x) \\
&\quad - \{\alpha+1\}_q\, q^n\mathrm{L}_{n,q}^{(\alpha)}(x) + \{\alpha+1\}_q\, q^n\mathrm{L}_{n-1,q}^{(\alpha)}(x) + \{n-1\}_q\, q\mathrm{L}_{n-1,q}^{(\alpha)}(x) \\
&\quad - xq^{n+\alpha+1}(1-q^{n+\alpha})\mathrm{L}_{n-1,q}^{(\alpha)}(qx) = 0,
\end{aligned} \tag{9.160}$$

$$\begin{aligned}
&\{n+1\}_q\mathrm{L}_{n+1,q}^{(\alpha)}(x) + xq^{1+\alpha+n}\mathrm{L}_{n,q}^{(\alpha)}(qx) - \{n\}_q(1+q)\mathrm{L}_{n,q}^{(\alpha)}(x) \\
&\quad - \{\alpha+1\}_q\, q^n\mathrm{L}_{n,q}^{(\alpha)}(qx) + \{\alpha+1\}_q\, q^n\mathrm{L}_{n-1,q}^{(\alpha)}(qx) + \{n-1\}_q\, q\mathrm{L}_{n-1,q}^{(\alpha)}(x) \\
&\quad - xq^{n+\alpha+1}(1-q^{n-1})\mathrm{L}_{n-1,q}^{(\alpha)}(qx) = 0,
\end{aligned} \tag{9.161}$$

$$\begin{aligned}
&\{n+1\}_q\mathrm{L}_{n+1,q}^{(\alpha)}(x) + xq^{1+\alpha+2n}\mathrm{L}_{n,q}^{(\alpha)}(x) - \{n\}_q(1+q)\mathrm{L}_{n,q}^{(\alpha)}(x) \\
&\quad - \{\alpha+1\}_q\, q^n\mathrm{L}_{n,q}^{(\alpha)}\left(q^{-1}x\right) + \{\alpha+1\}_q\, q^n\mathrm{L}_{n-1,q}^{(\alpha)}\left(q^{-1}x\right) \\
&\quad + \{n-1\}_q\, q\mathrm{L}_{n-1,q}^{(\alpha)}(x) - xq^{\alpha+2n-1}\left(1-q^{\alpha+1}\right)\mathrm{L}_{n-1,q}^{(\alpha)}(x) = 0.
\end{aligned} \tag{9.162}$$

Proof Compute the q-difference of $F(x,t,q,\alpha)$ with respect to t in two different ways. We start with (9.160):

$$\begin{aligned}
\mathrm{D}_{q,t}F &= \sum_{n=0}^{\infty}\frac{q^{n^2+n\alpha}(-x)(-xt)^{n-1}\{n\}_q}{(t;q)_{\alpha+1+n}\{n\}_q!} + \sum_{n=0}^{\infty}\frac{q^{n^2+n(\alpha+1)}(-xt)^n\{\alpha+1+n\}_q}{(t;q)_{\alpha+2+n}\{n\}_q!} \\
&= \sum_{n=0}^{\infty}\frac{q^{(n+1)^2+(n+1)\alpha}(-x)(-xt)^n}{(tq;q)_{\alpha+1+n}(1-t)\{n\}_q!} + \sum_{n=0}^{\infty}\frac{q^{n^2+n(\alpha+1)}(-xt)^n\{\alpha+1\}_q}{(tq;q)_{\alpha+1+n}(1-t)\{n\}_q!}
\end{aligned}$$

$$+\sum_{n=0}^{\infty}\frac{q^{(n+1)^2+(n+2)(\alpha+1)}(-xt)(-xt)^n}{(tq^2;q)_{\alpha+1+n}(1-t)(1-tq)\{n\}_q!}. \tag{9.163}$$

Then

$$\sum_{n=0}^{\infty}\frac{-x\mathrm{L}_{n,q}^{(\alpha)}(qx)(tq)^n q^{\alpha+1}}{1-t}+\sum_{n=0}^{\infty}\frac{\{1+\alpha\}_q\mathrm{L}_{n,q}^{(\alpha)}(x)(tq)^n}{1-t}$$
$$+\sum_{n=0}^{\infty}\frac{-xt\mathrm{L}_{n,q}^{(\alpha)}(qx)(tq^2)^n q^{2\alpha+3}}{(1-t)(1-tq)}=\sum_{n=0}^{\infty}\{1+n\}_q\mathrm{L}_{n+1,q}^{(\alpha)}(x)t^n. \tag{9.164}$$

Multiply with $(1-t)(1-tq)$ and equate the coefficients of t^n.

The proof of (9.162) goes as follows:

$$\mathrm{D}_{q,t}F=\sum_{n=0}^{\infty}\frac{q^{n^2+n\alpha}(-x)^n((t;q)_{\alpha+1+n}\{n\}_q t^{n-1}-t^n\mathrm{D}_q(t;q)_{\alpha+1+n})}{(tq;q)_{\alpha+1+n}(t;q)_{\alpha+1+n}\{n\}_q!}$$
$$=\sum_{n=0}^{\infty}\frac{q^{n^2+n\alpha}(-x)^n t^n((t;q)_{\alpha+1+n}\{n\}_q t^{-1}+\{\alpha+1+n\}_q(tq;q)_{\alpha+n})}{(tq;q)_{\alpha+1+n}(t;q)_{\alpha+1+n}\{n\}_q!}$$
$$=\sum_{n=0}^{\infty}\frac{q^{(n+1)^2+(n+1)\alpha}(-x)(-xt)^n}{(tq^2;q)_{\alpha+1+n}(1-tq)\{n\}_q!}+\sum_{n=0}^{\infty}\frac{q^{n^2+n\alpha}(-xt)^n\{\alpha+1\}_q}{(tq;q)_{\alpha+1+n}(1-t)\{n\}_q!}$$
$$+\sum_{n=0}^{\infty}\frac{q^{(n+1)^2+(n+1)\alpha}(-xt)(-xt)^n q^{\alpha+1}}{(tq^2;q)_{\alpha+1+n}(1-t)(1-tq)\{n\}_q!}. \tag{9.165}$$

Then

$$\sum_{n=0}^{\infty}\frac{-x\mathrm{L}_{n,q}^{(\alpha)}(x)(tq^2)^n q^{\alpha+1}}{1-tq}+\sum_{n=0}^{\infty}\frac{\{1+\alpha\}_q\mathrm{L}_{n,q}^{(\alpha)}(q^{-1}x)(tq)^n}{1-t}$$
$$+\sum_{n=0}^{\infty}\frac{-xt\mathrm{L}_{n,q}^{(\alpha)}(x)(tq^2)^n q^{2\alpha+2}}{(1-t)(1-tq)}=\sum_{n=0}^{\infty}\{1+n\}_q\mathrm{L}_{n+1,q}^{(\alpha)}(x)t^n. \tag{9.166}$$

Now multiply with $(1-t)(1-tq)$ and equate the coefficients of t^n. □

Remark 47 This proof is an example of a general method in q-calculus to obtain multiple q-analogues. In our case, $\{1+\alpha+n\}_q$ can be expressed as $\{1+\alpha\}_q+q^{1+\alpha}\{n\}_q$ or $\{1+\alpha\}_q q^n+\{n\}_q$. Compare with the similar reasoning using the q-Leibniz theorem.

Remark 48 The three formulas in the previous theorem have polynomials with both x and qx or $\frac{x}{q}$ as variables. They are analogues of different results for Laguerre polynomials, formulas which involve both Laguerre polynomials and their derivatives.

These formulas for q-Laguerre polynomials have some uses, but are of secondary importance, while the three term recurrence relation is of primary importance.

Theorem 9.2.21 *Yet another generating function for the q-Laguerre polynomials*:

$$\sum_{n=0}^{\infty}\frac{\mathrm{L}_{n,q}^{(\alpha)}(xq^{-n})t^n q^{\binom{n}{2}}}{\{1+\alpha\}_{n,q}}=\mathrm{E}_{\frac{1}{q}}(t){}_0\phi_2\big(-;1+\alpha,\infty|q;q^{1+\alpha}(1-q)^2(xt)\big). \quad (9.167)$$

Proof

$$\sum_{n=0}^{\infty}\frac{\mathrm{L}_{n,q}^{(\alpha)}(xq^{-n})}{\{1+\alpha\}_{n,q}}t^n q^{\binom{n}{2}}$$
$$=\sum_{k,n=0}^{\infty}\frac{t^n q^{\binom{n}{2}}(1-q)^{n+k}}{\langle 1+\alpha;q\rangle_k}\frac{\langle -n\rangle_k}{\langle 1\rangle_k}\frac{q^{\frac{k^2+k}{2}}}{\langle 1;q\rangle_n}q^{\alpha k}x^k$$
$$=\sum_{k,n=0}^{\infty}\frac{t^{n+k}q^{\frac{n^2+2nk+k^2-n-k}{2}}(1-q)^{2k+n}}{\langle 1,1+\alpha;q\rangle_k\langle 1;q\rangle_{n+k}}\langle -n-k;q\rangle_k q^{\frac{k^2+k}{2}+\alpha k}x^k$$
$$=\sum_{n}\frac{t^n q^{\binom{n}{2}}(1-q)^n}{\langle 1;q\rangle_n}\sum_{k}\frac{t^k(-1)q^{\binom{k}{2}+k^2+\alpha k}(1-q)^{2k}x^k}{\langle 1,1+\alpha;q\rangle_k}$$
$$=\mathrm{E}_{\frac{1}{q}}(t){}_0\phi_2\big(-;1+\alpha,\infty|q,xtq^{1+\alpha}(1-q)^2\big). \quad (9.168)$$

□

The generating functions for the q-Laguerre polynomials are special cases of

Theorem 9.2.22 *A q-analogue of Srivastava* [477, 1.2, p. 328], *Brafman* [80, (25), p. 947] *and Chaundy* [124, p. 62].

$$\sum_{n=0}^{\infty}\frac{\langle a;q\rangle_n}{\langle 1;q\rangle_n}t^n_{p+1}\phi_p\big(-n,(A);(C)|q;x\big)=\frac{1}{(t;q)_a}\sum_{m=0}^{\infty}\frac{(-xt)^m q^{\binom{m}{2}-m^2}\langle (A),a;q\rangle_m}{\langle 1,(C);q\rangle_m(tq^{-m};q)_m}. \quad (9.169)$$

Proof

$$\sum_{n=0}^{\infty}\frac{\langle a;q\rangle_n}{\langle 1;q\rangle_n}t^n{}_{p+1}\phi_p\big(-n,(A);(C)|q;x\big)$$
$$=\sum_{m=0}^{\infty}\frac{(-xt)^m q^{\binom{m}{2}-m^2}\langle (A),a;q\rangle_m}{\langle 1,(C);q\rangle_m}\sum_{n=0}^{\infty}\frac{\langle a+m;q\rangle_n}{\langle 1;q\rangle_n}t^n q^{-mn}$$

$$= \sum_{m=0}^{\infty} \frac{(-xt)^m q^{\binom{m}{2}-m^2} \langle (A), a; q\rangle_m}{\langle 1, (C); q\rangle_m (tq^{-m}; q)_{a+m}} = \text{LHS}. \tag{9.170}$$

In the last step we have used the q-binomial theorem. □

Corollary 9.2.23 *A confluent form and a q-analogue of Srivastava* [477, 1.7, p. 329] *and Brafman* [80, (27), p. 947]

$$\sum_{n=0}^{\infty} \frac{t^n}{\{n\}_q!} {}_{p+1}\phi_p\big(-n, (A); (C)|q; x\big)$$
$$= \mathrm{E}_q(t) \sum_{m=0}^{\infty} \frac{(-xt(1-q))^m q^{\binom{m}{2}-m^2} \langle (A); q\rangle_m}{\langle 1, (C); q\rangle_m (t(1-q)q^{-m}; q)_m}. \tag{9.171}$$

Proof Let $a \to \infty$ in (9.169). □

Theorem 9.2.24 *A q-extension of* (9.20):

$$\sum_{n=0}^{\infty} \frac{\langle d; q\rangle_n}{\langle 1+\alpha; q\rangle_n} t^n \mathrm{P}_{n,q}^{(\alpha,\beta)}(x) = \sum_{m=0}^{\infty} \frac{(-xtq^{-\beta+\alpha+1})^m q^{\binom{m}{2}-m^2} \langle \beta+m, d; q\rangle_m}{\langle 1, 1+\alpha; q\rangle_m}$$
$$\times \sum_{n=0}^{\infty} \frac{\langle d+m, \beta+2m; q\rangle_n}{\langle 1, \beta+m; q\rangle_n} t^n q^{-mn}. \tag{9.172}$$

Proof

$$\text{LHS} = \sum_{n=0}^{\infty} \frac{\langle d; q\rangle_n}{\langle 1; q\rangle_n} t^n \, {}_2\phi_1\big(-n, \beta+n; \alpha+1|q; xq^{-\beta+\alpha+1}\big)$$
$$= \sum_{n=0}^{\infty} \frac{\langle d; q\rangle_n}{\langle 1; q\rangle_n} t^n \sum_{m=0}^{n} \frac{(-xq^{-\beta+\alpha+1})^m \langle 1; q\rangle_n q^{\binom{m}{2}-mn} \langle \beta+n; q\rangle_m}{\langle 1+\alpha; q\rangle_m} \binom{n}{m}_q$$
$$= \text{RHS}. \tag{9.173}$$

□

Theorem 9.2.25 *A q-extension of* (9.36):

$$\sum_{n=0}^{\infty} t^n \mathrm{P}_{n,q}^{(\alpha-n,\beta)}(x) q^{\binom{n}{2}-n\alpha} = \sum_{m=0}^{\infty} \frac{(-xtq^{-\beta})^m \langle \beta+m; q\rangle_m}{\langle 1; q\rangle_m}$$
$$\times \sum_{n=0}^{\infty} \frac{\langle -\alpha, \beta+2m; q\rangle_n}{\langle 1, \beta+m; q\rangle_n} (-t)^n q^{-mn}. \tag{9.174}$$

Proof

$$\begin{aligned}
\text{LHS} &= \sum_{n=0}^{\infty} \frac{\langle 1+\alpha-n;q\rangle_n}{\langle 1;q\rangle_n} t^n q^{\binom{n}{2}-n\alpha}\ {}_2\phi_1\big(-n,\beta+n;\alpha+1-n|q;xq^{-\beta+\alpha+1-n}\big)\\
&\equiv \sum_{n=0}^{\infty} \frac{\langle -\alpha;q\rangle_n}{\langle 1;q\rangle_n} (-t)^n\ {}_2\phi_1\big(-n,\beta+n;\alpha+1-n|q;xq^{-\beta+\alpha+1-n}\big)\\
&= \sum_{n=0}^{\infty} \langle -\alpha;q\rangle_n (-t)^n \sum_{m=0}^{n} \frac{(-xq^{-\beta+\alpha+1-n})^m q^{\binom{m}{2}-mn} \langle \beta+n;q\rangle_m}{\langle 1,1+\alpha-n;q\rangle_m \langle 1;q\rangle_{n-m}}\\
&= \sum_{n,m=0}^{\infty} \frac{(xq^{1+\alpha-\beta-n-m})^m t^{n+m} (-1)^n \langle -\alpha;q\rangle_{n+m} \langle \beta+n+m;q\rangle_m q^{\binom{m}{2}-mn}}{\langle 1;q\rangle_n \langle 1,1+\alpha-n-m;q\rangle_m}\\
&= \text{RHS}.
\end{aligned} \tag{9.175}$$

□

9.2.4 q-orthogonality

The rule

$$pV = nRT \tag{9.176}$$

corresponds to the q-orthogonality relation (9.177).

Theorem 9.2.26

$$\begin{aligned}
&\int_0^{q^{\beta-\alpha-1}} \mathrm{P}_{n,q}^{(\alpha,\beta)}(x)\mathrm{P}_{m,q}^{(\alpha,\beta)}(x)x^\alpha \{n\}_q!\big(xq^{-\beta+\alpha+1};q\big)_{\beta-\alpha-1}\,d_q(x)\\
&\quad= \delta(m,n)\frac{\langle \beta+n;q\rangle_n}{(1-q)^n}\mathrm{QE}\big((1+\alpha)(-\alpha+\beta+n)\big)\mathrm{B}_q(\beta-\alpha+n,\alpha+1+n).
\end{aligned} \tag{9.177}$$

Proof q-Integration by parts gives, for $n \geq m$:

$$\begin{aligned}
&\int_0^{q^{\beta-\alpha-1}} \mathrm{P}_{n,q}^{(\alpha,\beta)}(x)\mathrm{P}_{m,q}^{(\alpha,\beta)}(x)x^\alpha \{n\}_q!\big(xq^{-\beta+\alpha+1};q\big)_{\beta-\alpha-1}\,d_q(x)\\
&= \int_0^{q^{\beta-\alpha-1}} \mathrm{D}_q^n\bigg(\frac{x^{\alpha+n}}{(x;q)_{\alpha+1-\beta-n}}\bigg)\mathrm{P}_{m,q}^{(\alpha,\beta)}(x)\,d_q(x)\\
&= \bigg[\mathrm{D}_q^{n-1}\bigg(\frac{x^{\alpha+n}}{(x;q)_{\alpha+1-\beta-n}}\bigg)\epsilon^{-1}\mathrm{P}_{m,q}^{(\alpha,\beta)}(x)\bigg]_0^{q^{\beta-\alpha-1}}\\
&\quad - \int_0^{q^{\beta-\alpha-1}} \mathrm{D}_q^{n-1}\bigg(\frac{x^{\alpha+n}}{(x;q)_{\alpha+1-\beta-n}}\bigg)\mathrm{D}_q\epsilon^{-1}\mathrm{P}_{m,q}^{(\alpha,\beta)}(x)\,d_q(x)
\end{aligned}$$

$$
\begin{aligned}
&= \cdots \\
&= \sum_{l=1}^{n} (-1)^{l+1} \left[\mathrm{D}_q^{n-l} \left(\frac{x^{\alpha+n}}{(x;q)_{\alpha+1-\beta-n}} \right) \left(\epsilon^{-1}\mathrm{D}_q\right)^{l-1} \epsilon^{-1} \mathrm{P}_{m,q}^{(\alpha,\beta)}(x) \right]_0^{q^{\beta-\alpha-1}} \\
&\quad + (-1)^n \int_0^{q^{\beta-\alpha-1}} x^{\alpha+n} \left(xq^{-\beta+\alpha+1-n};q\right)_{\beta+n-\alpha-1} \left(\mathrm{D}_q \epsilon^{-1}\right)^n \left[\mathrm{P}_{m,q}^{(\alpha,\beta)}(x)\right] d_q(x) \\
&= \sum_{l=1}^{n} \left[(-1)^{l+1} \sum_{k=0}^{n-l} \left(\binom{n-l}{k}_q \right.\right. \\
&\quad \left. \times \frac{\{-\beta-n+\alpha+1\}_{k,q} \{\alpha+1+k+l\}_{n-l-k,q} x^{\alpha+k+l}}{(xq^{n-k-l};q)_{\alpha+1-\beta+k-n}} \right) \\
&\quad \left. \times \left(\epsilon^{-1}\mathrm{D}_q\right)^{l-1} \epsilon^{-1} \mathrm{P}_{m,q}^{(\alpha,\beta)}(x) \right]_0^{q^{\beta-\alpha-1}} \\
&\quad + (-1)^n \int_0^{q^{\beta-\alpha-1}} x^{\alpha+n} \left(xq^{-\beta-n+\alpha+1};q\right)_{\beta-\alpha-1} \left(\mathrm{D}_q \epsilon^{-1}\right)^n \left[\mathrm{P}_{m,q}^{(\alpha,\beta)}(x)\right] d_q(x) \\
&= \mathrm{RHS}.
\end{aligned}
$$

The q-integral can be computed as follows.

$$
\begin{aligned}
&\int_0^{q^{\beta-\alpha-1}} x^{\alpha+n} \left(xq^{-\beta+\alpha+1-n};q\right)_{\beta-\alpha-1+n} d_q(x) \\
&\quad = q^{\beta-\alpha-1}(1-q) \sum_{m=n+1}^{\infty} \langle m-n;q\rangle_{n-\alpha+\beta-1} q^{(n+\alpha)(-\alpha+\beta+m-1)+m} \\
&\quad = q^{\beta-\alpha-1}(1-q) \sum_{m=n+1}^{\infty} \frac{\langle m-n;q\rangle_\infty}{\langle m-\alpha+\beta-1;q\rangle_\infty} q^{(n+\alpha)(-\alpha+\beta+m-1)+m} \\
&\quad = q^{\beta-\alpha-1}(1-q) \sum_{l=0}^{\infty} \frac{\langle l+1;q\rangle_\infty}{\langle l+n-\alpha+\beta;q\rangle_\infty} q^{(n+\alpha)(-\alpha+\beta+l+n)+l+1+n} \\
&\quad = q^{\beta-\alpha-1}(1-q) \sum_{l=0}^{\infty} \frac{\langle 1;q\rangle_\infty \langle n-\alpha+\beta;q\rangle_l}{\langle n-\alpha+\beta;q\rangle_\infty \langle 1;q\rangle_l} q^{(n+\alpha)(-\alpha+\beta+l+n)+l+1+n} \\
&\quad = q^{(\beta-\alpha+n)(n+\alpha+1)}(1-q) \frac{\langle 1, 2n+\beta+1;q\rangle_\infty}{\langle n-\alpha+\beta, n+\alpha+1;q\rangle_\infty} \\
&\quad = \mathrm{B}_q(\beta-\alpha+n, \alpha+1+n) q^{(\beta-\alpha+n)(n+\alpha+1)}.
\end{aligned}
\tag{9.178}
$$

□

There is a discrete version of this too.

Theorem 9.2.27 *Put* $x_k = q^{\beta-\alpha-1+k}$. *Then*

$$\sum_{k=0}^{\infty} \mathrm{P}_{n,q}^{(\alpha,\beta)}(x_k)\mathrm{P}_{m,q}^{(\alpha,\beta)}(x_k)x_k^{\alpha}\{n\}_q!\big(x_k q^{-\beta+\alpha+1};q\big)_{\beta-\alpha-1}(q-1)x_k$$
$$= \delta(m,n)\sum_{k=0}^{\infty}\frac{\langle\beta+n;q\rangle_n}{(1-q)^n}$$
$$\times\, \mathrm{QE}\big(n(\alpha-\beta-n)\big)x_k^{\alpha+n}\big(x_k q^{-\beta-n+\alpha+1};q\big)_{\beta+n-\alpha-1}(q-1)x_k. \quad (9.179)$$

Proof Discrete q-integration by parts gives

$$\sum_{k=0}^{\infty} \mathrm{P}_{n,q}^{(\alpha,\beta)}(x_k)\mathrm{P}_{m,q}^{(\alpha,\beta)}(x_k)x_k^{\alpha}\{n\}_q!\big(x_k q^{-\beta+\alpha+1};q\big)_{\beta-\alpha-1}(q-1)x_k$$
$$= \sum_{k=0}^{\infty}\mathrm{D}_q^n\left(\frac{x_k^{\alpha+n}}{(x_k;q)_{\alpha+1-\beta-n}}\right)\mathrm{P}_{m,q}^{(\alpha,\beta)}(x_k)(q-1)x_k$$
$$= \left[\mathrm{D}_q^{n-1}\left(\frac{x_k^{\alpha+n}}{(x_k;q)_{\alpha+1-\beta-n}}\right)\epsilon^{-1}\mathrm{P}_{m,q}^{(\alpha,\beta)}(x_k)\right]_{k=0}^{\infty}$$
$$- \sum_{k=0}^{\infty}\mathrm{D}_q^{n-1}\left(\frac{x_k^{\alpha+n}}{(x_k;q)_{\alpha+1-\beta-n}}\right)\big(\mathrm{D}_q\epsilon^{-1}\big)\big(\mathrm{P}_{m,q}^{(\alpha,\beta)}(x_k)\big)(q-1)x_k$$
$$= \cdots$$
$$= \sum_{l=1}^{n}(-1)^{l+1}\left[\mathrm{D}_q^{n-l}\left(\frac{x_k^{\alpha+n}}{(x_k;q)_{\alpha+1-\beta-n}}\right)\big(\epsilon^{-1}\mathrm{D}_q\big)^{l-1}\epsilon^{-1}\mathrm{P}_{m,q}^{(\alpha,\beta)}(x_k)\right]_{k=0}^{\infty}$$
$$+ (-1)^n\sum_{k=0}^{\infty}\big(\mathrm{D}_q\epsilon^{-1}\big)^n\big[\mathrm{P}_{m,q}^{(\alpha,\beta)}(x_k)\big]x_k^{\alpha+n}\big(x_k q^{-\beta-n+\alpha+1};q\big)_{\beta+n-\alpha-1}(q-1)x_k$$
$$= \sum_{l=1}^{n}\Bigg[(-1)^{l+1}\sum_{i=0}^{n-l}\left(\binom{n-l}{i}_q\right.$$
$$\left.\times\frac{\{-\beta-n+\alpha+1\}_{i,q}\{\alpha+1+i+l\}_{n-l-i,q}x_k^{\alpha+i+l}}{(x_k q^{n-i-l};q)_{\alpha+1-\beta-n+i}}\right)$$
$$\times\big(\epsilon^{-1}\mathrm{D}_q\big)^{l-1}\epsilon^{-1}\mathrm{P}_{m,q}^{(\alpha,\beta)}(x_k)\Bigg]_{k=0}^{\infty}$$
$$+ (-1)^n\sum_{k=0}^{\infty}\big(\mathrm{D}_q\epsilon^{-1}\big)^n\big[\mathrm{P}_{m,q}^{(\alpha,\beta)}(x_k)\big]x_k^{\alpha+n}\big(x_k q^{-\beta-n+\alpha+1};q\big)_{\beta-\alpha-1}(q-1)x_k$$
$$= \text{RHS}.$$

□

9.3 q-Legendre polynomials and Carlitz-Al-Salam polynomials

The temperature corresponds to the q-Legendre polynomials.

1 K	q-Legendre polynomial

In this section we treat the two orthogonal polynomials q-Legendre and Carlitz-Al-Salam.

9.3.1 *q-Legendre polynomials*

The q-Legendre polynomials were first introduced at the meeting *Progress on Difference Equations*, 1–5 April, 2007, in Laufen/Salzburg. We also found q-difference equations for these polynomials. Further q-orthogonal polynomials will be treated in a forthcoming book. Our q-Legendre polynomials are not the same as in Jackson [294] and [295].

The following polynomial is defined by the Rodrigues formula to enable an easy orthogonality relation. q-Legendre polynomials have been found before, but these do not have the same orthogonality range in the limit $q \to 1$ as in the classical case.

Definition 159 The q-Legendre polynomials are defined by

$$\mathrm{P}_{n,q}(x) \equiv \frac{q^{-\binom{n}{2}}(-1)^n}{\{n\}_q!(1 \boxplus_q q^{-n})^n} \mathrm{D}_q^n\big((1 \boxminus_q x)^n (1 \boxplus_q x)^n\big). \tag{9.180}$$

This implies

Theorem 9.3.1 *An explicit combinatorial formula for q-Legendre polynomials*:

$$\begin{aligned} \mathrm{P}_{n,q}(x) = {} & \frac{q^{-\binom{n}{2}}(-1)^n}{\{n\}_q!(1 \boxplus_q q^{-n})^n} \sum_{k=0}^{n} \binom{n}{k}_q \{n-k+1\}_{k,q}\, q^{\binom{k}{2}}(-1)^k \\ & \times \big(1 \boxminus_q q^n x\big)^{n-k} q^{\binom{n-k}{2}} \big(1 \boxplus_q q^{n-k} x\big)^k \{k+1\}_{n-k,q}. \end{aligned} \tag{9.181}$$

Proof We use the following lemma: □

Lemma 9.3.2

$$\mathrm{D}_q^k(1 \boxplus_q x)^l = \{l-k+1\}_{k,q} q^{\binom{k}{2}} \big(1 \boxplus_q q^k x\big)^{l-k}, \quad l \geq k. \tag{9.182}$$

The rule

$$1K = 1C + 273.16 \tag{9.183}$$

corresponds to the orthogonality relation (9.185).

Theorem 9.3.3 *For simplicity we put*

$$\widetilde{\mathrm{P}_{n,q}(x)} \equiv \mathrm{D}_q^n\big((1 \boxminus_q x)^n (1 \boxplus_q x)^n\big). \tag{9.184}$$

Orthogonality relation for q-Legendre polynomials:

$$\int_{-q^{1-m}}^{q^{1-m}} \widetilde{\mathrm{P}_{m,q}(x)}\widetilde{\mathrm{P}_{n,q}(x)}\, d_q(x) = \delta(m,n)(-1)^n \int_{-q^{1-m}}^{q^{1-m}} (1 \boxminus_q x)^m (1 \boxplus_q x)^n \times \big(\mathrm{D}_q\epsilon^{-1}\big)^n \widetilde{\mathrm{P}_{n,q}(x)}\, d_q(x), \quad n \ge m. \tag{9.185}$$

Proof q-integration by parts gives

$$\begin{aligned}
&\int_{-q^{1-m}}^{q^{1-m}} \widetilde{\mathrm{P}_{m,q}(x)}\widetilde{\mathrm{P}_{n,q}(x)}\, d_q(x) \\
&= \int_{-q^{1-m}}^{q^{1-m}} \mathrm{D}_q^m\big((1 \boxminus_q x)^m (1 \boxplus_q x)^m\big)\widetilde{\mathrm{P}_{n,q}(x)}\, d_q(x) \\
&= \big[\mathrm{D}_q^{m-1}\big((1 \boxminus_q x)^m (1 \boxplus_q x)^m\big)\widetilde{\mathrm{P}_{n,q}\big(xq^{-1}\big)}\big]_{-q^{1-m}}^{q^{1-m}} \\
&\quad - \int_{-q^{1-m}}^{q^{1-m}} \mathrm{D}_q^{m-1}\big((1 \boxminus_q x)^m (1 \boxplus_q x)^m\big)\mathrm{D}_q\widetilde{\mathrm{P}_{n,q}\big(xq^{-1}\big)}\, d_q(x) \\
&= \cdots \\
&= \sum_{k=1}^{n}(-1)^{k+1}\Bigg[\sum_{l=0}^{m-k}\binom{m-k}{l}_q \prod_{j=0}^{l-1}\{m-j\}_q \\
&\quad \times q^{\binom{l}{2}}(-1)^l\big(1 \boxminus_q q^{m-k}x\big)^{m-l} q^{\binom{m-l-k}{2}} \\
&\quad \times \big(1 \boxplus_q q^{m-l-k}x\big)^{k+l} \prod_{j=0}^{m-l-k-1}\{m-j\}_q\big(\epsilon^{-1}\mathrm{D}_q\big)^{k-1}\epsilon^{-1}\widetilde{\mathrm{P}_{n,q}(x)}\Bigg]_{-q^{1-m}}^{q^{1-m}} \\
&\quad + \delta(m,n)(-1)^n \\
&\quad \times \int_{-q^{1-m}}^{q^{1-m}} \sum_{k=0}^{m-n}\binom{m-n}{k}_q \prod_{l=0}^{k-1}\{m-l\}_q\, q^{\binom{k}{2}}(-1)^k\big(1 \boxminus_q q^{m-n}x\big)^{m-k} \\
&\quad \times q^{\binom{m-n-k}{2}}\big(1 \boxplus_q q^{m-n-k}x\big)^{k+n} \prod_{l=0}^{m-n-k-1}\{m-l\}_q\big(\mathrm{D}_q\epsilon^{-1}\big)^n\widetilde{\mathrm{P}_{n,q}(x)}\, d_q(x).
\end{aligned}$$

All terms disappear if $n > m$. □

Theorem 9.3.4 *The q-Legendre polynomials $P_{n,q}(x)$ for small indices are solutions of the following q-difference equations with initial value condition $f(q^{-n})=1$:*

$$(1-x^2)\mathrm{D}_q^2 f(x,q)-\{2\}_q x\mathrm{D}_q f(x,q)+\{2\}_q f(x,q)=0 \tag{9.186}$$

has solution $f(x,q)=\mathrm{P}_{1,q}(x)$,

$$(1-x^2q^2)\mathrm{D}_q^2 f(x,q)-q^3\{2\}_q x\mathrm{D}_q f(x,q)+q^2\{3\}_q! f(x,q)=0 \tag{9.187}$$

has solution $f(x,q)=\mathrm{P}_{2,q}(x)$,

$$\begin{aligned}&(1-x^2q^6)\mathrm{D}_q^2 f(x,q)-q^3\{2\}_q x\mathrm{D}_q f(x,q)\\&\quad+q^3\{3\}_q\big(\{2\}_q\big)^2\big(q^2-q+1\big)f(x,q)=0\end{aligned} \tag{9.188}$$

has solution $f(x,q)=\mathrm{P}_{3,q}(x)$.

Theorem 9.3.5 *The function $\mathrm{P}_{n,q}(x)$ is the solution of the following linear second-order q-difference equation with initial value $f(q^{-n})=1$:*

$$\begin{aligned}&\big(x^2q^{2n+2}-1\big)\mathrm{D}_q^2 f(x,q)+q^n\big(\{2\}_q\{n+1\}_q-q\{2n\}_q\big)x\mathrm{D}_q f(x,q)\\&\quad-q^n\{n\}_q\{n+1\}_q f(x,q)=0.\end{aligned} \tag{9.189}$$

Proof A q-analogue of Courant and Hilbert [143, p. 73]. Let

$$u=(-1)^n\big(x^2;q^2\big)_n. \tag{9.190}$$

Then

$$\big(x^2-1\big)\mathrm{D}_q u=\{2n\}_q xu. \tag{9.191}$$

Apply D_q^{n+1} on this and use the q-Leibniz theorem to obtain (9.189). □

9.3.2 Carlitz-Al-Salam polynomials

The Carlitz-Al-Salam polynomials $\mathrm{F}_{n,q}(x)$ [17], [154], with generating function which consists of q-exponential functions, are examples of q-analogues of x^n. These polynomials have applications in stochastics. The Carlitz-Al-Salam polynomials are closely related to two orthogonal polynomials U, V, which are orthogonal with respect to a discrete mass.

These polynomials are also known under the name discrete q-Hermite polynomials.

Definition 160

$$\sum_{\nu=0}^{\infty}\frac{t^\nu}{\{\nu\}_q!}\mathrm{F}_{\nu,q}(x)\equiv\frac{\mathrm{E}_q(xt)}{\mathrm{E}_q(t)\mathrm{E}_q(-t)}. \tag{9.192}$$

This special type of x-dependence on the right is typical for so-called q-Appell polynomials, see Section 4.3. The vector of $F_n(x)$ can be written as a q-Pascal matrix times $F_n(0)$.

The recurrence relation for $\mathrm{F}_{n,q}(x)$ is

$$\begin{aligned} &\mathrm{F}_{0,q}(x)=1, \qquad \mathrm{F}_{1,q}(x)=x, \\ &\mathrm{F}_{n+1,q}(x)=x\mathrm{F}_{n,q}(x)-\left(1-q^n\right)q^{n-1}\mathrm{F}_{n-1,q}(x), \quad n\geq 1. \end{aligned} \tag{9.193}$$

This is an example of a recurrence relation which has a very simple form for $q=1$.

The denominator in (9.192) can be written $\mathrm{E}_q(t \ominus_q t)$.

These polynomials $\mathrm{F}_{n,q}(x)$ are connected to the Cigler q-Hermite polynomials [135] by the following substitution, which is valid only for $q\neq 1$.

Definition 161

$$h_{\nu,q}(x)\equiv\left(\frac{q}{1-q}\right)^{\frac{\nu}{2}}\mathrm{F}_{\nu,q}\left(x\sqrt{\frac{1-q}{q}}\right). \tag{9.194}$$

We now find the following generating function for the Cigler q-Hermite polynomials [135, p. 42]:

$$\frac{\mathrm{E}_q(xt)}{\mathrm{E}_{2,q}(\frac{qt^2}{\{2\}_q})}=\sum_{\nu=0}^{\infty}\frac{t^\nu}{\{\nu\}_q!}h_{\nu,q}(x). \tag{9.195}$$

Chapter 10
q-functions of several variables

Symmetries correspond to the q-functions of many variables.

$SO(3)$	q-Appell functions
$SU(2)$	q-Lauricella functions

As molecules strive for the configuration with the greatest symmetry, our transformations are often symmetrical.

The multiple q-hypergeometric functions are defined by the q-shifted factorial and the tilde operator. By our method we are able to find q-analogues of corresponding formulas for the multiple hypergeometric case, which elucidate the integration properties of q-calculus and helps in trying to make a first systematic attempt to find summation and reduction theorems for multiple basic series, including the Jackson Γ_q function.

We give two definitions of the generalized q-Kampé de Fériet function, in the spirit of Karlsson and Srivastava [482], which are symmetric in the variables and allow us to treat q as a vector. Various connections between transformation formulas for multiple q-hypergeometric series and Lie algebras and finite groups known from the literature are cited.

10.1 The corresponding vector notation

Definition 162 We define vector versions of powers, q-shifted factorials, q-Pochhammer symbols and JHC q-additions, as follows:

$$\vec{x}^{\vec{\alpha}} \equiv \prod_{j=1}^{n} x_j^{\alpha_j}, \tag{10.1}$$

$$\frac{1}{(\vec{x};\vec{q})_{\vec{\beta}}} \equiv \prod_{j=1}^{n} \frac{1}{(x_j;q_j)_{\beta_j}}, \tag{10.2}$$

T. Ernst, *A Comprehensive Treatment of q-Calculus*,
DOI 10.1007/978-3-0348-0431-8_10, © Springer Basel 2012

$$\{\vec{l}\}_{\vec{q}}! \equiv \prod_{j=1}^{n} \{l_j\}_{q_j}!, \tag{10.3}$$

$$\{\vec{\alpha}\}_{\vec{m},\vec{q}} \equiv \prod_{j=1}^{n} \{\alpha_j\}_{m_j,q_j}, \tag{10.4}$$

$$\vec{q}^{\binom{\vec{k}}{2}} \equiv \prod_{j=1}^{n} q_j{}^{\binom{k_j}{2}}, \tag{10.5}$$

$$(-1)^{\vec{k}} \equiv (-1)^{|k|}, \qquad \vec{\epsilon} \equiv \prod_{j=1}^{n} \epsilon_j, \tag{10.6}$$

$$\mathrm{P}_{\vec{k},\vec{q}}(\vec{x}, \vec{y}) \equiv \prod_{j=1}^{n} \mathrm{P}_{k_j,q_j}(x_j, y_j), \tag{10.7}$$

$$\langle \vec{\alpha}; \vec{q} \rangle_{\vec{k}} \equiv \prod_{j=1}^{n} \langle \alpha_j; q_j \rangle_{k_j}. \tag{10.8}$$

The q-factorial (10.8) is usually written $\langle (\alpha); \vec{q} \rangle_{\vec{k}}$.

The partial q-derivative of a function of n variables is defined as an operator in the spirit of Hörmander [285, p. 12]:

$$\mathrm{D}_{\vec{q},\vec{x}}^{\vec{l}} F(\vec{x}, \vec{q}) \equiv \prod_{j=1}^{n} \big(\mathrm{D}_{q_j,x_j}^{l_j}\big) F(\vec{x}, \vec{q}), \tag{10.9}$$

$$\big(\mathrm{D}_{\vec{q},\vec{x}} \vec{\epsilon}^{\vec{t}}\big)^{\vec{l}} F(\vec{x}, \vec{q}) \equiv \prod_{j=1}^{n} \big(\mathrm{D}_{q_j,x_j} \epsilon_j^{t_j}\big)^{l_j} F(\vec{x}, \vec{q}). \tag{10.10}$$

Closed and open intervals are defined by

$$[\vec{a}, \vec{b}] \equiv \prod_{j=1}^{n} [a_j, b_j], \qquad (\vec{a}, \vec{b}) \equiv \prod_{j=1}^{n} (a_j, b_j). \tag{10.11}$$

Jackson [299, p. 145], [300], [304, p. 146] has shown certain connections between power series in x and series of the form $\sum_{k=0}^{\infty} \frac{a_k x^{f(k)}}{(x;q)_k}$, where $f(k)$ is an integer-valued function.

In 1921 Ryde [444] showed that certain linear homogeneous q-difference equations have series

$$\sum_{k=0}^{\infty} \frac{a_k}{(x; q^{-1})_k}$$

as solutions.

We need a q-analogue of holomorphic functions, which will be useful to characterize the functions to the far right in operator expressions.

Definition 163 Let $\vec{\mathrm{H}}_{\vec{q},n}$ denote the class of functions $F(\vec{x})$ of n variables which can be written in the form

$$F(\vec{x}) \equiv \sum_{\vec{k}=\vec{0}}^{\vec{\infty}} \vec{x}^{\vec{\alpha}_{\vec{k}}} \frac{\vec{\gamma}_{\vec{k}}}{(\vec{x}\vec{q}^{\vec{\delta}_{\vec{k}}};\vec{q}^{\pm 1})_{\vec{\beta}_{\vec{k}}}},$$
$$\text{where } \vec{\alpha}_{\vec{k}} \in \mathbb{N}^n,\ \vec{\beta}_{\vec{k}} \in \mathbb{R}^n,\ \vec{\gamma}_{\vec{k}} \in \mathbb{C}^n,\ \vec{\delta}_{\vec{k}} \in \mathbb{N}^n. \tag{10.12}$$

If the function is defined in an open region $\mathrm{O} \subset \mathbb{C}^n$, we write $\vec{\mathrm{H}}_{\vec{q},n}(\mathrm{O})$.

Example 34

$$\left(\mathrm{D}_q \epsilon^t\right)^k \sum_{m=0}^{\infty} \frac{a_m x^m}{\{m\}_q!} = \sum_{m=0}^{\infty} \frac{a_{k+m} x^m}{\{m\}_q!} \mathrm{QE}\left(tk(m+1) + t\binom{k}{2}\right), \tag{10.13}$$

$$\begin{aligned}
&\prod_{j=1}^{2} \left(\mathrm{D}_{q_j,x_j} \epsilon_j^{t_j}\right)^{k_j} \Phi_4\left(a;b;c,c'|q;x_1,x_2\right) \\
&\quad = \frac{\langle a,b;q\rangle_N \langle c+N-k_2;q\rangle_{k_2} \langle c'+N-k_1;q\rangle_{k_1}}{\langle c,c';q\rangle_N (1-q)^N} \\
&\qquad \times \sum_{m_1,m_2=0}^{\infty} \frac{\langle a+N,b+N;q\rangle_{m_1+m_2}}{\langle 1,c+k_1;q\rangle_{m_1} \langle 1,c'+k_2;q\rangle_{m_2}} x_1^{m_1} x_2^{m_2} \mathrm{QE}\left(\vec{t}\vec{m}\vec{k} + \vec{t}\binom{\vec{k}}{2} + \vec{t}\,\vec{k}\right),
\end{aligned} \tag{10.14}$$

where $N \equiv k_1 + k_2$, $\vec{t}\vec{m}\vec{k} \equiv t_1 m_1 k_1 + t_2 m_2 k_2$, and

$$\Phi_4\left(a;b;c,c'|q;x_1,x_2\right) \equiv \sum_{m_1,m_2=0}^{\infty} \frac{\langle a;q\rangle_{m_1+m_2} \langle b;q\rangle_{m_1+m_2}}{\langle 1;q\rangle_{m_1} \langle 1;q\rangle_{m_2} \langle c;q\rangle_{m_1} \langle c';q\rangle_{m_2}} x_1^{m_1} x_2^{m_2}. \tag{10.15}$$

Definition 164 The following notation for power series,

$$F(\vec{x}) \equiv \sum_{\vec{k}} \vec{x}^{\vec{k}} \vec{\gamma}_{\vec{k}},$$

will also be useful:

$$F(\vec{x} \oplus_{\vec{q}} \vec{y}) \equiv F\left((x_1 \oplus_{q_1} y_1), \ldots, (x_n \oplus_{q_n} y_n)\right), \tag{10.16}$$

$$F(\vec{x} \boxplus_{\vec{q}} \vec{y}) \equiv F\left((x_1 \boxplus_{q_1} y_1), \ldots, (x_n \boxplus_{q_n} y_n)\right), \tag{10.17}$$

$$F(\vec{x} \oplus_{\vec{q},\vec{t}} \vec{y}) \equiv F\left((x_1 \oplus_{q_1,t_1} y_1), \ldots, (x_n \oplus_{q_n,t_n} y_n)\right). \tag{10.18}$$

Theorem 10.1.1 *The Ward-q-Taylor formula for functions of n variables. Let $F(\vec{x})$ be a q-Kampé de Fériet function* [166], *or more generally, a formal power series $\sum_{\vec{m}} \vec{a}_{\vec{m}} \vec{x}^{\vec{m}}$. Then (compare with Hörmander* [285, 1.1.7′, p. 13]):

$$F(\vec{x} \oplus_{\vec{q}} \vec{y}) = \sum_{\vec{k}} \frac{\vec{y}^{\vec{k}}}{\{\vec{k}\}_{\vec{q}}!} \mathrm{D}_{\vec{q},\vec{x}}^{\vec{k}} F(\vec{x}), \tag{10.19}$$

$$F(\vec{x}) = \sum_{\vec{k}} \frac{(\vec{x} \boxminus_{\vec{q}} \vec{y})^{\vec{k}}}{\{\vec{k}\}_{\vec{q}}!} \mathrm{D}_{\vec{q}}^{\vec{k}} F(\vec{y}), \tag{10.20}$$

$$F(\vec{x} \boxplus_{\vec{q}} \vec{y}) = \sum_{\vec{k}} \frac{\vec{y}^{\vec{k}} \vec{q}^{\binom{\vec{k}}{2}}}{\{\vec{k}\}_{\vec{q}}!} \mathrm{D}_{\vec{q},\vec{x}}^{\vec{k}} F(\vec{x}). \tag{10.21}$$

We will use a kind of umbral calculus.

Definition 165 The two q-additions for q-shifted factorials are defined in the following way. If $F(x)$ is of the form $F(x) = \sum_{n=0}^{\infty} \frac{a_n}{(x;q)_{-n}}$, then

$$F(x \oplus_q y) \equiv \sum_{n=0}^{\infty} a_n \sum_{k=0}^{n} \binom{n}{k}_q \frac{1}{(x;q)_{k-n}(y;q)_{-k}}, \quad n = 0, 1, 2, \ldots, \tag{10.22}$$

$$F(x \boxplus_q y) \equiv \sum_{n=0}^{\infty} a_n \sum_{k=0}^{n} \binom{n}{k}_q q^{\binom{k}{2}} \frac{1}{(x;q)_{k-n}(y;q)_{-k}}, \quad n = 0, 1, 2, \ldots. \tag{10.23}$$

The following two notations for q-shifted factorials will be used:

$$F(\vec{x} \oplus_{\vec{q}} \vec{y}) \equiv F(x_1 \oplus_{q_1} y_1, \ldots, x_n \oplus_{q_n} y_n), \tag{10.24}$$

$$F(\vec{x} \boxplus_{\vec{q}} \vec{y}) \equiv F(x_1 \boxplus_{q_1} y_1, \ldots, x_n \boxplus_{q_n} y_n). \tag{10.25}$$

Theorem 10.1.2 *The Ward q-addition* (10.22) *for q-shifted factorials is associative and commutative.*

Proof Similar to the case power series. □

Theorem 10.1.3 *The Ward q-Taylor formula for q-shifted factorials. Let $\mathrm{F}_k(x)$ be of the form*

$$\mathrm{F}_k(x) \equiv \sum_{m=0}^{\infty} \frac{a_m q^{km}}{(x;q)_{-m}}. \tag{10.26}$$

Then

$$\mathrm{F}_0(x \oplus_q y) = \sum_{k=0}^{\infty} \frac{(-1)^k q^{-\binom{k}{2}}}{\{k\}_q!(y;q)_{-k}} \mathrm{D}_q^k \mathrm{F}_k(x). \tag{10.27}$$

Proof

$$\text{LHS}=\sum_{m=0}^{\infty}a_m\sum_{k=0}^{m}\binom{m}{k}_q\frac{1}{(x;q)_{k-m}(y;q)_{-k}}$$
$$=\sum_{k=0}^{\infty}\frac{(-1)^k q^{-\binom{k}{2}}}{(y;q)_{-k}\{k\}_q!}\sum_{m=k}^{\infty}\frac{a_m\{-m\}_{k,q}q^{km}}{(x;q)_{k-m}}=\text{RHS}. \tag{10.28}$$

□

Theorem 10.1.4 *Two q-Taylor formulas for functions in q-shifted factorials of n variables. Let* $\mathrm{F}_{\vec{k}}(\vec{x})$ *be of the form*

$$\mathrm{F}_{\vec{k}}(\vec{x})\equiv\sum_{\vec{m}}\vec{a}_{\vec{m}}\vec{q}^{\,\vec{m}k}\frac{1}{(\vec{x};\vec{q})_{-\vec{m}}}. \tag{10.29}$$

Then

$$\mathrm{F}_{\vec{0}}(\vec{x}\oplus_{\vec{q}}\vec{y})=\sum_{\vec{k}}\frac{(-1)^{\vec{k}}\vec{q}^{\,-\binom{\vec{k}}{2}}}{\{\vec{k}\}_{\vec{q}}!(\vec{y};\vec{q})_{-\vec{k}}}\mathrm{D}_{\vec{q},\vec{x}}^{\vec{k}}\mathrm{F}_{\vec{k}}(\vec{x}), \tag{10.30}$$

$$\mathrm{F}_{\vec{0}}(\vec{x}\boxplus_{\vec{q}}\vec{y})=\sum_{\vec{k}}\frac{(-1)^{\vec{k}}}{\{\vec{k}\}_{\vec{q}}!(\vec{y};\vec{q})_{-\vec{k}}}\mathrm{D}_{\vec{q},\vec{x}}^{\vec{k}}\mathrm{F}_{\vec{k}}(\vec{x}). \tag{10.31}$$

Proof We prove the first relation:

$$\text{LHS}=\sum_{\vec{m}}\vec{a}_{\vec{m}}\sum_{\vec{k}=\vec{0}}^{\vec{m}}\binom{\vec{m}}{\vec{k}}_{\vec{q}}\frac{1}{(\vec{x};\vec{q})_{\vec{k}-\vec{m}}(\vec{y};\vec{q})_{-\vec{k}}}$$
$$=\sum_{\vec{k}}\frac{(-1)^{\vec{k}}\vec{q}^{\,-\binom{\vec{k}}{2}}}{(\vec{y};\vec{q})_{-\vec{k}}\{\vec{k}\}_{\vec{q}}!}\sum_{\vec{m}=\vec{k}}^{\infty}\frac{\vec{a}_{\vec{m}}\{-\vec{m}\}_{\vec{k},\vec{q}}\vec{q}^{\,\vec{k}\vec{m}}}{(\vec{x};\vec{q})_{\vec{k}-\vec{m}}}=\text{RHS}. \tag{10.32}$$

□

Theorem 10.1.5 *The extended Ward q-Taylor formula for formal power series, where* (4.32) *have been extended in a natural way:*

$$F(x\oplus_{q,t}y)=\sum_{n=0}^{\infty}\frac{y^n}{\{n\}_q!}\left(\mathrm{D}_q\epsilon^t\right)^n F(x). \tag{10.33}$$

Theorem 10.1.6 *The extended Ward q-Taylor formula for functions of n variables. Let* $F(\vec{x})$ *be a q-Kampé de Fériet function* [166], *or more generally, a formal power* *series* $\sum_{\vec{m}} \vec{a}_{\vec{m}} \vec{x}^{\vec{m}}$. *Then (compare with* [285, 1.1.7′ p. 13])

$$F(\vec{x} \oplus_{\vec{q},t} \vec{y}) = \sum_{\vec{k}} \frac{\vec{y}^{\vec{k}}}{\{\vec{k}\}_{\vec{q}}!} (\mathrm{D}_{\vec{q},\vec{x}\vec{\epsilon}^t})^{\vec{k}} F(\vec{x}). \tag{10.34}$$

We now assume that the function $f(x, q)$ satisfies a certain restriction of growth given in Walliser [529] and [530], who improved a result of Gelfond from 1933. Then the q-Taylor formula (8.90) applies.

Theorem 10.1.7 *There is a unique* $\vec{q}' \in (\vec{0}, \vec{1})$, *such that for the function* $f(\vec{x}, \vec{q})$ *defined on* $[\vec{b}, \vec{c}] \times (\vec{q}', \vec{1})$, *and* $\vec{x}, \vec{a} \in (\vec{b}, \vec{c})$, $\vec{\xi} \in (\vec{b}, \vec{c})$ *can be found between* $\vec{x}$ *and* $\vec{a}$ *which satisfies*

$$f(\vec{x}, \vec{q}) = \sum_{\vec{k}=\vec{0}}^{\vec{n-1}} \frac{\mathrm{P}_{\vec{k},\vec{q}}(\vec{x}, \vec{a})}{\{\vec{k}\}_{\vec{q}}!} (D_{\vec{q}}^{\vec{k}}) f(\vec{a}, \vec{q}) + \frac{\mathrm{P}_{\vec{n},\vec{q}}(\vec{x}, \vec{a})}{\{\vec{n}\}_{\vec{q}}!} D_{\vec{q}}^{\vec{n}} f(\vec{\xi}, \vec{q}). \tag{10.35}$$

10.2 Historical introduction

We now go on to multiple hypergeometric series (MHS). This is perhaps the most beautiful and most difficult area in q-analysis, where many one-variable results are generalized. The number of formulas is large and, unlike the one-variable case, almost all formulas are new. We start with the first definitions and a historical introduction.

After Gauß's introduction of the hypergeometric series in 1812, various versions of MHS were developed by people whose names are now associated to the functions they introduced.

In 1880 Appell [39], [41] introduced some 2-variable hypergeometric series now called Appell functions.

Definition 166

$$\mathrm{F}_1\big(a; b, b'; c; x_1, x_2\big) \equiv \sum_{m_1,m_2=0}^{\infty} \frac{(a)_{m_1+m_2}(b)_{m_1}(b')_{m_2}}{m_1!m_2!(c)_{m_1+m_2}} x_1^{m_1} x_2^{m_2},$$

$$\max\big(|x_1|, |x_2|\big) < 1. \tag{10.36}$$

$$\mathrm{F}_2\big(a; b, b'; c, c'; x_1, x_2\big) \equiv \sum_{m_1,m_2=0}^{\infty} \frac{(a)_{m_1+m_2}(b)_{m_1}(b')_{m_2}}{m_1!m_2!(c)_{m_1}(c')_{m_2}} x_1^{m_1} x_2^{m_2},$$

$$|x_1| + |x_2| < 1. \tag{10.37}$$

$$\mathrm{F}_3\big(a,a';b,b';c;x_1,x_2\big)\equiv\sum_{m_1,m_2=0}^{\infty}\frac{(a)_{m_1}(a')_{m_2}(b)_{m_1}(b')_{m_2}}{m_1!m_2!(c)_{m_1+m_2}}x_1^{m_1}x_2^{m_2},$$

$$\max\big(|x_1|,|x_2|\big)<1. \tag{10.38}$$

$$\mathrm{F}_4\big(a;b;c,c';x_1,x_2\big)\equiv\sum_{m_1,m_2=0}^{\infty}\frac{(a)_{m_1+m_2}(b)_{m_1+m_2}}{m_1!m_2!(c)_{m_1}(c')_{m_2}}x_1^{m_1}x_2^{m_2},$$

$$|\sqrt{x_1}|+|\sqrt{x_2}|<1. \tag{10.39}$$

Jackson [307], [308] and Andrews [24] have q-deformed these functions.

In the formulas (10.40)–(10.43), (10.45), (10.46), (10.129), (11.67)–(11.69) below we assume that the variables x_i are letters in the alphabet (Chapter 4).

Definition 167 The q-analogues of the Appell functions are

$$\Phi_1\big(a;b,b';c|q;x_1,x_2\big)\equiv\sum_{m_1,m_2=0}^{\infty}\frac{\langle a;q\rangle_{m_1+m_2}\langle b;q\rangle_{m_1}\langle b';q\rangle_{m_2}}{\langle 1;q\rangle_{m_1}\langle 1;q\rangle_{m_2}\langle c;q\rangle_{m_1+m_2}}x_1^{m_1}x_2^{m_2}, \tag{10.40}$$

$$\Phi_2\big(a;b,b';c,c'|q;x_1,x_2\big)\equiv\sum_{m_1,m_2=0}^{\infty}\frac{\langle a;q\rangle_{m_1+m_2}\langle b;q\rangle_{m_1}\langle b';q\rangle_{m_2}}{\langle 1;q\rangle_{m_1}\langle 1;q\rangle_{m_2}\langle c;q\rangle_{m_1}\langle c';q\rangle_{m_2}}x_1^{m_1}x_2^{m_2}, \tag{10.41}$$

$$\Phi_3\big(a,a';b,b';c|q;x_1,x_2\big)\equiv\sum_{m_1,m_2=0}^{\infty}\frac{\langle a;q\rangle_{m_1}\langle a';q\rangle_{m_2}\langle b;q\rangle_{m_1}\langle b';q\rangle_{m_2}}{\langle 1;q\rangle_{m_1}\langle 1;q\rangle_{m_2}\langle c;q\rangle_{m_1+m_2}}x_1^{m_1}x_2^{m_2}, \tag{10.42}$$

$$\Phi_4\big(a;b;c,c'|q;x_1,x_2\big)\equiv\sum_{m_1,m_2=0}^{\infty}\frac{\langle a;q\rangle_{m_1+m_2}\langle b;q\rangle_{m_1+m_2}}{\langle 1;q\rangle_{m_1}\langle 1;q\rangle_{m_2}\langle c;q\rangle_{m_1}\langle c';q\rangle_{m_2}}x_1^{m_1}x_2^{m_2}. \tag{10.43}$$

Remark 49 For general $|q|<1$ the domains of convergence of Φ_2 and Φ_4 are slightly larger, as Mathematica computations show.

The domain of convergence for (10.37) and (10.39) given in [221, 10.2.6 and 10.2.8] were wrong.

Remark 50 By confluence, we obtain the Humbert functions from the above four formulas. These were studied in [287], where Laguerre polynomials of two variables were used.

After Appell had introduced his functions in 1880, it took nine years for J. Horn [286] to establish strict convergence concepts for hypergeometric series of two and three variables. Horn [286, p. 565] had realized that the sum of the numerator and denominator Pochhammer symbols (with sign) must be equal to ± 1 for convergence.

force equilibrium	balanced series

The four Lauricella functions of n variables were introduced 1893 [354]. The first systematic treatment of this subject was written in French by J. M. Kampé de Fériet (1893–1982) and Appell [41].

The Kampé de Fériet functions in their most general form are defined in [482, p. 27]. There are many generalizations of these functions in the literature, and we refer to the books by Srivastava and Karlsson [482] and Exton [193].

The results presented in this chapter are of the following form:

1. reduction formulas,
2. summation formulas,
3. transformation formulas.

The reduction formulas reduce the degree of the function in a certain sense. As in the case of (only) one variable, there are confluent forms of MHS, which are calculated by using limits.

Definition 168 The following operator [307] will also be useful (compare with (5.49)):

$$\theta_{q,j} \equiv x_j \mathrm{D}_{q,x_j}. \tag{10.44}$$

As was pointed out by Jackson [307, p. 78], the q-analogue of hypergeometric formulas with sums in the function arguments must contain a so-called q-addition. Jackson meant the Jackson-Hahn-Cigler q-addition, compare with [261, p. 362]. As we know from Chapter 4, there is also another q-addition, called NWA, which will prove more convenient for our purposes.

The paper [89] by Burchnall and Chaundy [89] was a breakthrough in the studies of expansions for MHS. By using an inverse pair of symbolic operators, these two mathematicians were able to prove a large number of expansion formulas for Appell functions. Their results were later generalized by Verma [520] and Srivastava [472] to Kampé de Fériet functions. These important works of Burchnall and Chaundy are of a different type than those of Erdelyi in this area.

In 1942, Jackson [307] made the first attempt to find a q-analogue of [89], see Section 10.9. Jackson [307, p. 70] and R. P. Agarwal [4, p. 187] defined four so-called normal q-Appell functions (10.40)–(10.43).

For the four so-called abnormal q-Appell functions, $x_2^{m_2} q^{k\binom{m_2}{2}}$ replaces $x_2^{m_2}$ in the general term. We will see that normal functions (in a broader sense) correspond to the Ward-Al-Salam q-addition and abnormal functions, with $k = 1$, (in a broader sense) correspond to the Jackson-Hahn q-addition.

q-Contiguity relations for ${}_2\phi_1$ were found by Heine [270], and for q-Appell functions by R. P. Agarwal [5]. The first balanced theorems for double series were published by Carlitz in 1963 [101] and 1967 [102]. q-analogues of [101] were found by

W. A. Al-Salam [14]. Bailey was greatly influenced by Ramanujan as an undergraduate at Cambridge in 1914, and he wrote the first systematic treatment of hypergeometric series [56]. Lucy Joan Slater attended Bailey's lectures on q-hypergeometric series in 1947–50 at London University, and wrote many important papers on this subject; among her pupils were Howard Exton and Margaret Jackson (Nottingham). R. P. Agarwal [4] also visited Bailey, in 1953, and made the aforementioned contributions to the subject. In 1963, after his return to Lucknow in India, Agarwal wrote the book [6], which summarized the contemporary knowledge of q-hypergeometric series.

In 1966, Slater stated [466, p. 234] that there seemed to be no systematic attempt to find summation theorems for basic Appell series, but Andrews [24] managed to prove some summation and transformation formulas for these series (Section 10.3). Some of these do not have a hypergeometric counterpart. This is typical for q-calculus, where we almost always get more equations than we had before.

In a series of papers [380], [381], Willard Miller Jr. proved that by using the maximal Lie algebra generated by the first-order differential recurrence relations satisfied by a MHS, we obtain addition theorems, transformations and generating functions. Then, in 1980, Kalnins, Manocha and Miller [321] developed a theory which enabled the powerful use of Lie algebraic methods for the solutions of partial differential equations by two-variable hypergeometric series. As was shown in [8], this technique can also be adapted to multiple q-hypergeometric series. This constitutes the Lie algebra approach.

There is also a finite group approach to transformation-formulas for q-MHS, which we will briefly summarize. In the footsteps of Rogers [434], [435] and Whipple [541], Lievens and Van der Jeugt [359], [513] proved some formulas for q-MHS by using the invariance group or symmetry group of the particular double series.

In 1996 Kuznetsov and Sklyanin [350] showed the connection between (multiple) Macdonald polynomials and q-Lauricella functions.

Obviously we have not seen the end of this story yet, as the subject gets more and more complex.

Various q-analogues of Kampé de Fériet functions occur in the literature, cf. Srivastava and Karlsson [482, p. 349–350] and [478, p. 50]. We will give a definition reminding of [220], which allows easy confluence to diminish the dimensions in (10.45) and (10.46) and has the advantage of being symmetric in the variables. Furthermore, q is allowed to be a vector, and the full machinery of tilde operators and q-additions can be used. The first definition is a q-analogue of [482, (24), p. 38], in the spirit of Srivastava. The second definition is a q-analogue of [482, (24), p. 38] with the constraint [482, (29), p. 38], due to Karlsson. It will be clear from the context which of the definitions we use.

Definition 169 The vectors

$$(a), (b), (g_i), (h_i), \left(a'\right), \left(b'\right), \left(g_i'\right), \left(h_i'\right)$$

have dimensions

$$A, B, G_i, H_i, A', B', G_i', H_i'.$$

Let

$$1+B+B'+H_i+H_i'-A-A'-G_i-G_i'\geq 0,\quad i=1,\ldots,n.$$

Then the generalized q-Kampé de Fériet function is defined by

$$\Phi^{A+A':G_1+G_1';\ldots;G_n+G_n'}_{B+B':H_1+H_1';\ldots;H_n+H_n'}\left[\begin{matrix}(\hat{a}):(\hat{g_1});\ldots;(\hat{g_n})\\(\hat{b}):(\hat{h_1});\ldots;(\hat{h_n})\end{matrix}|\vec{q};\vec{x}||\begin{matrix}(a'):(g_1');\ldots;(g_n')\\(b'):(h_1');\ldots;(h_n')\end{matrix}\right]$$
$$\equiv\sum_{\vec{m}}\frac{\langle(\hat{a});q_0\rangle_m(a')(q_0,m)\prod_{j=1}^n(\langle(\hat{g_j});q_j\rangle_{m_j}((g_j')(q_j,m_j)x_j^{m_j})}{\langle(\hat{b});q_0\rangle_m(b')(q_0,m)\prod_{j=1}^n(\langle(\hat{h_j});q_j\rangle_{m_j}(h_j')(q_j,m_j)\langle 1;q_j\rangle_{m_j})}$$
$$\times(-1)^{\sum_{j=1}^n m_j(1+H_j+H_j'-G_j-G_j'+B+B'-A-A')}$$
$$\times\operatorname{QE}\left((B+B'-A-A')\binom{m}{2},q_0\right)$$
$$\times\prod_{j=1}^n\operatorname{QE}\left((1+H_j+H_j'-G_j-G_j')\binom{m_j}{2},q_j\right).\tag{10.45}$$

It is assumed that there are no zero factors in the denominator. We assume that $(a')(q_0,m)$, $(g_j')(q_j,m_j)$, $(b')(q_0,m)$, $(h_j')(q_j,m_j)$ contain factors of the form $\langle a(\hat{k});q\rangle_k$, $(s;q)_k$, $(s(k);q)_k$ or $\operatorname{QE}(f(\vec{m}))$.

Definition 170 The vectors

$$(a),(b),(g_i),(h_i),(a'),(b'),(g_i'),(h_i')$$

have dimensions

$$A,B,G,H,A',B',G',H'.$$

Let

$$1+B+B'+H+H'-A-A'-G-G'\geq 0.$$

Then the generalized q-Kampé de Fériet function is defined by

$$\Phi^{A+A':G+G'}_{B+B':H+H'}\left[\begin{matrix}(\hat{a}):(\hat{g_1});\ldots;(\hat{g_n})\\(\hat{b}):(\hat{h_1});\ldots;(\hat{h_n})\end{matrix}|\vec{q};\vec{x}||\begin{matrix}(a'):(g_1');\ldots;(g_n')\\(b'):(h_1');\ldots;(h_n')\end{matrix}\right]$$
$$\equiv\sum_{\vec{m}}\frac{\langle(\hat{a});q_0\rangle_m(a')(q_0,m)\prod_{j=1}^n(\langle(\hat{g_j});q_j\rangle_{m_j}((g_j')(q_j,m_j)x_j^{m_j})}{\langle(\hat{b});q_0\rangle_m(b')(q_0,m)\prod_{j=1}^n(\langle(\hat{h_j});q_j\rangle_{m_j}(h_j')(q_j,m_j)\langle 1;q_j\rangle_{m_j})}$$
$$\times(-1)^{\sum_{j=1}^n m_j(1+H+H'-G-G'+B+B'-A-A')}$$
$$\times\operatorname{QE}\left((B+B'-A-A')\binom{m}{2},q_0\right)$$

$$\times \prod_{j=1}^{n} \mathrm{QE}\Bigg(\big(1+H+H'-G-G'\big)\binom{m_j}{2}, q_j \Bigg), \tag{10.46}$$

where

$$\hat{a} \equiv a \vee \tilde{a} \vee \widetilde{\tfrac{m}{n} a} \vee_k \tilde{a} \vee \Delta(q; l; \lambda). \tag{10.47}$$

It is assumed that there are no zero factors in the denominator. We assume that $(a')(q_0, m)$, $(g'_j)(q_j, m_j)$, $(b')(q_0, m)$, $(h'_j)(q_j, m_j)$ contain factors of the form $\langle a\hat{(k)}; q\rangle_k$, $(s; q)_k$, $(s(k); q)_k$ or $\mathrm{QE}(f(\vec{m}))$.

Remark 51 [479, p. 16]. The domain of convergence of the above series is extremely difficult to calculate. Consider the function (10.45) for $q = 1, n = 2$. If

$$A + A' + G_i + G'_i < B + B' + H_i + H'_i + 1, \quad i = 1, 2, \tag{10.48}$$

the series always converges. If

$$A + A' + G_i + G'_i = B + B' + H_i + H'_i + 1, \quad i = 1, 2, \tag{10.49}$$

the series converges for

$$\begin{cases} |x_1|^{\frac{1}{A+A'-B-B'}} + |x_2|^{\frac{1}{A+A'-B-B'}} < 1, & \text{if } A + A' > B + B', \\ \max\{|x_1|, |x_2|\} < 1, & \text{if } A + A' \leq B + B'. \end{cases} \tag{10.50}$$

Consider the function (10.46) for $q = 1, n = 2$. If

$$A + A' + G + G' < B + B' + H + H' + 1, \tag{10.51}$$

the series always converges. If

$$A + A' + G + G' = B + B' + H + H' + 1, \tag{10.52}$$

the series converges for

$$\begin{cases} |x_1|^{\frac{1}{A+A'-B-B'}} + |x_2|^{\frac{1}{A+A'-B-B'}} < 1, & \text{if } A + A' > B + B', \\ \max\{|x_1|, |x_2|\} < 1, & \text{if } A + A' \leq B + B'. \end{cases} \tag{10.53}$$

Remark 52 In case we want to get rid of the ∞ symbol, we just remove all the -1 and q-powers in the definitions. In the one-variable case, this was the original approach, see Thomae 1871 [499, p. 111], [467]. However, in that case we have to watch out for convergence and we lose the confluence property.

Example 35 Following the definitions (7.13) and (7.17) for balanced and well-poised q-hypergeometric series, we call the series

$$\Phi^{7:1;1}_{7:0;0} \left[\begin{matrix} (\alpha) : -M; -N \\ (\beta), 1 + a + M + N : -; - \end{matrix} \middle| q; q^{1-N}, q \right] \tag{10.54}$$

balanced with respect to both summation indices and very-well-poised, if we sum the two m and n q-factorials together with the last q-factorial in the denominator.

10.3 Transformations for basic double series

We start with a few examples where the Jackson [304], [220, p. 11, 1.5.4] q-analogue of the Euler-Pfaff-Kummer transformation (7.62) is used. If no convergence region is given for a certain formula, we first assume that the absolute value of the function arguments are small.

The following two reduction formulas are q-analogues of Appell and Kampé de Fériet [41, p. 23]:

$$\Phi_2\big(\alpha;\beta,\beta';\gamma,\beta'|q;x_1,x_2\big)\cong\frac{1}{(x_2;q)_\alpha}{}_3\phi_2\big(\alpha,\beta,\infty;\gamma|q;x_1||-;x_2q^\alpha\big), \quad (10.55)$$

$$\Phi_2\big(\alpha;\beta,\beta';\beta,\gamma'|q;x_1,x_2\big)$$
$$\cong\frac{1}{(x_1;q)_\alpha}{}_3\phi_2\big(\alpha,\beta',\infty;\gamma'|q;x_2||-;x_1q^\alpha\big). \quad (10.56)$$

The following two equations are q-analogues of [41, p. 23, (26)]:

$${}_2\phi_1(\alpha,\beta;\gamma|q;x_1\oplus_q x_2)$$
$$=\sum_{m=0}^{\infty}\frac{\langle\alpha,\beta;q\rangle_m x_1^m}{\langle 1,\gamma;q\rangle_m}{}_2\phi_1(\alpha+m,\beta+m;\gamma+m|q;x_2), \quad (10.57)$$

$${}_2\phi_1\big(\alpha,\beta;\gamma|q;(x_1\boxplus_q x_2)\big)$$
$$=\sum_{m=0}^{\infty}\frac{\langle\alpha,\beta;q\rangle_m x_1^m}{\langle 1,\gamma;q\rangle_m}{}_2\phi_2(\alpha+m,\beta+m;\gamma+m,\infty|q;-x_2). \quad (10.58)$$

This can be generalized to the following two equations, which are q-analogues of Exton [193, p. 24, 1.4.9]. We assume that the double sum is absolutely convergent:

$$\sum_{m_1,m_2=0}^{\infty}\frac{f(m_1+m_2)x_1^{m_1}x_2^{m_2}}{\langle 1;q\rangle_{m_1}\langle 1;q\rangle_{m_2}}=\sum_{N=0}^{\infty}\frac{f(N)}{\langle 1;q\rangle_N}(x_1\oplus_q x_2)^N, \quad (10.59)$$

$$\sum_{m_1,m_2=0}^{\infty}\frac{f(m_1+m_2)x_1^{m_1}x_2^{m_2}q^{\binom{m_1}{2}}}{\langle 1;q\rangle_{m_1}\langle 1;q\rangle_{m_2}}=\sum_{N=0}^{\infty}\frac{f(N)x_2^N}{\langle 1;q\rangle_N}\Big(-\frac{x_1}{x_2};q\Big)_N. \quad (10.60)$$

The following three expansions of Φ_1 in terms of one-variable ϕ are q-analogues of Appell and Kampé de Fériet [41, p. 24, (27), (27′), (27″)] formulas:

$$\begin{aligned}&\Phi_1(\alpha;\beta,\beta';\gamma|q;x_1,x_2)\\&\quad\cong\sum_{m=0}^{\infty}\frac{\langle\alpha,\beta;q\rangle_m x_1^m}{\langle 1,\gamma;q\rangle_m (x_2;q)_{\alpha+m}}\\&\qquad\times{}_2\phi_2(\alpha+m,\gamma-\beta'+m;\gamma+m|q;x_2q^{\beta'}||-;x_2q^{\alpha+m});\end{aligned}\tag{10.61}$$

$$\begin{aligned}&\Phi_1(\alpha;\beta,\beta';\gamma|q;x_1,x_2)\\&\quad\cong\sum_{m=0}^{\infty}\frac{\langle\alpha,\beta;q\rangle_m x_1^m}{\langle 1,\gamma;q\rangle_m (x_2;q)_{\beta'}}\\&\qquad\times{}_2\phi_2(\beta',\gamma-\alpha;\gamma+m|q;x_2q^{\alpha+m}||-;x_2q^{\beta'});\end{aligned}\tag{10.62}$$

$$\begin{aligned}&\Phi_1(\alpha;\beta,\beta';\gamma|q;x_1,x_2)\\&\quad\cong\sum_{m=0}^{\infty}\frac{\langle\alpha,\beta;q\rangle_m x_1^m}{\langle 1,\gamma;q\rangle_m (x_2;q)_{\alpha+\beta'-\gamma}}\\&\qquad\times{}_2\phi_1(\gamma-\alpha,\gamma-\beta'+m;\gamma+m|q;x_2q^{\alpha+\beta'-\gamma}).\end{aligned}\tag{10.63}$$

A q-analogue of [41, (28), p. 24]:

$$\begin{aligned}&\Phi_1(\alpha;\beta,\beta';\beta+\beta'|q;x_1,x_2)\\&\quad\cong\frac{1}{(x_2;q)_\alpha}\,{}_2\phi_2(\alpha,\beta;\beta+\beta'|q;-x_1\oplus_{q,\star}x_2q^{\beta'}||-;x_2q^{\alpha}),\end{aligned}\tag{10.64}$$

where the q-subtraction is defined by

$$(-x_1\oplus_{q,\star}x_2)^n\equiv\sum_{k=0}^{n}\binom{n}{k}_q(-x_1)^k x_2^{n-k}q^{\frac{k^2+k-2nk}{2}},\quad n=0,1,2,\ldots.\tag{10.65}$$

By (10.57) and Jackson [304], we obtain the following expansion of ${}_2\phi_1(|q;x_1\oplus_q x_2)$:

$$\begin{aligned}&{}_2\phi_1(\alpha,\beta;\gamma|q;x_1\oplus_q x_2)\\&\quad\cong\sum_{m=0}^{\infty}\frac{\langle\alpha,\beta;q\rangle_m x_1^m}{\langle 1,\gamma;q\rangle_m (x_2;q)_{\alpha+m}}\\&\qquad\times{}_2\phi_2(\alpha+m,\gamma-\beta;\gamma+m|q;x_2q^{\beta+m}||-;x_2q^{\alpha+m}).\end{aligned}\tag{10.66}$$

By [272, p. 115] we obtain

$$\begin{aligned}
&{}_2\phi_1\big(\alpha,\beta;\gamma|q;x_1\oplus_q x_2\big)\\
&\quad\cong\sum_{m=0}^{\infty}\frac{\langle\alpha,\beta;q\rangle_m x_1^m}{\langle 1,\gamma;q\rangle_m (x_2;q)_{\alpha+\beta-\gamma+m}}\\
&\qquad\times{}_2\phi_1\big(\gamma-\alpha,\gamma-\beta;\gamma+m|q;x_2q^{\alpha+\beta-\gamma+m}\big).
\end{aligned}\tag{10.67}$$

The following three expansions of Φ_2 in terms of one-variable ϕ are q-analogues of [41, p. 24, (30), (30′), (30″)]:

$$\begin{aligned}
&\Phi_2\big(\alpha;\beta,\beta';\gamma,\gamma'|q;x_1,x_2\big)\\
&\quad\cong\sum_{m=0}^{\infty}\frac{\langle\alpha,\beta;q\rangle_m x_1^m}{\langle 1,\gamma;q\rangle_m (x_2;q)_{\alpha+m}}\\
&\qquad\times{}_2\phi_2\big(\alpha+m,\gamma'-\beta';\gamma'|q;x_2q^{\beta'}||-;x_2q^{\alpha+m}\big).
\end{aligned}\tag{10.68}$$

$$\begin{aligned}
&\Phi_2\big(\alpha;\beta,\beta';\gamma,\gamma'|q;x_1,x_2\big)\\
&\quad\cong\sum_{m=0}^{\infty}\frac{\langle\alpha,\beta;q\rangle_m x_1^m}{\langle 1,\gamma;q\rangle_m (x_2;q)_{\beta'}}\\
&\qquad\times{}_2\phi_2\big(\beta',\gamma'-\alpha-m;\gamma'|q;x_2q^{\alpha+m}||-;x_2q^{\beta'}\big).
\end{aligned}\tag{10.69}$$

$$\begin{aligned}
&\Phi_2\big(\alpha;\beta,\beta';\gamma,\gamma'|q;x_1,x_2\big)\\
&\quad\cong\sum_{m=0}^{\infty}\frac{\langle\alpha,\beta;q\rangle_m x_1^m}{\langle 1,\gamma;q\rangle_m (x_2;q)_{\alpha+\beta'-\gamma'+m}}\\
&\qquad\times{}_2\phi_1\big(\gamma'-\alpha-m,\gamma'-\beta';\gamma'|q;x_2q^{\alpha+\beta'-\gamma'+m}\big).
\end{aligned}\tag{10.70}$$

By (7.62), we obtain a q-analogue of [41, p. 25, (31)]:

$$\begin{aligned}
&\Phi_2\big(\alpha;\beta,\beta';\gamma,\gamma'|q;x_1,x_2\big)\\
&\quad\cong\frac{1}{(x_2;q)_\alpha}\sum_{m_1=0}^{\infty}\frac{\langle\alpha,\beta;q\rangle_{m_1}x_1^{m_1}}{\langle 1,\gamma;q\rangle_{m_1}(x_2q^\alpha;q)_{m_1}}\\
&\qquad\times\sum_{m_2=0}^{\infty}\frac{\langle\alpha+m_1,\gamma'-\beta';q\rangle_{m_2}(x_2q^{\beta'})^{m_2}}{\langle 1,\gamma';q\rangle_{m_2}(x_2q^{\alpha+m_1};q)_{m_2}}(-1)^{m_2}q^{\binom{m_2}{2}}\\
&\quad\equiv\frac{1}{(x_2;q)_\alpha}\Phi^{1:1;1}_{1:1;2}\left[\begin{matrix}\alpha:\beta;\gamma'-\beta';\\ -:\gamma;\gamma',\infty;\end{matrix}|q;x_1,x_2q^{\beta'}||\begin{matrix}-:-;\\ x_2q^\alpha:-;\end{matrix}\right].
\end{aligned}$$

The following three expansions of Φ_3 in terms of one-variable ϕ are q-analogues of [41, p. 25, (33), (33′), (33″)]:

$$\Phi_3\big(\alpha, a'; \beta, \beta'; \gamma|q; x_1, x_2\big)$$
$$\cong \sum_{m=0}^{\infty} \frac{\langle \alpha, \beta; q\rangle_m x_1^m}{\langle 1, \gamma; q\rangle_m (x_2; q)_{\alpha'}}$$
$$\times\, {}_2\phi_2\big(\alpha', \gamma - \beta' + m; \gamma + m|q; x_2 q^{\beta'}||-; x_2 q^{\alpha'}\big), \tag{10.71}$$

$$\Phi_3\big(\alpha, \alpha'; \beta, \beta'; \gamma|q; x_1, x_2\big)$$
$$\cong \sum_{m=0}^{\infty} \frac{\langle \alpha, \beta; q\rangle_m x_1^m}{\langle 1, \gamma; q\rangle_m (x_2; q)_{\beta'}}$$
$$\times\, {}_2\phi_2\big(\beta', \gamma - \alpha' + m; \gamma + m|q; x_2 q^{\alpha'}||-; x_2 q^{\beta'}\big), \tag{10.72}$$

$$\Phi_3\big(\alpha, \alpha'; \beta, \beta'; \gamma|q; x_1, x_2\big)$$
$$\cong \sum_{m=0}^{\infty} \frac{\langle \alpha, \beta; q\rangle_m x_1^m}{\langle 1, \gamma; q\rangle_m (x_2; q)_{\alpha'+\beta'-\gamma-m}}$$
$$\times\, {}_2\phi_1\big(\gamma - \alpha' + m, \gamma - \beta' + m; \gamma + m|q; x_2 q^{\alpha'+\beta'-\gamma-m}\big). \tag{10.73}$$

By combining (10.56), Jackson [307, p. 75, (33)] and [307, p. 75, (34)], we obtain the expansions

$${}_3\phi_2\big(\alpha, \beta, \infty; \gamma|q; x_2||-; x_1 q^{\alpha}\big)$$
$$\cong \sum_{r=0}^{\infty} \frac{\langle \alpha, \beta; q\rangle_r}{\langle 1, \gamma; q\rangle_r} x_1^r x_2^r q^{r(\alpha+r-1)}$$
$$\times\, \frac{{}_2\phi_1(\alpha + r, \beta + r; \gamma + r|q; x_2)}{(x_1 q^{\alpha}; q)_r}, \tag{10.74}$$

$${}_2\phi_1(\alpha, \beta; \gamma|q; x_1)$$
$$\cong \sum_{r=0}^{\infty} \frac{(-1)^r \langle \alpha, \beta; q\rangle_r}{\langle 1, \gamma; q\rangle_r} x_1^r x_2^r q^{r\alpha + \binom{r}{2}}$$
$$\times\, \frac{1}{(x_2 q^{\alpha}; q)_r}\, {}_3\phi_2\big(\alpha + r, \beta + r, \infty; \gamma + r|q; x_1||-; x_2 q^{\alpha+r}\big). \tag{10.75}$$

10.3.1 Double q-balanced series

We continue with q-analogues of a couple of formulas by Carlitz [101] to show some balanced theorems for basic double series, which make sense when $q = 1$.

We then discuss the advantages and disadvantages of Watson's and Heine's notations, compare with Section 7.3.

Assume that

$$\gamma+\delta'\equiv\alpha+\beta'-n+1\left(\operatorname{mod}\frac{2\pi i}{\log q}\right), \tag{10.76}$$

$$\gamma\equiv\beta'+\beta\left(\operatorname{mod}\frac{2\pi i}{\log q}\right), \tag{10.77}$$

$$\delta\equiv\alpha-\beta'-m+1\left(\operatorname{mod}\frac{2\pi i}{\log q}\right). \tag{10.78}$$

This implies that

$$\gamma+\delta\equiv\alpha+\beta-m+1\left(\operatorname{mod}\frac{2\pi i}{\log q}\right). \tag{10.79}$$

Then according to Al-Salam [14, p. 456, (7)]

$$\begin{aligned}
S&\equiv\sum_{r=0}^{m}\sum_{s=0}^{n}\frac{\langle -m;q\rangle_r\langle -n;q\rangle_s\langle\alpha;q\rangle_{r+s}\langle\beta;q\rangle_r\langle\beta';q\rangle_s q^{r+s}}{\langle 1;q\rangle_r\langle 1;q\rangle_s\langle\gamma;q\rangle_{r+s}\langle\delta;q\rangle_r\langle\delta';q\rangle_s}\\
&=\sum_{r=0}^{m}\frac{\langle -m,\alpha,\beta;q\rangle_r}{\langle 1,\gamma,\delta;q\rangle_r}q^r\sum_{s=0}^{n}\frac{\langle -n,\alpha+r,\beta';q\rangle_s}{\langle 1,\gamma+r,\delta';q\rangle_s}q^s\\
&=\sum_{r=0}^{m}\frac{\langle -m,\alpha,\beta;q\rangle_r}{\langle 1,\gamma,\delta;q\rangle_r}q^r\frac{\langle\gamma-\alpha,\gamma-\beta'+r;q\rangle_n}{\langle\gamma+r,\gamma-\alpha-\beta';q\rangle_n}\\
&=\sum_{r=0}^{m}\frac{\langle -m,\alpha,\gamma-\beta'+n;q\rangle_r}{\langle 1,\gamma+n,\delta;q\rangle_r}q^r\frac{\langle\gamma-\alpha,\gamma-\beta';q\rangle_n}{\langle\gamma,\gamma-\alpha-\beta';q\rangle_n}\\
&=\frac{\langle\gamma-\alpha,\gamma-\beta';q\rangle_n}{\langle\gamma,\gamma-\alpha-\beta';q\rangle_n}\frac{\langle\gamma-\alpha+n,\beta';q\rangle_m}{\langle\gamma+n,\beta'-\alpha;q\rangle_m}\\
&=\frac{\langle\beta+\beta'-\alpha;q\rangle_{m+n}\langle\beta';q\rangle_m\langle\beta;q\rangle_n}{\langle\beta+\beta';q\rangle_{m+n}\langle\beta'-\alpha;q\rangle_m\langle\beta-\alpha;q\rangle_n}.
\end{aligned} \tag{10.80}$$

If instead of (10.77) we assume that

$$\delta'+\delta\equiv 1+\alpha\left(\operatorname{mod}\frac{2\pi i}{\log q}\right), \tag{10.81}$$

then (10.76) implies that

$$\gamma-\beta'+n\equiv\delta\left(\operatorname{mod}\frac{2\pi i}{\log q}\right). \tag{10.82}$$

In this case we have, according to Al-Salam [14, p. 456, (8)],

$$\begin{aligned}
S &= \sum_{r=0}^{m} \frac{\langle -m,\alpha,\beta;q\rangle_r}{\langle 1,\gamma,\delta;q\rangle_r} q^r \frac{\langle \gamma-\alpha,\gamma-\beta'+r;q\rangle_n}{\langle \gamma+r,\gamma-\alpha-\beta';q\rangle_n} \\
&= \frac{\langle \gamma-\alpha,\gamma-\beta';q\rangle_n}{\langle \gamma,\gamma-\alpha-\beta';q\rangle_n}\, {}_3\phi_2(-m,\alpha,\beta;\gamma+n,\gamma-\beta'|q;q) \\
&= \frac{\langle \gamma-\alpha,\gamma-\beta';q\rangle_n}{\langle \gamma,\gamma-\alpha-\beta';q\rangle_n} \frac{\langle \gamma+n-\alpha,\gamma+n-\beta;q\rangle_m}{\langle \gamma+n,\gamma+n-\alpha-\beta;q\rangle_m} \\
&= \frac{\langle \gamma-\alpha\rangle_{n+m}\langle \gamma-\beta'\rangle_n\langle \gamma-\alpha-\beta';q\rangle_m q^{m\alpha}}{\langle \gamma\rangle_{n+m}\langle \gamma-\alpha-\beta';q\rangle_n\langle \gamma-\beta'\rangle_m}. \qquad (10.83)
\end{aligned}$$

The summation formula (10.80) can be written in the form

$$\Phi^{1:2}_{1:1}\left[\begin{matrix}\alpha:-m,\beta;-n,\beta' \\ \beta+\beta':1+\alpha-\beta'-m;1+\alpha-\beta-n\end{matrix}|q;q,q\right] = \frac{\langle \beta+\beta'-\alpha;q\rangle_{m+n}\langle \beta';q\rangle_m\langle \beta;q\rangle_n}{\langle \beta+\beta';q\rangle_{m+n}\langle \beta'-\alpha;q\rangle_m\langle \beta-\alpha;q\rangle_n}. \qquad (10.84)$$

Definition 171 The elements of the vectors g_i are denoted by $g_{i,j}$. Consider Eq. (10.45), where all parameters with $'$ are put equal to zero and $A=B$, $1+H_i=G_i$, $i=1,\ldots,n$.

The q-Kampé de Fériet function (10.46) is called balanced, if

$$\begin{aligned}
&b_1+\cdots+b_B+h_{i,1}+\cdots+h_{i,H_i} \\
&\equiv 1+a_1+\cdots+a_A+g_{i,1}+\cdots+g_{i,G_i}\left(\operatorname{mod}\frac{2\pi i}{\log q}\right), \quad i=1,\ldots,n.
\end{aligned} \qquad (10.85)$$

Definition 172 The elements of the vectors g_i are denoted by $g_{i,j}$. Consider Eq. (10.46), where all parameters with $'$ are put equal to zero and $A=B$, $1+H=G$. The q-Kampé de Fériet function (10.46) is called balanced, if

$$\begin{aligned}
&b_1+\cdots+b_B+h_{i,1}+\cdots+h_{i,H} \\
&\equiv 1+a_1+\cdots+a_A+g_{i,1}+\cdots+g_{i,G}\left(\operatorname{mod}\frac{2\pi i}{\log q}\right), \quad i=1,\ldots,n. \qquad (10.86)
\end{aligned}$$

The series (10.84) is an example of a balanced q-Kampé de Fériet function.

10.3.2 Transformation formula of Carlitz-Srivastava

Assume again that $\gamma=\beta'+\beta$. We can now use (10.80) to prove the following transformation formula:

Theorem 10.3.1 *Srivastava* [478, p. 52, (3.3)].

$$\begin{aligned}
&\Phi_1\big(\beta'+\beta-\alpha;\beta',\beta;\beta+\beta'|q;x,y\big)\\
&\quad=\big(xq^{\beta'-\alpha};q\big)_{\alpha-\beta'}\big(yq^{\beta-\alpha};q\big)_{\alpha-\beta}\Phi_1\big(\alpha,\beta,\beta';\beta+\beta'|q;xq^{\beta'-\alpha},yq^{\beta-\alpha}\big).
\end{aligned} \tag{10.87}$$

Proof

$$\begin{aligned}
&\Phi_1\big(\beta'+\beta-\alpha;\beta',\beta;\beta+\beta'|q;x,y\big)\\
&=\sum_{m,n=0}^{\infty}\frac{x^m y^n\langle\beta'-\alpha;q\rangle_m\langle\beta-\alpha;q\rangle_n}{\langle 1;q\rangle_n\langle 1;q\rangle_m}\sum_{r=0}^{m}\sum_{s=0}^{n}\frac{\langle -m;q\rangle_r\langle -n;q\rangle_s}{\langle 1;q\rangle_r\langle 1;q\rangle_s}\\
&\quad\times\frac{q^{r+s}\langle\alpha;q\rangle_{r+s}\langle\beta;q\rangle_r\langle\beta';q\rangle_s}{\langle\gamma;q\rangle_{r+s}\langle\alpha-\beta'-m+1;q\rangle_r\langle\alpha-\beta-n+1;q\rangle_s}\\
&=\sum_{r=0}^{\infty}\sum_{s=0}^{\infty}\frac{x^r y^s q^{r+s}\langle\alpha;q\rangle_{r+s}\langle\beta;q\rangle_r\langle\beta';q\rangle_s}{\langle 1;q\rangle_r\langle 1;q\rangle_s\langle\gamma;q\rangle_{r+s}}\sum_{m=r}^{\infty}\sum_{n=s}^{\infty}\frac{\langle -m;q\rangle_r\langle\beta'-\alpha;q\rangle_{m-r}}{\langle 1;q\rangle_m\langle 1;q\rangle_n}\\
&\quad\times(-1)^{r+s}x^{m-r}\langle -n;q\rangle_s\langle\beta-\alpha;q\rangle_{n-s}y^{n-s}\\
&\quad\times q^{-\binom{s}{2}-\binom{r}{2}+s(\beta-\alpha+n-1)+r(\beta'-\alpha+m-1)}\\
&=\sum_{r=0}^{\infty}\sum_{s=0}^{\infty}\frac{x^r y^s\langle\alpha;q\rangle_{r+s}\langle\beta;q\rangle_r\langle\beta';q\rangle_s}{\langle 1;q\rangle_r\langle 1;q\rangle_s\langle\gamma;q\rangle_{r+s}}q^{s(\beta-\alpha)+r(\beta'-\alpha)}\\
&\quad\times\sum_{m=r}^{\infty}\frac{\langle\beta'-\alpha;q\rangle_{m-r}x^{m-r}}{\langle 1;q\rangle_{m-r}}\sum_{n=s}^{\infty}\frac{\langle\beta-\alpha;q\rangle_{n-s}y^{n-s}}{\langle 1;q\rangle_{n-s}}\\
&=\frac{(xq^{\beta'-\alpha};q)_\infty}{(x;q)_\infty}\frac{(yq^{\beta-\alpha};q)_\infty}{(y;q)_\infty}\Phi_1\big(\alpha,\beta,\beta';\beta+\beta'|q;xq^{\beta'-\alpha},yq^{\beta-\alpha}\big).
\end{aligned} \tag{10.88}$$

□

Remark 53 The original paper by Carlitz on the hypergeometric case, [101, p. 417, (12)], seems to contain a misprint. Also compare with [103, p. 138, (3)] and [104, 1, (1.5)]. In the first q-analogue (in Watson's notation) [14, p. 457, Eq. 9], a further misprint appears. This clearly shows the disadvantages of the Watson notation compared with the Heine notation.

For the special case $y=0$, (10.87) becomes a q-analogue of Euler's transformation:

Theorem 10.3.2 [220, p. 10, (1.4.3)], [272, p. 115] (*compare with* (7.52)):

$$ {}_2\phi_1(\gamma-\alpha;\beta;\gamma|q;x)=\left(xq^{\beta-\alpha};q\right)_{\alpha-\beta}\,{}_2\phi_1\left(\alpha,\gamma-\beta;\gamma|q;xq^{\beta-\alpha}\right). \quad (10.89) $$

10.3.3 Three formulas of Andrews

We continue with several formulas due to Andrews, which were originally presented in Watson's notation. We show that although these formulas have no hypergeometric counterpart, they can be presented in the current notation.

Theorem 10.3.3 *The following transformation* [24, p. 618] *holds*:

$$ \begin{aligned} &\Phi_1\left(\alpha;\beta,\beta';\gamma|q;q^x,q^y\right)\\ &\quad\cong\frac{\langle\alpha;q\rangle_\infty\langle\beta+x;q\rangle_\infty\langle\beta'+y;q\rangle_\infty}{\langle\gamma;q\rangle_\infty\langle x;q\rangle_\infty\langle y;q\rangle_\infty}\\ &\quad\times{}_3\phi_2\left(\gamma-\alpha,x,y;\beta+x,\beta'+y|q,q^\alpha\right). \end{aligned} \quad (10.90) $$

Proof We use the q-binomial theorem. Absolute convergence is assumed. Then

$$ \begin{aligned} &\Phi_1\left(\alpha;\beta,\beta';\gamma|q;q^x,q^y\right)\\ &=\sum_{m=0}^\infty\sum_{n=0}^\infty\frac{q^{xm}q^{yn}\langle\alpha;q\rangle_{m+n}\langle\beta;q\rangle_m\langle\beta';q\rangle_n}{\langle\gamma;q\rangle_{m+n}\langle 1;q\rangle_m\langle 1;q\rangle_n}\\ &=\frac{\langle\alpha;q\rangle_\infty}{\langle\gamma;q\rangle_\infty}\sum_{m=0}^\infty\sum_{n=0}^\infty\frac{q^{xm}q^{yn}\langle\gamma+m+n;q\rangle_\infty\langle\beta;q\rangle_m\langle\beta';q\rangle_n}{\langle\alpha+m+n;q\rangle_\infty\langle 1;q\rangle_m\langle 1;q\rangle_n}\\ &=\frac{\langle\alpha;q\rangle_\infty}{\langle\gamma;q\rangle_\infty}\sum_{m=0}^\infty\sum_{n=0}^\infty\sum_{r=0}^\infty\frac{q^{xm}q^{yn}q^{r(\alpha+m+n)}\langle\gamma-\alpha;q\rangle_r\langle\beta;q\rangle_m\langle\beta';q\rangle_n}{\langle 1;q\rangle_r\langle 1;q\rangle_m\langle 1;q\rangle_n}\\ &=\frac{\langle\alpha;q\rangle_\infty}{\langle\gamma;q\rangle_\infty}\sum_{r=0}^\infty\frac{\langle\gamma-\alpha;q\rangle_r}{\langle 1;q\rangle_r}q^{\alpha r}\frac{\langle\beta+x+r;q\rangle_\infty}{\langle x+r;q\rangle_\infty}\frac{\langle\beta'+y+r;q\rangle_\infty}{\langle y+r;q\rangle_\infty}\\ &=\frac{\langle\alpha;q\rangle_\infty}{\langle\gamma;q\rangle_\infty}\frac{\langle\beta+x;q\rangle_\infty}{\langle x;q\rangle_\infty}\frac{\langle\beta'+y;q\rangle_\infty}{\langle y;q\rangle_\infty}\sum_{r=0}^\infty\frac{\langle\gamma-\alpha;q\rangle_r}{\langle 1;q\rangle_r}q^{\alpha r}\\ &\times\frac{\langle x;q\rangle_r}{\langle\beta+x;q\rangle_r}\frac{\langle y;q\rangle_r}{\langle\beta'+x;q\rangle_r}\\ &=\frac{\langle\alpha;q\rangle_\infty\langle\beta+x;q\rangle_\infty\langle\beta'+y;q\rangle_\infty}{\langle\gamma;q\rangle_\infty\langle x;q\rangle_\infty\langle y;q\rangle_\infty}\,{}_3\phi_2\left(\gamma-\alpha,x,y;\beta+x,\beta'+y|q;q^\alpha\right). \end{aligned} \quad (10.91) $$

□

Furthermore

Theorem 10.3.4 [24, p. 619]

$$\Phi_1\big(\beta'-x;\beta,\beta';\beta+\beta'|q;q^x,q^y\big)\cong\frac{\langle\beta+x,\beta',\beta'+y-x;q\rangle_\infty}{\langle\beta+\beta',x,y;q\rangle_\infty}. \tag{10.92}$$

Proof

$$\begin{aligned}&\Phi_1\big(\beta'-x;\beta,\beta';\beta+\beta'|q;q^x,q^y\big)\\&=\frac{\langle\beta+x,\beta'-x,\beta'+y;q\rangle_\infty}{\langle\beta+\beta',x,y;q\rangle_\infty}{}_2\phi_1\big(x,y;\beta'+y|q;q^{\beta'-x}\big)\\&=\frac{\langle\beta+x,\beta',\beta'+y-x;q\rangle_\infty}{\langle\beta+\beta',x,y;q\rangle_\infty}.\end{aligned} \tag{10.93}$$

□

Finally, we have

Theorem 10.3.5 [24, p. 619].

$$\begin{aligned}&\Phi_1\big(\widetilde{1-y};\beta,x+1-2y;\beta+\widetilde{1+x}-y|q;q^x,q^y\big)\\&\cong\frac{\langle\beta+x,\widetilde{1};q\rangle_\infty\langle x+1,2(1-y)+x;q^2\rangle_\infty}{\langle\beta+\widetilde{1+x}-y,x,y;q\rangle_\infty}.\end{aligned} \tag{10.94}$$

Proof

$$\begin{aligned}&\Phi_1\big(\widetilde{1-y};\beta,x+1-2y;\beta+\widetilde{1+x}-y|q;q^x,q^y\big)\\&=\frac{\langle\widetilde{1-y},\beta+x,x+1-y;q\rangle_\infty}{\langle\beta+\widetilde{1+x}-y,x,y;q\rangle_\infty}{}_2\phi_1\big(x,y;x+1-y|q;-q^{1-y}\big)\\&=\frac{\langle\beta+x,\widetilde{1};q\rangle_\infty\langle x+1,2(1-y)+x;q^2\rangle_\infty}{\langle\beta+\widetilde{1+x}-y,x,y;q\rangle_\infty}.\end{aligned} \tag{10.95}$$

□

10.3.4 q-Analogues of Carlson's formulas

Our next task is to find q-analogues of some general formulas of Carlson [109]. We first recall the following two equivalent forms of the q-Vandermonde theorem. The second one also appeared in Jackson [307, (29), p. 74] and [308, (60), p. 56].

Lemma 10.3.6

$$\sum_{m+n=N}\frac{\langle b;q\rangle_m\langle b';q\rangle_n}{\langle 1;q\rangle_m\langle 1;q\rangle_n}q^{-nb'}=q^{-Nb'}\frac{\langle b+b';q\rangle_N}{\langle 1;q\rangle_N}, \tag{10.96}$$

$$\sum_{m+n=N} \frac{\langle b;q\rangle_m \langle b';q\rangle_n}{\langle 1;q\rangle_m \langle 1;q\rangle_n} q^{nb} = \frac{\langle b+b';q\rangle_N}{\langle 1;q\rangle_N}. \tag{10.97}$$

The order classification [286] for double hypergeometric series (DHS) was prevalent for a long time. Carlson [109] showed evidence indicating the need for a new classification of DHS.

The equations given in the next four theorems are q-analogues of Carlson's transformation [109, (8), p. 223]:

Theorem 10.3.7

$$\sum_{m+n=N} \frac{\langle b;q\rangle_m \langle b';q\rangle_n}{\langle 1;q\rangle_m \langle 1;q\rangle_n} q^{-nb'} (x_1 \oplus_q x_2)^m x_2^n$$
$$= \langle b+b';q\rangle_N \sum_{m+n=N} \frac{\langle b;q\rangle_m}{\langle b+b',1;q\rangle_m \langle 1;q\rangle_n} q^{-nb'} x_1^m x_2^n. \tag{10.98}$$

Proof

$$\text{LHS} = \sum_{n=0}^{\infty} \frac{\langle b;q\rangle_{N-n} \langle b';q\rangle_n}{\langle 1;q\rangle_{N-n} \langle 1;q\rangle_n} q^{-nb'} \sum_{m'=0}^{N-n} \binom{N-n}{m'}_q x_1^{m'} x_2^{N-m'}$$
$$= \sum_{m=0}^{N} \frac{x_1^m x_2^{N-m}}{\langle 1;q\rangle_m} \sum_{n=0}^{N-m} \frac{\langle b;q\rangle_{N-n} \langle b';q\rangle_n}{\langle 1;q\rangle_n \langle 1;q\rangle_{N-n-m}} q^{-nb'}$$
$$\overset{\text{by (10.96)}}{=} \sum_{m=0}^{N} \frac{x_1^m x_2^{N-m}}{\langle 1;q\rangle_m} q^{-b'(N-m)} \frac{\langle b;q\rangle_m \langle b+b'+m;q\rangle_{N-m}}{\langle 1;q\rangle_{N-m}}$$
$$= \sum_{m+n=N} \frac{x_1^m x_2^n}{\langle 1;q\rangle_m} q^{-b'n} \frac{\langle b;q\rangle_m \langle b+b'+m;q\rangle_{N-m}}{\langle 1;q\rangle_n} = \text{RHS}. \tag{10.99}$$

□

Theorem 10.3.8

$$\sum_{m+n=N} \frac{\langle b;q\rangle_m \langle b';q\rangle_n}{\langle 1;q\rangle_m \langle 1;q\rangle_n} q^{-nb'} (x_2 \boxplus_q x_1)^m x_2^n$$
$$= \langle b+b';q\rangle_N \sum_{m+n=N} \frac{\langle b;q\rangle_m}{\langle b+b',1;q\rangle_m \langle 1;q\rangle_n} q^{-nb'+\binom{m}{2}} x_1^m x_2^n. \tag{10.100}$$

Proof As above. □

Theorem 10.3.9

$$\sum_{m+n=N} \frac{\langle b;q\rangle_m \langle b';q\rangle_n}{\langle 1;q\rangle_m \langle 1;q\rangle_n} q^{nb} (x_1 \oplus_q x_2)^m x_2^n$$
$$= \langle b+b';q\rangle_N \sum_{m+n=N} \frac{\langle b;q\rangle_m}{\langle b+b',1;q\rangle_m \langle 1;q\rangle_n} x_1^m x_2^n. \tag{10.101}$$

Theorem 10.3.10

$$\sum_{m+n=N} \frac{\langle b;q\rangle_m \langle b';q\rangle_n}{\langle 1;q\rangle_m \langle 1;q\rangle_n} q^{nb} (x_2 \boxplus_q x_1)^m x_2^n$$
$$= \langle b+b';q\rangle_N \sum_{m+n=N} \frac{\langle b;q\rangle_m}{\langle b+b',1;q\rangle_m \langle 1;q\rangle_n} q^{\binom{m}{2}} x_1^m x_2^n. \tag{10.102}$$

Corollary 10.3.11 *Four q-analogues of Carlson* [109, (4), p. 222], *which are valid if*

$$|x_1| + |x_2| < 1, \qquad a \neq c, \qquad c \neq b+b', \qquad b \neq 1, \qquad b' \neq 1,$$

neither c *nor* $b+b' = 0$ *or* $-N$,

$$\begin{aligned}
&\sum_{m,n=0}^{\infty} \frac{\langle a;q\rangle_{m+n} \langle b;q\rangle_m \langle b';q\rangle_n}{\langle c;q\rangle_{m+n} \langle 1;q\rangle_m \langle 1;q\rangle_n} q^{-nb'} (x_1 \oplus_q x_2)^m x_2^n \\
&\quad = \sum_{m,n=0}^{\infty} \frac{\langle a, b+b';q\rangle_{m+n} \langle b;q\rangle_m}{\langle c;q\rangle_{m+n} \langle b+b',1;q\rangle_m \langle 1;q\rangle_n} q^{-nb'} x_1^m x_2^n, \\
&\sum_{m,n=0}^{\infty} \frac{\langle a;q\rangle_{m+n} \langle b;q\rangle_m \langle b';q\rangle_n}{\langle c;q\rangle_{m+n} \langle 1;q\rangle_m \langle 1;q\rangle_n} q^{nb} (x_1 \oplus_q x_2)^m x_2^n \\
&\quad = \sum_{m,n=0}^{\infty} \frac{\langle a, b+b';q\rangle_{m+n} \langle b;q\rangle_m}{\langle c;q\rangle_{m+n} \langle b+b',1;q\rangle_m \langle 1;q\rangle_n} x_1^m x_2^n, \\
&\sum_{m,n=0}^{\infty} \frac{\langle a;q\rangle_{m+n} \langle b;q\rangle_m \langle b';q\rangle_n}{\langle c;q\rangle_{m+n} \langle 1;q\rangle_m \langle 1;q\rangle_n} q^{-nb'} (x_2 \boxplus_q x_1)^m x_2^n \\
&\quad = \sum_{m,n=0}^{\infty} \frac{\langle a, b+b';q\rangle_{m+n} \langle b;q\rangle_m}{\langle c;q\rangle_{m+n} \langle b+b',1;q\rangle_m \langle 1;q\rangle_n} q^{-nb'+\binom{m}{2}} x_1^m x_2^n, \\
&\sum_{m,n=0}^{\infty} \frac{\langle a;q\rangle_{m+n} \langle b;q\rangle_m \langle b';q\rangle_n}{\langle c;q\rangle_{m+n} \langle 1;q\rangle_m \langle 1;q\rangle_n} q^{nb} (x_2 \boxplus_q x_1)^m x_2^n \\
&\quad = \sum_{m,n=0}^{\infty} \frac{\langle a, b+b';q\rangle_{m+n} \langle b;q\rangle_m}{\langle c;q\rangle_{m+n} \langle b+b',1;q\rangle_m \langle 1;q\rangle_n} q^{\binom{m}{2}} x_1^m x_2^n.
\end{aligned} \tag{10.103}$$

10.4 The q-Appell function Φ_1 as q-integral

Jackson has found a generalization of (7.50).

Theorem 10.4.1 [307], [5, p. 78], *a q-analogue of Picard* (1880) [416]:

$$\frac{\Gamma_q(a)\Gamma_q(c-a)}{\Gamma_q(c)}\Phi_1\big(a;b,b';c|q;x_1,x_2\big) \cong \int_0^1 t^{a-1}\frac{(qt;q)_{c-a-1}}{(x_1t;q)_b(x_2t;q)_{b'}}\,d_q(t). \tag{10.104}$$

Proof It turns out that formula (10.104) is equivalent to (10.90). Indeed, calculation shows that

$$\begin{aligned}
&\int_0^1 t^{a-1}\frac{(qt;q)_{c-a-1}}{(x_1t;q)_b(x_2t;q)_{b'}}\,d_q(t)\\
&= (1-q)\sum_{n=0}^{\infty}\frac{q^{an}\langle n+1;q\rangle_{c-a-1}}{\langle n+x_1;q\rangle_b\langle n+x_2;q\rangle_{b'}}\\
&= (1-q)\sum_{n=0}^{\infty}\frac{q^{an}\langle 1+n,n+x_1+b,n+x_2+b';q\rangle_\infty}{\langle n+x_1,n+x_2,n+c-a;q\rangle_\infty}\\
&= \frac{\langle 1,x_1+b,x_2+b';q\rangle_\infty}{\langle x_1,x_2,c-a;q\rangle_\infty}\\
&\quad\times(1-q)\sum_{n=0}^{\infty}\frac{q^{an}\langle 1+n,n+x_1+b,n+x_2+b',x_1,x_2,c-a;q\rangle_\infty}{\langle n+x_1,n+x_2,n+c-a,1,x_1+b,x_2+b';q\rangle_\infty}\\
&= (1-q)\frac{\langle 1,x_1+b,x_2+b';q\rangle_\infty}{\langle x_1,x_2,c-a;q\rangle_\infty}\sum_{n=0}^{\infty}\frac{q^{an}\langle x_1,x_2,c-a;q\rangle_n}{\langle 1,x_1+b,x_2+b';q\rangle_n}. \qquad (10.105)
\end{aligned}$$

□

Equation (10.104) is obviously symmetric in b and b', if we interchange x_1 and x_2. We therefore operate with $\mathrm{D}_{q,x}$ on both sides. If we denote the left and right sides of (10.104) by $A(a,b,b',c,x_1,x_2)$ and $B(a,b,b',c,x_1,x_2)$, respectively, we have

$$\begin{aligned}
\mathrm{D}_{q,x_1}A &= \frac{\Gamma_q(a)\Gamma_q(c-a)}{\Gamma_q(c)}\sum_{m_1,m_2=0}^{\infty}\frac{\langle a;q\rangle_{m_1+m_2+1}\langle b;q\rangle_{m_1+1}\langle b';q\rangle_{m_2}}{\langle 1;q\rangle_{m_1}\langle 1;q\rangle_{m_2}\langle c;q\rangle_{m_1+m_2+1}}\frac{x_1^{m_1}x_2^{m_2}}{1-q}\\
&= \{b\}_q\frac{\Gamma_q(a+1)\Gamma_q(c-a)}{\Gamma_q(c+1)}\Phi_1\big(a+1;b+1,b';c+1|q;x_1,x_2\big).
\end{aligned} \tag{10.106}$$

The RHS is

$$\mathrm{D}_{q,x_1} B = \{b\}_q \int_0^1 t^a \frac{(qt;q)_{c-a-1}}{(x_1t;q)_{b+1}(x_2t;q)_{b'}} d_q(t). \tag{10.107}$$

We get

$$\mathrm{D}_{q,x_1} A\big(a,b,b',c,x_1,x_2\big) = \{b\}_q A\big(a+1,b+1,b',c+1,x_1,x_2\big), \tag{10.108}$$

and the same for $B(a,b,b',c,x_1,x_2)$.

By the previous formula we get the next contiguity relation. The definitions are given in the article by Agarwal [5, p. 77], these definitions are straightforward generalizations of Section 7.7:

$$\{a\}_q \Phi_1\big(a^+\big) = \{1+a-c\}_q \Phi_1 + q^{1+a-c}\{c-1\}_q \Phi_1\big(c^-\big). \tag{10.109}$$

10.5 q-analogues of some of Srivastava's formulas

In this section we will derive some q-analogues more complex than before. Whenever certain special cases of these (non-q)-formulas have been published, we try to find q-analogues of them too. We will prove two general multiple formulas by the q-Vandermonde sums, and two general formulas by (8.41). The Bailey-Daum theorem will also be used. We always assume that $\{C_n\}_{n=0}^\infty$ is a sequence of bounded complex numbers.

We first derive two q-analogues of the general formula by Srivastava [473, (4), p. 295], (10.110) and (10.113).

Theorem 10.5.1 [483, p. 245, (16)], [166, p. 215, (62)]:

$$\sum_{m,n=0}^{\infty} \frac{C_{m+n} x^{m+n} q^{2\binom{n}{2}+\sigma n}}{\langle 1,\nu;q\rangle_m \langle 1,\sigma;q\rangle_n} = \sum_{N=0}^{\infty} \frac{C_N x^N}{\langle 1,\nu,\sigma;q\rangle_N} \langle \sigma-1+\nu+N;q\rangle_N. \tag{10.110}$$

Proof

$$\mathrm{LHS} = \sum_{N=0}^{\infty} \frac{C_N x^N}{\langle 1,\nu;q\rangle_N} \sum_{n=0}^{N} \frac{\langle -\nu+1-N,-N;q\rangle_n}{\langle 1,\sigma;q\rangle_n} q^{n(-1+\nu+2N+\sigma)} = \mathrm{RHS}. \tag{10.111}$$

□

Corollary 10.5.2 *Compare with* [481, p. 224, 4.23]. *A first q-analogue of* [473, (8), p. 296]. *If* $|x| < \frac{1}{4}$, *then*

$$\sum_{m,n=0}^{\infty} \frac{\langle \lambda,\mu;q\rangle_{m+n} x^{m+n} q^{2\binom{n}{2}+\sigma n}}{\langle \nu,1;q\rangle_m \langle \sigma,1;q\rangle_n} = {}_6\phi_5\left[\begin{matrix} \triangle(q;2;\nu+\sigma-1),\lambda,\mu \\ \nu,\sigma,\nu+\sigma-1,2\infty \end{matrix} \Big| q;x\right]. \tag{10.112}$$

Theorem 10.5.3 [166, p. 216, (66)]:

$$\sum_{m,n=0}^{\infty} \frac{C_{m+n}x^{m+n}q^{2\binom{n}{2}+n(-2(m+n)-\nu+2)}}{\langle 1,\nu;q\rangle_m\langle 1,\sigma;q\rangle_n}$$
$$=\sum_{N=0}^{\infty}\frac{C_N x^N}{\langle 1,\nu,\sigma;q\rangle_N}\langle \sigma-1+\nu+N;q\rangle_N q^{N(1-\nu-N)}. \tag{10.113}$$

Corollary 10.5.4 *A second q-analogue of* [473, (8), p. 296]:

$$\sum_{m,n=0}^{\infty} \frac{\langle \lambda,\mu;q\rangle_{m+n}x^{m+n}q^{2\binom{n}{2}+n(-2(m+n)-\nu+2)}}{\langle \nu,1;q\rangle_m\langle \sigma,1;q\rangle_n}$$
$$=\sum_{n=0}^{\infty}\frac{\langle \lambda,\mu,\triangle(q;2;\nu+\sigma-1);q\rangle_n x^n q^{n(1-\nu-n)}}{\langle \nu,1,\sigma,\nu+\sigma-1;q\rangle_n},\quad |x|<\frac{1}{4}. \tag{10.114}$$

Theorem 10.5.5 *A q-analogue of* [473, (5), p. 295].

$$\sum_{m,n=0}^{\infty} \frac{(-1)^n C_{m+n}x^{m+n}q^{-mn}}{\langle 1,\sigma;q\rangle_m\langle 1,\sigma;q\rangle_n}$$
$$=\sum_{n=0}^{\infty}(-1)^n\frac{C_{2n}x^{2n}}{\langle \sigma;q\rangle_{2n}}\frac{\langle -\sigma\widetilde{+1}-2n;q\rangle_n q^{\binom{n}{2}+n\sigma}}{\langle 1,\widetilde{1},\sigma;q\rangle_n}. \tag{10.115}$$

Proof

$$\text{LHS}=\sum_{N=0}^{\infty}\frac{C_N x^N}{\langle 1,\sigma;q\rangle_N}\sum_{n=0}^{N}(-1)^n\frac{\langle -\sigma+1-N,-N;q\rangle_n}{\langle 1,\sigma;q\rangle_n}q^{n(N+\sigma)}$$
$$=\sum_{N=0,N\text{ even}}^{\infty}\frac{C_N x^N}{\langle 1,\sigma;q\rangle_N}\frac{\langle -\sigma\widetilde{+1}-N;q\rangle_{\frac{N}{2}}}{\langle -\sigma+1-\frac{N}{2};q\rangle_{\frac{N}{2}}}\left\langle\frac{1}{2};q^2\right\rangle_{\frac{N}{2}}$$
$$=\sum_{n=0}^{\infty}\frac{C_{2n}x^{2n}}{\langle 1,\sigma;q\rangle_{2n}}\frac{\langle -\sigma\widetilde{+1}-2n;q\rangle_n}{\langle -\sigma+1-n;q\rangle_n}\left\langle\frac{1}{2};q^2\right\rangle_n=\text{RHS}. \tag{10.116}$$

□

Corollary 10.5.6 *A q-analogue of* [473, (9), p. 296]. *If* $|x| < \frac{1}{4}$, *then*

$$\sum_{m,n=0}^{\infty} (-1)^n \frac{\langle \lambda, \mu; q\rangle_{m+n} x^{m+n} q^{-mn}}{\langle \sigma, 1; q\rangle_m \langle \sigma, 1; q\rangle_n}$$

$$= {}_9\phi_9 \left[\begin{matrix} \Delta(q; 2; \lambda, \mu) \\ \Delta(q; 2; \sigma), \sigma, \widetilde{1}, 3\infty \end{matrix} |q; x^2 q^{\sigma} || \begin{matrix} \langle 1 - \widetilde{\sigma - 2k}; q\rangle_k \\ \cdot \end{matrix} \right]. \tag{10.117}$$

Remark 54 Formula (10.110) has some similarities with [482, p. 28, (33)] and formula (10.115) has some similarities with [482, p. 31, (49)]. In upcoming articles we will continue to work on this subject and the notation $\Delta(q; 2; \lambda)$ will play a special role.

The following corollary occurred in a different, veiled form in Jackson [308, (61), p. 56]. It is almost a q-analogue of Burchnall, Chaundy [90, p. 124].

Corollary 10.5.7 *A first q-analogue of a reduction formula for the Humbert function* [473, (12), p. 296]:

$$\Phi^{1:2;0}_{1:1;1} \left[\begin{matrix} \mu : \infty, \infty; -; \\ \infty : \nu; \sigma; \end{matrix} |q; x, xq^{\sigma} \right] = {}_5\phi_4 \left[\begin{matrix} \mu, \Delta(q; 2; \nu + \sigma - 1) \\ \nu, \sigma, \nu + \sigma - 1, \infty \end{matrix} |q; x \right]. \tag{10.118}$$

Proof

$$\Phi \left[\begin{matrix} \mu : \infty, \infty; -; \\ \infty : \nu; \sigma; \end{matrix} |q; x; xq^{\sigma} \right]$$

$$\overset{\text{by (10.110)}}{=} \sum_{N=0}^{\infty} \frac{\langle \mu; q\rangle_n \langle \sigma - 1 + \nu + N; q\rangle_N x^N}{\langle 1, \nu, \sigma; q\rangle_N}$$

$$= \sum_{N=0}^{\infty} \frac{\langle \mu, \frac{\nu+\sigma-1}{2}, \frac{\nu+\sigma}{2}, \widetilde{\frac{\nu+\sigma-1}{2}}, \widetilde{\frac{\nu+\sigma}{2}}; q\rangle_N x^N}{\langle 1, \nu, \sigma, \nu + \sigma - 1; q\rangle_N}$$

$$= {}_5\phi_4 \left[\begin{matrix} \mu, \frac{\nu+\sigma-1}{2}, \frac{\nu+\sigma}{2}, \widetilde{\frac{\nu+\sigma-1}{2}}, \widetilde{\frac{\nu+\sigma}{2}} \\ \nu, \sigma, \nu + \sigma - 1, \infty \end{matrix} |q, x \right]. \tag{10.119}$$

□

By (10.115) we obtain

Corollary 10.5.8 *A q-analogue of a reduction formula for the Humbert function by Srivastava* [473, (13), p. 297]:

$$\sum_{m,n=0}^{\infty} \frac{(-1)^n \langle \mu; q \rangle_{m+n} x^{m+n} q^{-mn}}{\langle 1, \nu; q \rangle_m \langle 1, \nu; q \rangle_n}$$
$$= {}_6\phi_6 \left[\begin{matrix} \triangle(q; 2; \mu), \infty \\ \triangle(q; 2; \nu), \nu, \widetilde{1} \end{matrix} | q; -x^2 q^{\nu} || \begin{matrix} \langle -\nu \widetilde{+1} - 2k; q \rangle_k \\ \cdot \end{matrix} \right]. \tag{10.120}$$

Corollary 10.5.9 [481, p. 225, 4.24]. *A q-analogue of* [473, (14), p. 297], *which leads to a relation for the product of the two Jackson q-Bessel functions* [220], [294], [296]:

$${}_2\phi_1(\infty, \infty; \nu | q; x)\, {}_0\phi_1\big(-; \sigma | q; x q^{\sigma}\big) = {}_4\phi_3 \left[\begin{matrix} \triangle(q; 2; \nu + \sigma - 1) \\ \nu, \sigma, \nu + \sigma - 1 \end{matrix} | q; x \right]. \tag{10.121}$$

Proof Put $\lambda = \mu = \infty$ in (10.112). □

Theorem 10.5.10 [166, p. 219, (74)]. *The first q-analogue of* [473, (16), p. 297]:

$$\sum_{m,n=0}^{\infty} \frac{C_{m+n} x^{m+n} \langle \nu; q \rangle_m \langle \sigma; q \rangle_n q^{-n\sigma}}{\langle 1; q \rangle_m \langle 1; q \rangle_n} = \sum_{n=0}^{\infty} \frac{C_n x^n \langle \nu + \sigma; q \rangle_n}{\langle 1; q \rangle_n} q^{-n\sigma}. \tag{10.122}$$

Proof

$$\text{LHS} = \sum_{N=0}^{\infty} \sum_{n=0}^{N} \frac{C_N x^N \langle -N, \sigma; q \rangle_n \langle \nu; q \rangle_N q^{n(-\nu+1-\sigma)}}{\langle 1; q \rangle_N \langle 1, -\nu + 1 - N; q \rangle_n}$$
$$= \sum_{N=0}^{\infty} \frac{C_N x^N \langle \nu, -\nu + 1 - \sigma - N; q \rangle_N}{\langle 1, -\nu + 1 - N; q \rangle_N} = \text{RHS}. \tag{10.123}$$

□

There is also a second q-analogue of [473, (16), p. 297]:

Theorem 10.5.11 [481, p. 229, 6.1], [166, p. 219, (76)], [483, p. 245]:

$$\sum_{m,n=0}^{\infty} \frac{C_{m+n} x^{m+n} \langle \nu; q \rangle_m \langle \sigma; q \rangle_n q^{n\nu}}{\langle 1; q \rangle_m \langle 1; q \rangle_n} = \sum_{n=0}^{\infty} \frac{C_n x^n \langle \nu + \sigma; q \rangle_n}{\langle 1; q \rangle_n}. \tag{10.124}$$

The maximal torus in $SO(3)$ corresponds to the summation formula (10.126).

By putting

$$C_n = \frac{\langle a; q\rangle_n}{\langle c; q\rangle_n}, \qquad \nu = b, \qquad \sigma = b', \qquad x = q^{c-a-b} \tag{10.125}$$

in the two equations above, we get the following two special cases, which are q-analogues of Appell and Kampé de Fériet [41, (24), p. 22]:

Corollary 10.5.12 [166, p. 220, (78)], [421, p. 70, (18)]:

$$\begin{aligned}&\Phi_1\big(a; b, b'; c|q; q^{c-a-b}, q^{c-a-b-b'}\big)\\ &\quad = \Phi_1\big(a; b, b'; c|q; q^{c-a-b-b'}, q^{c-a-b'}\big) = \Gamma_q\begin{bmatrix} c, c-a-b-b' \\ c-a, c-b-b' \end{bmatrix},\end{aligned} \tag{10.126}$$

where $|q^{c-a-b-b'}| < 1$.

Proof

$$\begin{aligned}&\Phi_1\big(a; b, b'; c|q; q^{c-a-b}, q^{c-a-b-b'}\big)\\ &\quad = \sum_{m_2=0}^{\infty} \frac{\langle a, b'; q\rangle_{m_2} q^{(c-a-b-b')m_2}}{\langle 1, c; q\rangle_{m_2}} \sum_{m_1=0}^{\infty} \frac{\langle a+m_2, b; q\rangle_{m_1} q^{(c-a-b)m_1}}{\langle 1, c+m_2; q\rangle_{m_1}}\\ &\quad = \sum_{m_2=0}^{\infty} \frac{\langle a, b'; q\rangle_{m_2} q^{(c-a-b-b')m_2}}{\langle 1, c; q\rangle_{m_2}} \Gamma_q\begin{bmatrix} c+m_2, c-a-b \\ c-a, c+m_2-b \end{bmatrix}\\ &\quad = \sum_{m_2=0}^{\infty} \frac{\langle a, b'; q\rangle_{m_2} q^{(c-a-b-b')m_2}}{\langle 1, c; q\rangle_{m_2}} \Gamma_q\begin{bmatrix} c, c-a-b \\ c-a, c-b \end{bmatrix} \frac{\langle c; q\rangle_{m_2}}{\langle c-b; q\rangle_{m_2}}\\ &\quad = \Gamma_q\begin{bmatrix} c, c-a-b, c-b, c-a-b-b' \\ c-a, c-b, c-b-a, c-b-b' \end{bmatrix} = \Gamma_q\begin{bmatrix} c, c-a-b-b' \\ c-a, c-b-b' \end{bmatrix}\\ &\quad = \Phi_1\big(a; b, b'; c|q, q^{c-a-b-b'}, q^{c-a-b'}\big).\end{aligned} \tag{10.127}$$

□

For $n = 2$, the above theorem implies the following reduction theorem for a q-Lauricella function from [167, p. 5, (27)]:

Theorem 10.5.13

$$\begin{aligned}&\Phi_D^{(n)}\big(a, b_1, \ldots, b_n; c|q; x, xq^{-b_2}, xq^{-b_2-b_3}, \ldots, xq^{-b_2-\cdots-b_n}\big)\\ &\quad = {}_2\phi_1\big(a, b_1+\cdots+b_n; c|q; xq^{-b_2-\cdots-b_n}\big),\end{aligned} \tag{10.128}$$

where

$$\Phi_{\mathrm{D}}^{(n)}(a,b_1,\ldots,b_n;c|q;x_1,\ldots,x_n) \equiv \sum_{\vec{m}} \frac{\langle a;q\rangle_m \prod_{j=1}^{n}\langle b_j;q\rangle_{m_j} x_j^{m_j}}{\langle c;q\rangle_m \prod_{j=1}^{n}\langle 1;q\rangle_{m_j}}, \quad \max\big(|x|,\ldots,|x_n|\big)<1. \tag{10.129}$$

Theorem 10.5.14 *A q-analogue of* [473, (17), p. 297]:

$$\sum_{m,n=0}^{\infty} \frac{(-1)^n C_{m+n} x^{m+n} \langle v;q\rangle_m \langle v;q\rangle_n}{\langle 1;q\rangle_m \langle 1;q\rangle_n} = \sum_{n=0}^{\infty} \frac{C_{2n} x^{2n} \langle v,\widetilde{v};q\rangle_n}{\langle 1;q^2\rangle_n}. \tag{10.130}$$

Proof See the proof of (10.174). □

By applying the Bailey-Daum theorem, we can prove the following

Theorem 10.5.15 *A q-analogue of* [471, p. 104]:

$$\Phi_1\big(a;b,b';1+a+b-b'|q;q^{1-b'},-q^{1-b'}\big) = \Gamma_q\begin{bmatrix}1+a+b-b',1-b',1+\frac{a}{2},1+\frac{a}{2}+\beta,1-b'+\beta\\ 1+a,1+b-b',1+\frac{a}{2}-b',1+\frac{a}{2}-b'+\beta,1+\beta\end{bmatrix}, \tag{10.131}$$

where $|q^{1-b'}|<1$ *and* $\beta\equiv\frac{\log(-1)}{\log q}$.

Proof

$$\begin{aligned}
\mathrm{LHS} &= \sum_{n=0}^{\infty} \frac{\langle a,b';q\rangle_n}{\langle 1,1+a+b-b';q\rangle_n}(-1)^n q^{n(1-b')} \\
&= \sum_{m=0}^{\infty} \frac{\langle a+n,b;q\rangle_m}{\langle 1,1+a+b-b'+n;q\rangle_m} q^{m(1-b')} \\
&= \sum_{n=0}^{\infty} \frac{\langle a,b';q\rangle_n}{\langle 1;q\rangle_n}(-1)^n q^{n(1-b')}\Gamma_q\begin{bmatrix}1+a+b-b',1-b'\\ 1+b-b',1+a-b'+n\end{bmatrix}(1-q)^{-n} \\
&= \sum_{n=0}^{\infty} \frac{\langle a,b';q\rangle_n}{\langle 1,1+a-b';q\rangle_n}(-1)^n q^{n(1-b')}\Gamma_q\begin{bmatrix}1+a+b-b',1-b'\\ 1+b-b',1+a-b'\end{bmatrix} \\
&= \Gamma_q\begin{bmatrix}1+a+b-b',1-b',1+\frac{a}{2}\\ 1+a,1+b-b',1+\frac{a}{2}-b'\end{bmatrix} \frac{\langle \widetilde{1+\frac{a}{2}-b'},\widetilde{1};q\rangle_\infty}{\langle \widetilde{1+\frac{a}{2}},\widetilde{1-b'};q\rangle_\infty} = \mathrm{RHS}.
\end{aligned} \tag{10.132}$$

□

When a is a negative integer, the above theorem may be rewritten in the form

Theorem 10.5.16

$$\Phi_1(-N;b,b';1+a+b-b'|q;q^{1-b'},-q^{1-b'})$$
$$=\begin{cases}\Gamma_q\begin{bmatrix}1-N+b-b',1-b'\\1-N-b',1+b-b'\end{bmatrix}\frac{1}{\langle\frac{N}{2}+b';q\rangle_{\frac{N}{2}}}\langle\widetilde{b'};q\rangle_{\frac{N}{2}}\langle\frac{1}{2};q^2\rangle_{\frac{N}{2}}, & N\text{ even},\\0, \quad N\text{ odd}.\end{cases}$$
(10.133)

Proof

$$\text{LHS}=\sum_{n=0}^{\infty}\frac{\langle-2M,b';q\rangle_n}{\langle1,1-2M-b';q\rangle_n}(-1)^nq^{n(1-b')}\Gamma_q\begin{bmatrix}1-2M+b-b',1-b'\\1+b-b',1-2M-b'\end{bmatrix}$$
$$=\text{RHS}.\tag{10.134}$$

□

The following four equations follow from (10.122). Each of them has a dual by (10.124).

Corollary 10.5.17 *A q-analogue of* [474, (2.3), p. 100]:

$$\Phi^{2:1;1}_{2:0;0}\left[\begin{matrix}\alpha,1+\frac{\alpha}{2}:\gamma;\delta;\\\frac{\alpha}{2},1+\alpha-\gamma-\delta:-;-;\end{matrix}|q;-1,-q^{-\delta}\right]$$
$$={}_3\phi_2\left[\begin{matrix}\alpha,1+\frac{\alpha}{2},\gamma+\delta\\\frac{\alpha}{2},1+\alpha-\gamma-\delta\end{matrix}|q;-q^{-\delta}\right].\tag{10.135}$$

Corollary 10.5.18 *A q-analogue of* [474, (2.5), p. 100]:

$$\Phi^{3:1;1}_{3:0;0}\left[\begin{matrix}\alpha,1+\frac{\alpha}{2},\beta:\gamma;\delta;\\\frac{\alpha}{2},1+\alpha-\beta,1+\alpha-\gamma-\delta:-;-;\end{matrix}|q;1,q^{-\delta}\right]$$
$$={}_4\phi_3\left[\begin{matrix}\alpha,1+\frac{\alpha}{2},\beta,\gamma+\delta\\\frac{\alpha}{2},1+\alpha-\beta,1+\alpha-\gamma-\delta\end{matrix}|q;q^{-\delta}\right].\tag{10.136}$$

In the same way we obtain the following

Corollary 10.5.19 *A q-analogue of* [474, (3.12), p. 104]:

$$\Phi^{2:1;1}_{2:0;0}\left[\begin{matrix}\alpha,-N:\gamma;\delta;\\2\alpha,1+\gamma+\delta-\alpha-N:-;-;\end{matrix}|q;q,q^{\gamma+1}\right]=\frac{\langle\alpha,2\alpha-\gamma-\delta;q\rangle_N}{\langle2\alpha,\alpha-\gamma-\delta;q\rangle_N}.\tag{10.137}$$

Corollary 10.5.20 *A q-analogue of* [474, (3.13), p. 105]:

$$\Phi^{2:1;1}_{2:0;0}\left[\begin{matrix}\alpha,\beta:-M;-N;\\ 2\alpha,1+\beta-\alpha-M-N:-;-;\end{matrix}|q;q,q^{-M+1}\right]=\frac{\langle\alpha,2\alpha-\beta;q\rangle_{M+N}}{\langle 2\alpha,\alpha-\beta;q\rangle_{M+N}}. \tag{10.138}$$

Proof

$$\text{LHS}={}_3\phi_2\left[\begin{matrix}\alpha,-M-N,\beta\\ 2\alpha,1+\beta-\alpha-M-N\end{matrix}|q;q\right]=\text{RHS}. \tag{10.139}$$

□

The following theorem is a q-analogue of Srivastava's generalization of Carlson's identity from [476, (3), p. 139]. Notice the difference in character from the equations in Section 3.

Theorem 10.5.21 *If $\{C_n\}_{n=0}^{\infty}$ is a bounded sequence of complex numbers,*

$$x_1\neq 0,\qquad v\neq 1,\qquad \sigma\neq 1,\qquad v+\sigma\neq 0,-1,-2,\ldots$$

and the two double series below are absolutely convergent, then

$$\begin{aligned}&\sum_{m,n=0}^{\infty}\frac{C_{m+n}x_2^{m+n}(-\frac{x_1}{x_2}q^{-\sigma};q)_m\langle v;q\rangle_m\langle\sigma;q\rangle_n}{\langle 1;q\rangle_m\langle 1;q\rangle_n}\\&\quad\times\text{QE}\left(-\binom{m}{2}+\binom{n}{2}+\sigma m+n(1-m-n)\right)\\&\cong\sum_{m,n=0}^{\infty}\frac{C_{m+n}x_1^m x_2^n\langle v;q\rangle_m\langle v+\sigma+m;q\rangle_n}{\langle 1;q\rangle_m\langle 1;q\rangle_n}\text{QE}\left(-\binom{n}{2}-mn\right).\end{aligned} \tag{10.140}$$

Remark 55 The LHS of the above relation can also be written as

$$\sum_{m,n=0}^{\infty}\frac{C_{m+n}x_2^n(x_2q^{\sigma}\boxplus_q x_1)^m\langle v;q\rangle_m\langle\sigma;q\rangle_n}{\langle 1;q\rangle_m\langle 1;q\rangle_n}\text{QE}\left(-\binom{m}{2}+\binom{n}{2}+n(1-m-n)\right). \tag{10.141}$$

Proof We use the q-Euler-Pfaffian equation (7.62):

$$\begin{aligned}\text{RHS}&=\sum_{N=0}^{\infty}\sum_{n=0}^{N}\frac{(-1)^n C_N x_1^N(\frac{x_2}{x_1})^n\langle v;q\rangle_N\langle -N,1-v-\sigma-N;q\rangle_n}{\langle 1;q\rangle_N\langle 1,1-v-N;q\rangle_n}q^{n(1+\sigma)}\\&=\sum_{N=0}^{\infty}\frac{C_N x_1^N\langle v;q\rangle_N}{\langle 1;q\rangle_N(-\frac{x_2}{x_1}q^{1+\sigma};q)_{-N}}\end{aligned}$$

$$
\begin{aligned}
&\times {}_2\phi_2\left[\begin{matrix}\sigma,-N\\ 1-v-N\end{matrix}|q;-\frac{x_2}{x_1}q^{2-v-N}||_{-\frac{x_2}{x_1}q^{1+\sigma-N}}\right]\\
&=\sum_{N=0}^{\infty}\frac{C_N x_1^N\langle v;q\rangle_N}{\langle 1;q\rangle_N}\sum_{n=0}^{N}\frac{\langle -N,\sigma;q\rangle_n}{\langle 1,1-v-N;q\rangle_n}\frac{(\frac{x_2}{x_1}q^{2-v-N})^n q^{\binom{n}{2}}}{(-\frac{x_2}{x_1}q^{1+\sigma};q)_{n-N}}\\
&=\sum_{N=0}^{\infty}\frac{C_N x_1^N\langle v;q\rangle_N}{\langle 1;q\rangle_N}\sum_{n=0}^{N}\frac{\langle -N,\sigma;q\rangle_n}{\langle 1,1-v-N;q\rangle_n}\left(\frac{y}{x_1}q^{2-v-N}\right)^n\\
&\times\left(-\frac{x_1}{x_2}q^{-\sigma};q\right)_{N-n}\mathrm{QE}\left(-\binom{N-n}{2}+\binom{n}{2}+\sigma(N-n)\right)\left(\frac{x_2}{x_1}\right)^{N-n}\\
&=\sum_{N=0}^{\infty}\sum_{n=0}^{N}\frac{C_N x_1^N\langle\sigma;q\rangle_n\langle v;q\rangle_{N-n}}{\langle 1;q\rangle_n\langle 1;q\rangle_{N-n}}\left(-\frac{x_1}{x_2}q^{-\sigma};q\right)_{N-n}\\
&\times\mathrm{QE}\left(-\binom{N-n}{2}+\binom{n}{2}+\sigma(N-n)+n(1-N)\right)\left(\frac{x_2}{x_1}\right)^{N}\\
&=\mathrm{LHS}.
\end{aligned}\tag{10.142}
$$

□

The following generalization of (10.122) is a q-analogue of Panda [410] and Singhal [465]. Compare with Srivastava [474, (4.3), p. 107].

Theorem 10.5.22

$$
\begin{aligned}
&\sum_{\vec{m}}\frac{C_{m_1+\cdots+m_n}\prod_{j=1}^{n}x^{m_j}\langle\alpha_j;q\rangle_{m_j}}{\prod_{j=1}^{n}\langle 1;q\rangle_{m_j}}\mathrm{QE}\left(-\sum_{k=1}^{n}m_k\sum_{l=2}^{k}\alpha_l\right)\\
&\quad=\sum_{N=0}^{\infty}\frac{C_N x^N\langle\sum_{k=1}^{n}\alpha_k;q\rangle_N}{\langle 1;q\rangle_N}\mathrm{QE}\left(-N\sum_{l=2}^{n}\alpha_l\right).
\end{aligned}\tag{10.143}
$$

Proof We use induction. Suppose that (10.143) is true for $n>1$ and denote the LHS by Δ_n; then

$$
\begin{aligned}
\Delta_{n+1}=&\sum_{m_{n+1}=0}^{\infty}\langle\alpha_{n+1};q\rangle_{m_{n+1}}\frac{x^{m_{n+1}}}{\langle 1;q\rangle_{m_{n+1}}}\mathrm{QE}\big(-m_{n+1}(\alpha_2+\cdots+\alpha_{n+1})\big)\\
&\times\sum_{m_1,\ldots,m_n}C_{m_1+\cdots+m_{n+1}}\frac{x^{m_1+\cdots+m_n}}{\prod_{i=1}^{n}\langle 1;q\rangle_{m_i}}\prod_{k=1}^{n}\langle\alpha_k;q\rangle_{m_k}\mathrm{QE}\left(-m_k\sum_{l=2}^{k}\alpha_l\right)\\
=&\sum_{m_{n+1}=0}^{\infty}\langle\alpha_{n+1};q\rangle_{m_{n+1}}\frac{x^{m_{n+1}}}{\langle 1;q\rangle_{m_{n+1}}}\mathrm{QE}\left(-m_{n+1}\sum_{l=2}^{n+1}\alpha_l\right)
\end{aligned}
$$

$$\times \sum_{N=0}^{\infty} C_{N+m_{n+1}} \left\langle \sum_{k=1}^{n} \alpha_k; q \right\rangle_N \frac{x^N}{\langle 1; q \rangle_N} \mathrm{QE}\left(-N \sum_{l=2}^{n} \alpha_l \right)$$

$$= \sum_{N=0}^{\infty} C_N x^N \mathrm{QE}\left(-N \sum_{l=2}^{n} \alpha_l \right) \frac{\langle \sum_{k=1}^{n+1} \alpha_k; q \rangle_N}{\langle 1; q \rangle_N}, \tag{10.144}$$

where in the last step we used the induction hypothesis for $n = 2$, with the following values of the parameters in (10.143):

$$m_1 \to N, \qquad m_2 \to m_n + 1,$$
$$\alpha_2 \to \alpha_n + 1, \qquad x \to x\mathrm{QE}\left(-\sum_{l=2}^{n} \alpha_l \right). \tag{10.145}$$

□

There is a dual statement too.

10.6 Some q-formulas of Srivastava

In this section we mention a selection of well-known q-formulas with many variables from the literature.

10.6.1 Generating functions

The following generating functions are extensions of the formulas in Section 9.2.3, (9.169) and (9.171):

Theorem 10.6.1 [477, 3.8, p. 333]. *The following generating function relation holds, where $\mu, \nu \in \mathbb{N}$ and $\gamma_n \in \mathbb{C}$ are arbitrary and c and d are vectors of size l and m:*

$$\sum_{n=0}^{\infty} \frac{\langle (c); q \rangle_n}{\langle (d), 1; q \rangle_n} \gamma_n t^n \sum_{k=0}^{[\frac{n}{N}]} \frac{x^k \langle (g+n); q \rangle_{\mu k} \langle (1-n-d), -n; q \rangle_{Nk} \delta_k}{\langle (h+n); q \rangle_{\nu k} \langle (1-n-c); q \rangle_{Nk} \langle 1; q \rangle_k} q^{(1-(l-m-1)n)Nk}$$

$$= \sum_{n,k=0}^{\infty} \frac{\langle (c); q \rangle_n \langle (g+n+Nk); q \rangle_{\mu k} \gamma_{n+Nk} t^n \delta_k}{\langle (d), 1; q \rangle_n \langle (h+n+Nk); q \rangle_{\nu k} \langle 1; q \rangle_k} \left(x \left((-1)^{l-m+1} t \right)^N \right)^k$$

$$\times \mathrm{QE}\left(-\frac{l-m-1}{2} Nk(Nk+1) + Nk\left(\sum_{j=1}^{l} c_j - \sum_{j=1}^{m} d_j \right) \right). \tag{10.146}$$

Proof

$$\text{LHS} = \sum_{n=0}^{\infty} \gamma_n t^n \sum_{k=0}^{[\frac{n}{N}]} \frac{x^k \langle (g+n); q\rangle_{\mu k} \langle (c); q\rangle_{n-Nk} \delta_k (-1)^{(l-m+1)Nk}}{\langle 1; q\rangle_k \langle (h+n); q\rangle_{\nu k} \langle (d), 1; q\rangle_{n-Nk}}$$
$$\times \mathrm{QE}\Bigg(\binom{Nk}{2}(m-l+1)$$
$$+ Nk\Bigg(1-(l-m)n + \sum_{j=1}^{m} 1-n-d_j - \sum_{j=1}^{l} 1-n-c_j\Bigg)\Bigg)$$
$$= \sum_{k=0}^{\infty} (-1)^{(l-m+1)Nk} \frac{\delta_k x^k}{\langle 1; q\rangle_k} \sum_{n=Nk}^{\infty} \frac{\gamma_n t^n \langle (g+n); q\rangle_{\mu k} \langle (c); q\rangle_{n-Nk}}{\langle (h+n); q\rangle_{\nu k} \langle (d), 1; q\rangle_{n-Nk}}$$
$$\times \mathrm{QE}\Bigg(\binom{Nk}{2}\Bigg((m-l+1)$$
$$+ Nk\Bigg(1-(l-m)n + \sum_{j=1}^{m} 1-n-d_j - \sum_{j=1}^{l} 1-n-c_j\Bigg)\Bigg)\Bigg). \quad (10.147)$$

□

Theorem 10.6.2 [477, 3.9, p. 334]. *The numbers $\triangle_{n,k} \in \mathbb{C}$ are arbitrary.*

$$\sum_{n=0}^{\infty} \frac{\langle (c); q\rangle_n}{\langle (d), 1; q\rangle_n} \gamma_n t^n \sum_{k=0}^{[\frac{n}{N}]} \frac{x^k \langle (1-n-d), -n; q\rangle_{Nk} \triangle_{n,k}}{\langle (1-n-c); q\rangle_{Nk} \langle 1; q\rangle_k} q^{(1-(l-m-1)n)Nk}$$
$$= \sum_{n,k=0}^{\infty} \frac{\langle (c); q\rangle_n \gamma_{n+Nk} t^n \triangle_{n+Nk,k} (x((-1)^{l-m+1} t)^N)^k}{\langle (d), 1; q\rangle_n \langle 1; q\rangle_k}$$
$$\times \mathrm{QE}\Bigg(-\frac{l-m-1}{2} Nk(Nk+1) + Nk\Bigg(\sum_{j=1}^{l} c_j - \sum_{j=1}^{m} d_j\Bigg)\Bigg). \quad (10.148)$$

Proof

$$\text{LHS} = \sum_{n=0}^{\infty} \gamma_n t^n \sum_{k=0}^{[\frac{n}{N}]} \frac{x^k \langle (c); q\rangle_{n-Nk} \triangle_{n,k} (-1)^{(l-m+1)Nk}}{\langle 1; q\rangle_k \langle (d), 1; q\rangle_{n-Nk}} \mathrm{QE}\Bigg(\binom{Nk}{2}(m-l+1)$$
$$+ Nk\Bigg(1-(l-m)n + \sum_{j=1}^{m} 1-n-d_j - \sum_{j=1}^{l} 1-n-c_j\Bigg)\Bigg)$$
$$= \sum_{k=0}^{\infty} (-1)^{(l-m+1)Nk} \frac{x^k}{\langle 1; q\rangle_k} \sum_{n=Nk}^{\infty} \frac{\gamma_n t^n \triangle_{n,k} \langle (c); q\rangle_{n-Nk}}{\langle (d), 1; q\rangle_{n-Nk}}$$

$$\times \mathrm{QE}\left(\binom{Nk}{2}(m-l+1)\right)$$
$$\times \mathrm{QE}\left(Nk\left(1-(l-m)n+\sum_{j=1}^{m}1-n-d_j-\sum_{j=1}^{l}1-n-c_j\right)\right)$$
$$= \mathrm{RHS}. \tag{10.149}$$

□

10.6.2 Transformations

Theorem 10.6.3 [478, p. 50 1.8]. *The sequences $F(z_1,\ldots,z_n)$ and $\Omega(m_1,\ldots,m_n)$ are connected as follows:*

$$F(z_1,\ldots,z_n)=\sum_{\vec{m}}\Omega(m_1,\ldots,m_n)\frac{\vec{z}^{\vec{m}}}{\langle 1;q\rangle_{\vec{m}}}. \tag{10.150}$$

The vectors $\vec{m}$, $\vec{z}$ have dimension n and the vectors $\vec{\lambda}$, $\vec{\mu}$ have dimension p. Then the following multiple transformation applies:

$$\sum_{\vec{m}}\frac{\langle\vec{\lambda};q\rangle_m}{\langle\vec{\mu};q\rangle_m}\Omega(m_1,\ldots,m_n)\frac{\vec{z}^{\vec{m}}}{\langle 1;q\rangle_{\vec{m}}}$$
$$=\frac{\langle\vec{\lambda};q\rangle_\infty}{\langle\vec{\mu};q\rangle_\infty}\sum_{\vec{k}}F\left(z_1q^k,\ldots,z_nq^k\right)\langle\vec{\mu}-\vec{\lambda};q\rangle_{\vec{k}}\frac{q^{\vec{\lambda}k}}{\langle 1;q\rangle_{\vec{k}}}. \tag{10.151}$$

Proof Use the q-binomial theorem. □

Theorem 10.6.4 [478, p. 51, 2.4]. *The vectors $\vec{\lambda}$, $\vec{\mu}$ have dimension p and the vectors $\vec{\alpha}$, $\vec{\beta}$ have dimension n. Then*

$$\Phi^{p:2;\ldots,2}_{p:1;\ldots,1}\left[\begin{matrix}(\lambda):(\alpha),(\beta)\\(\mu):-\end{matrix}\Big|q;\vec{z}\,\Big\|\,\frac{\cdot}{(q^{\vec{\alpha+\beta}}\vec{z};q)_{\vec{m}}}\right]$$
$$=\frac{\langle\vec{\lambda};q\rangle_\infty}{\langle\vec{\mu};q\rangle_\infty}\frac{(q^{\vec{\alpha}}\vec{z},q^{\vec{\beta}}\vec{z};q)_\infty}{(\vec{z},q^{\vec{\alpha+\beta}}\vec{z};q)_\infty}$$
$$\times\Phi^{2n:1;\ldots,1}_{2n:0;\ldots,0}\left[\begin{matrix}-:(\vec{\mu}-\vec{\lambda})\\-:-\end{matrix}\Big|q;q^{\vec{\lambda}}\,\Big\|\,\frac{(q^{\vec{\alpha+\beta}}\vec{z},\vec{z};q)_m}{(q^{\vec{\alpha}}\vec{z},q^{\vec{\beta}}\vec{z};q)_m}\right]. \tag{10.152}$$

Proof Put $\Omega(m_1,\ldots,m_n)=\frac{\langle\vec{\alpha},\vec{\beta};q\rangle_{\vec{m}}}{(q^{\vec{\alpha}}\vec{z},q^{\vec{\beta}}\vec{z};q)_{\vec{m}}}$, use the q-Gaussian theorem and (10.151). □

Theorem 10.6.5 [478, p. 53, 3.5]:

$$\Phi_1\big(a; b, b'; c|q; x_1, x_2\big) = \frac{1}{(x_1; q)_{a+b-c}} \Phi^{1:1;2}_{1:0;1}\left[\begin{matrix} c-b : c-a; a, b' \\ c : -; c-b \end{matrix} |q; x_1 q^{a+b-c}, x_2\right]. \quad (10.153)$$

Proof Use formula (7.52). □

10.6.3 Double sum identities (Srivastava and Jain)

We now come to double sum identities of Srivastava and Jain [481]. This section is divided into two parts. In the first part the q-Saalschütz theorem is used; in the second part the q-Watson theorem (8.9) is used; there also occur reduction formulas and a quadratic transformation. In this and the following section $\{C_{m,n}\}_{m,n=0}^{\infty}$ denotes a bounded sequence of complex numbers.

Theorem 10.6.6 [481, p. 217, (2.5)]:

$$\sum_{m,n=0}^{\infty} \frac{C_{m+n} x^{m+n} q^{m(c-a-b)} \langle a, b; q\rangle_m \langle c-a-b; q\rangle_n}{\langle 1, c; q\rangle_m \langle 1; q\rangle_n} = \sum_{N=0}^{\infty} C_N x^N \frac{\langle c-a, c-b; q\rangle_N}{\langle 1, c; q\rangle_N}. \quad (10.154)$$

Proof

$$\text{LHS} \overset{\text{by (6.15), (6.20)}}{=} \sum_{N=0}^{\infty} \frac{C_N x^N \langle c-a-b; q\rangle_N}{\langle 1; q\rangle_N} \sum_{n=0}^{N} \frac{\langle a, b, -N; q\rangle_n}{\langle 1, c, a+b-c+1-N; q\rangle_n} q^n$$

$$= \sum_{N=0}^{\infty} \frac{C_N x^N \langle c-a-b; q\rangle_N}{\langle 1; q\rangle_N} \frac{\langle c-a, c-b; q\rangle_N}{\langle c, c-a-b; q\rangle_N} = \text{RHS}. \quad (10.155)$$

□

Theorem 10.6.7 [481, p. 217, (2.6)]:

$$\sum_{m,n=0}^{\infty} \frac{C_{m+n} (-1)^m x^{m+n} q^{m(b-a)+\binom{m}{2}} \langle a; q\rangle_m \langle b+c-a-1; q\rangle_{m+2n}}{\langle 1, b; q\rangle_m \langle 1, c; q\rangle_n} = \sum_{N=0}^{\infty} C_N x^N \frac{\langle b-a, b+c-a-1, b+c+N-1; q\rangle_N}{\langle 1, b, c; q\rangle_N}. \quad (10.156)$$

Proof

$$\text{LHS} \overset{\text{by } 2\times(6.15),\,(6.20)}{=} \sum_{N=0}^{\infty} \frac{C_N x^N \langle b+c-a-1;q\rangle_{2N}}{\langle c,1;q\rangle_N} \times \sum_{n=0}^{N} \frac{\langle a,1-N-c,-N;q\rangle_n}{\langle 1,b,a-b-c+2-2N;q\rangle_n} q^n$$

$$= \sum_{N=0}^{\infty} \frac{C_N x^N \langle b+c-a-1;q\rangle_{2N}}{\langle c,1;q\rangle_N} \frac{\langle b-a,b+N+c-1;q\rangle_N}{\langle b,b+c-a+N-1;q\rangle_N} = \text{RHS}. \tag{10.157}$$

□

Corollary 10.6.8 [481, p. 220, (3.5)]:

$$\Phi^{p:2;1}_{p:1;0}\left[\begin{matrix}\alpha_1,\ldots,\alpha_p : a,b; c-a-b \\ \beta_1,\ldots,\beta_p : c; -\end{matrix} |q; xq^{c-a-b}, x\right] = \sum_{N=0}^{\infty} x^N \frac{\langle \alpha_1,\ldots,\alpha_p, c-a, c-b;q\rangle_N}{\langle \beta_1,\ldots,\beta_p,1,c;q\rangle_N}. \tag{10.158}$$

Proof Use (10.154). □

Corollary 10.6.9 [481, p. 220, (3.6)]:

$$\sum_{m,n=0}^{\infty} \frac{\langle \alpha_1,\ldots,\alpha_p;q\rangle_{m+n}(-1)^m x^{m+n} q^{m(b-a)+\binom{m}{2}} \langle a;q\rangle_m \langle b+c-a-1;q\rangle_{m+2n}}{\langle \beta_1,\ldots,\beta_p;q\rangle_{m+n}\langle 1,b;q\rangle_m \langle 1,c;q\rangle_n}$$

$$= \sum_{N=0}^{\infty} x^N \frac{\langle \alpha_1,\ldots,\alpha_p, b-a, b+c-a-1, b+c+N-1;q\rangle_N}{\langle \beta_1,\ldots,\beta_p,1,b,c;q\rangle_N}. \tag{10.159}$$

Proof Use (10.156). □

Corollary 10.6.10 [481, p. 223, (4.17)]:

$$\Phi^{p:1;1}_{p:0;0}\left[\begin{matrix}\alpha_1,\ldots,\alpha_p : a; c \\ \beta_1,\ldots,\beta_p : -; -\end{matrix} |q; xq^{c}, x\right] = \sum_{N=0}^{\infty} x^N \frac{\langle \alpha_1,\ldots,\alpha_p, a+c;q\rangle_N}{\langle \beta_1,\ldots,\beta_p,1;q\rangle_N}. \tag{10.160}$$

Proof Use (10.158), put $c \to a+b+c$ and let $b \to \infty$. □

In the second part we use a different double sum identity. We will for example find a q-analogue of the following quadratic transformation:

Corollary 10.6.11

$$
{}_2\mathrm{F}_1\left(\lambda, a; 2a; 2\frac{x}{1+x}\right) = (1+x)^{\lambda}\,{}_2\mathrm{F}_1\left(\frac{\lambda}{2}, \frac{\lambda+1}{2}; a+\frac{1}{2}; x^2\right). \tag{10.161}
$$

Proof Put $\frac{y}{1-y} \to x$ in (3.36). □

Theorem 10.6.12 [481, p. 217, 2.2]. *A q-analogue of* [91, 3.10, p. 440]:

$$
\sum_{m,n=0}^{\infty} \frac{C_{m+n}x^{m+n}(-1)^n \langle a, \widetilde{a}; q\rangle_m \langle b, \widetilde{b}; q\rangle_n}{\langle 1, 2a; q\rangle_m \langle 1, 2b; q\rangle_n} = \sum_{N=0}^{\infty} \frac{C_{2N}x^{2N}\langle a+b, \widetilde{a+b}; q\rangle_{2N}}{\langle 1, a+\frac{1}{2}, b+\frac{1}{2}, a+b; q^2\rangle_N}. \tag{10.162}
$$

Proof

$$
\begin{aligned}
\text{LHS} &= \sum_{N=0}^{\infty}\sum_{n=0}^{N} \frac{C_N x^N (-1)^n \langle a, \widetilde{a}; q\rangle_{N-n} \langle b, \widetilde{b}; q\rangle_n}{\langle 1, 2a; q\rangle_{N-n} \langle 1, 2b; q\rangle_n} \\
&= \sum_{N=0}^{\infty} \frac{C_N x^N \langle a, \widetilde{a}; q\rangle_N}{\langle 1, 2a; q\rangle_N} \sum_{n=0}^{N} \frac{\langle b, \widetilde{b}, 1-2a-N, -N; q\rangle_n q^n}{\langle 1, 2b, 1-a-N, \widetilde{1-a-N}; q\rangle_n} \\
&\overset{\text{by (8.9)}}{=} \sum_{N=0}^{\infty} \frac{C_{2N}\langle a, \widetilde{a}; q\rangle_{2N} x^{2N} \langle \frac{1}{2}, a+b+N; q^2\rangle_N}{\langle 1, 2a; q\rangle_{2N} \langle \frac{1}{2}+b, a+N; q^2\rangle_N} = \text{RHS}.
\end{aligned} \tag{10.163}
$$

□

Theorem 10.6.13 [481, p. 220, 3.2]:

$$
\begin{aligned}
&\Phi^{p:2;2}_{p:1;1}\left[\begin{matrix} \vec{\lambda} : a, \widetilde{a}; b, \widetilde{b} \\ \vec{\mu} : 2a; 2b \end{matrix} \middle| q; x, -x \right] \\
&= {}_{8+4p}\phi_{7+4p}\left[\begin{matrix} \triangle(q; 2; \vec{\lambda}, a+b, \widetilde{a+b}) \\ \triangle(q; 2; \vec{\mu}), a+b, a+\frac{1}{2}, b+\frac{1}{2}, \widetilde{a+\frac{1}{2}}, \widetilde{b+\frac{1}{2}}, \widetilde{1}, \widetilde{a+b} \end{matrix} \middle| q; x^2 \right].
\end{aligned} \tag{10.164}
$$

Proof Use (10.162). □

Theorem 10.6.14 [481, p. 222, 4.5]:

$$\Phi^{1:2;2}_{1:1;1}\left[\begin{array}{c}\lambda : a, \widetilde{a}; b, \widetilde{b}\\ \infty : 2a; 2b\end{array}|q; x, -x\right]$$
$$= {}_6\phi_5\left[\begin{array}{c}\frac{\lambda}{2}, \frac{\lambda+1}{2}, \triangle(q^2; 2; a+b)\\ a+b, a+\frac{1}{2}, b+\frac{1}{2}, 2\infty\end{array}|q^2; x^2\right]. \tag{10.165}$$

Proof Use (10.164). □

Corollary 10.6.15 [481, p. 222, 4.6]. *A q-analogue of* (10.161):

$${}_3\phi_2\left[\begin{array}{c}\lambda, a, \widetilde{a}\\ 2a\end{array}|q; x||\begin{array}{c}\cdot\\ (-xq^{\lambda}; q)_k\end{array}\right] = (-x; q)_\lambda\, {}_2\phi_1\left[\begin{array}{c}\frac{\lambda}{2}, \frac{\lambda+1}{2}\\ a+\frac{1}{2}\end{array}|q^2; x^2\right]. \tag{10.166}$$

Proof Put $b = \infty$ in (10.165) and use the q-binomial theorem. □

Corollary 10.6.16 [481, p. 222, 4.9]. *A q-analogue of* [53, p. 246, (2.11)]:

$${}_2\phi_1\left[\begin{array}{c}a, \widetilde{a}\\ 2a\end{array}|q; x\right] {}_2\phi_1\left[\begin{array}{c}b, \widetilde{b}\\ 2b\end{array}|q; -x\right] = {}_4\phi_3\left[\begin{array}{c}\triangle(q^2; a+b)\\ a+b, a+\frac{1}{2}, b+\frac{1}{2}\end{array}|q^2; x^2\right]. \tag{10.167}$$

Proof Use (10.164). □

10.7 Two reduction formulas of Karlsson and Srivastava

This section is based in part on the work of Srivastava and Jain [481]. Karlsson and Srivastava, who wrote a joint book, also wrote an article [483] about transformations of multiple q-series. We will prove only two formulas of [483] by using the finite Bailey–Daum formula.

Theorem 10.7.1 [483, p. 246, (23)]:

$$\sum_{i,j} \frac{C_{i+j} x^{i+j} \mathrm{QE}(-ij)(-1)^j}{\langle 1, \nu; q\rangle_i \langle 1, \nu; q\rangle_j}$$
$$= \sum_k \frac{C_{2k} x^{2k} (-1)^k \langle \widetilde{-\nu+1-2k}; q\rangle_k \mathrm{QE}(\nu k + \binom{k}{2})}{\langle \triangle(q; 2; \nu), \nu, \widetilde{1}, 1; q\rangle_k}. \tag{10.168}$$

Proof We prove the equivalent formula

$$\sum_{i,j} \frac{C_{i+j} x^{i+j} \mathrm{QE}((\nu)(i+j)-ij)(-1)^j}{\langle 1,\nu;q\rangle_i \langle 1,\nu;q\rangle_j}$$

$$= \sum_k \frac{C_{2k} x^{2k} (-1)^k \langle \widetilde{-\nu+1}-2k;q\rangle_k \mathrm{QE}\big(3\nu k+\binom{k}{2}\big)}{\langle \frac{\nu}{2}, \frac{\widetilde{\nu}}{2}, \frac{\nu+1}{2}, \widetilde{\frac{\nu+1}{2}}, \nu, \widetilde{1}, 1;q\rangle_k}. \tag{10.169}$$

$$\mathrm{LHS} = \sum_k \sum_{j=0}^{k} \frac{C_k x^k (-1)^k \langle -\nu+1-k;q\rangle_j \mathrm{QE}\big((\nu-k)j+j^2-kj-\binom{k-j}{2}\big)}{\langle 1;q\rangle_{k-j} \langle 1,\nu;q\rangle_j \langle -\nu+1-k;q\rangle_k}$$

$$= \sum_k \frac{C_k x^k (-1)^k \mathrm{QE}(-\binom{k}{2})}{\langle 1,-\nu+1-k;q\rangle_k} \sum_{j=0}^{k} \binom{k}{j}_q \mathrm{QE}\bigg(\nu j+\binom{j}{2}\bigg) \frac{\langle -\nu+1-k;q\rangle_j}{\langle \nu;q\rangle_j}$$

$$\stackrel{\text{by (8.41)}}{=} \sum_k \frac{C_{2k} x^{2k} \langle \widetilde{-\nu+1}-2k;q\rangle_k \langle \frac{1}{2};q^2\rangle_k \mathrm{QE}\big(-\binom{2k}{2}\big)}{\langle 1-\nu-k;q\rangle_k \langle 1-\nu-2k,1;q\rangle_{2k}} = \mathrm{RHS}. \tag{10.170}$$

□

Theorem 10.7.2 [483, p. 246, (22)]:

$$\sum_{i,j} \frac{C_{i+j} \langle \nu;q\rangle_i \langle \nu;q\rangle_j x^{i+j} (-1)^j}{\langle 1;q\rangle_i \langle 1;q\rangle_j} = \sum_k \frac{C_{2k} x^{2k} \langle \widetilde{\nu},\nu;q\rangle_k}{\langle \widetilde{1},1;q\rangle_k}. \tag{10.171}$$

Proof We prove the equivalent formula

$$\sum_{i,j} \frac{C_{i+j} \langle \nu;q\rangle_i \langle \nu;q\rangle_j x^{i+j} \mathrm{QE}(-\nu(i+j)-\binom{i+j}{2})(-1)^j}{\langle 1;q\rangle_i \langle 1;q\rangle_j}$$

$$= \sum_k \frac{C_{2k} x^{2k} \langle \widetilde{\nu},\nu;q\rangle_k \mathrm{QE}(-2\nu k-4\binom{k}{2}-k)}{\langle \widetilde{1},1;q\rangle_k}. \tag{10.172}$$

$$\mathrm{LHS} = \sum_k \frac{C_k \langle -\nu+1-k;q\rangle_k x^k (-1)^k}{\langle 1;q\rangle_k}$$

$$\times \sum_{j=0}^{k} \binom{k}{j}_q \mathrm{QE}\bigg(-\nu j-kj+\frac{j^2+j}{2}\bigg) \frac{\langle \nu;q\rangle_j}{\langle -\nu+1-k;q\rangle_j}$$

$$\stackrel{\text{by (8.41)}}{=} \sum_k \frac{C_{2k} x^{2k} \langle \widetilde{\nu};q\rangle_k \langle -\nu+1-2k;q\rangle_{2k} \langle \frac{1}{2};q^2\rangle_k}{\langle \nu+k;q\rangle_k \langle 1;q\rangle_{2k}} = \mathrm{RHS}. \tag{10.173}$$

□

10.8 q-analogues of reducibility theorems of Karlsson

Karlsson [328] has derived two interesting reduction formulas of general character for triple sums, which can be used to deduce many reduction formulas for hypergeometric functions in three variables [482, p. 311]. The proofs use formulas (7.113) and (8.41). Clearly many more reducibility formulas of this type can be deduced. We illustrate with a q-analogue of one interesting special case known from the literature.

Theorem 10.8.1 *A q-analogue of Karlsson* [328, p. 200].
If $\{C_{m,n}\}_{m,n=0}^{\infty}$ is a bounded sequence of complex numbers, then

$$\sum_{m,n,p=0}^{\infty}\frac{C_{m,n+p}\langle b;q\rangle_n\langle b;q\rangle_p x_1^m x_2^n(-x_2)^p}{\langle 1;q\rangle_m\langle 1;q\rangle_n\langle 1;q\rangle_p}$$
$$=\sum_{m=0}^{\infty}\sum_{k=0}^{\infty}\frac{C_{m,2k}\langle b,\widetilde{b};q\rangle_k x_1^m x_2^{2k}}{\langle 1;q\rangle_m\langle 1,\widetilde{1};q\rangle_k}. \tag{10.174}$$

Proof

$$\mathrm{LHS}=\sum_{m,t=0}^{\infty}\sum_{p=0}^{t}\frac{C_{m,t}\langle b;q\rangle_t\langle b,-t;q\rangle_p x_1^m x_2^t(-1)^p q^{p(1-b)}}{\langle 1;q\rangle_m\langle 1;q\rangle_t\langle 1,1-t-b;q\rangle_p}$$
$$=\sum_{m=0}^{\infty}\sum_{t=0,t\text{ even}}^{\infty}\frac{C_{m,t}\langle b;q\rangle_t x_1^m x_2^t}{\langle 1;q\rangle_m\langle 1;q\rangle_t}\frac{\langle\widetilde{b};q\rangle_{\frac{t}{2}}\langle\frac{1}{2};q^2\rangle_{\frac{t}{2}}}{\langle b+\frac{t}{2};q\rangle_{\frac{t}{2}}}=\mathrm{RHS}. \tag{10.175}$$

□

The formula (10.130) is the special case $x_1=0$ in the theorem above.

Corollary 10.8.2 *A q-analogue of* [328, 2.5, p. 201]:

$$\Phi_{\mathrm{D}}^{(3)}\left(a+1,1+a-c,b,b;c|q;-q^{c-a},x,-x\right)$$
$$=\frac{1}{1+q^{\frac{a}{2}}}\Gamma_q\begin{bmatrix}c,\frac{a}{2},c-a+\beta,1+\frac{a}{2}+\beta\\ a,c-\frac{a}{2},1+\beta,c-\frac{a}{2}+\beta\end{bmatrix}{}_2\phi_1\begin{bmatrix}b,\frac{a}{2}\\ c-\frac{a}{2}\end{bmatrix}|q^2;x^2\Bigg],$$
$$\beta\equiv\frac{\log(-1)}{\log q}. \tag{10.176}$$

Proof Put $C_{m,n}\equiv\frac{\langle a+1;q\rangle_{m+n}\langle 1+a-c;q\rangle_m}{\langle c;q\rangle_{m+n}}$ and $x_1=-q^{c-a}$ in (10.174). We use Bailey-Daum's formula together with the relation between Γ_q-functions and q-shifted factorials. Then we have

$$\mathrm{RHS}=\sum_{m,k=0}^{\infty}\frac{\langle a;q\rangle_{2k}\langle 1+a-c,a+2k;q\rangle_m}{\langle c;q\rangle_{2k}\langle 1,c+2k;q\rangle_m}\frac{\langle b,\widetilde{b};q\rangle_k}{\langle 1,\widetilde{1};q\rangle_k}x^{2k}(-1)^m q^{m(c-a)}$$

$$
\begin{aligned}
&= \sum_{k=0}^{\infty} \frac{\langle a;q\rangle_{2k}}{\langle c;q\rangle_{2k}} \frac{\langle b,\widetilde{b};q\rangle_k}{\langle 1,\widetilde{1};q\rangle_k} x^{2k} \Gamma_q\left[\begin{matrix} c+2k, \frac{a}{2}+1+k \\ 1+a+2k, c-\frac{a}{2}+k \end{matrix}\right] \\
&\times \frac{\langle \widetilde{-\frac{a}{2}+c+k}, \widetilde{1};q\rangle_\infty}{\langle \widetilde{\frac{a}{2}+1+k}, \widetilde{c-a};q\rangle_\infty} = \sum_{k=0}^{\infty} \frac{\langle \frac{a}{2}, b, \widetilde{b};q\rangle_k}{\langle c-\frac{a}{2}, \widetilde{1}, 1;q\rangle_k} x^{2k} \\
&\times \Gamma_q\left[\begin{matrix} c, \frac{a}{2} \\ a, c-\frac{a}{2} \end{matrix}\right] \frac{\langle \widetilde{-\frac{a}{2}+c+k};q\rangle_{a+1-c}}{\langle \widetilde{c-a};q\rangle_{a+1-c}(1+q^{\frac{a}{2}+k})} \\
&= \Gamma_q\left[\begin{matrix} c, \frac{a}{2} \\ a, c-\frac{a}{2} \end{matrix}\right] \\
&\times \frac{\langle \widetilde{1}, \widetilde{c-\frac{a}{2}};q\rangle_\infty}{\langle \widetilde{c-a}, \widetilde{1+\frac{a}{2}};q\rangle_\infty (1+q^{\frac{a}{2}})} {}_2\phi_1\left[\begin{matrix} b, \frac{a}{2} \\ c-\frac{a}{2} \end{matrix} | q^2; x^2\right] = \text{LHS}. \qquad (10.177)
\end{aligned}
$$

□

We use the q-Dixon formula for the proof of the next theorem.

Theorem 10.8.3 *A q-analogue of* [328, 4.1, p. 202].

If $\{C_{m,n}\}_{m,n=0}^{\infty}$ is a bounded sequence of complex numbers then we have the reduction formula

$$
\begin{aligned}
&\sum_{m,n,p=0}^{\infty} \frac{C_{m,n+p}\langle a,b;q\rangle_n \langle a,b,\widetilde{1-\frac{n+p}{2}};q\rangle_p x_1^m x_2^n (-x_2)^p q^{\frac{p}{2}(n-1)}}{\langle 1;q\rangle_m \langle 1;q\rangle_n \langle 1, \widetilde{-\frac{n+p}{2}};q\rangle_p} \\
&= \sum_{m,k=0}^{\infty} \frac{C_{m,2k}\langle a,b,\widetilde{1-k},\widetilde{b};q\rangle_k \langle a+b;q\rangle_{2k} x_1^m x_2^{2k} q^{-kb}}{\langle 1;q\rangle_m \langle 1, a+b, \widetilde{1-b-k}, \widetilde{1};q\rangle_k}. \qquad (10.178)
\end{aligned}
$$

Proof Compare with Rainville [425, p. 56].

$$
\begin{aligned}
&\sum_{m,n=0}^{\infty}\sum_{p=0}^{n} \frac{C_{m,n}\langle a,b;q\rangle_n \langle a,b,-n,\widetilde{1-\frac{n}{2}};q\rangle_p x_1^m x_2^n q^{\binom{p}{2}+p(2-n-a-b)+\frac{p}{2}(n-p-1)}}{\langle 1;q\rangle_m \langle 1;q\rangle_n \langle 1, 1-n-b, 1-n-a, \widetilde{-\frac{n}{2}};q\rangle_p} \\
&\overset{\text{by (7.113)}}{=} \sum_{m,n=0}^{\infty} \frac{C_{m,n}\langle a,b;q\rangle_n x_1^m x_2^n}{\langle 1;q\rangle_m \langle 1;q\rangle_n} \\
&\times \Gamma_q\left[\begin{matrix} 1-b-n, 1-a-n, 1-\frac{n}{2}, 1-a-b-\frac{n}{2} \\ 1-n, 1-a-b-n, 1-b-\frac{n}{2}, 1-a-\frac{n}{2} \end{matrix}\right] \\
&= \sum_{m=0}^{\infty} \sum_{n=0, n \text{ even}}^{\infty} \frac{C_{m,n}\langle a,b;q\rangle_n x_1^m x_2^n}{\langle 1;q\rangle_m \langle 1;q\rangle_n} \Gamma_q\left[\begin{matrix} 1-a-n, 1-a-b-\frac{n}{2} \\ 1-a-b-n, 1-a-\frac{n}{2} \end{matrix}\right]
\end{aligned}
$$

$$
\times \frac{\langle \widetilde{1-\frac{n}{2}}, \widetilde{b}; q\rangle_{\frac{n}{2}}}{\langle \widetilde{1-\frac{n}{2}-b}, b+\frac{n}{2}; q\rangle_{\frac{n}{2}}} \left\langle \frac{1}{2}; q^2 \right\rangle_{\frac{n}{2}}
$$

$$
= \sum_{m,k=0}^{\infty} \frac{C_{m,2k}\langle a,b;q\rangle_{2k} x_1^m x_2^{2k}}{\langle 1;q\rangle_m \langle 1;q\rangle_{2k}}
$$

$$
\times \frac{\langle 1-a-b-2k, \widetilde{1-k}, \widetilde{b}; q\rangle_k}{\langle 1-a-2k, \widetilde{1-k-b}, b+k; q\rangle_k} \left\langle \frac{1}{2}; q^2 \right\rangle_k = \text{RHS}. \tag{10.179}
$$

□

10.9 q-Analogues of Burchnall-Chaundy expansions

This section is mainly based on the following inverse pair of symbolic operators defined (for $q = 1$) in [89]. We will get an improved version of [307] in the process, compare with [4].

Definition 173 Set

$$
\begin{aligned}
\nabla_q(h) &\equiv \Gamma_q \begin{bmatrix} h, h+\theta_{q,1}+\theta_{q,2} \\ h+\theta_{q,1}, h+\theta_{q,2} \end{bmatrix}, \\
\Delta_q(h) &\equiv \Gamma_q \begin{bmatrix} h+\theta_{q,1}, h+\theta_{q,2} \\ h+\theta_{q,1}+\theta_{q,2}, h \end{bmatrix}.
\end{aligned} \tag{10.180}
$$

Then

$$
\nabla_q(h)\langle h;q\rangle_m \langle h;q\rangle_n x_1^m x_2^n = \langle h;q\rangle_{m+n} x_1^m x_2^n. \tag{10.181}
$$

We limit our considerations to the case $h \neq -n$, $n \in \mathbb{N}$.

We have

$$
{}_2\phi_1(a,b;c|q;x_1)\,{}_2\phi_1(a',b';c'|q;x_2) = \sum_{m,n=0}^{\infty} \frac{\langle a,b;q\rangle_m \langle a',b';q\rangle_n}{\langle 1,c;q\rangle_m \langle 1,c';q\rangle_n} x_1^m x_2^n, \tag{10.182}
$$

$$
{}_2\phi_1(a,b;c|q;x_1 \oplus_q x_2) = \sum_{m,n=0}^{\infty} \frac{\langle a;q\rangle_{m+n} \langle b;q\rangle_{m+n}}{\langle 1;q\rangle_m \langle 1;q\rangle_n \langle c;q\rangle_{m+n}} x_1^m x_2^n. \tag{10.183}
$$

We can write symbolically (the last three equations first appeared in Jackson [307, (14), p. 72]):

$$
\begin{aligned}
&\Phi_2(a;b,b';c,c'|q;x_1,x_2) \\
&\quad = \nabla_q(a)\,{}_2\phi_1(a,b;c|q;x_1)\,{}_2\phi_1(a,b';c'|q;x_2),
\end{aligned} \tag{10.184}
$$

$$\begin{aligned}&\Phi_3\big(a,a';b,b';c|q;x_1,x_2\big)\\&\quad=\triangle_q(c)\,{}_2\phi_1(a,b;c|q;x_1)\,{}_2\phi_1\big(a',b';c|q;x_2\big),\end{aligned} \tag{10.185}$$

$$\begin{aligned}&\Phi_1\big(a;b,b';c|q;x_1,x_2\big)\\&\quad=\nabla_q(a)\triangle_q(c)\,{}_2\phi_1(a,b;c|q;x_1)\,{}_2\phi_1\big(a,b';c|q;x_2\big),\end{aligned} \tag{10.186}$$

$$\Phi_1\big(a;b,b';c|q;x_1,x_2\big)=\nabla_q(a)\Phi_3\big(a,a;b,b';c|q;x_1,x_2\big), \tag{10.187}$$

$$\Phi_1\big(a;b,b';c|q;x_1,x_2\big)=\triangle_q(c)\Phi_2\big(a;b,b';c,c|q;x_1,x_2\big), \tag{10.188}$$

$$\Phi_4\big(a,b;c,c'|q;x_1,x_2\big)=\nabla_q(b)\Phi_2\big(a;b,b;c,c'|q;x_1,x_2\big), \tag{10.189}$$

$${}_2\phi_1(a,b;c|q;x_1\oplus_q x_2)=\nabla_q(b)\Phi_1(a;b,b;c|q;x_1,x_2), \tag{10.190}$$

$${}_2\phi_1(a,b;c|q;x_1\oplus_q x_2)=\triangle_q(c)\Phi_4(a,b;c,c|q;x_1,x_2), \tag{10.191}$$

$${}_2\phi_1(a,b;c|q;x_1\oplus_q x_2)=\nabla_q(b)\triangle_q(c)\Phi_2(a;b,b;c,c|q;x_1,x_2), \tag{10.192}$$

$${}_2\phi_1\big(a,b;c|q;(x_1\boxplus_q x_2)\big)=\nabla_q(b)\Phi^{1:1;1}_{1:0;1}\left[\begin{matrix}a:b;b\\c:-;\infty\end{matrix}|q;x_1,-x_2\right], \tag{10.193}$$

$$\begin{aligned}&{}_2\phi_1\big(a,b;c|q;(x_1\boxplus_q x_2)\big)\\&\quad=\triangle_q(c)\Phi^{2:2;1}_{2:1;1}\left[\begin{matrix}a,b:\infty,\infty;\infty\\\infty,\infty:c;c\end{matrix}|q;x_1,-x_2\right],\end{aligned} \tag{10.194}$$

$$\begin{aligned}&{}_2\phi_1\big(a,b;c|q;(x_1\boxplus_q x_2)\big)\\&\quad=\nabla_q(b)\triangle_q(c)\Phi^{1:2;1}_{1:1;1}\left[\begin{matrix}a:b,\infty;b\\\infty:c;c\end{matrix}|q;x_1,-x_2\right].\end{aligned} \tag{10.195}$$

For a simple q-hypergeometric series we obtain

$$\frac{1}{\langle\theta_{q,1}+c;q\rangle_r}\,{}_2\phi_1(a,b;c|q;x_1)=\frac{1}{\langle c;q\rangle_r}\,{}_2\phi_1(a,b;c+r|q;x_1), \tag{10.196}$$

and analogous formulae for double series.

Similarly,

$$\begin{aligned}&\frac{1}{\langle -a-\theta_{q,1}-\theta_{q,2}+1;q\rangle_r}\langle a;q\rangle_{m+n}x_1^m x_2^n\\&\quad=\frac{(-1)^r\langle a;q\rangle_r\langle a-r;q\rangle_{m+n}}{\langle a-r;q\rangle_{2r}}q^{\binom{r}{2}+r(a+m+n-r)}.\end{aligned} \tag{10.197}$$

Theorem 10.9.1 *The following formulas for $\theta_{q,1}$ and $\theta_{q,2}$ hold:*

$$\frac{1}{\langle -a-\theta_{q,1}-\theta_{q,2}+1;q\rangle_r}\Phi_1\big(a;b,b';c|q;x_1,x_2\big) = \frac{(-1)^r\langle a;q\rangle_r}{\langle a-r;q\rangle_{2r}}\Phi_1\big(a-r;b,b';c|q;x_1q^r,x_2q^r\big)q^{-\binom{r}{2}+r(a-1)}, \tag{10.198}$$

$$\langle -\theta_{q,1};q\rangle_r {}_2\phi_1(a,b;c|q;x_1) = \frac{(-1)^r\langle a,b;q\rangle_r x_1^r}{\langle c;q\rangle_r}{}_2\phi_1\big(a+r,b+r;c+r|q;x_1q^{-r}\big)q^{\binom{r}{2}-r^2}, \tag{10.199}$$

$$\begin{aligned}\langle -\theta_{q,1},-\theta_{q,2};q\rangle_r \Phi_1\big(a;b,b';c|q;x_1,x_2\big) &= \frac{\langle a;q\rangle_{2r}\langle b,b';q\rangle_r}{\langle c;q\rangle_{2r}}x_1^r x_2^r \\ &\quad\times \Phi_1\big(a+2r;b+r,b'+r;c+2r|q;x_1q^{-r},x_2q^{-r}\big)q^{-r^2-r},\end{aligned} \tag{10.200}$$

$$\frac{\langle -\theta_{q,1},-\theta_{q,2};q\rangle_r}{\langle -a-\theta_{q,1}-\theta_{q,2}+1;q\rangle_r}\Phi_1\big(a;b,b';c|q;x_1,x_2\big) = \frac{(-1)^r\langle a,b,b';q\rangle_r}{\langle c;q\rangle_{2r}}x_1^r x_2^r\Phi_1\big(a+r;b+r,b'+r;c+2r|q;x_1,x_2\big)q^{\binom{r}{2}+r(a-1)}. \tag{10.201}$$

Proof Combine (10.198) and (10.200) to obtain the last formula. □

Lemma 10.9.2 *Various formulas for Γ_q functions as infinite sums of q-factorials:*

$$\Gamma_q\begin{bmatrix} h, m+n+h \\ m+h, n+h\end{bmatrix} = \sum_{r=0}^{\infty}\frac{\langle -m,-n;q\rangle_r}{\langle 1,h;q\rangle_r}q^{r(m+n+h)}, \tag{10.202}$$

$$\Gamma_q\begin{bmatrix} m+k, n+k \\ k, m+n+k\end{bmatrix} = \sum_{r=0}^{\infty}\frac{\langle -m,-n;q\rangle_r}{\langle 1,-k-m-n+1;q\rangle_r}q^{r} \tag{10.203}$$

$$= \sum_{r=0}^{\infty}(-1)^r\frac{\langle -m,-n;q\rangle_r\langle k;q\rangle_{2r}}{\langle 1,k+m,k+n,k+r-1;q\rangle_r}q^{\binom{r}{2}+r(m+n+k)}, \tag{10.204}$$

$$\Gamma_q\begin{bmatrix} k+m, k+n, h, m+n+h \\ k, m+n+k, m+h, n+h\end{bmatrix} = \sum_{r=0}^{\infty}\frac{\langle -m,-n,k-h;q\rangle_r\langle k;q\rangle_{2r}}{\langle 1,h,k+r-1,m+k,n+k;q\rangle_r}q^{r(m+n+h)} \tag{10.205}$$

$$= \sum_{r=0}^{\infty} \frac{\langle -m, -n, h-k; q\rangle_r}{\langle 1, h, -k-m-n+1; q\rangle_r} q^r. \tag{10.206}$$

Keep in mind that (10.203) follows by letting $h \to \infty$ in (10.206), (10.204) follows by letting $h \to \infty$ in (10.205) and (10.202) follows by letting $k \to \infty$ in (10.205).

In (10.202) replace m, n by θ_1, θ_2 to obtain

$$\nabla_q(h) = \sum_{r=0}^{\infty} \frac{\langle -\theta_{q,1}, -\theta_{q,2}; q\rangle_r}{\langle 1, h; q\rangle_r} q^{rh} \epsilon_1^r \epsilon_2^r. \tag{10.207}$$

Similarly, we obtain

$$\Delta_q(h) = \sum_{r=0}^{\infty} \frac{\langle -\theta_{q,1}, -\theta_{q,2}; q\rangle_r}{\langle 1, 1-h-\theta_{q,1}-\theta_{q,2}; q\rangle_r} q^r, \tag{10.208}$$

$$\Delta_q(h) = \sum_{r=0}^{\infty} \frac{(-1)^r \langle -\theta_{q,1}, -\theta_{q,2}; q\rangle_r \langle h; q\rangle_{2r}}{\langle 1, h+r-1, h+\theta_{q,1}, h+\theta_{q,2}; q\rangle_r} q^{\binom{r}{2}+rh} \epsilon_1^r \epsilon_2^r, \tag{10.209}$$

and

$$\nabla_q(h)\, \Delta_q\,(k) = \sum_{r=0}^{\infty} \frac{\langle -\theta_{q,1}, -\theta_{q,2}, k-h; q\rangle_r \langle k; q\rangle_{2r}}{\langle 1, h, k+r-1, k+\theta_{q,1}, k+\theta_{q,2}; q\rangle_r} q^{rh} \epsilon_1^r \epsilon_2^r, \tag{10.210}$$

$$\nabla_q(h)\, \Delta_q\,(k) = \sum_{r=0}^{\infty} \frac{\langle h-k, -\theta_{q,1}, -\theta_{q,2}; q\rangle_r}{\langle 1, h, 1-k-\theta_{q,1}-\theta_{q,2}; q\rangle_r} q^r, \tag{10.211}$$

where the last formula is the q-Pfaff-Saalschütz theorem. By adopting these operators, we obtain the following expansions, which are q-analogues of [89, (26)–(43)].

The proofs are always done in the same way. First we write the ∇_q and Δ_q in the form of infinite sums. Then we replace $\theta_{q,1}$ and $\theta_{q,2}$ by m and n in the double sum (say m and n are the summation indices).

Theorem 10.9.3 *A q-analogue of Burchnall, Chaundy* [89, (26)]:

$$\Phi_2\big(a; b, b'; c, c'|q; x_1, x_2\big) = \sum_{r=0}^{\infty} \frac{\langle a, b, b'; q\rangle_r}{\langle 1, c, c'; q\rangle_r} x_1^r x_2^r q^{r(a+r-1)} \times {}_2\phi_1(a+r, b+r; c+r|q; x_1)\, {}_2\phi_1\big(a+r, b'+r; c'+r|q; x_2\big). \tag{10.212}$$

This corresponds to a form of the first q-Vandermonde theorem:

$$\frac{\langle a;q\rangle_{m+n}}{\langle 1;q\rangle_m\langle 1;q\rangle_n}=\sum_{r=0}^{\min(m,n)}\frac{\langle a;q\rangle_r}{\langle 1;q\rangle_r}\frac{\langle a+r;q\rangle_{m-r}}{\langle 1;q\rangle_{m-r}}\frac{\langle a+r;q\rangle_{n-r}}{\langle 1;q\rangle_{n-r}}q^{r(a+r-1)}. \tag{10.213}$$

Theorem 10.9.4 *A q-analogue of* [89, (27)]:

$$\begin{aligned}&{}_2\phi_1(a,b;c|q;x_1)\,{}_2\phi_1\big(a,b';c'|q;x_2\big)\\&=\sum_{r=0}^{\infty}\frac{(-1)^r\langle a,b,b';q\rangle_r}{\langle 1,c,c';q\rangle_r}x_1^r x_2^r q^{ra+\binom{r}{2}}\\&\quad\times\Phi_2\big(a+r;b+r,b'+r;c+r,c'+r|q;x_1,x_2\big).\end{aligned} \tag{10.214}$$

Proof

$$\begin{aligned}&{}_2\phi_1(a,b;c|q;x_1)\,{}_2\phi_1\big(a,b';c'|q;x_2\big)\\&=\sum_{r=0}^{\infty}\frac{\langle -\theta_{q,1},-\theta_{q,2};q\rangle_r}{\langle 1,1-a-\theta_{q,1}-\theta_{q,2};q\rangle_r}q^r\,\Phi_2\big(a;b,b';c,c'|q;x_1,x_2\big)=\cdots.\end{aligned} \tag{10.215}$$

□

This corresponds to

$$\frac{\langle a;q\rangle_m\langle a;q\rangle_n}{\langle 1;q\rangle_m\langle 1;q\rangle_n}=\sum_{r=0}^{\min(m,n)}\frac{\langle a;q\rangle_{m+n-r}}{\langle 1;q\rangle_r}\frac{(-1)^r}{\langle 1;q\rangle_{m-r}\langle 1;q\rangle_{n-r}}q^{ra+\binom{r}{2}}. \tag{10.216}$$

Theorem 10.9.5 *A q-analogue of* [89, (28)]. *Compare with R. P. Agarwal* [4, p. 194]:

$$\begin{aligned}&\Phi_3\big(a,a';b,b';c|q;x_1,x_2\big)\\&=\sum_{r=0}^{\infty}\frac{(-1)^r\langle a,a',b,b';q\rangle_r}{\langle 1,c+r-1;q\rangle_r\langle c;q\rangle_{2r}}x_1^r x_2^r q^{rc+\frac{3}{2}r(r-1)}\\&\quad\times{}_2\phi_1(a+r,b+r;c+2r|q;x_1)\,{}_2\phi_1\big(a'+r,b'+r;c+2r|q;x_2\big).\end{aligned} \tag{10.217}$$

Proof

$$\begin{aligned}&\Phi_3\big(a,a';b,b';c|q;x_1,x_2\big)\\&=\sum_{r=0}^{\infty}\frac{(-1)^r\langle -\theta_{q,1},-\theta_{q,2};q\rangle_r\langle c;q\rangle_{2r}}{\langle 1,c+r-1,c+\theta_{q,1},c+\theta_{q,2};q\rangle_r}q^{\binom{r}{2}+rc}\epsilon_1^r\epsilon_2^r\end{aligned}$$

$$\times\, {}_2\phi_1(a,b;c|q;x_1)\,{}_2\phi_1\big(a',b';c|q;x_2\big)=\cdots. \tag{10.218}$$

□

This corresponds to

$$\begin{aligned}&\frac{1}{\langle 1;q\rangle_m\langle 1;q\rangle_n\langle a;q\rangle_{m+n}}\\&\quad=\sum_{r=0}^{\min(m,n)}\frac{(-1)^r}{\langle 1,a+r-1;q\rangle_r\langle 1;q\rangle_{m-r}}\frac{q^{ra+\frac{3}{2}r(r-1)}}{\langle a;q\rangle_{m+r}\langle 1,a+2r;q\rangle_{n-r}}.\end{aligned} \tag{10.219}$$

Theorem 10.9.6 *A q-analogue of* [89, (29)]:

$$\begin{aligned}&{}_2\phi_1(a,b;c|q;x_1)\,{}_2\phi_1\big(a',b';c|q;x_2\big)\\&\quad=\sum_{r=0}^{\infty}\frac{\langle a,a',b,b';q\rangle_r}{\langle 1,c;q\rangle_r\langle c;q\rangle_{2r}}x_1^r x_2^r q^{rc+r(r-1)}\\&\qquad\times\,\Phi_3\big(a+r,a'+r;b+r,b'+r;c+2r|q;x_1,x_2\big).\end{aligned} \tag{10.220}$$

This corresponds to

$$\frac{\langle a;q\rangle_{m+n}}{\langle 1,a;q\rangle_m\langle 1,a;q\rangle_n}=\sum_{r=0}^{\min(m,n)}\frac{q^{ra+r(r-1)}}{\langle 1,a;q\rangle_r\langle 1;q\rangle_{m-r}\langle 1;q\rangle_{n-r}}, \tag{10.221}$$

which for $m=n$ is a special case of the first q-Vandermonde theorem.

Theorem 10.9.7 *A q-analogue of* [89, (30)]. *The first version of this equation occurred in Jackson* [307, (37), p. 75]. *The same improved version appeared in R. P. Agarwal* [4, 6.8, p. 193].

$$\begin{aligned}&\Phi_1\big(a;b,b';c|q;x_1,x_2\big)\\&\quad=\sum_{r=0}^{\infty}\frac{\langle c-a,a,b,b';q\rangle_r}{\langle 1,c+r-1;q\rangle_r\langle c;q\rangle_{2r}}x_1^r x_2^r q^{ra+r(r-1)}\\&\qquad\times\,{}_2\phi_1(a+r,b+r;c+2r|q;x_1)\,{}_2\phi_1\big(a+r,b'+r;c+2r|q;x_2\big).\end{aligned} \tag{10.222}$$

Proof

$$\begin{aligned}&\Phi_1\big(a;b,b';c|q;x_1,x_2\big)\\&\quad\overset{\text{by (10.205)}}{=}\sum_{r=0}^{\infty}\frac{\langle -\theta_{q,1},-\theta_{q,2},c-a;q\rangle_r\langle c;q\rangle_{2r}}{\langle 1,a,c+r-1,c+\theta_{q,1},c+\theta_{q,2};q\rangle_r}\end{aligned}$$

$$\times q^{ra}\epsilon_1^r\epsilon_2^r\,{}_2\phi_1(a,b;c|q;x_1)\,{}_2\phi_1\big(a,b';c|q;x_2\big)$$

$$\stackrel{\text{by (10.196)}}{=}\sum_{r=0}^{\infty}\frac{\langle-\theta_{q,1},-\theta_{q,2},c-a;q\rangle_r\langle c;q\rangle_{2r}}{\langle 1,c+r-1,a,c,c;q\rangle_r}q^{ra}\epsilon_1^r\epsilon_2^r\,{}_2\phi_1(a,b;c+r|q;x_1)$$

$$\times\,{}_2\phi_1\big(a,b';c+r|q;x_2\big)$$

$$\stackrel{\text{by (10.199)}}{=}\sum_{r=0}^{\infty}\frac{\langle b,a,b',c-a;q\rangle_r\langle c;q\rangle_{2r}}{\langle 1,c+r-1,c,c,c+r,c+r;q\rangle_r}q^{ra+r(r-1)}$$

$$\times x_1^r x_2{}^r\,{}_2\phi_1(a+r,b+r;c+2r|q;x_1)$$

$$\times\,{}_2\phi_1\big(a+r,b'+r;c+2r|q;x_2\big)=\cdots. \tag{10.223}$$

□

This corresponds to a form of the q-Whipple theorem:

$$\frac{\langle a;q\rangle_{m+n}}{\langle 1;q\rangle_m\langle 1;q\rangle_n\langle c;q\rangle_{m+n}}$$
$$=\sum_{r=0}^{\min(m,n)}\frac{\langle c-a,a;q\rangle_r q^{ra+r(r-1)}}{\langle 1,c+r-1;q\rangle_r\langle 1;q\rangle_{m-r}}\frac{\langle a+r;q\rangle_{m-r}}{\langle c;q\rangle_{m+r}}\frac{\langle a+r;q\rangle_{n-r}}{\langle 1,c+2r;q\rangle_{n-r}}. \tag{10.224}$$

Theorem 10.9.8 *A q-analogue of* [89, (31)]:

$$ {}_2\phi_1(a,b;c|q;x_1)\,{}_2\phi_1\big(a,b';c|q;x_2\big)$$
$$=\sum_{r=0}^{\infty}\frac{(-1)^r\langle a,b,b',c-a;q\rangle_r}{\langle 1,c;q\rangle_r\langle c;q\rangle_{2r}}q^{ra+\binom{r}{2}}$$
$$\times x_1^r x_2{}^r\,\Phi_1\big(a+r;b+r,b'+r;c+2r|q;x_1,x_2\big). \tag{10.225}$$

Proof

$${}_2\phi_1(a,b;c|q;x_1)\,{}_2\phi_1\big(a,b';c|q;x_2\big)$$
$$=\sum_{r=0}^{\infty}\frac{\langle c-a,-\theta_{q,1},-\theta_{q,2};q\rangle_r}{\langle 1,c,1-a-\theta_{q,1}-\theta_{q,2};q\rangle_r}q^r\,\Phi_1\big(a;b,b';c|q;x_1,x_2\big). \tag{10.226}$$

□

This corresponds to a form of the q-Pfaff-Saalschütz theorem:

$$\frac{\langle a;q\rangle_m\langle a;q\rangle_n}{\langle 1,c;q\rangle_m\langle 1,c;q\rangle_n}=\sum_{r=0}^{\min(m,n)}\frac{(-1)^r\langle c-a;q\rangle_r}{\langle 1,c;q\rangle_r\langle c;q\rangle_{m+n}}\frac{\langle a;q\rangle_{m+n-r}}{\langle 1;q\rangle_{m-r}\langle 1;q\rangle_{n-r}}q^{ra+\binom{r}{2}}. \tag{10.227}$$

Theorem 10.9.9 *A q-analogue of* [89, (32)]:

$$\begin{aligned}&\Phi_1\big(a;b,b';c|q;x_1,x_2\big)\\&\quad=\sum_{r=0}^{\infty}\frac{\langle a,b,b';q\rangle_r}{\langle 1;q\rangle_r\langle c;q\rangle_{2r}}q^{ra+r(r-1)}x_1^r x_2^r\\&\qquad\times\Phi_3\big(a+r,a+r;b+r,b'+r;c+2r|q;x_1,x_2\big).\end{aligned} \tag{10.228}$$

Proof

$$\begin{aligned}\Phi_1\big(a;b,b';c|q;x_1,x_2\big)&=\sum_{r=0}^{\infty}\frac{\langle -\theta_{q,1},-\theta_{q,2};q\rangle_r}{\langle 1,a;q\rangle_r}q^{ra}\epsilon_1^r\epsilon_2^r\\&\quad\times\Phi_3\big(a,a;b,b';c|q;x_1,x_2\big)\cdots.\end{aligned} \tag{10.229}$$

□

This corresponds to (10.213).

Theorem 10.9.10 *A q-analogue of* [89, (33)]:

$$\begin{aligned}&\Phi_3\big(a,a;b,b';c|q;x_1,x_2\big)\\&\quad=\sum_{r=0}^{\infty}\frac{(-1)^r\langle a,b,b';q\rangle_r}{\langle 1;q\rangle_r\langle c;q\rangle_{2r}}q^{ra+\binom{r}{2}}x_1^r x_2^r\\&\qquad\times\Phi_1\big(a+r;b+r,b'+r;c+2r|q;x_1,x_2\big).\end{aligned} \tag{10.230}$$

Proof

$$\begin{aligned}&\Phi_3\big(a,a;b,b';c|q;x_1,x_2\big)\\&\quad=\sum_{r=0}^{\infty}\frac{\langle -\theta_{q,1},-\theta_{q,2};q\rangle_r}{\langle 1,1-a-\theta_{q,1}-\theta_{q,2};q\rangle_r}q^r\Phi_1\big(a;b,b';c|q;x_1,x_2\big)=\cdots.\end{aligned} \tag{10.231}$$

□

This corresponds to (10.216).

Theorem 10.9.11 *A q-analogue of* [89, (34)]:

$$\begin{aligned}&\Phi_1\big(a; b, b'; c|q; x_1, x_2\big)\\&\quad=\sum_{r=0}^{\infty}\frac{(-1)^r\langle b, b'; q\rangle_r\langle a; q\rangle_{2r}}{\langle 1, c+r-1; q\rangle_r\langle c; q\rangle_{2r}}q^{rc+\frac{3}{2}r(r-1)}x_1^r x_2^r\\&\quad\times\Phi_2\big(a+2r; b+r, b'+r; c+2r, c+2r|q; x_1, x_2\big).\end{aligned}\tag{10.232}$$

Proof

$$\begin{aligned}&\Phi_1\big(a; b, b'; c|q; x_1, x_2\big)\\&\quad=\sum_{r=0}^{\infty}\frac{(-1)^r\langle -\theta_{q,1}, -\theta_{q,2}; q\rangle_r\langle c; q\rangle_{2r}}{\langle 1, c+r-1, c+\theta_{q,1}, c+\theta_{q,2}; q\rangle_r}q^{\binom{r}{2}+rc}\epsilon_1^r\epsilon_2^r\\&\quad\times\Phi_2\big(a; b, b'; c, c|q; x_1, x_2\big)=\cdots.\end{aligned}\tag{10.233}$$

□

This corresponds to (10.219).

Theorem 10.9.12 *A q-analogue of* [89, (35)]:

$$\begin{aligned}&\Phi_2\big(a; b, b'; c, c|q; x_1, x_2\big)\\&\quad=\sum_{r=0}^{\infty}\frac{\langle b, b'; q\rangle_r\langle a; q\rangle_{2r}}{\langle 1, c; q\rangle_r\langle c; q\rangle_{2r}}q^{rc+r(r-1)}x_1^r x_2^r\\&\quad\times\Phi_1\big(a+2r; b+r, b'+r; c+2r|q; x_1, x_2\big).\end{aligned}\tag{10.234}$$

Proof

$$\begin{aligned}\Phi_2\big(a; b, b'; c, c|q; x_1, x_2\big)&=\sum_{r=0}^{\infty}\frac{\langle -\theta_{q,1}, -\theta_{q,2}; q\rangle_r}{\langle 1, c; q\rangle_r}q^{rc}\epsilon_1^r\epsilon_2^r\\&\quad\times\Phi_1\big(a; b, b'; c|q; x_1, x_2\big)=\cdots.\end{aligned}\tag{10.235}$$

□

This corresponds to (10.221).

Theorem 10.9.13 *A q-analogue of* [89, (36)]:

$$\begin{aligned}&\Phi_4\big(a;b;c,c'|q;x_1,x_2\big)\\&\quad=\sum_{r=0}^{\infty}\frac{\langle b;q\rangle_r\langle a;q\rangle_{2r}}{\langle 1,c,c';q\rangle_r}q^{rb+r(r-1)}x_1^rx_2^r\\&\quad\times\Phi_2\big(a+2r;b+r,b+r;c+r,c'+r|q;x_1,x_2\big).\end{aligned}\tag{10.236}$$

Proof

$$\begin{aligned}&\Phi_4\big(a;b;c,c'|q;x_1,x_2\big)\\&\quad=\sum_{r=0}^{\infty}\frac{\langle -\theta_{q,1},-\theta_{q,2};q\rangle_r}{\langle 1,b;q\rangle_r}q^{rb}\epsilon_1^r\epsilon_2^r\\&\quad\times\Phi_2\big(a;b,b;c,c'|q;x_1,x_2\big)=\cdots.\end{aligned}\tag{10.237}$$

□

This corresponds to (10.213).

Theorem 10.9.14 *A q-analogue of* [89, (37)]:

$$\begin{aligned}&\Phi_2\big(a;b,b;c,c'|q;x_1,x_2\big)\\&\quad=\sum_{r=0}^{\infty}\frac{(-1)^r\langle b;q\rangle_r\langle a;q\rangle_{2r}}{\langle 1,c,c';q\rangle_r}q^{\binom{r}{2}+rb}x_1^rx_2^r\\&\quad\times\Phi_4\big(a+2r;b+r;c+r,c'+r|q;x_1,x_2\big).\end{aligned}\tag{10.238}$$

Proof

$$\begin{aligned}&\Phi_2\big(a;b,b;c,c'|q;x_1,x_2\big)\\&\quad=\sum_{r=0}^{\infty}\frac{(-1)^r\langle -\theta_{q,1},-\theta_{q,2};q\rangle_r\langle c;q\rangle_{2r}}{\langle 1,c+r-1,c+\theta_{q,1},c+\theta_{q,2};q\rangle_r}q^{\binom{r}{2}+rc}\epsilon_1^r\epsilon_2^r\\&\quad\times\Phi_4\big(a;b;c,c'|q;x_1,x_2\big)=\cdots.\end{aligned}\tag{10.239}$$

□

This corresponds to (10.216).

Theorem 10.9.15 *A q-analogue of* [89, (38)]. *The first version of this equation, with Jackson q-addition, occurred in Jackson* [307, (45), p. 76]:

$$\begin{aligned}
&{}_2\phi_1(a,b;c|q;x_1 \oplus_q x_2)\\
&\quad=\sum_{r=0}^{\infty}\frac{\langle b;q\rangle_r\langle a;q\rangle_{2r}}{\langle 1;q\rangle_r\langle c;q\rangle_{2r}}q^{rb+r(r-1)}x_1^r x_2^r\\
&\quad\times \Phi_1(a+2r;b+r,b+r;c+2r|q;x_1,x_2).
\end{aligned}\tag{10.240}$$

Proof

$$\begin{aligned}
{}_2\phi_1(a,b;c|q;x_1 \oplus_q x_2)&=\sum_{r=0}^{\infty}\frac{\langle -\theta_{q,1},-\theta_{q,2};q\rangle_r}{\langle 1,b;q\rangle_r}q^{rb}\epsilon_1^r\epsilon_2^r\\
&\quad\times \Phi_1(a;b,b;c|q;x_1,x_2)=\cdots.
\end{aligned}\tag{10.241}$$

□

This corresponds to (10.221).

Theorem 10.9.16 *A q-analogue of* [89, (39)]. *The first version of this equation, with Jackson q-addition, occurred in Jackson* [307, (46), p. 77].

$$\begin{aligned}
&\Phi_1(a;b,b;c|q;x_1,x_2)\\
&\quad=\sum_{r=0}^{\infty}\frac{(-1)^r\langle b;q\rangle_r\langle a;q\rangle_{2r}}{\langle 1;q\rangle_r\langle c;q\rangle_{2r}}q^{\binom{r}{2}+rb}x_1^r x_2^r\\
&\quad\times {}_2\phi_1(a+2r,b+r;c+2r|q;x_1 \oplus_q x_2).
\end{aligned}\tag{10.242}$$

Proof

$$\begin{aligned}
\Phi_1(a;b,b;c|q;x_1,x_2)&=\sum_{r=0}^{\infty}\frac{\langle -\theta_{q,1},-\theta_{q,2};q\rangle_r}{\langle 1,1-b-\theta_{q,1}-\theta_{q,2};q\rangle_r}q^r\\
&\quad\times {}_2\phi_1(a,b;c|q;x_1 \oplus_q x_2)=\cdots.
\end{aligned}\tag{10.243}$$

□

This corresponds to (10.216).

Theorem 10.9.17 *A q-analogue of* [89, (40)]. *The first version of this equation, with Jackson q-addition, occurred in Jackson* [307, (47), p. 77]:

$$\begin{aligned}
&{}_2\phi_1(a, b; c|q; x_1 \oplus_q x_2) \\
&= \sum_{r=0}^{\infty} \frac{(-1)^r \langle a, b; q\rangle_{2r}}{\langle 1, c+r-1; q\rangle_r \langle c; q\rangle_{2r}} q^{\frac{3}{2}r(r-1)+rc} x_1^r x_2^r \\
&\times \Phi_4(a+2r; b+2r; c+2r, c+2r|q; x_1, x_2).
\end{aligned} \tag{10.244}$$

This corresponds to (10.219).

Theorem 10.9.18 *A q-analogue of* [89, (41)]. *The first version of this equation, with Jackson q-addition, occurred in Jackson* [307, (48), p. 77]:

$$\begin{aligned}
&\Phi_4(a; b; c, c|q; x_1, x_2) \\
&= \sum_{r=0}^{\infty} \frac{\langle a, b; q\rangle_{2r}}{\langle 1, c; q\rangle_r \langle c; q\rangle_{2r}} q^{r(r-1)+rc} x_1^r x_2^r \\
&\times {}_2\phi_1(a+2r, b+2r; c+2r|q; x_1 \oplus_q x_2).
\end{aligned} \tag{10.245}$$

Proof

$$\begin{aligned}
\Phi_4(a; b; c, c|q; x_1, x_2) &= \sum_{r=0}^{\infty} \frac{\langle -\theta_{q,1}, -\theta_{q,2}; q\rangle_r}{\langle 1, c; q\rangle_r} q^{rc} \epsilon_1^r \epsilon_2^r \\
&\times {}_2\phi_1(a, b; c|q; x_1 \oplus_q x_2) = \cdots .
\end{aligned} \tag{10.246}$$

□

This corresponds to the limit $a \to \infty$ in (10.224).

Theorem 10.9.19 *A q-analogue of* [89, (42)]:

$$\begin{aligned}
&{}_2\phi_1(a, b; c|q; x_1 \oplus_q x_2) \\
&= \sum_{r=0}^{\infty} \frac{\langle a; q\rangle_{2r} \langle b, c-b; q\rangle_r}{\langle c; q\rangle_{2r} \langle 1, c+r-1; q\rangle_r} q^{r(b+r-1)} x_1^r x_2^r \\
&\times \Phi_2(a+2r; b+r, b+r; c+2r, c+2r|q; x_1, x_2).
\end{aligned} \tag{10.247}$$

This corresponds to (10.221).

Theorem 10.9.20 *A q-analogue of* [89, (43)]:

$$\Phi_2(a; b, b; c, c|q; x_1, x_2)$$

$$= \sum_{r=0}^{\infty} (-1)^r \frac{\langle a; q\rangle_{2r} \langle b, c-b; q\rangle_r}{\langle c; q\rangle_{2r} \langle 1, c; q\rangle_r} q^{\binom{r}{2}+rb} x_1^r x_2^r$$
$$\times {}_2\phi_1(a+2r, b+r; c+2r|q; x_1 \oplus_q x_2). \tag{10.248}$$

Proof

$$\Phi_2(a; b, b; c, c|q; x_1, x_2)$$
$$\overset{\text{by (10.206)}}{=} \sum_{r=0}^{\infty} \frac{\langle -\theta_{q,1}, -\theta_{q,2}, c-b; q\rangle_r q^r}{\langle 1, c, -b-\theta_{q,1}-\theta_{q,2}+1; q\rangle_r} {}_2\phi_1(a, b; c|q; x_1 \oplus_q x_2)$$
$$\overset{\text{by (10.201)}}{=} \text{RHS}. \tag{10.249}$$

□

This corresponds to (10.224).

10.9.1 q-analogues of Verma expansions

By the same method which was used in [307], we obtain the following q-analogues of the corrected version of Verma [520, (10)–(14)]. The formulas also closely resemble q-analogues of Srivastava [472, (11), (12), (8), (13), (7)].

When there will be no danger of confusion, we will write G and H for $G_1; G_2$ and $H_1; H_2$.

Theorem 10.9.21

$$\Phi^{A:G_1;G_2}_{B:H_1;H_2}\left[\begin{matrix}(a):(g_1);(g_2);\\(b):(h_1);(h_2);\end{matrix}|q; x_1, x_2\right]$$
$$= \sum_{r=0}^{\infty} \frac{\langle (g_1), (g_2), \alpha; q\rangle_r \langle (a); q\rangle_{2r}}{\langle 1, (h_1), (h_2); q\rangle_r \langle (b), \alpha; q\rangle_{2r}} q^{2\binom{r}{2}+r\alpha} x_1^r x_2^r \Phi^{A+1:G_1+1;G_2+1}_{B+1:H_1+1;H_2+1}$$
$$\times \left[\begin{matrix}(a+2r), \infty : (g_1+r), \alpha+r; (g_2+r), \alpha+r;\\ (b+2r), \alpha+2r : (h_1+r), \infty; (h_2+r), \infty;\end{matrix}|q; x_1, x_2\right] \tag{10.250}$$
$$= \sum_{r=0}^{\infty} \frac{(-1)^r \langle (g_1), (g_2); q\rangle_r \langle (a); q\rangle_{2r}}{\langle 1, (h_1), (h_2), \alpha; q\rangle_r \langle (b); q\rangle_{2r}} q^{\binom{r}{2}+r\alpha} x_1^r x_2^r \Phi^{A+1:G+1}_{B+1:H+1}$$
$$\times \left[\begin{matrix}(a+2r), \alpha+r : (g_1+r), \infty; (g_2+r), \infty;\\ (b+2r), \infty : (h_1+r), \alpha+r; (h_2+r), \alpha+r;\end{matrix}|q; x_1, x_2\right] \tag{10.251}$$
$$= \sum_{r=0}^{\infty} \frac{\langle (g_1), (g_2), \gamma-\alpha, \alpha; q\rangle_r \langle (a); q\rangle_{2r}}{\langle 1, (h_1), (h_2), \gamma+r-1; q\rangle_r \langle (b), \alpha; q\rangle_{2r}} q^{2\binom{r}{2}+r\alpha} x_1^r x_2^r \Phi^{A+1:G+1}_{B+1:H+1}$$

$$\times \left[\begin{matrix} (a+2r), \gamma+2r : (g_1+r), \alpha+r; (g_2+r), \alpha+r; \\ (b+2r), \alpha+2r : (h_1+r), \gamma+2r; (h_2+r), \gamma+2r; \end{matrix} \middle| q; x_1, x_2 \right] \tag{10.252}$$

$$= \sum_{r=0}^{\infty} \frac{(-1)^r \langle (g_1), (g_2); q\rangle_r \langle (a); q\rangle_{2r}}{\langle 1, (h_1), (h_2), \alpha+r-1; q\rangle_r \langle (b); q\rangle_{2r}} q^{3\binom{r}{2}+r\alpha} x_1^r x_2^r \Phi^{A+1:G+1}_{B+1:H+1}$$

$$\times \left[\begin{matrix} (a+2r), \alpha+2r : (g_1+r), \infty; (g_2+r), \infty; \\ (b+2r), \infty : (h_1+r), \alpha+2r; (h_2+r), \alpha+2r; \end{matrix} \middle| q; x_1, x_2 \right] \tag{10.253}$$

$$= \sum_{r=0}^{\infty} \frac{(-1)^r \langle (g_1), (g_2), \alpha-\gamma, \alpha; q\rangle_r \langle (a); q\rangle_{2r}}{\langle 1, (h_1), (h_2), \gamma; q\rangle_r \langle (b), \alpha; q\rangle_{2r}} q^{\binom{r}{2}+r\gamma} x_1^r x_2^r \Phi^{A+1:G+1}_{B+1:H+1}$$

$$\times \left[\begin{matrix} (a+2r), \gamma+r : (g_1+r), \alpha+r; (g_2+r), \alpha+r; \\ (b+2r), \alpha+2r : (h_1+r), \gamma+r; (h_2+r), \gamma+r; \end{matrix} \middle| q; x_1, x_2 \right]. \tag{10.254}$$

These formulas hold when $A+G_1 \le B+H_1+1$, $A+G_2 \le B+H_2+1$ *and* $|x_1|$, $|x_2|$ *are chosen in such a way that both sides converge* [472, p. 49].

Proof We will use the abbreviation

$$\Theta \equiv (-1)^{m(1+H_1-G_1+B-A)+n(1+H_2-G_2+B-A)}$$
$$\times \operatorname{QE}\left((B-A)\binom{m+n}{2} + (1+H_1-G_1)\binom{m}{2} + (1+H_2-G_2)\binom{n}{2} \right).$$

First we prove (10.252).

$$\text{LHS} = \sum_{r=0}^{\infty} \frac{\langle -\theta_{q,1}, -\theta_{q,2}, \gamma-\alpha; q\rangle_r \langle \gamma; q\rangle_{2r}}{\langle 1, \alpha, \gamma+r-1, \gamma+\theta_{q,1}, \gamma+\theta_{q,2}; q\rangle_r} q^{r\alpha} \epsilon_1^r \epsilon_2^r$$
$$\times \Phi^{A+1:G+1}_{B+1:H+1} \left[\begin{matrix} (a), \gamma : (g_1), \alpha; (g_2), \alpha; \\ (b), \alpha : (h_1), \gamma; (h_2), \gamma; \end{matrix} \middle| q; x_1, x_2 \right]$$
$$= \sum_{r,m,n=0}^{\infty} \frac{q^{2\binom{r}{2}+r\alpha} \langle \gamma-\alpha; q\rangle_r \langle \gamma; q\rangle_{2r}}{\langle \alpha, \gamma+r-1, \gamma, \gamma; q\rangle_r}$$
$$\times \Phi^{A+1:G+1}_{B+1:H+1} \left[\begin{matrix} (a), \gamma : (g_1), \alpha; (g_2), \alpha; \\ (b), \alpha : (h_1), \gamma+r; (h_2), \gamma+r; \end{matrix} \middle| q; x_1, x_2 \right] = \text{RHS}. \tag{10.255}$$

Equations (10.250) and (10.253) follow from (10.252) by letting $\gamma \to +\infty$ and $\alpha \to +\infty$ respectively. Equation (10.254) is proved in a similar way:

$$\text{LHS} = \sum_{r=0}^{\infty} \frac{\langle -\theta_{q,1}, -\theta_{q,2}, \alpha-\gamma; q\rangle_r}{\langle 1, \alpha, 1-\gamma-\theta_{q,1}-\theta_{q,2}; q\rangle_r} q^r$$

$$\times \Phi^{A+1:G+1}_{B+1:H+1}\left[\begin{matrix}(a),\gamma:(g_1),\alpha;(g_2),\alpha;\\(b),\alpha:(h_1),\gamma;(h_2),\gamma;\end{matrix}|q;x_1,x_2\right]$$

$$=\sum_{r,m,n=0}^{\infty}\frac{(-1)^r q^{\binom{r}{2}+r\gamma}\langle\alpha-\gamma;q\rangle_r\langle\gamma;q\rangle_{m+n-r}}{\langle 1,\alpha;q\rangle_r}\Theta$$

$$\times\frac{\langle(a);q\rangle_{m+n}\langle(g_1),\alpha;q\rangle_m\langle(g_2),\alpha;q\rangle_n x_1^m x_2^n}{\langle(b),\alpha;q\rangle_{m+n}\langle(h_1),\gamma;q\rangle_m\langle(h_2),\gamma;q\rangle_n\langle 1;q\rangle_{m-r}\langle 1;q\rangle_{n-r}}$$

$$=\sum_{r,m,n=0}^{\infty}\frac{(-1)^r q^{\binom{r}{2}+r\gamma}x_1^r x_2^r\langle\alpha-\gamma,(g_1),\alpha,(g_2);q\rangle_r\langle(a);q\rangle_{2r}}{\langle 1,(h_1),\gamma,(h_2);q\rangle_r\langle(b),\alpha;q\rangle_{2r}}\Theta$$

$$\times\frac{x_1^m x_2^n\langle(a+2r),\gamma+r;q\rangle_{m+n}\langle(g_1+r),\alpha+r;q\rangle_m\langle(g_2+r),\alpha+r;q\rangle_n}{\langle(b+2r),\alpha+2r;q\rangle_{m+n}\langle(h_1+r),\gamma+r;q\rangle_m\langle(h_2+r),\gamma+r;q\rangle_n}$$

$$=\text{RHS}.$$

Equation (10.251) follows from (10.254) by letting $\alpha\to+\infty$. □

Remark 56 As pointed out in [520], formula (10.250) gives on specialization of parameters (10.212), (10.220), (10.228), (10.234), (10.236), (10.240), (10.245). Formula (10.251) gives (10.214), (10.230), (10.238), (10.242); formula (10.252) gives (10.222) and (10.247); (10.253) gives (10.217), (10.232) and (10.244); formula (10.254) gives (10.225) and (10.248).

10.9.2 A similar formula

We now turn to a different problem, which can be solved by a similar technique. By Jackson's theorem, we obtain the following generalization of certain expansions in [166].

$$\nabla_q(h_1)\nabla_q(h_2)\triangle_q(h_3)\triangle_q(h_4)$$
$$=\sum_{k=0}^{\infty}\frac{\langle h_4-1,h_4-h_1,h_4-h_2,-\theta_{q,1},-\theta_{q,2},\frac{h_4+1}{2},\widetilde{\frac{h_4+1}{2}},e;q\rangle_k}{\langle 1,h_1,h_2,h_4+\theta_{q,1},h_4+\theta_{q,2},h_4-e,\frac{h_4-1}{2},\widetilde{\frac{h_4-1}{2}};q\rangle_k}q^k, \tag{10.256}$$

where

$$e\equiv-1+\theta_{q,1}+\theta_{q,2}+h_1+h_2\left(\bmod\frac{2\pi i}{\log q}\right),\qquad h_1+h_3\equiv h_3+h_4\left(\bmod\frac{2\pi i}{\log q}\right).$$

This implies the following theorem:

Theorem 10.9.22

$$\Phi_{2:0}^{2:1}\left[\begin{matrix} a,b:a+b-c;a+b-c \\ c,a+b-c:- \end{matrix}|q;x,y\right]$$
$$=\sum_{r,m,n=0}^{\infty}\frac{\langle c-1,c-a,c-b,a,b,\frac{c+1}{2},\widetilde{\frac{c+1}{2}},a+b+2r+m+n-1;q\rangle_r}{\langle 1,c,c,c+r,c+r,a+b+r+m+n-c,\frac{c-1}{2},\widetilde{\frac{c-1}{2}};q\rangle_r}$$
$$\times\frac{\langle a+r,b+r;q\rangle_m\langle a+r,b+r;q\rangle_n}{\langle 1,c+2r;q\rangle_m\langle 1,c+2r;q\rangle_n}(-1)^r x^{r+m}y^{r+n}q^{\binom{r}{2}+r(a+b-c)}. \tag{10.257}$$

Proof Denote the LHS of (10.257) by X

$$X=\nabla_q(a)\,\nabla_q(b)\,\Delta_q(c)\,\Delta_q(a+b-c)\,{}_2\phi_1(a,b;c|q;x)\,{}_2\phi_1(a,b;c|q;y)$$
$$=\sum_{r=0}^{\infty}\frac{\langle c-1,c-a,c-b,-\theta_{q,1},-\theta_{q,2},\frac{c+1}{2},\widetilde{\frac{c+1}{2}},a+b+\theta_{q,1}+\theta_{q,2}-1;q\rangle_r}{\langle 1,a,b,c+\theta_{q,1},c+\theta_{q,2},c+1-a-b-\theta_{q,1}-\theta_{q,2},\frac{c-1}{2},\widetilde{\frac{c-1}{2}};q\rangle_r}q^r$$
$$\times\sum_{m,n=0}^{\infty}\frac{\langle a,b;q\rangle_m\langle a,b;q\rangle_n}{\langle 1,c;q\rangle_m\langle 1,c;q\rangle_n}x^m y^n$$
$$=\sum_{r=0}^{\infty}\frac{\langle c-1,c-a,c-b,-\theta_{q,1},-\theta_{q,2},\frac{c+1}{2},\widetilde{\frac{c+1}{2}},a+b+\theta_{q,1}+\theta_{q,2}-1;q\rangle_r}{\langle 1,a,b,c,c,c+1-a-b-\theta_{q,1}-\theta_{q,2},\frac{c-1}{2},\widetilde{\frac{c-1}{2}};q\rangle_r}x^r y^r q^r$$
$$\times\sum_{m,n=0}^{\infty}\frac{\langle a,b;q\rangle_m\langle a,b;q\rangle_n}{\langle 1,c+r;q\rangle_m\langle 1,c+r;q\rangle_n}x^m y^n$$
$$=\sum_{r=0}^{\infty}\frac{\langle c-1,c-a,c-b,a,a,b,b,\frac{c+1}{2},\widetilde{\frac{c+1}{2}},a+b+\theta_{q,1}+\theta_{q,2}-1;q\rangle_r}{\langle 1,a,b,c+1-a-b-\theta_{q,1}-\theta_{q,2},\frac{c-1}{2},\widetilde{\frac{c-1}{2}};q\rangle_r\langle c,c;q\rangle_{2r}}x^r y^r$$
$$\times\sum_{m,n=0}^{\infty}\frac{\langle a+r,b+r;q\rangle_m\langle a+r,b+r;q\rangle_n}{\langle 1,c+2r;q\rangle_m\langle 1,c+2r;q\rangle_n}x^m y^n q^{-r(r+m+n)}$$
$$=\sum_{r=0}^{\infty}\frac{\langle c-1,c-a,c-b,a,b,\frac{c+1}{2},\widetilde{\frac{c+1}{2}},a+b+2r+m+n-1;q\rangle_r}{\langle 1,a+b+r+m+n-c,\frac{c-1}{2},\widetilde{\frac{c-1}{2}};q\rangle_r\langle c,c;q\rangle_{2r}}x^r y^r$$
$$\times\sum_{m,n=0}^{\infty}\frac{\langle a+r,b+r;q\rangle_m\langle a+r,b+r;q\rangle_n}{\langle 1,c+2r;q\rangle_m\langle 1,c+2r;q\rangle_n}x^m y^n q^{\binom{r}{2}+r(a+b-c)}(-1)^r. \tag{10.258}$$

□

This implies the following identity:

$$\frac{\langle a,b;q\rangle_{m+n}\langle a+b-c;q\rangle_m\langle a+b-c;q\rangle_n}{\langle c,a+b-c;q\rangle_{m+n}\langle 1;q\rangle_m\langle 1;q\rangle_n}$$
$$=\sum_{r=0}^{\min(m,n)}\frac{\langle c-1,c-a,c-b,a,b,\frac{c+1}{2},\widetilde{\frac{c+1}{2}},a+b+m+n-1;q\rangle_r}{\langle 1,a+b-r+m+n-c,\frac{c-1}{2},\widetilde{\frac{c-1}{2}};q\rangle_r\langle c,c;q\rangle_{2r}}$$
$$\times\frac{\langle a+r,b+r;q\rangle_{m-r}\langle a+r,b+r;q\rangle_{n-r}}{\langle 1,c+2r;q\rangle_{m-r}\langle 1,c+2r;q\rangle_{n-r}}q^{\binom{r}{2}+r(a+b-c)}(-1)^r. \tag{10.259}$$

10.10 Multiple extensions of the Rothe-von Grüson-Gauß formula

The Rothe-von Grüson-Gauß formula (2.19) forms the basis of q-analysis.

We can generalize this to n dimensions:

Theorem 10.10.1

$$\sum_{\vec{m}=\vec{0}}^{\vec{k}}(-1)^{\vec{m}}\binom{\vec{k}}{\vec{m}}_q q^{\binom{\vec{m}}{2}}\vec{u}^{\vec{m}}=(\vec{u};q)_{\vec{k}}. \tag{10.260}$$

Corollary 10.10.2

$$\sum_{\vec{m}=\vec{0}}^{\vec{k}}(-1)^{\vec{m}}\binom{\vec{k}}{\vec{m}}_q q^{\binom{\vec{m}}{2}}=\delta_{\vec{k},\vec{0}}. \tag{10.261}$$

Proof Put $\vec{u}=\vec{1}$ above. □

The proof of the next lemma is typical for q-calculus, compare with the proof in Section 8.4; the Γ_q function plays a major part, the occurrence of a $\Gamma_q(0)$ in the denominator implies a Kronecker delta.

We find some more multi-dimensional versions of (10.261) and some associated formulas. For the proof we need the following lemma, a q-analogue of Chaundy [123, p. 164]. We have kept the original notation.

Lemma 10.10.3

$$\sum_{r=0}^{R}\sum_{s=0}^{S}\frac{(-1)^{r+s}\langle 1;q\rangle_R\langle 1;q\rangle_S}{\langle 1;q\rangle_r\langle 1;q\rangle_{R-r}\langle 1;q\rangle_s\langle 1;q\rangle_{S-s}}$$

$$\times\frac{\langle c-1;q\rangle_{r+s}\langle c;q\rangle_{2r+2s}}{\langle c-1;q\rangle_{2r+2s}\langle c;q\rangle_{R+S+r+s}}\mathrm{QE}\left(\binom{r}{2}+\binom{s}{2}+Sr\right)$$

$$=\delta_{R,0}\delta_{S,0}. \tag{10.262}$$

Proof We will use q-Dixons formula in the end.

$$\mathrm{LHS}=\sum_{r=0}^{R}\sum_{s=0}^{S}\frac{\langle -R;q\rangle_r\langle -S;q\rangle_s}{\langle 1;q\rangle_r\langle 1;q\rangle_s\langle\frac{c-1}{2},\frac{c}{2};q^2\rangle_{r+s}}$$

$$\times\frac{\langle c-1;q\rangle_{r+s}\langle\frac{c+1}{2},\frac{c}{2};q^2\rangle_{r+s}}{\langle c;q\rangle_{R+S}\langle c+R+S;q\rangle_{r+s}}\mathrm{QE}\big(Rr+S(r+s)\big)$$

$$=\frac{1}{\langle c;q\rangle_{R+S}}\sum_{k=0}^{R+S}\frac{\langle c-1;q\rangle_k\langle\frac{c+1}{2},\frac{c}{2};q^2\rangle_k}{\langle c+R+S;q\rangle_k\langle\frac{c-1}{2},\frac{c}{2};q^2\rangle_k}$$

$$\times\sum_{r+s=k}\frac{\langle -R;q\rangle_r\langle -S;q\rangle_s}{\langle 1;q\rangle_r\langle 1;q\rangle_s}\mathrm{QE}(Rr+Sk)$$

$$=\frac{1}{\langle c;q\rangle_{R+S}}\sum_{k=0}^{R+S}\frac{\langle c-1;q\rangle_k\langle\frac{c+1}{2},\frac{c}{2};q^2\rangle_k}{\langle c+R+S;q\rangle_k\langle\frac{c-1}{2},\frac{c}{2};q^2\rangle_k}$$

$$\times\sum_{r=0}^{k}\frac{\langle -R,-k;q\rangle_r\langle -S;q\rangle_k}{\langle 1,1+S-k;q\rangle_r\langle 1;q\rangle_k}\mathrm{QE}\big(Rr+Sk+r(1+S)\big)$$

$$=\frac{1}{\langle c;q\rangle_{R+S}}\sum_{k=0}^{R+S}\frac{\langle c-1;q\rangle_k\langle\frac{c+1}{2},\frac{c}{2};q^2\rangle_k}{\langle c+R+S;q\rangle_k\langle\frac{c-1}{2},\frac{c}{2};q^2\rangle_k}$$

$$\times\frac{\langle -S;q\rangle_k}{\langle 1;q\rangle_k}{}_2\phi_1\big(-k,-R;1+S-k|q;q^{1+S+R}\big)$$

$$=\frac{1}{\langle c;q\rangle_{R+S}}\sum_{k=0}^{R+S}\frac{\langle c-1;q\rangle_k\langle\frac{c+1}{2},\frac{c}{2};q^2\rangle_k}{\langle c+R+S;q\rangle_k\langle\frac{c-1}{2},\frac{c}{2};q^2\rangle_k}$$

$$\times\frac{\langle -S;q\rangle_k}{\langle 1;q\rangle_k}\Gamma_q\begin{bmatrix}1+S-k,1+S+R\\1+S,1+S-k+R\end{bmatrix}$$

$$=\frac{1}{\langle c;q\rangle_{R+S}}\sum_{k=0}^{R+S}\frac{\langle c-1;q\rangle_k\langle\frac{c+1}{2},\frac{c}{2};q^2\rangle_k}{\langle c+R+S;q\rangle_k\langle\frac{c-1}{2},\frac{c}{2};q^2\rangle_k}$$

$$\times \frac{\langle -S;q\rangle_k}{\langle 1;q\rangle_k}\frac{\langle 1+S;q\rangle_{-k}}{\langle 1+R+S;q\rangle_{-k}}$$

$$=\frac{1}{\langle c;q\rangle_{R+S}}\sum_{k=0}^{R+S}\frac{\langle c-1,-S-R,\frac{c+1}{2},\widetilde{\frac{c+1}{2}};q\rangle_k q^{k(S+R)}}{\langle c+R+S,1,\frac{c-1}{2},\widetilde{\frac{c-1}{2}};q\rangle_k}$$

$$=\frac{1}{\langle c;q\rangle_{R+S}}\,{}_4\phi_3\left(c-1,-R-S,\frac{c+1}{2},\widetilde{\frac{c+1}{2}};\right.$$

$$\left. c+R+S,\frac{c-1}{2},\widetilde{\frac{c-1}{2}}|q;q^{S+R}\right)$$

$$=\frac{1}{\langle c;q\rangle_{R+S}}\Gamma_q\begin{bmatrix} R+S,c+R+S,\frac{c+1}{2},\frac{c-1}{2} \\ \frac{c+1}{2}+R+S,\frac{c-1}{2}+R+S,0,c\end{bmatrix}. \tag{10.263}$$

The result is 0, except for $R+S=0$, since we divide by $\Gamma_q(0)=\infty$. Then the ${}_4\phi_3$ equals 1, and we get the product of Kronecker deltas in the lemma. □

Corollary 10.10.4

$$\sum_{m=0}^{s}\sum_{n=0}^{t}(-1)^{m+n}\binom{s}{m}_q\binom{t}{n}_q q^{\binom{m}{2}+\binom{n}{2}+mt}=\delta_{s,0}\delta_{t,0}. \tag{10.264}$$

We now put $b=-s$, $b'=-t$ in (10.126) to obtain formula (10.264):

$$\sum_{m=0}^{s}\sum_{n=0}^{t}(-1)^{m+n}\binom{s}{m}_q\binom{t}{n}_q \langle a;q\rangle_{m+n}\langle c+m+n;q\rangle_{s+t-m-n}$$

$$\times\,\mathrm{QE}\left(\binom{m}{2}+\binom{n}{2}+m(c-a)+n(c-a+s)\right)=\langle c-a;q\rangle_{s+t}. \tag{10.265}$$

Now multiply with $q^{a(s+t)}$ and apply D_{q,q^a}^{s+t} to both sides to obtain

$$\sum_{m=0}^{s}\sum_{n=0}^{t}\binom{s}{m}_q\binom{t}{n}_q \langle c+m+n;q\rangle_{s+t-m-n}$$

$$\times\,\mathrm{QE}\left(\binom{m+n}{2}\binom{m}{2}+\binom{n}{2}+c(m+n)+ns\right)=1. \tag{10.266}$$

Finally, apply D_{q,q^c}^{s+t} to both sides to obtain

$$\sum_{m=0}^{s}\sum_{n=0}^{t}(-1)^{m+n}\binom{s}{m}_q\binom{t}{n}_q \mathrm{QE}\left(\binom{m+n}{2}+\binom{m}{2}+\binom{n}{2}+ns\right)$$

$$\times\,\mathrm{QE}\left(\binom{s+t-m-n}{2}+(s+t-m-n)(m+n)\right)=0. \tag{10.267}$$

(The maximal torus in $SU(2)$ corresponds to the summation formulas (10.268) and (10.269).)

Corollary 10.10.5 *A generalization of q-Vandermonde II*:

$$\Phi_{\mathrm{D}}^{n}\big(b,-k_1,\ldots,-k_n;c|q;q^{1-k_2-\cdots-k_n},q^{1-k_3-\cdots-k_n},\ldots,q\big)$$
$$=\frac{\langle c-b;q\rangle_K}{\langle c;q\rangle_K}q^{bK},\quad \sum_{i=1}^{n}k_i=K. \tag{10.268}$$

A generalization of q-Vandermonde I:

$$\Phi_{\mathrm{D}}^{n}\big(b,-k_1,\ldots,-k_n;c|q;q^{c-b+K},q^{c-b+K-k_1},\ldots,q^{c-b+k_n}\big)$$
$$=\frac{\langle c-b;q\rangle_K}{\langle c;q\rangle_K}. \tag{10.269}$$

Proof Put

$$x=\mathrm{QE}(1-k_2-\cdots-k_n),\qquad C_N=\frac{\langle b;q\rangle_N}{\langle c;q\rangle_N},\qquad \alpha_l=-k_l \tag{10.270}$$

in (10.143) to obtain (10.268). To obtain (10.269), invert the basis q and use (6.24). □

Definition 174 If $\{x_j\}_{j=1}^n$ and $\{y_j\}_{j=1}^n$ are two arbitrary sequences of complex numbers, then their scalar product is defined by

$$\overline{xy}\equiv\sum_{j=1}^{n}x_jy_j. \tag{10.271}$$

Corollary 10.10.6

$$\sum_{\vec{m}=\vec{0}}^{\vec{k}}\binom{\vec{k}}{\vec{m}}_q(-1)^{\vec{m}}q^{\binom{\vec{m}}{2}}\mathrm{QE}\left(-\overline{km}+\sum_{j=1}^{n}m_j\left(1-\sum_{l=j+1}^{n}k_l\right)\right)=\delta_{\vec{k},\vec{0}}. \tag{10.272}$$

Proof Put $c=b$ in (10.268). □

Corollary 10.10.7 *Andrews* [47, p. 211], [171, p. 185, (108)], [421, p. 73, (31)], *a q-analogue of Lauricella* [354, p. 150]:

$$\Phi_{\mathrm{D}}^{n}\big(\beta,\alpha_1,\ldots,\alpha_n;\gamma|q;q^{\gamma-\alpha-\beta+\alpha_2+\cdots+\alpha_n},q^{\gamma-\alpha-\beta+\alpha_3+\cdots+\alpha_n},\ldots,q^{\gamma-\alpha-\beta}\big)$$
$$=\Gamma_q\begin{bmatrix}\gamma,\gamma-\alpha-\beta\\ \gamma-\alpha,\gamma-\beta\end{bmatrix},\quad \Re(\gamma-\alpha-\beta)>0,\ \sum_{i=1}^{n}\alpha_i=\alpha. \tag{10.273}$$

Proof Put

$$x = \mathrm{QE}(\gamma - \alpha_1 - \beta), \qquad C_N = \frac{\langle\beta; q\rangle_N}{\langle\gamma; q\rangle_N} \tag{10.274}$$

in (10.143). □

Corollary 10.10.8 *A generalization of* (10.264) *and* (10.267):

$$\sum_{\vec{m}=\vec{0}}^{\vec{k}} \binom{\vec{k}}{\vec{m}}_q (-1)^{\vec{m}} q^{\binom{\vec{m}}{2}} \mathrm{QE}\left(-\overline{km} + \sum_{j=1}^{n} m_j \left(\sum_{l=1}^{j} k_l\right)\right) = \delta_{\vec{k},\vec{0}}. \tag{10.275}$$

Proof Put $\alpha_l = -k_l$, $\gamma = \beta$ in (10.273). □

The formulas (10.272) and (10.275) are the two multiple extensions of the Gauß formula.

10.11 An expansion formula in the spirit of Chaundy

We are now going to prove an expansion formula in the spirit of Chaundy with the aid of (10.262).

Theorem 10.11.1

$$\begin{aligned}
&\Phi_3\big(A, A'; B, B'; C|q; x, y\big) \\
&\quad = \sum_{r,s=0}^{\infty} (-1)^{r+s} q^{\binom{s}{2}+\binom{r}{2}} \\
&\qquad \times \frac{\langle a, b; q\rangle_r \langle a', b'; q\rangle_s \mathrm{QE}(sr - ms - nr + mn)}{\langle 1; q\rangle_r \langle 1; q\rangle_s \langle c + r + s - 1; q\rangle_{r+s}} \\
&\qquad \times \Phi_{1:2}^{1:3}\left[\begin{matrix} c + r + s - 1 : A, B, -r; A', B', -s \\ C : a, b; a', b' \end{matrix} \Big| q; q, q\right] x^r y^s \\
&\qquad \times \Phi_3\big(a + r, a' + s; b + r, b' + s; c + 2r + 2s|q; x, yq^{r-m}\big).
\end{aligned} \tag{10.276}$$

Proof The coefficient for

$$\frac{\langle A, B; q\rangle_m \langle A', B'; q\rangle_n}{\langle C; q\rangle_{m+n} \langle 1, a, b; q\rangle_m \langle 1, a', b'; q\rangle_n}$$

on the RHS is granted absolute convergence:

$$\sum_{r=m}^{\infty}\sum_{s=n}^{\infty}(-1)^{r+s}\mathrm{QE}\left(\binom{r}{2}+\binom{s}{2}+sr-ms-nr+mn+m+n\right)$$

$$\times\frac{\langle c+r+s-1;q\rangle_{m+n}\langle a,b;q\rangle_r\langle a',b';q\rangle_s}{\langle 1;q\rangle_r\langle 1;q\rangle_s\langle c+r+s-1;q\rangle_{r+s}}\langle -r;q\rangle_m$$

$$\times\langle -s;q\rangle_n x^r y^s\Phi_3\big(a+r,a'+s;b+r,b'+s;c+2r+2s|q;x,yq^{r-m}\big)$$

$$=\sum_{R_1,R_2=0}^{\infty}\sum_{r=m}^{R_1}\sum_{s=n}^{R_2}\frac{\langle a,b;q\rangle_{R_1}\langle a',b';q\rangle_{R_2}}{\langle c+r+s-1;q\rangle_{r+s}}x^{R_1}y^{R_2}\frac{\langle c+r+s-1;q\rangle_{m+n}}{\langle c+2r+2s;q\rangle_{R_1+R_2-r-s}}$$

$$\times\frac{(-1)^{r+s+m+n}\mathrm{QE}(\binom{r}{2}+\binom{s}{2}+\binom{m}{2}+\binom{n}{2}+m(1-r-R_2+n)+n(1-s-r)+R_2r)}{\langle 1;q\rangle_{r-m}\langle 1;q\rangle_{s-n}\langle 1;q\rangle_{R_1-r}\langle 1;q\rangle_{R_2-s}}$$

$$=\sum_{R_1,R_2=0}^{\infty}\sum_{r'=0}^{R_1-m}\sum_{s'=0}^{R_2-n}\langle a,b;q\rangle_{R_1}\langle a',b';q\rangle_{R_2}\langle c+r'+m+s'+n-1;q\rangle_{m+n}x^{R_1}y^{R_2}$$

$$\times\frac{(-1)^{r'+s'}\mathrm{QE}(\binom{r'}{2}+\binom{s'}{2}+r'(R_2-n))}{\langle 1;q\rangle_{r'}\langle 1;q\rangle_{s'}\langle 1;q\rangle_{R_1-r'-m}\langle 1;q\rangle_{R_2-s'-n}\langle c+2r'+2m+2s'+2n;q\rangle_{R_1+R_2-r'-s'-m-n}}$$

$$\times\frac{1}{\langle c+r'+m+s'+n-1;q\rangle_{r'+s'+m+n}}$$

$$\stackrel{\text{by }(10.262)}{=}\langle a,b;q\rangle_m\langle a',b';q\rangle_n x^m y^n. \tag{10.277}$$

□

10.12 Formulas according to Burchnall-Chaundy and Jackson

We now present some q-analogues of hypergeometric equations in Burchnall-Chaundy II [88]. These are mostly formal identities, but we still write equality signs. Whenever possible, the new notation for q-hypergeometric series will be used. The first formula is a q-analogue of Burchnall and Chaundy [90, p. 114, (15)]:

Theorem 10.12.1 [167]

$$\Phi_1\big(a;b,b';c|q;x,xq^{-b'}\big)={}_2\phi_1\big(a,b+b';c|q;xq^{-b'}\big). \tag{10.278}$$

By (10.222) *we obtain a corrected version of* [307, (55), p. 79]. *This is a q-analogue of an addition formula in the arguments b, b' and a q-analogue of* [90, p. 114, (16)]:

$$ {}_2\phi_1\big(a,b+b';c|q;xq^{-b'}\big)$$

$$=\sum_{r=0}^{\infty}\frac{\langle c-a,a,b,b';q\rangle_r}{\langle 1,c+r-1;q\rangle_r\langle c;q\rangle_{2r}}x^{2r}q^{-rb'+ra+r(r-1)}$$

$$\times{}_2\phi_1(a+r,b+r;c+2r|q;x)\,{}_2\phi_1\big(a+r,b'+r;c+2r|q;xq^{-b'}\big). \tag{10.279}$$

Theorem 10.12.2 *Jackson* [307, (56), p. 79], *a q-analogue of* [90, p. 114, (17)]. *By* (10.225) *we obtain*

$$
\begin{aligned}
&{}_2\phi_1(a,b;c|q;x)\,{}_2\phi_1\big(a,b';c|q;xq^{-b'}\big)\\
&\quad=\sum_{r=0}^{\infty}\frac{(-1)^r\langle a,b,b',c-a;q\rangle_r}{\langle 1,c;q\rangle_r\langle c;q\rangle_{2r}}x^{2r}\\
&\quad\times q^{-rb'+ra+\binom{r}{2}}\,{}_2\phi_1\big(a+r,b+b'+2r;c+2r|q;xq^{-b'}\big).
\end{aligned}
\tag{10.280}
$$

In this equation we put successively

1. $c=b+b'$,
2. $b=\frac{c+1}{2}$, $b'=\frac{c-1}{2}$,
3. $b=b'=\frac{c}{2}$.

This yields the following three equations. Compare Jackson [307, (57)–(58), p. 79].

Theorem 10.12.3

$$
\begin{aligned}
&{}_2\phi_1\big(a,b;b+b'|q;x\big){}_2\phi_1\big(a,b';b+b'|q;xq^{-b'}\big)\\
&\quad=\sum_{r=0}^{\infty}\frac{(-1)^r\langle a,b,b',b+b'-a;q\rangle_r}{\langle 1,b+b';q\rangle_r\langle b+b';q\rangle_{2r}}x^{2r}\frac{q^{-rb'+ra+\binom{r}{2}}}{(xq^{-b'};q)_{a+r}}\\
&\quad=\frac{1}{(xq^{-b'};q)_a}\sum_{r=0}^{\infty}\frac{(-1)^r\langle a,b,b',b+b'-a;q\rangle_r}{\langle 1,b+b',\frac{b+b'}{2},\frac{b+b'+1}{2},\widetilde{\frac{b+b'}{2}},\widetilde{\frac{b+b'+1}{2}};q\rangle_r}\frac{x^{2r}q^{-rb'+ra+\binom{r}{2}}}{(xq^{a-b'};q)_r}\\
&\quad\equiv\frac{1}{(xq^{-b'};q)_a}{}_6\phi_6\left[\begin{matrix}a,b,b',b+b'-a,\infty,\infty\\ b+b',\triangle(q;2;b+b')\end{matrix}\,|q;x^2q^{a-b'}\,||\,\frac{\cdot}{(xq^{a-b'};q)_k}\right],
\end{aligned}
\tag{10.281}
$$

$$
\begin{aligned}
&{}_2\phi_1\left(a,\frac{c+1}{2};c|q;x\right){}_2\phi_1\left(a,\frac{c-1}{2};c|q;xq^{\frac{1-c}{2}}\right)\\
&\quad=\sum_{r=0}^{\infty}\frac{(-1)^r\langle a,c-a,\frac{c+1}{2},\frac{c-1}{2};q\rangle_r x^{2r}q^{r\frac{1-c}{2}+ra+\binom{r}{2}}}{\langle 1,c;q\rangle_r\langle c;q\rangle_{2r}(xq^{\frac{1-c}{2}};q)_{a+r}}\\
&\quad=\frac{1}{(xq^{\frac{1-c}{2}};q)_a}{}_5\phi_5\left[\begin{matrix}a,c-a,\frac{c-1}{2},\infty,\infty\\ c,\frac{c}{2},\widetilde{\frac{c+1}{2}},\widetilde{\frac{c}{2}}\end{matrix}\,|q;x^2q^{a+\frac{1-c}{2}}\,||\,\frac{\cdot}{(xq^{a+\frac{1-c}{2}};q)_k}\right],
\end{aligned}
\tag{10.282}
$$

$$
{}_2\phi_1\left(a,\frac{c}{2};c|q;x\right){}_2\phi_1\left(a,\frac{c}{2};c|q;xq^{-\frac{c}{2}}\right)
$$
$$
=\sum_{r=0}^{\infty}\frac{(-1)^r\langle a,c-a,\frac{c}{2},\frac{c}{2};q\rangle_r x^{2r}q^{-r\frac{c}{2}+ra+\binom{r}{2}}}{\langle 1,c;q\rangle_r\langle c;q\rangle_{2r}(xq^{-\frac{c}{2}};q)_{a+r}}
$$
$$
=\frac{1}{(xq^{\frac{-c}{2}};q)_a}{}_5\phi_5\left[\begin{matrix}a,c-a,\frac{c}{2},\infty,\infty\\ c,\frac{c+1}{2},\widetilde{\frac{c}{2}},\widetilde{\frac{c+1}{2}}\end{matrix}|q;x^2q^{a-\frac{c}{2}}||_{(xq^{a-\frac{c}{2}};q)_k}^{\cdot}\right]. \tag{10.283}
$$

As a special case consider $b=b'$, and put $y=xq^{-b'}$ in (10.240), and use (10.278) to get

$$
{}_2\phi_1\big(a,b;c|q;x\big(1\oplus_q q^{-b}\big)\big)
=\sum_{r=0}^{\infty}\frac{\langle b;q\rangle_r\langle a;q\rangle_{2r}}{\langle 1;q\rangle_r\langle c;q\rangle_{2r}}x^{2r}q^{r(r-1)}
\times {}_2\phi_1\big(a+2r,2b+2r;c+2r|q;xq^{-b}\big). \tag{10.284}
$$

We can easily find q-analogues of some further expansions in Burchnall-Chaundy I.

Theorem 10.12.4 *A q-analogue of* [89, p. 256, (44)].

$$
{}_2\phi_1(a,b;c|q;(x\oplus_q y\boxminus_q xy)
=\sum_{r=0}^{\infty}\frac{(-xy)^r\langle a,b;q\rangle_r}{\langle 1,c;q\rangle_r}q^{\binom{r}{2}}\,{}_2\phi_1\big(a+r,b+r;c+r|q;q^{-r}(x\oplus_q y)\big). \tag{10.285}
$$

Proof This can also be written in the form

$$
\sum_{r=0}^{\infty}(-1)^r\frac{\langle a,b;q\rangle_r x^r y^r q^{\binom{r}{2}}}{\langle 1,c;q\rangle_r}\sum_{k=0}^{\infty}\frac{\langle a+r,b+r;q\rangle_k}{\langle 1,c+r;q\rangle_k}\sum_{s=0}^{k}\binom{k}{s}_q x^s y^{k-s}
$$
$$
=\sum_{l=0}^{\infty}\frac{\langle a,b;q\rangle_l}{\langle 1,c;q\rangle_l}\sum_{t=0}^{l}x^t y^{l-t}\binom{l}{t}_q
\times\sum_{m=0}^{\infty}\frac{\langle 1-c-l,-t,t-l;q\rangle_m}{\langle 1,1-a-l,1-b-l;q\rangle_m}q^{m(-a+1-b+c)}, \tag{10.286}
$$

$$\text{LHS} = \sum_{r,k=0}^{\infty}\sum_{s=0}^{k}(-1)^r \frac{\langle a,b;q\rangle_{r+k}}{\langle 1;q\rangle_r \langle c;q\rangle_{r+k}} \frac{x^{r+s}y^{r+k-s}q^{\binom{r}{2}}}{\langle 1;q\rangle_{k-s}\langle 1;q\rangle_s},$$

$$\text{RHS} = \sum_{m,l=0}^{\infty}\sum_{t=0}^{l} \frac{\langle a,b;q\rangle_{l-m}}{\langle c;q\rangle_{l-m}} \frac{\langle -t,t-l;q\rangle_m x^t y^{l-t}(-1)^m q^{-\binom{m}{2}}}{\langle 1;q\rangle_t \langle 1;q\rangle_m \langle 1;q\rangle_{l-t}} q^{ml}$$

$$= \sum_{m,l=0}^{\infty}\sum_{t=0}^{l} \frac{\langle a,b;q\rangle_{l-m}}{\langle c;q\rangle_{l-m}} \frac{x^t y^{l-t}(-1)^m q^{\binom{m}{2}}}{\langle 1;q\rangle_m \langle 1;q\rangle_{t-m}\langle 1;q\rangle_{l-t-m}}. \tag{10.287}$$

□

Let us show that the previous proof can be made in a more systematic way. We can write the two first q-Taylor series in the forms

$$\begin{aligned}
&{}_2\phi_1(a,b;c|q;(x \ominus_q h)) \\
&= \sum_{r=0}^{\infty} \frac{(-h)^r \langle a,b;q\rangle_r}{\langle 1,c;q\rangle_r} {}_2\phi_1(a+r,b+r;c+r|q;x) \\
&= \sum_{r=0}^{\infty} \frac{h^r x^{-r}\langle -\theta_{q,1};q\rangle_r q^{-\binom{r}{2}}}{\langle 1;q\rangle_r} {}_2\phi_1\big(a,b;c|q;xq^r\big),
\end{aligned} \tag{10.288}$$

$$\begin{aligned}
&{}_2\phi_1(a,b;c|q;(x \boxminus_q h)) \\
&= \sum_{r=0}^{\infty} \frac{(-h)^r \langle a,b;q\rangle_r q^{\binom{r}{2}}}{\langle 1,c;q\rangle_r} {}_2\phi_1(a+r,b+r;c+r|q;x) \\
&= \sum_{r=0}^{\infty} \frac{h^r x^{-r}\langle -\theta_{q,1};q\rangle_r}{\langle 1;q\rangle_r} {}_2\phi_1\big(a,b;c|q;xq^r\big).
\end{aligned} \tag{10.289}$$

The following two formulas occurred in slightly different form in Jackson [307, p. 78].

By replacing the function arguments x and h by $x \oplus_q y$ and xy, we get the following second q-analogue of [89, (44), p. 256]:

$$\begin{aligned}
&{}_2\phi_1(a,b;c|q;(x \oplus_q y \ominus_q xy) \\
&= \sum_{r=0}^{\infty} \frac{(-xy)^r \langle a,b;q\rangle_r}{\langle 1,c;q\rangle_r} {}_2\phi_1(a+r,b+r;c+r|q;x \oplus_q y).
\end{aligned} \tag{10.290}$$

We can rewrite (10.290) symbolically as

$$
\begin{aligned}
&{}_2\phi_1(a,b;c|q;(x \oplus_q y \ominus_q xy)\\
&\quad = {}_4\phi_2\left[\begin{matrix} 1-c-\theta_{q,1}-\theta_{q,2}, -\theta_{q,1}, -\theta_{q,2}, \infty \\ 1-a-\theta_{q,1}-\theta_{q,2}, 1-b-\theta_{q,1}-\theta_{q,2} \end{matrix} |q; -q^{-a-b+c+1}\right]\\
&\qquad \times {}_2\phi_1\big(a,b;c|q;(x \oplus_q y)\big). \qquad (10.291)
\end{aligned}
$$

By replacing the function arguments x and h by $x \oplus_q y \ominus_q xy$ and $-xy$, we get the following q-analogue of [89, (45), p. 256]:

$$
\begin{aligned}
&{}_2\phi_1\big(a,b;c|q;(x \oplus_q y)\big)\\
&\quad = \sum_{r=0}^{\infty} \frac{(xy)^r \langle a,b;q\rangle_r q^{\binom{r}{2}}}{\langle 1,c;q\rangle_r} {}_2\phi_1(a+r,b+r;c+r|q;x \oplus_q y \ominus_q xy). \qquad (10.292)
\end{aligned}
$$

According to Jackson [307, p. 78], it is not possible to find q-analogues of the formulas [89, (46)–(51), p. 256].

Chapter 11
Linear partial q-difference equations

The differential equations correspond to the q-difference equations.

$A(D) = 0$	$A(\mathrm{D}_q) = 0$

11.1 Introduction

The purpose of this chapter is to introduce some new notions which are used to solve linear partial q-difference equations of n variables. The functions of interest will be denoted by

$$f(x_1, \ldots, x_n). \tag{11.1}$$

We introduce the following operators, $j, k = 1, \ldots, n$.

$$[\epsilon_j f](x_1, \ldots, x_n) \equiv f(x_1, \ldots, q_j x_j, \ldots, x_n), \tag{11.2}$$

$$[\mathrm{M}_{x_j} f](x_1, \ldots, x_n) \equiv x_j f(x_1, \ldots, x_n), \tag{11.3}$$

$$[\partial_{x_j, q_j} f](x_1, \ldots, x_n) \equiv \frac{f(x_1, \ldots, x_n) - f(x_1, \ldots, q_j x_j, \ldots, x_n)}{(1 - q_j)x_j}, \tag{11.4}$$

with the following relations:

$$\epsilon_j = \mathrm{I} - (1 - q_j)\mathrm{M}_{x_j} \partial_{x_j, q_j}, \tag{11.5}$$

$$\partial_{x_j, q_j} \mathrm{M}_{x_j} = \mathrm{I} + q_j \mathrm{M}_{x_j} \partial_{x_j, q_j}. \tag{11.6}$$

We deduce that

$$\epsilon_j = \partial_{x_j, q_j} \mathrm{M}_{x_j} - \mathrm{M}_{x_j} \partial_{x_j, q_j}. \tag{11.7}$$

It is obvious that

$$\begin{aligned} \left[\epsilon_j^m \epsilon_k^n f\right](x_1, \ldots, x_n) &= \left[\epsilon_k^n \epsilon_j^m f\right](x_1, \ldots, x_n) \\ &= f\left(x_1, \ldots, q_j^m x_j, \ldots, q_k^n x_k, \ldots, x_n\right), \quad j < k, \end{aligned} \tag{11.8}$$

T. Ernst, *A Comprehensive Treatment of q-Calculus*,
DOI 10.1007/978-3-0348-0431-8_11, © Springer Basel 2012

or

$$\begin{aligned}&\big[(\partial_{x_j,q_j}\mathrm{M}_{x_j}-\mathrm{M}_{x_j}\partial_{x_j,q_j})^m(\partial_{x_k,q_k}\mathrm{M}_{x_k}-\mathrm{M}_{x_k}\partial_{x_k,q_k})^n f\big](x_1,\ldots,x_n)\\&\quad=f\big(x_1,\ldots,q_j^m x_j,\ldots,q_k^n x_k,\ldots,x_n\big),\quad j<k.\end{aligned}\tag{11.9}$$

11.2 Canonical equations and symmetry techniques for q-series (Kalnins, Miller)

The q-algebra approach, due to Floreanini and Vinet, Kalnins and Miller, and Feinsilver uses only the q-analogue of the Lie algebra. The following sections are based on the article by A. K. Agarwal, Kalnins and Miller [8]. We have changed the notation slightly in accordance with the new notation, which is introduced in this paper. Based on a work by Thomae of 1871 [499], we define a ${}_r\Phi_s$ q-hypergeometric series by

Definition 175

$${}_r\Phi_s(a_1,\ldots,a_r;b_1,\ldots,b_s|q;z)\equiv\sum_{n=0}^{\infty}\frac{\langle a_1;q\rangle_n\cdots\langle a_r;q\rangle_n}{\langle 1;q\rangle_n\langle b_1;q\rangle_n\cdots\langle b_s;q\rangle_n}z^n.\tag{11.10}$$

Remark 57 This definition for a q-hypergeometric series differs from the earlier one (7.1), which was based on [220]. This is the reason for changing notation from ϕ to Φ!

Let ϵ_u be the q-dilation operator corresponding to the variable u, i.e., ϵ_u maps a function f of the variables $u,v,w,\ldots$ into the function

$$\epsilon_u f(u,v,w,\ldots)\equiv f(qu,v,w,\ldots).\tag{11.11}$$

We can verify the following recurrence relations for the q-series (11.10):

$$\begin{aligned}&\big(\mathrm{I}-q^{a_k}\big(\mathrm{I}-(1-q)\mathrm{M}_z\partial_{z,q}\big)\big)\,{}_r\Phi_s(a_i;b_j|q;z)\\&\quad=\big(1-q^{a_k}\big)\,{}_r\Phi_s\big(e^k a_i;b_j|q;z\big),\quad 1\le k\le r,\end{aligned}\tag{11.12}$$

$$\begin{aligned}&\big(\mathrm{I}-q^{b_l-1}\big(\mathrm{I}-(1-q)\mathrm{M}_z\partial_{z,q}\big)\big)\,{}_r\Phi_s(a_i;b_j|q;z)\\&\quad=\big(1-q^{b_l-1}\big)\,{}_r\Phi_s(a_i;e_l b_j|q;z),\quad 1\le l\le s,\end{aligned}\tag{11.13}$$

$$\begin{aligned}&z^{-1}\big(\mathrm{I}-\big(\mathrm{I}-(1-q)\mathrm{M}_z\partial_{z,q}\big)\big)\,{}_r\Phi_s(a_i;b_j|q;z)\\&\quad=\frac{\prod_{j=1}^r(1-q^{a_j})}{\prod_{j=1}^s(1-q^{b_j})}\,{}_r\Phi_s(a_i+1;b_j+1|q;z),\end{aligned}\tag{11.14}$$

where

$$e^k a_i = \begin{cases} a_i, & \text{if } i \neq k; \\ a_i + 1, & \text{if } i = k, \end{cases} \tag{11.15}$$

$$e_l b_j = \begin{cases} b_j, & \text{if } j \neq l; \\ b_j - 1, & \text{if } j = l. \end{cases} \tag{11.16}$$

These relations have the following form in [8]:

$$\left(\mathrm{I} - q^{a_k}\epsilon_z\right) {}_r\Phi_s(a_i; b_j|q; z) = \left(1 - q^{a_k}\right) {}_r\Phi_s\left(e^k a_i; b_j|q; z\right),$$
$$1 \le k \le r, \tag{11.17}$$

$$\left(\mathrm{I} - q^{b_l - 1}\epsilon_z\right) {}_r\Phi_s(a_i; b_j|q; z) = \left(1 - q^{b_l - 1}\right) {}_r\Phi_s(a_i; e_l b_j|q; z),$$
$$1 \le l \le s, \tag{11.18}$$

$$z^{-1}(\mathrm{I} - \epsilon_z) {}_r\Phi_s(a_i; b_j|q; z) = \frac{\prod_{j=1}^r (1 - q^{a_j})}{\prod_{j=1}^s (1 - q^{b_j})} {}_r\Phi_s(a_i + 1; b_j + 1|q; z). \tag{11.19}$$

The relations (11.12)–(11.14) imply the fundamental q-difference equation satisfied by the ${}_r\Phi_s$:

$$\left(z \prod_{j=1}^{r} \left(\mathrm{I} - q^{a_j}\left(\mathrm{I} - (1-q)\mathrm{M}_z \partial_{z,q}\right)\right) - \left((1-q)\mathrm{M}_z \partial_{z,q}\right) \right.$$
$$\left. \times \prod_{j=1}^{s} \left(\mathrm{I} - q^{b_j - 1}\left(\mathrm{I} - (1-q)\mathrm{M}_z \partial_{z,q}\right)\right)\right) {}_r\Phi_s(a_i; b_j|q; z) = 0. \tag{11.20}$$

For $b_j \neq 0, -1, -2, \ldots$ the only solution of this equation which is analytic in z at $z = 0$ is ${}_r\Phi_s(a_i; b_j|q; z)$. This equation has the following form in [8]:

$$\left(z \prod_{j=1}^{r} \left(\mathrm{I} - q^{a_j}\epsilon_z\right) - (\mathrm{I} - \epsilon_z) \prod_{j=1}^{s} \left(\mathrm{I} - q^{b_j - 1}\epsilon_z\right)\right) {}_r\Phi_s(a_i; b_j|q; z) = 0. \tag{11.21}$$

We want to show that the formulas (6.185) ($p = 2$) and (11.21) ($r = 2$, $s = 1$) give the same result. Indeed, (6.185) gives

$$(1 - x) + \left(-q^{b_1 - 1} - 1 + \left(q^{a_1} + q^{a_2}\right)x\right)\epsilon + \left(q^{b_1 - 1} - q^{a_1 + a_2}x\right)\epsilon^2 = 0, \tag{11.22}$$

while (11.21) gives

$$-x\left(1 - q^{a_1}\epsilon\right)\left(1 - q^{a_2}\epsilon\right) + (1 - \epsilon)\left(1 - q^{b_1 - 1}\epsilon\right) = 0. \tag{11.23}$$

The two expressions coincide.

Definition 176 Next we define the function ${}_r\Psi_s$ of $2(r+s)+1$ variables by

$$ {}_r\Psi_s(a_i;b_j|q;u_p) \equiv {}_r\Phi_s\left(a_i;b_j|q;\frac{\prod_{j=0}^{s} u_{r+1+j}}{\prod_{j=1}^{r} u_j}\right)\prod_{j=1}^{r} u_j^{-a_j}\prod_{j=1}^{s} u_{r+j}^{b_j-1}. \tag{11.24} $$

Let $\Delta_p^{\pm}$ be the q-difference operators

$$ \begin{aligned} \Delta_p^{+} &\equiv u_p^{-1}(\mathrm{I}-\epsilon_{u_p}) = (1-q)\partial_{u_p,q}, \\ \Delta_p^{-} &\equiv u_p^{-1}(\mathrm{I}-\epsilon_{u_p}^{-1}) = (1-q^{-1})\partial_{u_p,q^{-1}}, \quad 1 \le p \le r+s+1. \end{aligned} \tag{11.25} $$

In terms of these operators, Eqs. (11.12)–(11.14) take the simple forms

$$ \Delta_k^{-}\,{}_r\Psi_s = (1-q^{-1})\partial_{u_k,q^{-1}}\,{}_r\Psi_s = (1-q^{a_k})\,{}_r\Psi_s(e^k a_i;b_j|q;z), \quad 1 \le k \le r, \tag{11.26} $$

$$ \Delta_{r+l}^{+}\,{}_r\Psi_s = (1-q)\partial_{u_{r+l},q}\,{}_r\Psi_s = (1-q^{b_l-1})\,{}_r\Psi_s(a_i;e_l b_j|q;z), \quad 1 \le l \le s, \tag{11.27} $$

$$ \begin{aligned} \Delta_{r+s+1}^{+}\,{}_r\Psi_s &= (1-q)\partial_{u_{r+s+1},q}\,{}_r\Psi_s \\ &= \frac{\prod_{j=1}^{r}(1-q^{a_j})}{\prod_{j=1}^{s}(1-q^{b_j})}\,{}_r\Psi_s(a_i+1;b_j+1|q;z). \end{aligned} \tag{11.28} $$

The following canonical partial q-difference equation holds:

$$ \begin{aligned} &\left(\prod_{j=1}^{r}\Delta_j^{-} - \prod_{k=r+1}^{r+s+1}\Delta_k^{+}\right){}_r\Psi_s \\ &= \left(\prod_{j=1}^{r}(1-q^{-1})\partial_{u_j,q^{-1}} - \prod_{k=r+1}^{r+s+1}(1-q)\partial_{u_k,q}\right){}_r\Psi_s = 0. \end{aligned} \tag{11.29} $$

Furthermore, ${}_r\Psi_s$ satisfies the eigenvalue equations

$$ \begin{aligned} &\left(\mathrm{I}-(1-q)\mathrm{M}_{u_{r+s+1}}\partial_{u_{r+s+1},q}\right)^{-1}\left(\mathrm{I}-(1-q)\mathrm{M}_{u_k}\partial_{u_k,q}\right)^{-1}{}_r\Psi_s \\ &= \epsilon_{r+s+1}^{-1}\epsilon_k^{-1}\,{}_r\Psi_s = q^{a_k}\,{}_r\Psi_s, \quad 1 \le k \le r, \end{aligned} \tag{11.30} $$

$$ \begin{aligned} &\left(\mathrm{I}-(1-q)\mathrm{M}_{u_{r+s+1}}\partial_{u_{r+s+1},q}\right)^{-1}\left(\mathrm{I}-(1-q)\mathrm{M}_{u_{r+l}}\partial_{u_{r+l},q}\right){}_r\Psi_s \\ &= \epsilon_{r+s+1}^{-1}\epsilon_{r+l}\,{}_r\Psi_s = q^{b_l-1}\,{}_r\Psi_s, \quad 1 \le l \le s. \end{aligned} \tag{11.31} $$

The function ${}_r\Psi_s$ is characterized by (11.29), (11.30) and (11.31): it is the only solution of these equations analytic in the u_p at $u_{r+s+1}=0$, up to a constant multiple.

An analytic solution $\Theta(u_p)$ of the canonical equation

$$\left(\prod_{j=1}^{r}\Delta_j^- - \prod_{k=r+1}^{r+s+1}\Delta_k^+\right)\Theta$$
$$=\left(\prod_{j=1}^{r}(1-q^{-1})\partial_{u_j,q^{-1}} - \prod_{k=r+1}^{r+s+1}(1-q)\partial_{u_k,q}\right)\Theta=0 \qquad (11.32)$$

is a generating function for q-hypergeometric functions. We will compute such a generating function Θ by characterizing it as a joint eigenfunction of a set of $r+s$ commuting symmetry operators for (11.32). A symmetry operator for the canonical equation is a linear operator L, which maps any local analytic solution Θ for (11.32) into another local analytic solution $L\Theta$.

The dilation operators $\epsilon_{r+s+1}^{-1}\epsilon_k^{-1}$, $1\leq k\leq r$, $\epsilon_{r+s+1}^{-1}\epsilon_{r+l}$, $1\leq l\leq s$ are commuting symmetries and the eigenvalue equations (11.30), (11.31) characterize the basis solutions ${}_r\Psi_s$ in terms of these symmetries. Furthermore the q-difference operators Δ_j^-, $1\leq j\leq r$ and Δ_{r+h}^+, $1\leq h\leq s+l$ are commuting symmetries. Any permutation of the variables $u_j : 1\leq j\leq r$ is a symmetry of (11.32), as is any permutation of the variables $\{u_{r+h}\} : 1\leq h\leq s+1$.

The canonical equation (11.32) for q-hypergeometric functions is a q-analogue of the canonical equation for the hypergeometric functions ${}_r\mathrm{F}_s$ (see Miller [379]).

We conclude by giving some examples of q-algebras and their representations. All these are q-analogues of the Lie algebra representations from Miller [382].

The q-algebra $Su_q(1,1)$ has a representation whose matrix elements can be expressed by q-Gegenbauer polynomials [213]. The two-dimensional Euclidean quantum algebra, the symmetry algebra of the q-Helmholtz equation, has a representation whose matrix elements can be expressed by q-Bessel functions [211], [212]. The q-oscillator algebra has a representation whose matrix elements can be expressed by q-Laguerre polynomials and q-Hermite polynomials [322], [323], [212]. The q-algebra $sl_q(2)$ has a representation whose matrix elements can be expressed by Stieltjes–Wigert polynomials [212]. The four-dimensional Euclidean quantum algebra $U_q(sl(2))$ has a representation whose matrix elements can be expressed by continuous q-Jacobi polynomials [216]. The 6-dimensional q-Schrödinger algebra, the symmetry algebra of the 2-dimensional q-heat equation, has a representation whose matrix elements can be expressed by q-Bessel functions and q-Hermite polynomials [215]. The 9-dimensional q-Schrödinger algebra, the symmetry algebra of the 3-dimensional q-heat equation written in complex light-cone coordinates in the spatial directions, has a representation whose matrix elements can be expressed by q-Laguerre polynomials [215]. The quantum algebra $U_q(so(5))$, the symmetry algebra of the 3-dimensional q-wave equation written in light-cone coordinates, has a representation whose matrix elements can be expressed by ${}_3\phi_2$ series [214]. The quantum algebra $U_q(sl(4))$, the symmetry algebra of the 4-dimensional q-wave equation written in light-cone coordinates, has a representation whose matrix elements can be expressed by ${}_2\phi_1$ series [214].

11.3 q-difference equations for q-Appell and q-Lauricella functions

In this section we are going to find q-difference equations for q-Appell and q-Lauricella functions. So as a preliminary lemma we need the Heine q-difference equation for a ${}_2\phi_1$ q-hypergeometric series (6.182).

Some of the following equations have appeared in different form and different notation in Jackson [307, p. 79–80]. These equations have multiple forms, but there is always a simplest form, which is called canonical.

Theorem 11.3.1 *In corrected form, the partial q-difference equations for the q-Appell functions*

$$\Phi_1(a;b,b';c|q;x_1,x_2),\qquad \Phi_2(a;b,b';c,c'|q;x_1,x_2),$$
$$\Phi_3(a,a';b,b';c|q;x_1,x_2),\qquad \Phi_4(a;b;c,c'|q;x_1,x_2)$$

read

$$\begin{aligned}&x_1\big(q^c-x_1q^{a+b+1}\big)\epsilon_2\mathrm{D}^2_{q,1,1}\\&\quad+x_2\big[q^c+x_1\big(q^a-q^{a+b}-q^{a+b+1}\big)\big]\mathrm{D}^2_{q,1,2}-\{b\}_q q^a x_2\mathrm{D}_{q,2}\\&\quad+\big[\{c\}_q-\big(\{a\}_q q^b+\{b\}_q q^a+q^{a+b}\big)x_1\big]\mathrm{D}_{q,1}-\{a\}_q\{b\}_q\mathrm{I}=0,\end{aligned}\tag{11.33}$$

$$\begin{aligned}&x_1\big(q^c-x_1q^{a+b+1}\epsilon_2\big)\mathrm{D}^2_{q,1,1}+x_1x_2\big(q^a-q^{a+b}-q^{a+b+1}\big)\mathrm{D}^2_{q,1,2}-\{b\}_q q^a x_2\mathrm{D}_{q,2}\\&\quad+\big[\{c\}_q-\big(\{a\}_q q^b+\{b\}_q q^a+q^{a+b}\big)x_1\big]\mathrm{D}_{q,1}-\{a\}_q\{b\}_q\mathrm{I}=0,\end{aligned}\tag{11.34}$$

$$\begin{aligned}&x_1\big(q^c\epsilon_2-x_1q^{a+b+1}\big)\mathrm{D}^2_{q,1,1}+x_2q^c\mathrm{D}^2_{q,1,2}\\&\quad+\big[\{c\}_q-\big(\{a\}_q q^b+\{b\}_q q^a+q^{a+b}\big)x_1\big]\mathrm{D}_{q,1}-\{a\}_q\{b\}_q\mathrm{I}=0,\end{aligned}\tag{11.35}$$

$$\begin{aligned}&x_1\big(q^c-x_1q^{a+b+1}\epsilon_2^2\big)\mathrm{D}^2_{q,1,1}-2q^{a+b}\epsilon_2x_1x_2\mathrm{D}^2_{q,1,2}\\&\quad-\big[\{a\}_q q^b+\{b\}_q q^a+q^{a+b-1}\epsilon_2\big]x_2\mathrm{D}_{q,2}\\&\quad+\big[\{c\}_q-\epsilon_2\big(\{a\}_q q^b+\{b\}_q q^a+\epsilon_2q^{a+b}\big)x_1\big]\mathrm{D}_{q,1}\\&\quad-q^{a+b}x_2^2\mathrm{D}^2_{q,2,2}-\{a\}_q\{b\}_q\mathrm{I}=0.\end{aligned}\tag{11.36}$$

Proof The proof of (11.33) goes as follows: Write the first q-Appell function in the form

$$\Phi_1(a;b,b';c|q;x_1,x_2)=\sum_{m_2=0}^{\infty}\frac{\langle a,b';q\rangle_{m_2}}{\langle 1,c;q\rangle_{m_2}}\sum_{m_1=0}^{\infty}\frac{\langle a+m_2,b;q\rangle_{m_1}}{\langle 1,c+m_2;q\rangle_{m_1}}x_1^{m_1}x_2^{m_2}.\tag{11.37}$$

Then the inner sum satisfies the q-difference equation

$$\sum_{m_2=0}^{\infty} \frac{\langle a, b'; q\rangle_{m_2}}{\langle 1, c; q\rangle_{m_2}} \big[x_1\big(q^{c+m_2} - x_1 q^{a+b+1+m_2}\big)\mathrm{D}_{q,1,1}^2 \\ + \big[\{c+m_2\}_q - \big(\{a+m_2\}_q q^b + \{b\}_q q^{a+m_2} + q^{a+b+m_2}\big)x_1\big]\mathrm{D}_{q,1} \\ - \{a+m_2\}_q\{b\}_q \mathrm{I}\big] \sum_{m_1=0}^{\infty} \frac{\langle a+m_2, b; q\rangle_{m_1}}{\langle 1, c+m_2; q\rangle_{m_1}} x_1^{m_1} x_2^{m_2} = 0. \tag{11.38}$$

We have

$$\{c+m_2\}_q = \{c\}_q + q^c\{m_2\}_q \vee \{c+m_2\}_q = \{m_2\}_q + q^{m_2}\{c\}_q. \tag{11.39}$$

Therefore we get the terms

$$\begin{cases} \{c\}_q \mathrm{D}_{q,1} + q^c x_2 \mathrm{D}_{q,1,2}^2, \\ x_2 \mathrm{D}_{q,1,2}^2 + \epsilon_2\{c\}_q \mathrm{D}_{q,1}. \end{cases} \tag{11.40}$$

In the same way we have

$$- \big(\{a+m_2\}_q q^b + \{b\}_q q^{a+m_2} + q^{a+b+m_2}\big) \\ = -\big(q^b\{a\}_q + q^a\{b\}_q + q^{a+b} + \{m_2\}_q\big(q^{a+b+1} + q^{a+b} - q^a\big)\big) \\ \vee -\big(\{a+m_2\}_q q^b + \{b\}_q q^{a+m_2} + q^{a+b+m_2}\big) \\ = -\big(\{m_2\}_q q^{a+b+1} + q^b\{a+1\}_q + q^{m_2}\{b\}_q\big). \tag{11.41}$$

Therefore we get the terms

$$-\big(\{a\}_q q^b + \{b\}_q q^a + q^{a+b}\big)x_1 \mathrm{D}_{q,1} - x_1 x_2\big(-q^a + q^{a+b} + q^{a+b+1}\big)\mathrm{D}_{q,1,2}^2 \tag{11.42}$$

or

$$-\big(x_1 x_2 q^{a+b+1}\mathrm{D}_{q,1,2}^2 - \big[\{a+1\}_q q^b + \epsilon_2\{b\}_q\big]\mathrm{D}_{q,1}\big). \tag{11.43}$$

This gives us eight equivalent q-difference equations for Φ_1.

The proof of (11.34) goes as follows: Write the second q-Appell function in the form

$$\Phi_2\big(a; b, b'; c, c'|q; x_1, x_2\big) = \sum_{m_2=0}^{\infty} \frac{\langle a, b'; q\rangle_{m_2}}{\langle 1, c'; q\rangle_{m_2}} \sum_{m_1=0}^{\infty} \frac{\langle a+m_2, b; q\rangle_{m_1}}{\langle 1, c; q\rangle_{m_1}} x_1^{m_1} x_2^{m_2}. \tag{11.44}$$

Then the inner sum satisfies the q-difference equation

$$\sum_{m_2=0}^{\infty}\frac{\langle a,b';q\rangle_{m_2}}{\langle 1,c';q\rangle_{m_2}}\big[x_1\big(q^c-x_1q^{a+b+1+m_2}\big)\mathrm{D}^2_{q,1,1}$$
$$+\big[\{c\}_q-\big(\{a+m_2\}_q q^b+\{b\}_q q^{a+m_2}+q^{a+b+m_2}\big)x_1\big]\mathrm{D}_{q,1}$$
$$-\{a+m_2\}_q\{b\}_q\mathrm{I}\big]\sum_{m_1=0}^{\infty}\frac{\langle a+m_2,b;q\rangle_{m_1}}{\langle 1,c;q\rangle_{m_1}}x_1^{m_1}x_2^{m_2}=0. \tag{11.45}$$

We have

$$-\big(\{a+m_2\}_q q^b+\{b\}_q q^{a+m_2}+q^{a+b+m_2}\big)$$
$$=-\big(q^b\{a\}_q+q^a\{b\}_q+q^{a+b}+\{m_2\}_q\big(q^{a+b+1}+q^{a+b}-q^a\big)\big)$$
$$\vee-\big(\{a+m_2\}_q q^b+\{b\}_q q^{a+m_2}+q^{a+b+m_2}\big)$$
$$=-\big(\{m_2\}_q q^{a+b+1}+q^b\{a+1\}_q+q^{m_2}\{b\}_q\big). \tag{11.46}$$

Therefore we get the terms

$$-\big(\{a\}_q q^b+\{b\}_q q^a+q^{a+b}\big)x_1\mathrm{D}_{q,1}-x_1x_2\big(-q^a+q^{a+b}+q^{a+b+1}\big)\mathrm{D}^2_{q,1,2} \tag{11.47}$$

or

$$-\big(x_1x_2q^{a+b+1}\mathrm{D}^2_{q,1,2}-\big[\{a+1\}_q q^b+\epsilon_2\{b\}_q\big]\mathrm{D}_{q,1}\big). \tag{11.48}$$

This gives us four equivalent q-difference equations for Φ_2.

The proof of (11.35) goes as follows: Write the third q-Appell function in the form

$$\Phi_3\big(a,a';b,b';c|q;x_1,x_2\big)=\sum_{m_2=0}^{\infty}\frac{\langle a',b';q\rangle_{m_2}}{\langle 1,c;q\rangle_{m_2}}\sum_{m_1=0}^{\infty}\frac{\langle a,b;q\rangle_{m_1}}{\langle 1,c+m_2;q\rangle_{m_1}}x_1^{m_1}x_2^{m_2}. \tag{11.49}$$

Then the inner sum satisfies the q-difference equation

$$\sum_{m_2=0}^{\infty}\frac{\langle a',b';q\rangle_{m_2}}{\langle 1,c;q\rangle_{m_2}}\big[x_1\big(q^{c+m_2}-x_1q^{a+b+1}\big)\mathrm{D}^2_{q,1,1}$$
$$+\big[\{c+m_2\}_q-\big(\{a\}_q q^b+\{b\}_q q^a+q^{a+b}\big)x_1\big]\mathrm{D}_{q,1}$$
$$-\{a\}_q\{b\}_q\mathrm{I}\big]\sum_{m_1=0}^{\infty}\frac{\langle a,b;q\rangle_{m_1}}{\langle 1,c+m_2;q\rangle_{m_1}}x_1^{m_1}x_2^{m_2}=0. \tag{11.50}$$

We have

$$\{c+m_2\}_q=\{c\}_q+q^c\{m_2\}_q\vee\{c+m_2\}_q=\{m_2\}_q+q^{m_2}\{c\}_q. \tag{11.51}$$

Therefore we get the terms

$$\begin{cases}\{c\}_q \mathrm{D}_{q,1} + q^c x_2 \mathrm{D}^2_{q,1,2}, \\ x_2 \mathrm{D}^2_{q,1,2} + \epsilon_2 \{c\}_q \mathrm{D}_{q,1}.\end{cases} \tag{11.52}$$

This gives us two equivalent q-difference equations for Φ_3.

The proof of (11.36) goes as follows: Write the fourth q-Appell function in the form

$$\Phi_4(a;b;c,c'|q;x_1,x_2) = \sum_{m_2=0}^{\infty} \frac{\langle a,b;q\rangle_{m_2}}{\langle 1,c';q\rangle_{m_2}} \sum_{m_1=0}^{\infty} \frac{\langle a+m_2,b+m_2;q\rangle_{m_1}}{\langle 1,c;q\rangle_{m_1}} x_1^{m_1} x_2^{m_2}. \tag{11.53}$$

Then the inner sum satisfies the q-difference equation

$$\begin{aligned}&\sum_{m_2=0}^{\infty} \frac{\langle a,b;q\rangle_{m_2}}{\langle 1,c';q\rangle_{m_2}} \big[x_1\big(q^c - x_1 q^{a+b+1+2m_2}\big)\mathrm{D}^2_{q,1,1} \\ &\quad + \big[\{c\}_q - \big(\{a+m_2\}_q q^{b+m_2} + \{b+m_2\}_q q^{a+m_2} + q^{a+b+2m_2}\big)x_1\big]\mathrm{D}_{q,1} \\ &\quad - \{a+m_2\}_q\{b+m_2\}_q \mathrm{I}\big] \sum_{m_1=0}^{\infty} \frac{\langle a+m_2,b+m_2;q\rangle_{m_1}}{\langle 1,c;q\rangle_{m_1}} x_1^{m_1} x_2^{m_2} = 0.\end{aligned} \tag{11.54}$$

We have

$$\{b+m_2\}_q = \{b\}_q + q^b\{m_2\}_q \vee \{b+m_2\}_q = \{m_2\}_q + q^{m_2}\{b\}_q. \tag{11.55}$$

Therefore we get the terms

$$\begin{cases}\{b\}_q \mathrm{D}_{q,1} + q^b x_2 \mathrm{D}^2_{q,1,2}, \\ x_2 \mathrm{D}^2_{q,1,2} + \epsilon_2 \{b\}_q \mathrm{D}_{q,1}.\end{cases} \tag{11.56}$$

In the same way we have

$$\begin{aligned}&-\big(\{a+m_2\}_q q^b + \{b\}_q q^{a+m_2} + q^{a+b+m_2}\big) \\ &= -\big(q^b\{a\}_q + q^a\{b\}_q + q^{a+b} + \{m_2\}_q\big(q^{a+b+1} + q^{a+b} - q^a\big)\big) \\ &\quad \vee -\big(\{a+m_2\}_q q^b + \{b\}_q q^{a+m_2} + q^{a+b+m_2}\big) \\ &= -\big(\{m_2\}_q q^{a+b+1} + q^b\{a+1\}_q + q^{m_2}\{b\}_q\big).\end{aligned} \tag{11.57}$$

Therefore we get the terms

$$-\big(\{a\}_q q^b + \{b\}_q q^a + q^{a+b}\big)x_1 \mathrm{D}_{q,1} - x_1 x_2\big(-q^a + q^{a+b} + q^{a+b+1}\big)\mathrm{D}^2_{q,1,2} \tag{11.58}$$

or

$$-\big(x_1 x_2 q^{a+b+1} \mathrm{D}^2_{q,1,2} - \big[\{a+1\}_q q^b + \epsilon_2\{b\}_q\big]\mathrm{D}_{q,1}\big). \tag{11.59}$$

This gives us 16 equivalent q-difference equations for Φ_4. □

These equations are stated in a different form in [221, p. 299].

Theorem 11.3.2 *The q-difference equation for Φ_1 can be written in the following canonical form, a q-analogue of* [375, p. 146], [466, p. 213, 8.1.2.4]:

$$-x_1\{\theta_{q,1}+b\}_q\{\theta_{q,1}+\theta_{q,2}+a\}_q+\{\theta_{q,1}\}_q\{\theta_{q,1}+\theta_{q,2}+c-1\}_q=0. \tag{11.60}$$

The q-difference equation for Φ_2 can be written in the following canonical form:

$$-x_1\{\theta_{q,1}+a\}_q\{\theta_{q,1}+\theta_{q,2}+b\}_q+\{\theta_{q,1}\}_q\{\theta_{q,1}+c-1\}_q=0. \tag{11.61}$$

The q-difference equation for Φ_3 can be written in the following canonical form:

$$-x_1\{\theta_{q,1}+a\}_q\{\theta_{q,1}+b\}_q+\{\theta_{q,1}\}_q\{\theta_{q,1}+\theta_{q,2}+c-1\}_q=0. \tag{11.62}$$

The q-difference equation for Φ_4 can be written in the following canonical form:

$$-x_1\{\theta_{q,1}+\theta_{q,2}+a\}_q\{\theta_{q,1}+\theta_{q,2}+\theta_{q,2}+b\}_q+\{\theta_{q,1}\}_q\{\theta_{q,1}+c-1\}_q=0. \tag{11.63}$$

The q-difference equation for Φ_1 can be rewritten in the operator form:

$$\begin{aligned}&\left(q^c-x_1q^{a+b+1}\right)\frac{\epsilon_2}{(1-q)^2}q^{-1}\left(\epsilon_1^2-(1+q)\epsilon_1+q\right)\\&+\left[q^c+x_1\left(q^a-q^{a+b}-q^{a+b+1}\right)\right]\frac{1}{(1-q)^2}[\mathrm{I}-\epsilon_1][\mathrm{I}-\epsilon_2]\\&-\{b\}_q\frac{x_1q^a}{1-q}[\mathrm{I}-\epsilon_2]-x_1\{a\}_q\{b\}_q\\&+\left[\{c\}_q-\left(\{a\}_qq^b+\{b\}_qq^a+q^{a+b}\right)x_1\right]\frac{1}{1-q}[\mathrm{I}-\epsilon_1]=0.\end{aligned} \tag{11.64}$$

Another q-difference equation satisfied by Φ_1 is (special thanks to Axel Riese for finding this equation by using Mathematica)

$$x_2\{b'\}_qx_1\mathrm{D}_{q,x_1}f-x_1\{b\}_qx_2\mathrm{D}_{q,x_2}f+\left(-x_1q^b+x_2q^{b'}\right)x_2\mathrm{D}_{q,x_2}x_1\mathrm{D}_{q,x_1}f=0. \tag{11.65}$$

Theorem 11.3.3 *Equation* (11.34) *is also satisfied* (*compare with* [467, p. 34, (65)], *where all the solutions of a homogeneous second order q-difference equation was found*) *by*

$$x_1^{1-c}\Phi_2\left(a-c+1;b-c+1,b';2-c,c'|q;x_1,x_2\right).$$

Assume a solution to (11.34) of the form

$$\sum_{m_1,m_2=0}^{\infty} a_{m_1,m_2} x_1^{m_1+\mu_1} x_2^{m_2+\mu_2}.$$

Then the method of Frobenius gives the indicial equation for the term $a_{0,0}x_1^{\mu_1-1}x_2^{\mu_2}$:

$$\{\mu_1\}_q\big(\{c\}_q + q^c\{\mu_1-1\}_q\big) = \{\mu_1\}_q\{\mu_1+c-1\}_q. \tag{11.66}$$

We now go over to another symmetric function.
(The symmetry group $SU(2)$ corresponds to the q-Lauricella function.)

Definition 177 The q-Lauricella functions are (for $\Phi_{\mathrm{D}}^{(n)}(a,\vec{b};c|q;\vec{x})$ see (10.129)):

$$\Phi_{\mathrm{A}}^{(n)}\big(a,\vec{b};\vec{c}|q;\vec{x}\big) \equiv \sum_{\vec{m}} \frac{\langle a;q\rangle_m \langle\vec{b};q\rangle_{\vec{m}} \vec{x}^{\vec{m}}}{\langle\vec{c},\vec{1};q\rangle_{\vec{m}}}, \tag{11.67}$$

$$\Phi_{\mathrm{B}}^{(n)}\big(\vec{a},\vec{b};c|q;\vec{x}\big) \equiv \sum_{\vec{m}} \frac{\langle\vec{a},\vec{b};q\rangle_{\vec{m}} \vec{x}^{\vec{m}}}{\langle c;q\rangle_m \langle\vec{1};q\rangle_{\vec{m}}}, \tag{11.68}$$

$$\Phi_{\mathrm{C}}^{(n)}\big(a,b;\vec{c}|q;\vec{x}\big) \equiv \sum_{\vec{m}} \frac{\langle a,b;q\rangle_m \vec{x}^{\vec{m}}}{\langle\vec{c},\vec{1};q\rangle_{\vec{m}}}. \tag{11.69}$$

The domains of convergence for the above functions ($q=1$) are for

$$\mathrm{F}_{\mathrm{A}}^{(n)}\big(a,\vec{b};\vec{c}|\vec{x}\big): \quad |x_1|+\cdots+|x_n|<1. \tag{11.70}$$

For

$$\mathrm{F}_{\mathrm{B}}^{(n)}\big(\vec{a},\vec{b};c|\vec{x}\big): \quad \max\big(|x_1|,\ldots,|x_n|\big)<1. \tag{11.71}$$

For

$$\mathrm{F}_{\mathrm{C}}^{(n)}\big(a,b;\vec{c}|\vec{x}\big): \quad |\sqrt{x_1}|+\cdots+|\sqrt{x_n}|<1. \tag{11.72}$$

For

$$\mathrm{F}_{\mathrm{D}}^{(n)}\big(a,\vec{b};c|\vec{x}\big): \max\big(|x_1|,\ldots,|x_n|\big)<1. \tag{11.73}$$

For general $|q|<1$ the domains of convergence for F_{A} and F_{C} are somewhat larger, as Mathematica computations show.

Theorem 11.3.4 *The partial q-difference equations for the three q-Lauricella functions Φ_{A}, Φ_{B}, Φ_{D} as functions of three variables are*:

$$\begin{aligned}
&x_1\big(q^{c_1}-x_1q^{a+b_1+1}\epsilon_2\epsilon_3\big)\mathrm{D}^2_{q,1,1}+\big(q^a-q^{a+b_1}-q^{a+b_1+1}\big)\epsilon_3\theta_{q,1}\theta_{q,2}\\
&\quad-\{b_1\}_q q^a\epsilon_3\theta_{q,2}-q^{a+b_1}\theta_{q,1}\theta_{q,3}-\{b_1\}_q q^a\theta_{q,3}\\
&\quad+\big[\{c_1\}_q-\big(\{a\}_q q^{b_1}+\epsilon_3\big(\{b_1\}_q q^a+q^{a+b_1}\big)\big)x_1\big]\mathrm{D}_{q,1}\\
&\quad-\{a\}_q\{b_1\}_q\mathrm{I}=0,
\end{aligned}\tag{11.74}$$

$$\begin{aligned}
&x_1\big(q^c\epsilon_2\epsilon_3-x_1q^{a_1+b_1+1}\big)\mathrm{D}^2_{q,1,1}+x_2q^c\mathrm{D}^2_{q,1,2}\\
&\quad+\big[\{c\}_q+q^c\theta_{q,3}\epsilon_2-\big(\{a_1\}_q q^{b_1}+\{b_1\}_q q^{a_1}+q^{a_1+b_1}\big)x_1\big]\mathrm{D}_{q,1}\\
&\quad-\{a_1\}_q\{b_1\}_q\mathrm{I}=0,
\end{aligned}\tag{11.75}$$

$$\begin{aligned}
&x_1\big(q^c-x_1q^{a+b_1+1}\big)\epsilon_2\epsilon_3\mathrm{D}^2_{q,1,1}-\theta_{q,3}q^a\big(q^{b_1}\theta_{q,1}+\{b_1\}_q\big)\\
&\quad+x_2\big[q^c+x_1\big(q^a-q^{a+b_1}-q^{a+b_1+1}\big)\big]\epsilon_3\mathrm{D}^2_{q,1,2}-\epsilon_3\{b_1\}_q q^a\theta_{q,2}\\
&\quad+\big[\{c\}_q+q^c\theta_{q,3}-\{a\}_q q^{b_1}x_1-\epsilon_3\big(\{b_1\}_q q^a+q^{a+b_1}\big)x_1\big]\mathrm{D}_{q,1}\\
&\quad-\{a\}_q\{b_1\}_q\mathrm{I}=0.
\end{aligned}\tag{11.76}$$

Proof The proof of (11.74) goes as follows: Write the first q-Lauricella function in the form

$$\Phi_{\mathrm{A}}^{(3)}\big(a,\vec{b};\vec{c}|q;\vec{x}\big)=A\Phi_2(a+m_3;b_1,b_2;c_1,c_2|q;x_1,x_2)x_3^{m_3},\tag{11.77}$$

where

$$A\equiv\sum_{m_3=0}^{\infty}\frac{\langle a,b_3;q\rangle_{m_3}}{\langle 1,c_3;q\rangle_{m_3}}.\tag{11.78}$$

Then the q-difference equation (11.34) holds for the inner sum:

$$\begin{aligned}
0=A\big(&x_1\big(q^{c_1}-x_1q^{a+m_3+b_1+1}\epsilon_2\big)\mathrm{D}^2_{q,1,1}\\
&+x_1x_2\big(q^{a+m_3}-q^{a+m_3+b_1}-q^{a+m_3+b_1+1}\big)\mathrm{D}^2_{q,1,2}-\{b_1\}_q q^{a+m_3}x_2\mathrm{D}_{q,2}\\
&+\big[\{c_1\}_q-\big(\{a+m_3\}_q q^{b_1}+\{b_1\}_q q^{a+m_3}+q^{a+m_3+b_1}\big)x_1\big]\mathrm{D}_{q,1}\\
&-\{a+m_3\}_q\{b_1\}_q\mathrm{I}\big)\Phi_2x_3^{m_3}.
\end{aligned}\tag{11.79}$$

This can be simplified to

$$\begin{aligned}
0=A\big(&x_1\big(q^{c_1}-x_1q^{a+b_1+1}\epsilon_2\epsilon_3\big)\mathrm{D}^2_{q,1,1}+\big(q^a-q^{a+b_1}-q^{a+b_1+1}\big)\epsilon_3\theta_{q,1}\theta_{q,2}\\
&-\{b_1\}_q q^a\epsilon_3\theta_{q,2}-q^{a+b_1}\theta_{q,1}\theta_{q,3}-\{b_1\}_q q^a\theta_{q,3}
\end{aligned}$$

$$+\big[\{c_1\}_q-\big(\{a\}_q q^{b_1}+\epsilon_3\big(\{b_1\}_q q^a+q^{a+b_1}\big)\big)x_1\big]\mathrm{D}_{q,1}$$
$$-\{a\}_q\{b_1\}_q\mathrm{I}\big)\Phi_2 x_3^{m_3}. \tag{11.80}$$

This gives us 16 equivalent q-difference equations for $\Phi_{\mathrm{A}}^{(3)}$.

The proof of (11.75) goes as follows: Write the second q-Lauricella function in the form

$$\Phi_{\mathrm{B}}^{(3)}(\vec{a},\vec{b};c|q;\vec{x})=B\Phi_3(a_1,a_2;b_1,b_2;c+m_3|q;x_1,x_2)x_3^{m_3}, \tag{11.81}$$

where

$$B\equiv\sum_{m_3=0}^{\infty}\frac{\langle a_3,b_3;q\rangle_{m_3}}{\langle 1,c;q\rangle_{m_3}}. \tag{11.82}$$

Then the q-difference equation (11.35) holds for the inner sum:

$$0=B\big(x_1\big(q^{c+m_3}\epsilon_2-x_1q^{a_1+b_1+1}\big)\mathrm{D}_{q,1,1}^2+x_2q^{c+m_3}\mathrm{D}_{q,1,2}^2$$
$$+\big[\{c+m_3\}_q-\big(\{a_1\}_q q^{b_1}+\{b_1\}_q q^{a_1}+q^{a_1+b_1}\big)x_1\big]\mathrm{D}_{q,1}$$
$$-\{a_1\}_q\{b_1\}_q\mathrm{I}\big)\Phi_3 x_3^{m_3}. \tag{11.83}$$

The formula (11.83) is in simplified form:

$$0=B\big(x_1\big(q^c\epsilon_2\epsilon_3-x_1q^{a_1+b_1+1}\big)\mathrm{D}_{q,1,1}^2+x_2q^c\mathrm{D}_{q,1,2}^2\epsilon_3$$
$$+\big[\{c\}_q+q^c\theta_{q,3}-\big(\{a_1\}_q q^{b_1}+\{b_1\}_q q^{a_1}+q^{a_1+b_1}\big)x_1\big]\mathrm{D}_{q,1}$$
$$-\{a_1\}_q\{b_1\}_q\mathrm{I}\big)\Phi_3 x_3^{m_3}. \tag{11.84}$$

This gives us four equivalent q-difference equations for $\Phi_{\mathrm{B}}^{(3)}$.

The proof of (11.76) goes as follows: Write the fourth q-Lauricella function in the form

$$\Phi_{\mathrm{D}}^{(3)}(a,\vec{b};c|q;\vec{x})=D\Phi_1(a+m_3;b_1,b_2;c+m_3|q;x_1,x_2)x_3^{m_3}, \tag{11.85}$$

where

$$D\equiv\sum_{m_3=0}^{\infty}\frac{\langle a,b_3;q\rangle_{m_3}}{\langle 1,c;q\rangle_{m_3}}. \tag{11.86}$$

Then the q-difference equation (11.33) holds for the inner sum:

$$\begin{aligned}0= D\big(x_1\big(q^{c+m_3}-x_1q^{a+m_3+b_1+1}\big)\epsilon_2\mathrm{D}_{q,1,1}^2\\ +x_2\big[q^{c+m_3}+x_1\big(q^{a+m_3}-q^{a+m_3+b_1}-q^{a+m_3+b_1+1}\big)\big]\mathrm{D}_{q,1,2}^2\\ -\{b_1\}_q q^{a+m_3}x_2\mathrm{D}_{q,2}\\ +\big[\{c+m_3\}_q-\big(\{a+m_3\}_q q^{b_1}+\{b_1\}_q q^{a+m_3}+q^{a+m_3+b_1}\big)x_1\big]\mathrm{D}_{q,1}\\ -\{a+m_3\}_q\{b_1\}_q\big)\Phi_1 x_3^{m_3}.\end{aligned} \tag{11.87}$$

This can be simplified to

$$\begin{aligned}0= D\big(x_1\big(q^{c}-x_1q^{a+b_1+1}\big)\epsilon_2\epsilon_3\mathrm{D}_{q,1,1}^2-\theta_{q,3}q^a\big(q^{b_1}\theta_{q,1}+\{b_1\}_q\big)\\ +x_2\big[q^{c}+x_1\big(q^{a}-q^{a+b_1}-q^{a+b_1+1}\big)\big]\epsilon_3\mathrm{D}_{q,1,2}^2-\epsilon_3\{b_1\}_q q^a\theta_{q,2}\\ +\big[\{c\}_q+q^c\theta_{q,3}-\{a\}_q q^{b_1}x_1-\epsilon_3\big(\{b_1\}_q q^{a}+q^{a+b_1}\big)x_1\big]\mathrm{D}_{q,1}\\ -\{a\}_q\{b_1\}_q\mathrm{I}\big)\Phi_1 x_3^{m_3}.\end{aligned} \tag{11.88}$$

This gives us 64 equivalent q-difference equations for $\Phi_{\mathrm{D}}^{(3)}$. □

One can note here that in 1893 Raymond Le Vavasseur (1862–1930) has investigated the system of partial differential equations of the Appell functions.

Chapter 12
q-Calculus and physics

Applications of q-calculus to problems in physics abound. We have tried to describe this briefly in nine separate sections.

12.1 The q-Coulomb problem and the q-hydrogen atom

In 1935 V. Fock [217] studied the $O(4)$ Symmetriegruppe (the group of motions of $O(3)$) of the hydrogen atom.

In 1967 R. J. Finkelstein [203] showed how the complete dynamics of the hydrogen atom is related to $O(3)$ and the Schrödinger equation in the momentum representation may be interpreted as an integral equation on $O(3)$. The solutions of this integral equation are the Wigner functions D^j_{mn}. The author [179] has introduced a q-analogue of D^j_{mn} with the help of Ward numbers.

Definition 178 In q-field theory the momentum operator is denoted by p_q and the wave function is denoted by $\Psi(x)$. We then have

$$p_q \Psi(x) \equiv \frac{\hbar}{i} \mathrm{D}_q \Psi(x). \tag{12.1}$$

The replacement of a differential operator by a difference operator then suggests the replacement of the usual continuum by an underlying lattice or, alternatively, by the use of a space with noncommuting coordinates whose spectra define this lattice [204].

The q-Coulomb-problem has been studied by Finkelstein in [117], [92], [205], Feigenbaum and Freund [195]. It turns out that the new wave functions are the matrix elements of the irreducible representations of the quantum group $SU_q(2)$. The symmetric q-derivative was also used. The resulting new integral equation is formulated in terms of the Woronowicz integral. The eigenvalues are given by a modified Balmer formula, which replaces n by $\{n\}_q$. This model is called the q-hydrogen atom.

T. Ernst, *A Comprehensive Treatment of q-Calculus*,
DOI 10.1007/978-3-0348-0431-8_12, © Springer Basel 2012

In [431] representations of $SU_q(2)$ were constructed on a two-dimensional sphere, in the form of q-special functions, which include q-spherical harmonics.

In [208] a q-delta distribution was obtained by a deformation of quantum mechanics.

12.2 Connections to knot theory

Knot theory is a branch of topology. The so-called quantum groups play an important role; the representation theory of quantum groups can be used to model elementary particles.

There are various forms of q-Wigner coefficients in the literature. The reason for this is that several q-analogues of $SU(2)$ exist. In some works of Finkelstein, a connection between gravitation, $SU_q(2)$ and certain knots is presented. The price for this is that the q-analogue of the Wigner coefficients [495, p. 112] must be sacrificed.

Recently, q-calculus has found applications in the theory of coloured Jones polynomials.

12.3 General relativity

In [126], a canonical quantization of Yang-Mills theories was made. In 1996 Finkelstein [206] constructed a q-deformation of general relativity by replacing the Lorentz group by the q-Lorentz group. The *q-Yang-Mills equation* has been studied by Kamata and Nakamula in 1999 [324].

12.4 Molecular and nuclear spectroscopy

In 1990 the quantum algebra $SU_q(2)$ was applied to study rotational spectra of deformed nuclei [429].

In 1991 it was also shown that spectra of superdeformed bands in even-even nuclei, as well as rotational bands with normal deformation are described approximately by $SU_q(2)$ [77]. In the same year a quantum group theoretic application to vibrating and rotating diatomic molecules was made [553], [551], [552]. This theoretical model coincides approximately with the vibrational Raman spectra. In 1992 a q-deformed Aufbau principle working for both atoms and monoatomic ions was based on the q-deformed chain $SO(4) > SO(3)_q$ [393].

Two years later, in [370] a q-algebra technique was applied to molecular backbending in AgH. In 1996 a q-Heisenberg algebra technique was applied to the superfluidity of ^{4}He [388].

We can describe these mathematical models briefly as follows: The classical q-oscillator

$$\mathrm{D}_q^2 f(x) + \omega^2 f(x) = 0 \tag{12.2}$$

has the solution

$$f(x) = C_1 \operatorname{Sin}_q(\omega x) + C_2 \operatorname{Cos}_q(\omega x). \tag{12.3}$$

The q-commutator is defined as follows (compare (1.59)):

$$[A, B]_q \equiv qAB - BA. \tag{12.4}$$

The Heisenberg q-uncertainty relation can be proved as follows:

$$\begin{aligned}
[x, p_q]_q \Psi(x) &= (qxp_q - p_q x)\Psi(x) \\
&= qx\frac{\hbar}{i}\mathrm{D}_q \Psi(x) - \frac{\hbar}{i}\mathrm{D}_q\big(x\Psi(x)\big) \\
&= qx\frac{\hbar}{i}\mathrm{D}_q \Psi(x) - \frac{\hbar}{i}\big(\Psi(x) + qx\mathrm{D}_q\Psi(x)\big) = i\hbar\Psi(x).
\end{aligned} \tag{12.5}$$

Let the eigenfunctions of the q-Hamiltonian H_q be $\Psi_{n,q}$ and let $\overline{a}$ and a denote the creation and annihilation operators:

$$a\Psi_{n,q} = \big(\{n\}_q\big)^{\frac{1}{2}} \Psi_{n-1,q}, \tag{12.6}$$

$$\overline{a}\Psi_{n,q} = \big(\{n+1\}_q\big)^{\frac{1}{2}} \Psi_{n+1,q}. \tag{12.7}$$

Then

$$a\overline{a} - q\overline{a}a = 1. \tag{12.8}$$

12.5 Elementary particle physics and chemical physics

The study of the Arik–Coon algebra started in 1968 when Veneziano [519] invented the Veneziano model. In 1970 [60], in 1971 [61] and in 1972 [62], Baker and Coon used q-calculus in computations with Feynman diagrams. This work continued in the papers [44] and [43], where the harmonic oscillator model was generalized to the Arik–Coon algebra, a quantum oscillator algebra with three generators. An application of this was the 1992 paper of Van der Jeugt [512], where the boson canonical commutation relations were used to find representations by q-Hermite polynomials. Yet another application was the 1998 paper by Katriel [330], where various q-analogues of Stirling number identities where derived.

In 1994 [226] q-analogues of hadron mass sum rules were obtained through Alexander polynomials of certain knots. In 1996 [207] a q-gauge theory was used to modify the Möller formula for electron-positron annihilation and a q-analogue of Planck's blackbody spectrum was anticipated (compare with [52]).

In 1997 [327] q-deformed bosons and fermions were used to derive a generalized Fokker–Planck equation, which can be easily integrated under stationary conditions and which reproduces the statistical distributions for these particles for both real and complex q-values. Also in 1997 [288] a consecutive description of functional methods of quantum field theory for systems of interacting q-particles was given. These particles obey exotic statistics and appear in many problems of condensed matter physics, magnetism and quantum optics.

An introduction to the Lipkin model was given in [337]. The q-deformed Lipkin model was presented in 1995 [51]. In 1999 [502] a q-deformation of the NJL model for quantum chromodynamics (QCD) was made.

12.6 Electroweak interaction

The q-electroweak theory proposes a description of elementary particles characterized as solitons by using the irreducible representations of $SU_q(2)$, compare with [495, p. 112]. Alain Connes [142] has given a geometrical model for the Weinberg-Salam theory of electroweak interactions.

12.7 String theory

In the 1940s it became obvious that the proton and the neutron are not point-like particles like the electron. Influenced by the Swedish physicist Oskar Klein (1894–1977), string theory started in 1943, when Werner Heisenberg proposed that the strongly interacting particles have extended dimension. This program was then further developed by prominent theoretical physicists starting in the late 1950s and continuing throughout the 1960s. Heisenberg even suggested a fundamental field equation, which has a coupling constant l with dimension length. This length, which has similarities with the so-called string length, should be a fundamental unit as important as the speed of light and the Planck constant. String theory should be a theory containing all physics. String theory was discarded and marginalized in the 1970s and finally disappeared by the 1980s. In the last two decades the subjects topology and string theory have come together in topological string theory. This was made possible after Alice Rogers' introduction of the concept of supermanifold (1980), which allows the most general topology.

In 1983 this was followed by the introduction of the Wess-Zumino-Witten model, a simple model of conformal field theory whose solutions are realized by affine Kac-Moody algebras. This was the same year as Wess and Bagger wrote their book about supersymmetry and supergravity. About 1988 Woronowicz and Manin put the quantum group concept on a sound mathematical basis. There is also the Wess-Zumino-Witten-Novikov model, which for genus zero gives the Knizhnik-Zamolodchikov equation (K-equation). The WZWN model contains a spectral parameter, which lives on a Riemann surface. The K-equation is connected to holonomic q-difference equations. The coefficients of the K-equation for genus one are found in terms of Weierstraß σ function. String theory works with a certain string length. This is somehow connected to the Planck length $L_{Planck} = \sqrt{\frac{hG}{2\pi c^3}}$. This is merely a kind of intelligent guessing, since this size is about 20 times smaller than the characteristic size of known hadrons.

Superstring theory tries to unify quantum theory and gravitation.

During the second superstring revolution, 1994–1997, the number of dimensions increased as will now be described. Originally, superstring theory was made up of five similar 10-dimensional theories. The goal of the 11-dimensional M-theory (master theory) of 1995 was to unify the three string theories. The M-theory, supposedly the correct theory, is based on conjectures that contradict physical experience; there are no experiments that can examine this M-theory. As a resort, a 26-dimensional membrane (brane) theory was introduced. The 26 dimensions are so complicated that nobody can see through this theory. There are four interactions: electrodynamics, strong interaction, weak interaction and gravitation. The aim of the grand unification theory is to describe these four interactions as one interaction. The standard model describes the first three unifying interactions as a product of three gauge groups (Lie groups). But this theory does not work well regarding gravitation. One of the leading advocates for supergravity and string theory is Hermann Nicolai in Potsdam-Golm, who got his PhD in 1978 under the supervision of Julius Wess in Karlsruhe.

12.8 Wess-Zumino model

Bruno Zumino wrote his first article about Dirac particles in 1954. Julius Wess wrote his first article about the meson-baryon scattering in 1964. In 1967 Wess and Zumino worked together at New York University on a paper about the Lagrangian Method for Chiral Symmetries. A model containing only pions and nucleons was also described. Finally, the way to generalize the method to $SU(3) \times SU(3)$ was indicated.

Wess returned to Karlsruhe in 1966, to Munich in 1990 and finally moved to DESY. He was above all a physicist. He cooperated with Witten and Zumino; Witten was good at explaining their theories at meetings and conferences and Zumino was also a mathematician. The paper that started the Wess-Zumino model was written in 1971. Therein the strong and electroweak forces are connected with gravity and require for their realization a doubling of the number of elementary particles of the Standard Model. The Wess-Zumino model is the simplest example of supersymmetric field theory. An experimental verification of the Wess-Zumino model could be provided by the current experiments at CERN.

In the 1960s D. Ruelle proved that the only group making the Lagrangian invariant is the direct product of the Lorentz group and a compact group.

A Lie group can be generalized in at least two ways: quantum group ($SU_q(2)$) and super group. In supersymmetric gauge theory, to every boson, there is a corresponding fermion; however, the number of predicted fermions is much larger than the number of predicted bosons. Bosons and fermions both go to infinity. Hence, in the correction of the Higgs mass, both bosonic and fermionic loops enter but contribute with opposite sign, and the sum will be finite. The Lie algebra is easier to handle than the Lie group, according to the experts on this subject. A super Lie algebra is a vector space A over $\mathbb{C}$ or $\mathbb{R}$ which splits into two subspaces G and U, called respectively the even and odd subspace. This is ideal for q-deformation, since we only have to apply the q-exponential function.

Both the Santilli and the Wess theory use q-calculus as a mathematical model. But it was Wess and his group in Munich who managed to q-deform the Maxwell equations, the Poincaré group and the Poincaré algebra. The Poincaré group is the semidirect product of the Lorentz group and the translation group. Previously, attempts had been made to study q-Maxwell equations by Dobrev in Clausthal using conformal symmetry. This was a very algebraic approach.

12.9 Quantum Chromodynamics

q-Calculus has wide-ranging applications in theoretical physics and analytic number theory. One example in theoretical physics is hadronic mechanics. Let us also hint at a possible connection between some algebras of A. A. Albert and the two types of q-addition mentioned before above. Adrian A. Albert (1905–1972) in [18], introduced a nonassociative algebra together with two products, one a simple composition and the second an antisymmetric product. The work of Albert also promoted the investigation of a q-analogue of the Heisenberg commutator in the form $[A, B] = pAB - qBA$. Then quantum groups were introduced in the mid-eighties by Drinfeld and Jimbo. Nowadays, quantum groups and their representations form an important part of both mathematics and theoretical physics. The representations of quantum groups involve the well-known q-special functions.

In 1973, the pure Yang-Mills theory with gauge group $SU(3)$ finally became accepted as the most likely explanation for the strong interactions, and it received the beautiful name "Quantum Chromodynamics" (QCD). QCD effectively describes the fundamental degrees of freedom for quarks and gluons, whereas the Santilli theory works only for hadrons. Experiments have shown that anti-matter is affected by gravity in the same way as matter, so no repulsion. In QCD the wave function of the gluons described as a matrix transforms by matrix multiplication of a certain Lie group. Bruno Zumino and Julius Wess were producing highly intriguing papers and in 1984–86 the first superstring revolution took place.

References

1. Abel, N.H.: Oeuvres Completes de Niels Henrik Abel. Imprimerie de Grondahl and Son, Christiana; Johnson Reprint Corporation, New York and London, VIII, 621 p. (1965)
2. Aceto, L., Trigiante, D.: The matrices of Pascal and other greats. Am. Math. Mon. **108**(3), 232–245 (2001)
3. Aczel, A.D.: Descartes Hemliga Anteckningar (Swedish). English Title: Descartes's Secret Notebook. Fahrenheit, Stockholm (2006)
4. Agarwal, R.P.: Some basic hypergeometric identities. Ann. Soc. Sci. Brux. Ser. I **67**, 186–202 (1953)
5. Agarwal, R.P.: Some relations between basic hypergeometric functions of two variables. Rend. Circ. Mat. Palermo (2) **3**, 76–82 (1954)
6. Agarwal, R.P.: Generalized Hypergeometric Series. Uttar Pradesh Scientific Research Committee Allahabad, India Asia Publishing House, Bombay (1963)
7. Agarwal, A.K., Srivastava, H.M.: Generating functions for a class of q-polynomials. Ann. Mat. Pura Appl. (4) **154**, 99–109 (1989)
8. Agarwal, A.K., Kalnins, E.G., Miller, W. Jr.: Canonical equations and symmetry techniques for q-series. SIAM J. Math. Anal. **18**(6), 1519–1538 (1987)
9. Aigner, M.: Combinatorial Theory. Grundlehren der Mathematischen Wissenschaften, vol. 234. Springer, Berlin (1979)
10. Aiton, E.J.: Leibniz. A Biography. Adam Hilger, Bristol (1985)
11. Åkerman, S.: Rose Cross over the Baltic. Brill, Leiden (1998)
12. Al-Salam, W.A.: q-Bernoulli numbers and polynomials. Math. Nachr. **17**, 239–260 (1959)
13. Al-Salam, W.A.: Operational representations for the Laguerre and other polynomials. Duke Math. J. **31**, 127–142 (1964)
14. Al-Salam, W.A.: Saalschützian theorems for basic double series. J. Lond. Math. Soc. **40**, 455–458 (1965)
15. Al-Salam, W.A.: q-Appell polynomials. Ann. Mat. Pura Appl. (4) **77**, 31–45 (1967)
16. Al-Salam, N.A.: On some q-operators with applications. Nederl. Akad. Wetensch. Indag. Math. **51**(1), 1–13 (1989)
17. Al-Salam, W.A., Carlitz, L.: Some orthogonal q-polynomials. Math. Nachr. **30**, 47–61 (1965)
18. Albert, A.A.: On Jordan algebras of linear transformations. Trans. Am. Math. Soc. **59**, 524–555 (1946)
19. Alzer, H.: Sharp bounds for the ratio of q-gamma functions. Math. Nachr. **222**, 5–14 (2001)
20. Alzer, H., Grinshpan, A.Z.: Inequalities for the gamma and q-gamma functions. J. Approx. Theory **144**(1), 67–83 (2007)
21. André, D.: Développements de $\sec x$ et de $\tan x$. C. R. Acad. Sci. Paris **88**, 965–967 (1879)

T. Ernst, *A Comprehensive Treatment of q-Calculus*,
DOI 10.1007/978-3-0348-0431-8, © Springer Basel 2012

22. André, Y.: Différentielles non commutatives et théorie de Galois différentielle ou aux différences. Ann. Sci. Éc. Norm. Super. (4) **34**(5), 685–739 (2001)
23. Andrews, G.E.: A simple proof of Jacobi's triple product identity. Proc. Am. Math. Soc. **16**, 333–334 (1965)
24. Andrews, G.E.: Summations and transformations for basic Appell series. J. Lond. Math. Soc. (2) **4**, 618–622 (1972)
25. Andrews, G.E.: On the q-analog of Kummer's theorem and applications. Duke Math. J. **40**, 525–528 (1973)
26. Andrews, G.E.: Applications of basic hypergeometric functions. SIAM Rev. **16**, 441–484 (1974)
27. Andrews, G.E.: Eulerian series—history and introduction. In: Eulerian Series and Applications. Proceedings from the Conference at Penn State University (1974)
28. Andrews, G.E.: On q-analogues of the Watson and Whipple summations. SIAM J. Math. Anal. **7**(3), 332–336 (1976)
29. Andrews, G.E.: Pfaff's method. II. Diverse applications. J. Comput. Appl. Math. **68**(1, 2), 15–23 (1996)
30. Andrews, G.E.: The Theory of Partitions. Reprint of the 1976 Original. Cambridge Mathematical Library. Cambridge University Press, Cambridge (1998)
31. Andrews, G.E., Askey, R.: Enumeration of partitions: the role of Eulerian series and q-orthogonal polynomials. In: Higher Combinatorics. Proc. NATO Advanced Study Inst., Berlin, 1976. NATO Adv. Study Inst. Ser., Ser. C: Math. Phys. Sci., vol. 31, pp. 3–26. Reidel, Dordrecht (1977)
32. Andrews, G.E., Askey, R.: A simple proof of Ramanujan's summation of the ${}_1\psi_1$. Aequ. Math. **18**(3), 333–337 (1978)
33. Andrews, G.E., Foata, D.: Congruences for the q-secant numbers. Eur. J. Comb. **1**(4), 283–287 (1980)
34. Andrews, G.E., Gessel, I.: Divisibility properties of the q-tangent numbers. Proc. Am. Math. Soc. **68**(3), 380–384 (1978)
35. Andrews, G.E., Askey, R., Roy, R.: Special Functions. Encyclopedia of Mathematics and Its Applications, vol. 71. Cambridge University Press, Cambridge (1999), xvi+664 pp.
36. Angelesco, P.A.: Sur les polynômes hypergéométriques. Ann. Math. **3**, 161–177 (1925)
37. Apostol, T.M.: Introduction to Analytic Number Theory. Undergraduate Texts in Mathematics. Springer, Berlin (1976)
38. Appell, P.: Sur la serie hypergéométrique et les polynômes de Jacobi. C. R. Acad. Sci. **89**, 31–33 (1879)
39. Appell, P.: Sur des séries hypergéométriques de deux variables, et sur des équations différentielles linéaires aux dérivées partielles. C. R. Acad. Sci. **90**, 296–298, 731–734 (1880)
40. Appell, P.: Sur une classe de polynômes. Ann. Sci. Éc. Norm. Super. (2) **9**, 119–144 (1880)
41. Appell, P., Kampé de Fériet, J.: Fonctions Hypergéométriques et Hypersphériques. Paris (1926)
42. Arbogast, L.: Du Calcul des Dérivation. Strasbourg (1800)
43. Arik, M., Coon, D.D.: Hilbert spaces of analytic functions and generalized coherent states. J. Math. Phys. **17**(4), 524–527 (1976)
44. Arik, M., Coon, D.D., Lam, Y.M.: Operator algebra of dual resonance models. J. Math. Phys. **16**(9), 1776–1779 (1975)
45. Ashton, C.H.: Die Heineschen O-Funktionen. Dissertation, Muenchen (1909)
46. Askey, R.: A note on the history of series. Mathematical Research Center Technical Report 1532, University of Wisconsin, Madison (1975)
47. Askey, R. (ed.): Theory and Application of Special Functions. Proceedings of an Advanced Seminar Sponsored by the Mathematics Research Center, The University of Wisconsin-Madison, March 21–April 2. Academic Press, New York (1975)
48. Askey, R.: The q-gamma and q-beta functions. Appl. Anal. **8**(2), 125–141 (1978/79)
49. Askey, R.: Review of H. Exton's book q-Hypergeometric Functions and Applications (see [194]). Zbl. Math. 0514.33001

50. Atakishiyev, N.M., Atakishiyeva, M.K.: A q-analogue of the Euler gamma integral. Theor. Math. Phys. **129**(1), 1325–1334 (2001) (Russian, English)
51. Avancini, S.S., Eiras, A., Galetti, D., Pimentel, B.M., Lima, C.L.: Phase transition in a q-deformed Lipkin model. J. Phys. A **28**(17), 4915–4923 (1995)
52. Babinec, P.: On the q-analogue of a black body radiation. Acta Phys. Pol. A **82**(6), 957–960 (1992)
53. Bailey, W.N.: Products of generalized hypergeometric series. Proc. Lond. Math. Soc. (2) **28**, 242–254 (1928)
54. Bailey, W.N.: An identity involving Heine's basic hypergeometric series. J. Lond. Math. Soc. **4**, 254–257 (1929)
55. Bailey, W.N.: Some identities involving generalized hypergeometric series. Proc. Lond. Math. Soc. (2) **29**, 503–516 (1929)
56. Bailey, W.N.: Generalized Hypergeometric Series. Cambridge (1935), reprinted by Stechert-Hafner, New York (1964)
57. Bailey, W.N.: A note on certain q-identities. Q. J. Math. (Oxford) **12**, 173–175 (1941)
58. Bailey, W.N.: Identities of the Rogers-Ramanujan type. Proc. Lond. Math. Soc. (2) **50**, 1–10 (1949)
59. Bailey, W.N.: On the sum of a terminating ${}_3F_2(1)$. Q. J. Math. Oxf. Ser. (2) **4**, 237–240 (1953)
60. Baker, M., Coon, D.: Dual resonance theory with nonlinear trajectories. Phys. Rev. D **2**, 2349–2358 (1970)
61. Baker, M., Coon, D.: Loop diagrams in dual resonance theories with non-linear trajectories. Phys. Rev. D **3**, 2478–2485 (1971)
62. Baker, M., Coon, D., Yu, S.: Operator formulation of a dual multiparticle theory with non-linear trajectories. Phys. Rev. D **5**, 1429–1433 (1972)
63. Barnes, E.W.: A new development of the theory of the hypergeometric functions. Proc. Lond. Math. Soc. (2) **6**, 141–177 (1908)
64. Barzun, J.: From Dawn to Decadence: 500 Years of Western Cultural Life. HarperCollins, London (2001)
65. Becker, O., Hofmann, J.E.: Geschichte der Mathematik. Athenäum-Verlag, Bonn (1951)
66. Bell, E.T.: Men of Mathematics. Simon and Schuster, New York (1937)
67. Bender, E.A.: A generalized q-binomial Vandermonde convolution. Discrete Math. **1**(2), 115–119 (1971/1972)
68. Berman, G., Fryer, K.D.: Introduction to Combinatorics. Academic Press, New York (1972)
69. Berndt, B.: Ramanujan's Notebooks. Part II. Springer, New York (1989)
70. Bernoulli, Ja.: Ars Conjectandi. Basel (1713)
71. Bessel, F.: Über die Theorie der Zahlenfacultäten. Königsberger Archiv f. Naturwiss. und Mathem. (1812)
72. Bhatnagar, G., Schlosser, M.: C_n and D_n very-well-poised ${}_{10}\phi_9$ transformations. Constr. Approx. **14**(4), 531–567 (1998)
73. Birkhoff, G., Rota, G.-C.: Ordinary Differential Equations. Ginn and Company, Boston (1962)
74. Björling, E.G.: In determinationem coefficientium Ckn in pag. 247 seqq. T. XXV. hujus Diarii relatarum. J. Reine Angew. Math. **28**, 284–288 (1844)
75. Blissard, J.: Theory of generic equations. Q. J. Math. **4**, 279–305 (1861)
76. Bombelli, R.: L'Algebra. G. Rossi, Bologna (1579)
77. Bonatsos, D., Dreska, S.B., Raychev, P.P., Roussev, R.P., Smirnov, F.: Description of superdeformed bands by the quantum algebra $SU_q(2)$. J. Phys. G **17**, L67–L73 (1991)
78. Boole, G.: On a general method in analysis. Philos. Trans. **134**, 225–282 (1844)
79. Boole, G.: A Treatise on the Calculus of Finite Differences. 1st edn. 1860, 2nd edn. 1872
80. Brafman, F.: Generating functions of Jacobi and related polynomials. Proc. Am. Math. Soc. **2**, 942–949 (1951)
81. Bressoud, D.M.: Almost poised basic hypergeometric series. Proc. Indian Acad. Sci. Math. Sci. **97**(1–3), 61–66 (1987)

82. Brown, B.M., Eastham, M.S.P.: A note on the Dixon formula for a finite hypergeometric series. J. Comput. Appl. Math. **194**, 173–175 (2006)
83. Brychkov, Y.: Handbook of Special Functions. Derivatives, Integrals, Series and Other Formulas. CRC Press, Boca Raton (2008)
84. Burchnall, J.L.: A note on the polynomials of Hermite. Q. J. Math. Oxf. Ser. **12**, 9–11 (1941)
85. Burchnall, J.L.: Differential equations associated with hypergeometric functions. Q. J. Math. **13**, 90–106 (1942)
86. Burchnall, J.L., Chaundy, T.W.: Commutative ordinary differential operators. Proc. Lond. Math. Soc. **21**, 420–440 (1923)
87. Burchnall, J.L., Chaundy, T.W.: Commutative ordinary differential operators. Proc. R. Soc. Lond. **118**, 557–583 (1928)
88. Burchnall, J.L., Chaundy, T.W.: Commutative ordinary differential operators II. The identity $P^n = Q^m$. Proc. R. Soc. Lond. **134**, 471–485 (1931)
89. Burchnall, J.L., Chaundy, T.W.: Expansions of Appell's double hypergeometric functions. Q. J. Math. **11**, 249–270 (1940)
90. Burchnall, J.L., Chaundy, T.W.: Expansions of Appell's double hypergeometric functions. II. Q. J. Math. **12**, 112–128 (1941)
91. Buschman, R.G., Srivastava, H.M.: Series identities and reducibility of Kampé de Fériet functions. Math. Proc. Camb. Philos. Soc. **91**(3), 435–440 (1982)
92. Cadavid, A.C., Finkelstein, R.J.: The q-Coulomb problem in configuration space. J. Math. Phys. **37**(8), 3675–3683 (1996)
93. Cajori, F.: A History of the Conceptions of Limits and Fluxions in Great Britain from Newton to Woodhouse. Open Court Company, Chicago (1919)
94. Cajori, F.: A History of Mathematical Notations, vol. 2. Open Court Company, Chicago (1929)
95. Cajori, F.: A History of Mathematics, 4th edn. Chelsea Publishing Co., Bronx (1985)
96. Cantor, M.: Vorlesungen über Geschichte der Mathematik. Zweiter Band. Von 1200–1668. Unveränderter Neudruck der Zweiten Auflage. B.G. Teubner, Leipzig (1913) (German)
97. Carlitz, L.: q-Bernoulli numbers and polynomials. Duke Math. J. **15**, 987–1000 (1948)
98. Carlitz, L.: Bernoulli and Euler numbers and orthogonal polynomials. Duke Math. J. **26**, 1–16 (1959)
99. Carlitz, L.: A note on the Laguerre polynomials. Mich. Math. J. **7**, 219–223 (1960)
100. Carlitz, L.: On the product of two Laguerre polynomials. J. Lond. Math. Soc. **36**, 399–402 (1961)
101. Carlitz, L.: A Saalschützian theorem for double series. J. Lond. Math. Soc. **38**, 415–418 (1963)
102. Carlitz, L.: A summation theorem for double hypergeometric series. Rend. Semin. Mat. Univ. Padova **37**, 230–233 (1967)
103. Carlitz, L.: Summation of a double hypergeometric series. Matematiche (Catania) **22**, 138–142 (1967)
104. Carlitz, L.: A transformation formula for multiple hypergeometric series. Monatshefte Math. **71**, 1–6 (1967)
105. Carlitz, L.: Some formulas of F.H. Jackson. Monatshefte Math. **73**, 193–198 (1969)
106. Carlitz, L.: Some inverse relations. Duke Math. J. **40**, 893–901 (1973)
107. Carlitz, L.: Stirling pairs. Rend. Semin. Mat. Univ. Padova **59**, 19–44 (1978)
108. Carlitz, L., Scoville, R.: Tangent numbers and operators. Duke Math. J. **39**, 413–429 (1972)
109. Carlson, B.C.: The need for a new classification of double hypergeometric series. Proc. Am. Math. Soc. **56**, 221–224 (1976)
110. Caspar, M.: Kepler. Transl. by C. Doris Hellman. With a New Introduction and References by Owen Gingerich. Dover, New York (1993)
111. Cauchy, A.: Exercices de Mathématique, 2nd edn. Paris (1827)
112. Cauchy, A.: Résumés analytiques. Turin (1833)
113. Cauchy, A.: Mémoire sur les fonctions dont plusiers valeurs sont liées entre elles par une équation linéaire, et sur diverses transformations de produits composés d'un nombre indéfinie

de facteurs. C. R. Acad Sci. Paris **17**, 526–534 (1843)
114. Cayley, A.: Table of $\Delta^m 0^n \div \Pi(m)$ up to $m = n = 20$. Trans. Camb. Philos. Soc. **XIII**, 1–4 (1881)
115. Cazzaniga, P.: Il calcolo dei simboli d'operazione elementaremente esposto. G. Mat. **20**, 48–77 (1880). 194–229
116. Chak, A.M.: A class of polynomials and a generalization of Stirling numbers. Duke Math. J. **23**, 45–55 (1956)
117. Chan, F.L., Finkelstein, R.J.: q-deformation of the Coulomb problem. J. Math. Phys. **35**(7), 3273–3284 (1994)
118. Chatterjea, S.K.: A generalization of Laguerre polynomials. Collect. Math. **15**, 285–292 (1963)
119. Chatterjea, S.K.: Operational formulae for certain classical polynomials. I. Q. J. Math. Oxf. Ser. (2) **14**, 241–246 (1963)
120. Chatterjea, S.K.: Operational representations for the Laguerre polynomials. Ann. Sc. Norm. Super. Pisa 3e Sér. **20**, 739–744 (1966)
121. Chatterjea, S.K.: Generalization of Hardy–Hille formula. Proc. Natl. Inst. Sci. India Part A **34**(suppl. 1), 87–90 (1968)
122. Chatterjea, S.K.: Some series involving products of Laguerre polynomials. Math. Balk. **5**, 55–59 (1975)
123. Chaundy, T.W.: Expansions of hypergeometric functions. Q. J. Math. Oxf. Ser. **13**, 159–171 (1942)
124. Chaundy, T.W.: An extension of hypergeometric functions. I. Q. J. Math. **14**, 55–78 (1943)
125. Chen, W.Y.C., Liu, Z.-G.: Parameter augmentation for basic hypergeometric series, I. In: Mathematical Essays in Honor of Gian-Carlo Rota, pp. 111–129. Birkhäuser, Basel (1998)
126. Cheng, H.: Canonical quantization of Yang-Mills theories. In: Perspectives in Mathematical Physics. International Press, Somerville (1994)
127. Chistiakov, I.I.: Bernoulli Numbers (Russian). Moscow (1895)
128. Choi, M.D., Elliott, G.A., Yui, N.: Gauss polynomials and the rotation algebra. Invent. Math. **99**, 225–246 (1990)
129. Chu, S.C.: Jade Mirror of the Four Unknowns (1303)
130. Chu, W.: Inversion techniques and combinatorial identities. Basic hypergeometric identities. Publ. Math. **44**, 301–320 (1994)
131. Chu, W.: Basic hypergeometric identities: an introductory revisiting through the Carlitz inversions. Forum Math. **7**(1), 117–129 (1995)
132. Chung, K.S., Chung, W.S., Nam, S.T., Kang, H.J.: New q-derivative and q-logarithm. Int. J. Theor. Phys. **33**(10), 2019–2029 (1994)
133. Cigler, J.: Operatormethoden für q-Identitäten. Monatshefte Math. **88**, 87–105 (1979)
134. Cigler, J.: Operatormethoden für q-Identitäten II: q-Laguerre-Polynome. Monatshefte Math. **91**, 105–117 (1981)
135. Cigler, J.: Elementare q-Identitäten. Publications Institut de Recherche Mathématique Avancée, pp. 23–57 (1982)
136. Cigler, J.: Operatormethoden für q-Identitäten VI: Geordnete Wurzelbäume und q-Catalan-Zahlen. Österreich. Akad. Wiss. Math.-Natur. Kl. Sitzungsber. II **206**(1997), 253–266 (1998)
137. Cigler, J.: Eine Charakterisierung der q-Exponentialpolynome. Österreich. Akad. Wiss. Math.-Natur. Kl. Sitzungsber. II **208**, 143–157 (1999)
138. Cigler, J.: Finite Differences (Differenzenrechnung). Vienna (2001) (German)
139. Cigler, J.: Elementare q-Identitäten. Skriptum, Wien (2006)
140. Clausen, T.: Über die Fälle, wenn die Reihe von der Form.... J. Reine Angew. Math. **3**, 89–91 (1828)
141. Comtet, L.: Advanced Combinatorics. Reidel, Dordrecht (1974)
142. Connes, A.: Noncommutative Geometry. Academic Press, Inc., San Diego (1994)
143. Courant, R., Hilbert, D.: Methoden der Mathematischen Physik I, 3rd edn. Heidelberger Taschenbücher, vol. 30, p. 469. Springer, Heidelberg (1968)

144. Crilly, T.: The Cambridge Mathematical Journal and its descendants: the linchpin of a research community in the early and mid-Victorian Age. Hist. Math. **31**, 455–497 (2004)
145. Crippa, D., Simon, K., Trunz, P.: Markov processes involving q-Stirling numbers. Comb. Probab. Comput. **6**(2), 165–178 (1997)
146. Curry, H.B.: Abstract differential operators and interpolation formulas. Port. Math. **10**, 135–162 (1951)
147. Das, M.K.: Operational representation for the Laguerre polynomials. Acta Math. Acad. Sci. Hung. **18**, 335–338 (1967)
148. Daum, J.A.: Basic hypergeometric series. Thesis, Lincoln, Nebraska (1941)
149. Daum, J.A.: The basic analog of Kummer's theorem. Bull. Am. Math. Soc. **48**, 711–713 (1942)
150. De Moivre, A.: Miscellanea Analytica. London (1730)
151. De Morgan, A.: Differential and Integral Calculus. Baldwin and Cradock, London (1842); Elibron Classics (2002)
152. De Sole, A., Kac, V.: On integral representations of q-gamma and q-beta functions. Atti Accad. Naz. Lincei Cl. Sci. Fis. Mat. Natur. Rend. Lincei (9) Mat. Appl. **16**(1), 11–29 (2005)
153. Debiard, A., Gaveau, B.: Hypergeometric symbolic calculus. I: systems of two symbolic hypergeometric equations. Bull. Sci. Math. **126**(10), 773–829 (2002)
154. Désarménien, J.: Les q-analogues des polynômes d'Hermite. Sémin. Lothar. Comb. B **6**, B06b (1982), 12 p.
155. Dixon, A.C.: Summation of a certain series. Proc. Lond. Math. Soc. (1) **35**, 285–289 (1903)
156. Dobinski, G.: Summierung der Reihe $\sum_{n=1}^{\infty} \frac{n^m}{n!}$ für $m = 1, 2, \ldots$. Arch. Math. Phys. **61**, 333–336 (1877)
157. d'Ocagne, M.: Sur une classe de nombres remarquables. Am. J. Math. **IX**, 353–380 (1887)
158. Dougall, J.: On Vandermonde's theorem and some more general expansions. Proc. Edinb. Math. Soc. **25**, 114–132 (1907)
159. Dutka, J.: The early history of the hypergeometric function. Arch. Hist. Exact Sci. **31**(1), 15–34 (1984)
160. Dutka, J.: The early history of the factorial function. Arch. Hist. Exact Sci. **43**(3), 225–249 (1991)
161. Edwards, A.W.F.: Pascal's Arithmetical Triangle. Charles Griffin & Co., Ltd., London; The Clarendon Press, Oxford University Press, New York (1987)
162. Erdelyi, A. (ed.): Higher Transcendental Functions. Vol. I. Based, in part, on notes left by Harry Bateman and compiled by the Staff of the Bateman Manuscript Project (Repr. of the original 1953 edition publ. by McGraw-Hill Book Company, Inc., New York).
163. Ernst, T.: The history of q-calculus and a new method. U.U.D.M. Report 2000:16, ISSN 1101-3591, Department of Mathematics, Uppsala University (2000)
164. Ernst, T.: A new method for q-calculus. Uppsala Dissertations (2002)
165. Ernst, T.: A method for q-calculus. J. Nonlinear Math. Phys. **10**(4), 487–525 (2003)
166. Ernst, T.: Some results for q-functions of many variables. Rend. Semin. Mat. Univ. Padova **112**, 199–235 (2004)
167. Ernst, T.: q-Generating functions for one and two variables. Simon Stevin **12**(4), 589–605 (2005)
168. Ernst, T.: q-Bernoulli and q-Euler polynomials, an umbral approach. Int. J. Differ. Equ. **1**(1), 31–80 (2006)
169. Ernst, T.: q-analogues of some operational formulas. Algebras Groups Geom. **23**(4), 354–374 (2006)
170. Ernst, T.: A renaissance for a q-umbral calculus. In: Proceedings of the International Conference, Munich, Germany, 25–30 July 2005, pp. 178–188. World Scientific, Singapore (2007)
171. Ernst, T.: Some new formulas involving Γ_q functions. Rend. Semin. Mat. Univ. Padova **118**, 159–188 (2007)
172. Ernst, T.: q-calculus as operational algebra. In: Ruffing, A., et al. (eds.) Communications of the Laufen Colloquium on Science. Laufen, Austria, April 1–5, 2007. Berichte aus der Mathematik, vol. 7, pp. 1–31. Shaker, Aachen (2007).

173. Ernst, T.: The different tongues of q-calculus. Proc. Est. Acad. Sci. **57**(2), 81–99 (2008)
174. Ernst, T.: q-Stirling numbers, an umbral approach. Adv. Dyn. Syst. Appl. **3**(2), 251–282 (2008)
175. Ernst, T.: Motivation for introducing q-complex numbers. Adv. Dyn. Syst. Appl. **3**(1), 107–129 (2008). Special Volume in Honor of Allan Peterson
176. Ernst, T.: Examples of a q-umbral calculus. Adv. Stud. Contemp. Math. **16**(1), 1–22 (2008)
177. Ernst, T.: Die Jacobi-Gudermann-Glaisherschen elliptischen Funktionen nach Heine. Hadron. J. **33**, 273–302 (2010)
178. Ernst, T.: Sur les polynômes q-Hermite de Cigler. Algebras Groups Geom. **27**, 121–142 (2010)
179. Ernst, T.: q-deformed matrix pseudo-groups. Royal Flemish Academy of Belgium, 151–162 (2010)
180. Ernst, T.: Zur Theorie der Γ_q-Funktion. Proc. Jangjeon Math. Soc. **14**, 91–113 (2011)
181. Ernst, T.: q-analogues of general reduction formulas by Buschman and Srivastava and an important q-operator reminding of MacRobert. Demonstr. Math. **44**, 285–296 (2011)
182. Ernst, T.: An umbral approach for q-Bernoulli and q-Euler matrices and its connection to the polynomial matrix approach (submitted)
183. Ernst, T.: The correspondence between two mathematicians: Niels Erik Nørlund and Gösta Mittag-Leffler (2010, submitted)
184. Ernst, T.: q-Pascal and q-Wronskian matrices with implications to q-Appell polynomials, in particular q-Bernoulli and q-Euler polynomial formulas (2011, submitted)
185. Ernst, T.: Introduction to q-complex analysis, together with the q-complex numbers $\mathbb{C}_{\oplus_q}$ (2011, submitted)
186. Etingof, P.I., Frenkel, I.B., Kirillov, A.A.: Lectures on Representation Theory and Knizhnik-Zamolodchikov Equations. Mathematical Surveys and Monographs, vol. 58. American Mathematical Society, Providence (1998)
187. Euler, L.: Methodus generalis summandi progressiones. Comment. Acad. Sci. Petropolitanae **6**, 68–97 (1732–1733)
188. Euler, L.: Inventio summae cuiusque seriei ex dato termino generali. Comment. Acad. Sci. Petropolitanae **8**, 9–22 (1736)
189. Euler, L.: Institutiones Calculi Differentialis (1755), new printing, Birkhäuser (1913)
190. Euler, L.: Introductio in Analysin Infinitorum, vol. 1. Lausanne (1748)
191. Euler, L.: Euler's hypergeometric transformation. Nova Acta Acad. Petropol. **7** (1778)
192. Euler, L.: Einleitung in die Analysis des Unendlichen / aus dem Lateinischen übersezt und mit Anmerkungen und Zusätzen begleitet von Johann Andreas Christian Michelsen. Berlin (1788)
193. Exton, H.: Multiple Hypergeometric Functions and Applications. Mathematics and Its Applications. Ellis Horwood Ltd., Chichester; Halsted Press [Wiley, Inc.], New York, 312 pp. (1976)
194. Exton, H.: Q-Hypergeometric Functions and Applications. Ellis Horwood, Chichester; Halsted Press [Wiley, Inc.], New York (1983)
195. Feigenbaum, J., Freund, P.G.O.: A q-deformation of the Coulomb problem. J. Math. Phys. **37**(4), 1602–1616 (1996)
196. Feldheim, E.: Quelques nouvelles relations pour les polynômes d'Hermite. J. Lond. Math. Soc. **13**, 22–29 (1938)
197. Feldheim, E.: Équations intégrales pour les polynomes d'Hermite à une et plusieurs variables, pour les polynômes de Laguerre, et pour les fonctions hypergéométriques les plus générales. Ann. Sc. Norm. Super. Pisa (2) **9**, 225–252 (1940)
198. Feldheim, E.: Formules d'inversion et autres rélations pour les polynômes orthogonaux classiques. Bull. Soc. Math. Fr. **68**, 199–228 (1940)
199. Feldheim, E.: Contributi alla teoria delle funzioni ipergeometriche di piu variabili. Ann. Sc. Norm. Super. Pisa (2) **12**, 17–59 (1943)
200. Feldheim, E.: Rélations entre les polynômes de Jacobi, Laguerre et Hermite. Acta Math. **75**, 117–138 (1943)

201. Ferraro, G.: Some aspects of Euler's theory of series: inexplicable functions and the Euler-Maclaurin summation formula. Hist. Math. **25**(3), 290–317 (1998)
202. Fine N. (Andrews, G.): Basic Hypergeometric Series and Applications. With a Foreword by George E. Andrews. Mathematical Surveys and Monographs, vol. 27. American Mathematical Society, Providence (1988)
203. Finkelstein, R.J.: Symmetry group of the hydrogen atom. J. Math. Phys. **8**(3), 443–449 (1967)
204. Finkelstein, R.J.: q-field theory. Lett. Math. Phys. **34**(2), 169–176 (1995)
205. Finkelstein, R.J.: The q-Coulomb problem. J. Math. Phys. **37**(6), 2628–2636 (1996)
206. Finkelstein, R.J.: q-gravity. Lett. Math. Phys. **38**(1), 53–62 (1996)
207. Finkelstein, R.J.: q-gauge theory. Int. J. Mod. Phys. A **11**(4), 733–746 (1996)
208. Finkelstein, R., Marcus, E.: Transformation theory of the q-oscillator. J. Math. Phys. **36**(6), 2652–2672 (1995)
209. Fitouhi, A., Brahim, K., Bettaibi, N.: Asymptotic approximations in quantum calculus. J. Nonlinear Math. Phys. **12**(4), 586–606 (2005)
210. Fleckenstein, J.O.: Die Taylorsche Formel bei Johann I. Bernoulli. Elem. Math. **1**, 13–17 (1946)
211. Floreanini, R., Vinet, L.: Representations of quantum algebras and q-special functions. In: Quantum Symmetries, Clausthal, 1991, pp. 264–284. World Sci. Publ., River Edge (1993)
212. Floreanini, R., Vinet, L.: Quantum algebras and q-special functions. Ann. Phys. **221**(1), 53–70 (1993)
213. Floreanini, R., Vinet, L.: Quantum algebra approach to q-Gegenbauer polynomials. In: Proceedings of the Fourth International Symposium on Orthogonal Polynomials and Their Applications, Evian-Les-Bains, 1992. J. Comput. Appl. Math. **57**(1–2), 123–133 (1995)
214. Floreanini, R., Vinet, L.: Quantum symmetries of q-difference equations. J. Math. Phys. **36**(6), 3134–3156 (1995)
215. Floreanini, R., Vinet, L.: Symmetries of the q-deformed heat equations. In: Nonlinear, Deformed and Irreversible Quantum Systems, Clausthal, 1994, pp. 385–400. World Sci. Publ., River Edge (1995)
216. Floreanini, R., LeTourneux, J., Vinet, L.: A q-deformed $e(4)$ and continuous q-Jacobi polynomials. J. Math. Phys. **37**(8), 4135–4149 (1996)
217. Fock, V.: Zur Theorie des Wasserstoffatoms. Z. Phys. **98**, 145–154 (1935)
218. Fröberg, C.-E.: Lärobok i Numerisk Analys. Svenska Bokförlaget/Bonniers, Stockholm (1962)
219. Fujiwara, I.: A unified presentation of classical orthogonal polynomials. Math. Jpn. **11**, 133–148 (1966)
220. Gasper, G., Rahman, M.: Basic Hypergeometric Series. Cambridge University Press, Cambridge (1990)
221. Gasper, G., Rahman, M.: Basic Hypergeometric Series, 2nd edn. Cambridge University Press, Cambridge (2004)
222. Gauß, C.F.: Werke 2, 9–45 (1876).
223. Gauß, C.F.: Werke 3 (1876)
224. Gauß, C.F.: Disquisitiones generales circa seriem infinitam.... (1876)
225. Gauß, C.F.: Summatio quarundam serierum singularium. Opera **2**, 16–17 (1876)
226. Gavrilik, A.M.: q-Serre relations in $U_q(u_n)$ and q-deformed meson mass sum rules. J. Phys. A **27**(3), L91–L94 (1994)
227. Gegenbauer, L.: Ein vergessener Österreicher. Jahresber. Dtsch. Math.-Ver. **12**, 324–344 (1903)
228. Gelfand, I.M., Graev, M.I., Retakh, V.S.: General hypergeometric systems of equations and series of hypergeometric type. Usp. Mat. Nauk **47**(4), 3–82 (1992) (Russian). Math. Surv. **47**(4), 1–88 (1992)
229. Glaisher, J.W.L.: General summation-formulae in finite differences. Q. J. Math. **29**, 303–328 (1898)

230. Glushkov, S.: An interpretation of Viète's 'Calculus of triangles' as a precursor of the algebra of complex numbers. Hist. Math. **4**, 127–136 (1977)
231. Goldman, J., Rota, G.C.: The number of subspaces of a vector space. In: Recent Progress in Combinatorics, pp. 75–83. Academic Press, New York (1969)
232. Goldman, J., Rota, G.C.: On the foundations of combinatorial theory. IV. Finite vector spaces and Eulerian generating functions. Stud. Appl. Math. **49**, 239–258 (1970)
233. Goldstine, H.H.: A History of Numerical Analysis from the 16th Through the 19th Century. Studies in the History of Mathematics and Physical Sciences, vol. 2. Springer, New York (1977)
234. Gould, H.W.: The q-series generalization of a formula of Sparre Andersen. Math. Scand. **9**, 90–94 (1961)
235. Gould, H.W.: The q-Stirling numbers of first and second kinds. Duke Math. J. **28**, 281–289 (1961)
236. Gould, H.W.: The operator $(a^x\Delta)^n$ and Stirling numbers of the first kind. Am. Math. Mon. **71**, 850–858 (1964)
237. Gould, H.W.: Note on a q-series transform. J. Comb. Theory **1**, 397–399 (1966)
238. Gould, H.W.: Combinatorial Identities. Henry W. Gould, Morgantown (1972)
239. Gould, H.W.: Evaluation of sums of convolved powers using Stirling and Eulerian numbers. Fibonacci Q. **16**(6), 488–497 (1978)
240. Gould, H.W.: Euler's formula for n-th differences of powers. Am. Math. Mon. **85**, 450–467 (1978)
241. Gould, H.W., Hsu, L.C.: Some new inverse series relations. Duke Math. J. **40**, 885–891 (1973)
242. Goulden, I.P., Jackson, D.M.: An inversion model for q-identities. Eur. J. Comb. **4**(3), 225–230 (1983)
243. Goursat, M.: Mémoiré sur les fonctions hypérgéometrique d'ordre supérieur. Ann. École Norm. II **XII**, 261–287, 395–430 (1883)
244. Graham, R.L., Knuth, D.E., Patashnik, O.: Concrete Mathematics. A Foundation for Computer Science, 2nd edn. Addison-Wesley, Reading (1994)
245. Grattan-Guinness, I.: From the Calculus to Set Theory, 1630–1910. An Introductory History. Gerald Duckworth & Co. Ltd., London (1980)
246. Gray, A., Matthews, G.B.A.: Treatise on Bessel Functions and Their Applications to Physics, 2nd edn. Dover, New York (1966)
247. Grigoriew, E.: Bernoullische Zahlen höherer Ordnungen. Kasan Ges. (2) **71**, 146–202 (1898) (Russian)
248. Grinshpan, A.Z., Ismail, M.E.H.: Completely monotonic functions involving the gamma and q-gamma functions. Proc. Am. Math. Soc. **134**(4), 1153–1160 (2006)
249. Grosjean, C.C., Sharma, R.K.: Transformation formulae for hypergeometric series in two variables II. Simon Stevin **62**(2), 97–125 (1988)
250. Grünert, J.A.: Mathematische Abhandlungen, Erste Sammlung. Hammerich, Altona (1822)
251. Grünert, J.A.: Summierung der Reihe .... Crelle J. **2**, 358–363 (1827)
252. Grünert, J.A.: Über die Summierung der Reihen von der Form .... Crelle J. **25**, 240–279 (1843)
253. Gudermann, C.: Allgemeiner Beweis des polynomischen Lehrsatzes ohne die Vorausstzung des binomischen und ohne die Hilfe der höheren Rechnung. Schulprogramm Cleve (1825)
254. Gudermann, C.: Grundriss der Analytischen Sphärik. Köln (1830)
255. Gudermann, C.: Theorie der Potenzial oder Cyklisch-Hyperbolische Functionen. G. Reimer, Berlin (1833)
256. Gudermann, C.: Theorie der Modular-Functionen und der Modular-Integrale. Reimer, Berlin (1844)
257. Guelfond, A.O.: Calcul des Différences Finies. Collection Universitaire de Mathématiques, XII. Traduit Par G. Rideau. Dunod, Paris (1963)
258. Guo, V.: Elementary proofs of some q-identities of Jackson and Andrews-Jain. Discrete Math. **295**(1–3), 63–74 (2005)

259. Gupta, A.: On a q-extension of "incomplete" beta function. J. Indian Math. Soc. (N.S.) **66**(1–4), 193–201 (1999)
260. Hahn, W.: Über Orthogonalpolynome, die q-Differenzengleichungen genügen. Math. Nachr. **2**, 4–34 (1949)
261. Hahn, W.: Beiträge zur Theorie der Heineschen Reihen. Math. Nachr. **2**, 340–379 (1949)
262. Hahn, W.: Über die höheren Heineschen Reihen und eine einheitliche Theorie der sogenannten speziellen Funktionen. Math. Nachr. **3**, 257–294 (1950)
263. Hahn, W.: Lineare geometrische Differenzengleichungen. Forschungszentrum Graz, Bericht 169 (1981)
264. Hall, N.A.: An algebraic identity. J. Lond. Math. Soc. **11**, 276 (1936)
265. Hardy, G.H.: Summation of a series of polynomials of Laguerre. J. Lond. Math. Soc. **7**, 138–139 (1932); Addendum. Ibid., **7**, 192
266. Hardy, G.H.: The Indian mathematician Ramanujan. Am. Math. Mon. **44**, 137–155 (1937)
267. Hardy, G.H.: Ramanujan. Cambridge University Press, Cambridge (1940). Reprinted by Chelsea, New York (1978)
268. Hardy, G.H., Wright, E.M.: An Introduction to the Theory of Numbers. Edited and revised by D.R. Heath-Brown and J.H. Silverman. With a foreword by Andrew Wiles, 6th edn. Oxford University Press, Oxford (2008)
269. Heine, E.: Über die Reihe $1 + [(1-q^{\alpha})(1-q^{\beta})]/[(1-q)(1-q^{\gamma})]x + [(1-q^{\alpha})(1-q^{\alpha+1})(1-q^{\beta})(1-q^{\beta+1})]/[(1-q)(1-q^2)(1-q^{\gamma})(1-q^{\gamma+1})]x^2 + \ldots$. J. Reine Angew. Math. **32**, 210–212 (1846)
270. Heine, E.: Untersuchungen über die Reihe $1 + [(1-q^{\alpha})(1-q^{\beta})]/[(1-q)(1-q^{\gamma})]x + [(1-q^{\alpha})(1-q^{\alpha+1})(1-q^{\beta})(1-q^{\beta+1})]/[(1-q)(1-q^2)(1-q^{\gamma})(1-q^{\gamma+1})]x^2 + \ldots$. J. Reine Angew. Math. **34**, 285–328 (1847)
271. Heine, E.: Über die Zähler und Nenner der Näherungswerte von Kettenbrüchen. J. Reine Angew. Math. **57**, 231–247 (1860)
272. Heine, E.: Handbuch der Kugelfunctionen, Theorie und Anwendungen, vol. 1. G. Reimer, Berlin (1878)
273. Heine, E.: der Kugelfunctionen Handbuch Theorie und Anwendungen, vol. 2. G. Reimer, Berlin (1878), reprinted Würzburg (1961)
274. Hellström, L., Silvestrov, S.D.: Commuting Elements in the q-Deformed Heisenberg Algebras. World Scientific, Singapore (2000)
275. Henrici, P.: Applied and Computational Complex Analysis, vol. 2. Special Functions–Integral Transforms–Asymptotics–Continued Fractions. Wiley Interscience, New York (1977)
276. Herschel, J.: On the development of exponential functions. Philos. Trans. R. Soc. Lond. 25–45 (1816)
277. Herschel, J.: A Collection of Examples of the Applications of Calculus of Finite Differences. J. Smith, Cambridge (1820)
278. Hille, E.: On Laguerre's series. I, II and III. Proc. Natl. Acad. Sci. USA **12**, 261–269, 348–352 (1926)
279. Hindenburg, C.F.: Über combinatorische Analysis und Derivations-Calcul. Schwickert, Leipzig (1803)
280. Hindenburg, C.F.: Novi Systematis Permutationem Combinationum ac Variationum. Primae Lineae et Logisticae Serierum Formulis Analytico-combinatoriis per Tabulas Exhibendae. Leipzig (1781)
281. Hindenburg, C.F.: Archiv der Reinen und Angewandten Mathematik. Leipzig (1795)
282. Hofbauer, J.: Beiträge zu Rota's Theorie der Folgen von Binomialtyp. Österreich. Akad. Wiss. Math.-Natur. Kl. Sitzungsber. II **187**(8–10), 437–489 (1978)
283. Hofmann, J.E.: Leibniz in Paris, 1672–1676. His Growth to Mathematical Maturity. Cambridge University Press, London (1974)
284. Hofmann, J.E.: Ausgewählte Schriften, vol. I. Georg Olms Verlag, Hildesheim (1990)
285. Hörmander, L.: The Analysis of Linear Partial Differential Operators 1. Springer, Berlin (1983)

286. Horn, J.: Über die Convergenz der hypergeometrischen Reihen zweier und dreier Veränderlichen. Math. Ann. **34**, 544–600 (1889)
287. Humbert, P.: Les polynômes de Sonine. J. Éc. Polytech. **2**(24), 59–75 (1925)
288. Ilinski, K.N., Kalinin, G.V., Stepanenko, A.S.: q-functional field theory for particles with exotic statistics. Phys. Lett. A **232**(6), 399–408 (1997)
289. Imchenetsky, V.G.: Sur la généralisation des fonctions de Jacques Bernoulli. Zap. Peterb. Akad. Nauk (7) **31**(11), 1–58 (1883)
290. Ismail, M.E.H., Muldoon, M.E.: Inequalities and monotonicity properties for gamma and q-gamma functions. In: Approximation and Computation, West Lafayette, IN, 1993. Internat. Ser. Numer. Math., vol. 119, pp. 309–323. Birkhäuser, Boston (1994)
291. Ismail, M.E.H., Lorch, L., Muldoon, M.E.: Completely monotonic functions associated with the gamma function and its q-analogues. J. Math. Anal. Appl. **116**(1), 1–9 (1986)
292. Itard, J.: Fermat précurseur du calcul différentiel. Arch. Int. Hist. Sci. **27**, 589–610 (1948)
293. Jackson, F.H.: A basic-sine and cosine with symbolical solution of certain differential equations. Proc. Edinb. Math. Soc. **22**, 28–39 (1904)
294. Jackson, F.H.: On generalized functions of Legendre and Bessel. Trans. R. Soc. Edinb. **41**, 1–28 (1904)
295. Jackson, F.H.: The application of basic numbers to Bessel's and Legendre's functions. Proc. Lond. Math. Soc. (2) **2**, 192–220 (1904)
296. Jackson, F.H.: Theorems relating to a generalisation of the Bessel-function. Trans. R. Soc. Edinb. **41**, 105–118 (1904)
297. Jackson, F.H.: Some properties of a generalized hypergeometric function. Am. J. Math. **27**, 1–6 (1905)
298. Jackson, F.H.: On q-functions and a certain difference operator. Trans. R. Soc. Edinb. **46**, 253–281 (1908)
299. Jackson, F.H.: Generalization of Montmort's formula for the transformation of power series. Messenger **37**, 145–147 (1908)
300. Jackson, F.H.: Note on a generalization of Montmort's series. Messenger **37**, 191–192 (1908)
301. Jackson, F.H.: A q-form of Taylor's theorem. Messenger **38**, 62–64 (1909)
302. Jackson, F.H.: On q-definite integrals. Q. J. Pure Appl. Math. **41**, 193–203 (1910)
303. Jackson, F.H.: On q-difference equations. Am. J. Math. **32**, 305–314 (1910)
304. Jackson, F.H.: Transformations of q-series. Messenger **39**, 145–153 (1910)
305. Jackson, F.H.: Summation of q-hypergeometric series. Messenger Math. **50**, 101–112 (1921)
306. Jackson, F.H.: Certain q-identities. Q. J. Math. Oxf. Ser. **12**, 167–172 (1941)
307. Jackson, F.H.: On basic double hypergeometric functions. Q. J. Math. **13**, 69–82 (1942)
308. Jackson, F.H.: Basic double hypergeometric functions. Q. J. Math. Oxf. Ser. **15**, 49–61 (1944)
309. Jackson, F.H.: Basic integration. Q. J. Math. Oxf. Ser. (2) **2**, 1–16 (1951)
310. Jacobi, C.G.J.: Fundamenta Nova. Königsberg (1829)
311. Jacobi, C.G.J.: Gesammelte Werke, vol. 6. Berlin (1891)
312. Jahnke, H.: Mathematik und Bildung in der Humboldtschen Reform. Vandenhoeck & Ruprecht, Göttingen (1990)
313. Jain, V.K., Srivastava, H.M.: Some families of multilinear q-generating functions and combinatorial q-series identities. J. Math. Anal. Appl. **192**(2), 413–438 (1995)
314. Järvheden, B.: Elementär Kombinatorik. Studentlitteratur, Lund (1976)
315. Jensen, J.L.W.: Studier over en Afhandling af Gauß (Studies on a paper by Gauß). Nyt. Tidsskr. Math. **29**, 29–36 (1918) (Danish)
316. Jing, S.-C., Fan, H.-Y.: q-Taylor's formula with its q-remainder. Commun. Theor. Phys. **23**(1), 117–120 (1995)
317. Johnson, W.P.: Some applications of the q-exponential formula. Proceedings of the 6th Conference on Formal Power Series and Algebraic Combinatorics, New Brunswick, NJ, 1994. Discrete Math. **157**(1–3), 207–225 (1996)
318. Johnson, W.P.: The curious history of Faà di Bruno's formula. Am. Math. Mon. **109**(3), 217–234 (2002).

319. Jones, G., Singerman, D.: Complex Functions. An Algebraic and Geometric Viewpoint. Cambridge University Press, Cambridge (1987)
320. Jordan, Ch.: Calculus of Finite Differences, 3rd edn. Chelsea Publishing Co., New York (1950)
321. Kalnins, E.G., Manocha, H.L., Miller, W. Jr.: The Lie theory of two-variable hypergeometric functions. Stud. Appl. Math. **62**, 143–173 (1980)
322. Kalnins, E.G., Manocha, H.L., Miller, W. Jr.: Models of q-algebra representations: tensor products of special unitary and oscillator algebras. J. Math. Phys. **33**(7), 2365–2383 (1992)
323. Kalnins, E.G., Miller, W. Jr., Mukherjee, S.: Models of q-algebra representations: matrix elements of the q-oscillator algebra. J. Math. Phys. **34**(11), 5333–5356 (1993)
324. Kamata, M., Nakamula, A.: One-parameter family of selfdual solutions in classical Yang-Mills theory. Phys. Lett. B **463**, 257–262 (1999)
325. Kamke, E.: Differentialgleichungen Lösungsmethoden und Lösungen. Chelsea Publishing Co., New York (1959)
326. Kaneko, M., Kurokawa, N., Wakayama, M.: A variation of Euler's approach to values of the Riemann zeta function. Kyushu J. Math. **57**(1), 175–192 (2003)
327. Kaniadakis, G., Lavagno, A., Quarati, P.: Kinetic model for q-deformed bosons and fermions. Phys. Lett. A **227**(3–4), 227–231 (1997)
328. Karlsson, P.W.: Reduction of certain hypergeometric functions of three variables. Glas. Mat. Ser. III 8 **28**, 199–204 (1973)
329. Kästner, A.: Anfangsgründe der Analysis des Unendlichen. Göttingen (1770)
330. Katriel, J.: Stirling number identities: interconsistency of q-analogues. J. Phys. A **31**(15), 3559–3572 (1998)
331. Khan, M.A.: q-analogues of certain operational formulae. Houst. J. Math. **13**(1), 75–82 (1987)
332. Khan, M.A.: On a calculus for the $T_{k,q,x}$-operator. Math. Balk. (N.S.) **6**(1), 83–90 (1992)
333. Kim, T.: Sums of powers of consecutive q-integers. Adv. Stud. Contemp. Math. (Kyungshang) **9**(1), 15–18 (2004)
334. Kim, T., Adiga, C.: On the q-analogue of gamma functions and related inequalities. J. Inequal. Pure Appl. Math. **6**(4), 118 (2005), 4 pp.
335. Kim, Y., Rathie, A., Lee, C.: On q-Gauss's second summation theorem. Far East J. Math. Sci. **17**(3), 299–303 (2005)
336. Klein, F.: Vorlesungen über die Entwickelung der Mathematik im 19. Jahrhundert, vols. I, II. Bearbeitet von R. Courant, St. Cohn-Vossen. Springer, Berlin (1979)
337. Klein, A., Marshalek, E.R.: Boson realizations of Lie algebras with applications to nuclear physics. Rev. Mod. Phys. **63**(2), 375–558 (1991)
338. Klügel, G.S.: Mathematisches Wörterbuch. Leipzig (1805)
339. Klügel, G.S.: Analytische Trigonometrie. Braunschweig (1770)
340. Knuth, D.: The Art of Computer Programming, Volume 4, Fascicle 2. Generating All Tuples and Permutations. Addison-Wesley, Reading (2005)
341. Koekoek, J., Koekoek, R.: A note on the q-derivative operator. J. Math. Anal. Appl. **176**(2), 627–634 (1993)
342. Koekoek, R., Swarttouw, R.F.: The Askey-scheme of hypergeometric orthogonal polynomials and its q-analogue. Delft University of Technology, Faculty of Technical Mathematics and Informatics, Report no. 98–17 (1998)
343. Koornwinder, T.: Jacobi functions as limit cases of q-ultraspherical polynomials. J. Math. Anal. Appl. **148**(1), 44–54 (1990)
344. Koornwinder, T.: Compact quantum groups and q-special functions. In: Baldoni, V., Picardello, M.A. (eds.) Representations of Lie Groups and Quantum Groups. Pitman Research Notes in Math., vol. 311. Longman Scientific & Technical, Harlow (1994)
345. Koppelman, E.: The calculus of operations and the rise of abstract algebra. J. Arch. Hist. Exact Sci. **8**, 155–242 (1971)
346. Kowalewski, G.: Große Mathematiker. Eine Wanderung durch die Geschichte der Mathematik vom Altertum bis zur Neuzeit. J.F. Lehmann, München (1938)

347. Kramp, C.: Analyse des réfractions astronomique et terrestres. Strasbourg/Leipzig (1798)
348. Kummer, E.E.: Über die hypergeometrische Reihe .... J. Reine Angew. Math. **15**, 39–83, 127–172 (1836)
349. Kupershmidt, B.A.: q-Newton binomial: from Euler to Gauß. J. Nonlinear Math. Phys. **7**(2), 244–262 (2000)
350. Kuznetsov, V.B., Sklyanin, E.K.: Factorisation of Macdonald polynomials. In: Symmetries and Integrability of Difference Equations. Canterbury, 1996. London Math. Soc. Lecture Note Ser., vol. 255, pp. 370–384. Cambridge Univ. Press, Cambridge (1999)
351. Landers, D., Rogge, L.: Nichtstandard Analysis. Springer, Heidelberg (1994)
352. Larcombe, P.J.: On the history of the Catalan numbers: a first record in China. Math. Today **35**(3), 89 (1999)
353. Larsson, D., Silvestrov, S.: Burchnall-Chaundy theory for q-difference operators and q-deformed Heisenberg algebras. J. Nonlinear Math. Phys. **10**(suppl. 2), 95–106 (2003)
354. Lauricella, G.: Sulle funzioni ipergeometriche a piu variabili. Rend. Circ. Mat. Palermo **7**, 111–158 (1893)
355. Lebedev, N.N.: Special Functions and Their Applications. Dover, New York (1972)
356. Lehmer, D.H.: Lacunary recurrence formulas for the numbers of Bernoulli and Euler. Ann. Math. (2) **36**(3), 637–649 (1935)
357. Leibniz, G.: Dissertatio de Arte Combinatoria. Leipzig (1666)
358. Lesky, P.: Orthogonalpolynome in x und q^{-x} als Lösungen von reellen q-Operatorgleichungen zweiter Ordnung. Monatshefte Math. **132**(2), 123–140 (2001)
359. Lievens, S., Van der Jeugt, J.: Transformation formulas for double hypergeometric series related to 9-j coefficients and their basic analogues. J. Math. Phys. **42**, 5417–5430 (2001)
360. Liu, Z.: Private communication
361. Loeffel, H.: Blaise Pascal: 1623–1662. Vita Mathematica, vol. 2. Birkhäuser, Basel (1987)
362. Lucas, E.: On the development of $(\frac{z}{1-e^{-z}})^{\alpha}$ in a series. Messenger (2) **VII**, 82–84 (1878)
363. Lucas, E.: Théorie des Nombres, vol. 1. Gauthier-Villars, Paris (1891)
364. Luke, Y.L.: The Special Functions and Their Approximations, vol. I. Mathematics in Science and Engineering, vol. 53. Academic Press, New York (1969)
365. MacRobert, T.M.: The multiplication formula for the gamma function and E-function series. Math. Ann. **139**, 133–139 (1959)
366. Mahoney, M.: The Mathematical Career of Pierre de Fermat, 1601–1665, 2nd edn. Princeton Paperbacks. Princeton University Press, Princeton (1994)
367. Manocha, H.L., Sharma, B.L.: Some formulae for Jacobi polynomials. Proc. Camb. Philos. Soc. **62**, 459–462 (1966)
368. Mansour, M.: An asymptotic expansion of the q-gamma function $\Gamma_q(x)$. J. Nonlinear Math. Phys. **13**(4), 479–483 (2006)
369. Mansour, T.: Some inequalities for the q-gamma function. J. Inequal. Pure Appl. Math. **9**(1), 18 (2008), 4 pp.
370. Marinova, L.P., Raychev, P.P., Maruani, J.: Molecular backbending in AgH and its description in terms of q-algebras. Mol. Phys. **82**(6), 1115–1129 (1994)
371. Maserse, F.: Scriptores Logarithmici, vols. 1–6. Inner Temple, London (1791–1806)
372. Mathews, J.: William Rowan Hamilton's paper of 1837 on the arithmetization of analysis. Arch. Hist. Exact Sci. **19**, 177–200 (1978)
373. McIntosh, C.: The Rosicrusians. Samuel Weiser Inc., Newburyport (1997)
374. Mehler, F.G.: Über die Entwicklung einer Funktion von beliebig vielen Variablen nach Laplaceschen Functionen höherer Ordnung. J. Reine Angew. Math. **66**, 161–176 (1866)
375. Mellin, Hj.: Über den Zusammenhang zwischen den linearen Differential- und Differenzengleichunge. Acta Math. **25**, 139–164 (1902)
376. Mellin, Hj.: Abriss einer einheitlichen Theorie der Gamma und der hypergeometrischen Funktionen. Math. Ann. **68**, 305–337 (1910)
377. Melzak, Z.A.: Companion to Concrete Mathematics, vol. 1. Mathematical Techniques and Various Applications. Pure and Applied Mathematics. Wiley, New York (1973)
378. Mikami, Y.: Mathematical papers from the far East. Abh. Gesch. Math. Wiss. **28** (1910)

379. Miller, W. Jr.: Lie theory and generalized hypergeometric functions. SIAM J. Math. Anal. **3**, 31–44 (1972)
380. Miller, W. Jr.: Lie theory and Lauricella functions F_D. J. Math. Phys. **13**, 1393–1399 (1972)
381. Miller, W. Jr.: Lie theory and the Appell functions F_1. SIAM J. Math. Anal. **4**, 638–655 (1973)
382. Miller, W. Jr.: Symmetry and Separation of Variables. Encyclopedia of Mathematics and Its Applications. Addison-Wesley, Reading (1977)
383. Milne, S.C.: q-analog of restricted growth functions, Dobinski's equality, and Charlier polynomials. Trans. Am. Math. Soc. **245**, 89–118 (1978)
384. Milne-Thomson, L.M.: Two classes of generalized polynomials. Proc. Lond. Math. Soc. (2) **35**, 514–522 (1933)
385. Milne-Thomson, L.M.: The Calculus of Finite Differences. Macmillan and Co., Ltd., London (1951)
386. Moak, D.S.: The q-gamma function for $q > 1$. Aequ. Math. **20**(2–3), 278–285 (1980)
387. Moak, D.S.: The q-analogue of the Laguerre polynomials. J. Math. Anal. Appl. **81**(1), 20–47 (1981)
388. Monteiro, M.R., Rodrigues, L.M.C.S.: Quantum algebraic nature of the phonon spectrum in ^{4}He. Phys. Rev. Lett. **76**(7), 1098–1100 (1996)
389. Muir, T.: Schweins, an overlooked discoverer in the theory of determinants. The London, Edinburgh, and Dublin Philosophical Magazine and Journal of Science. Glasgow (1884)
390. Munch, O.J.: On power product sums. Nordisk Mat. Tidskr. **7**, 5–19 (1959)
391. Nalli, P.: Sopra un procedimento di calcolo analogo all integrazione. Palermo Rend. **47**, 337–374 (1923)
392. Needham, J.: Science and Civilization in China, vol. 3, Mathematics and the Sciences of the Heavens and the Earth. Cambridge University Press, Cambridge (1959)
393. Négadi, T., Kibler, M.: A q-deformed Aufbau Prinzip. J. Phys. A **25**(4), L157–L160 (1992)
394. Netto, E.: Lehrbuch der Kombinatorik. Chelsea, New York (1958)
395. Neville E.H.: Jacobian Elliptic Functions. Clarendon Press, Oxford (1944)
396. Nielsen, N.: Handbuch der Theorie der Zylinderfunktionen. B.G. Teubner, Leipzig (1904)
397. Nielsen, N.: Handbuch der Theorie der Gammafunktion. Teubner, Leipzig (1906)
398. Nielsen, N.: Sur les fonctions de Bernoulli et des sommes de puissances numériques. Nieuw Arch. (2) **10**, 396–415 (1913)
399. Nielsen, N.: Traité Élémentaire des Nombres de Bernoulli. Gauthier Villars, Paris (1923)
400. Nishizawa, M.: Evaluation of a certain q-determinant. Linear Algebra Appl. **342**(1), 107–115 (2002)
401. Niven, I.: Formal power series. Am. Math. Mon. **76**, 871–889 (1969)
402. Nørlund, N.E.: Mémoire sur les polynômes de Bernoulli. Acta Math. **43**, 121–196 (1920)
403. Nørlund, N.E.: Vorlesungen über Differenzenrechnung. Springer, Berlin (1924)
404. Oldham, K.B., Spanier, J.: The Fractional Calculus. Theory and Applications of Differentiation and Integration to Arbitrary Order. With an annotated chronological bibliography by Bertram Ross. Mathematics in Science and Engineering, vol. 111. Academic Press [A subsidiary of Harcourt Brace Jovanovich, Publishers], New York (1974)
405. Orr, W.M.: Theorems relating to the products of two hypergeometric series. Trans. Camb. Philos. Soc. **17**, 1–15 (1899)
406. Oruc, H.: Generalized Bernstein polynomials and total positivity. Thesis, University of St. Andrews (1998)
407. Oruc, H., Akmaz, H.: Symmetric functions and the Vandermonde matrix. J. Comput. Appl. Math. **172**(1), 49–64 (2004)
408. Öttinger, L.: Forschungen in dem Gebiete der Höheren Analysis mit der Resultaten ihrer Anwendung. Osswald, Heidelberg (1831)
409. Ozhigova, E.P.: The origins of symbolic and combinatorial methods at the end of the 18th and at the beginning of the 19th century. Istor.-Mat. Issled. **24**, 121–157 (1979) (Russian)
410. Panda, R.: Some multiple series transformations. Jnānabha Sect. A **4**, 165–168 (1974)
411. Pearson, J.: The Elements of the Calculus of Finite Differences. Cambridge (1850)

412. Petkovsek, M., Wilf, H., Zeilberger, D.: $A = B$. A.K. Peters, Wellesley (1996)
413. Pfaff, J.F.: Disquisitiones Analyticae. Fleckeisen, Helmstedt (1797)
414. Pfaff, J.F.: Observationes analyticae ad L. Euleri Institutiones Calculi Integralis, vol. IV, Supplem. II et IV, Nova Acta Academiae Scientiarum Petropolitanae, Tome XI, Histoire, 38–57 (1797)
415. Phillips, G.M.: Interpolation and Approximation by Polynomials. CMS Books in Mathematics/Ouvrages de Mathématiques de la SMC, vol. 14. Springer, New York (2003)
416. Picard, E.: Sur une extension aux fonctions de deux variables du problème de Riemann relatif aux fonctions hypergéométriques. C. R. Acad. Sci. Paris **XC**, 1267–1269 (1880)
417. Pochhammer, L.: Über die Differentialgleichung der allgemeinen hypergeometrischen Reihe mit zwei endlichen singulären Punkten. J. Reine Angew. Math. **102**, 76–159 (1888)
418. Prabhakar, T.R., Reva, An.: Appell cross-sequence suggested by the Bernoulli and Euler polynomials of general order. Indian J. Pure Appl. Math. **10**(10), 1216–1227 (1979)
419. Pringsheim, A.: Vorlesungen über Zahlen- und Funktionenlehre. II, 2: Eideutige Analytische Funktionen, vol. XIV, pp. 625–1224. B.G. Teubner, Leipzig (1932)
420. Prudnikov, A.P., Brychkov, Yu.A., Marichev, O.I.: Integrals and Series. More Special Functions, vol. 3. Gordon and Breach Science Publishers, New York (1990). Translated from the Russian by G.G. Gould
421. Purohit, S.D.: Summation formulae for basic hypergeometric functions via fractional q-calculus. Matematiche **64**(1), 67–75 (2009)
422. Raabe, J.L.: Die Jacob Bernoullische Funktion. Orell Füssli und Cie, Zürich (1848)
423. Raabe, J.L.: Zurückführung einiger Summen und bestimmten Integrale auf die Jacob Bernoullischen Function. Crelle J. **42**, 348–376 (1851)
424. Rai, P.N., Singh, S.N.: On a class of generalized Hermite polynomials. Math. Stud. **46**(2–4), 374–378 (1978)
425. Rainville, E.D.: Special Functions. Chelsea Publishing Co., Bronx (1971). Reprint of 1960 first edition
426. Rajeswari, V., Srinivasa Rao, K.: Generalized basic hypergeometric functions and the q-analogues of 3-j and 6-j coefficients. J. Phys. A **24**(16), 3761–3780 (1991)
427. Rajković, P.M., Stanković, M.S., Marinković, S.D.: Mean value theorems in q-calculus. Mat. Vestn. **54**(3–4), 171–178 (2002). Proceedings of the 5th International Symposium on Mathematical Analysis and Its Applications, Niška Banja, 2002
428. Rauch, H.E., Lebowitz, A.: Elliptic Functions, Theta Functions and Riemann Surfaces. Williams & Wilkins, Baltimore (1973)
429. Raychev, P.P., Roussev, R.P., Smirnov, Yu.F.: The quantum algebra $SU_q(2)$ and rotational spectra of deformed nuclei. J. Phys. G **16**, L137–141 (1990)
430. Reid, C.: The Search for E.T. Bell. Also Known as John Taine. MAA Spectrum. Mathematical Association of America, Washington (1993)
431. Rideau, G., Winternitz, P.: Representations of the quantum algebra $\mathrm{su}_q(2)$ on a real two-dimensional sphere. J. Math. Phys. **34**(12), 6030–6044 (1993)
432. Riemann, B.: Elliptischen Funktionen. Ed. Stahl, H. Teubner, Leipzig (1899)
433. Riordan, J.: Combinatorial Identities. Robert E. Krieger Publishing Co., Huntington (1979). Reprint of the 1968 original
434. Rogers, L.J.: On a three-fold symmetry in the elements of Heine's series. Proc. Lond. Math. Soc. **24**, 171–179 (1893)
435. Rogers, L.J.: On the expansion of some infinite products. Proc. Lond. Math. Soc. **24**, 337–352 (1893)
436. Roman, S.: The theory of the umbral calculus I. J. Math. Anal. Appl. **87**, 58–115 (1982)
437. Roman, S.: The Umbral Calculus. Pure and Applied Mathematics, vol. 111. Academic Press, Inc. [Harcourt Brace Jovanovich, Publishers], New York (1984)
438. Roman, S.: More on the umbral calculus, with emphasis on the q-umbral calculus. J. Math. Anal. Appl. **107**, 222–254 (1985)
439. Rota, G.-C., Taylor, B.D.: The classical umbral calculus. SIAM J. Math. Anal. **25**(2), 694–711 (1994)

440. Rota, G.-C., Doubilet, P., Greene, C., Kahaner, D., Odlyzko, A., Stanley, R. (eds.): Finite Operator Calculus. Academic Press, Inc. [Harcourt Brace Jovanovich, Publishers], New York (1975)
441. Rothe, H.A.: Systematisches Lehrbuch der Arithmetik. Leipzig (1811)
442. Roy, R.: Review of the book Kac, V. and Cheung P., Quantum Calculus. Universitext. Springer-Verlag, New York, 2002. Am. Math. Mon. **110**(7), 652–657 (2002)
443. Rudin, W.: Real and Complex Analysis. McGraw-Hill, New York (1987)
444. Ryde, F.: A contribution to the theory of linear homogeneous geometric difference equations (q-difference equations). Dissertation, Lund (1921)
445. Saalschütz, L.: Eine Summationsformel. Z. Math. Phys. **35**, 186–188 (1890)
446. Saalschütz, L.: Vorlesungen über die Bernoullischen Zahlen, ihren Zusammenhang mit den Secanten-Coefficienten und ihre wichtigeren Anwendungen. Springer, Berlin (1893)
447. Salié, H.: Eulersche Zahlen. Sammelband zu Ehren des 250. In: Geburtstages Leonhard Eulers, pp. 293–310. Akademie-Verlag, Berlin (1959)
448. Sansone, G., Gerretsen, J.: Lectures on the Theory of Functions of a Complex Variable. II: Geometric Theory. Wolters-Noordhoff Publishing, Groningen (1969)
449. Schafheitlin, P.: Bemerkungen über die allgemeine hypergeometrische Reihe.... Arch. Math. Phys. (3) **19**, 20–25 (1912)
450. Schendel, L.: Zur Theorie der Functionen (x). J. Reine Angew. Math. **LXXXIV**, 80–84 (1877)
451. Schendel, L.: Beiträge zur Theorie der Functionen. Halle (1880)
452. Schlesinger, L.: Über Gauß' Arbeiten zur Funktionentheorie. Gött. Nachr. (Beiheft) **143** (1912)
453. Schlömilch, O.: Compendium, vol. II. Braunschweig (1895)
454. Schlosser, M.: q-analogues of the sums of consecutive integers, squares, cubes, quarts and quints. Electron. J. Comb. **11**(1) (2004)
455. Schneider, I.: Johannes Faulhaber, 1580–1635. Rechenmeister in einer Welt des Umbruchs. Vita Mathematica, vol. 7. Birkhäuser, Basel (1993)
456. Schwatt, I.J.: An Introduction to the Operations with Series. The Press of the University of Pennsylvania, Philadelphia (1924)
457. Schweins, F.: Analysis. Heidelberg (1820)
458. Schweins, F.: Theorie der Differenzen und Differentiale. Heidelberg (1825)
459. Scriba, C., Schreiber, P.: 5000 Jahre Geometrie: Geschichte Kulturen Menschen. Springer, Berlin (2005)
460. Sears, D.B.: On the transformation theory of hypergeometric functions. Proc. Lond. Math. Soc. (2) **52**, 14–35 (1950)
461. Sears, D.B.: On the transformation theory of basic hypergeometric functions. Proc. Lond. Math. Soc. (2) **53**, 158–180 (1951)
462. Sellami, M., Brahim, K., Bettaibi, N.: New inequalities for some special and q-special functions. J. Inequal. Pure Appl. Math. **8**(2), 47 (2007), 7 pp.
463. Sharma, A., Chak, A.M.: The basic analogue of a class of polynomials. Riv. Mat. Univ. Parma **5**, 325–337 (1954)
464. Singh, R.P.: Operational formulae for Jacobi and other polynomials. Rend. Semin. Mat. Univ. Padova **35**, 237–244 (1965)
465. Singhal, B.M.: On the reducibility of Lauricella's function F_D. Jnānabha A **4**, 163–164 (1974)
466. Slater, L.J.: Generalized Hypergeometric Functions. Cambridge (1966)
467. Smith, E.R.: Zur Theorie der Heineschen Reihe und ihrer Verallgemeinerung. Diss. Univ. München (1911)
468. Sonine, N.J.: Recherches sur les fonctions cylindriques et le développement des fonctions continues en séries. Clebsch Ann. **XVI**, 1–80 (1880)
469. Sonine, N.: Die Gammafunktion und die Omegafunktion von Heine. Warsaw (1884) (Russian)

470. Srinivasa Rao, K., Rajeswari, V.: Quantum Theory of Angular Momentum. Springer/Narosa, Berlin/New Delhi (1993)
471. Srivastava, H.M.: Hypergeometric functions of three variables. Ganita **15**, 97–108 (1964)
472. Srivastava, H.M.: Certain pairs of inverse series relations. J. Reine Angew. Math. **245**, 47–54 (1970)
473. Srivastava, H.M.: On the reducibility of Appell's function F_4. Can. Math. Bull. **16**, 295–298 (1973)
474. Srivastava, H.M.: A note on certain summation theorems for multiple hypergeometric series. Simon Stevin **52**(3), 97–109 (1978)
475. Srivastava, H.M.: A note on certain identities involving generalized hypergeometric series. Akad. Wetensch. Indag. Math. **41**(2), 191–201 (1979)
476. Srivastava, H.M.: Some generalizations of Carlson's identity. Boll. Unione Mat. Ital. A (5) **18**(1), 138–143 (1981)
477. Srivastava, H.M.: A family of q-generating functions. Bull. Inst. Math. Acad. Sin. **12**, 327–336 (1984)
478. Srivastava, H.M.: Some transformations and reduction formulas for multiple q-hypergeometric series. Ann. Mat. Pura Appl. (4) **144**, 49–56 (1986)
479. Srivastava, H.M.: A class of finite q-series. Rend. Semin. Mat. Univ. Padova **75**, 15–24 (1986)
480. Srivastava, H.M.: Sums of a certain class of q-series. Proc. Jpn. Acad., Ser. A, Math. Sci. **65**(1), 8–11 (1989)
481. Srivastava, H.M., Jain, V.K.: q-series identities and reducibility of basic double hypergeometric functions. Can. J. Math. **38**(1), 215–231 (1986)
482. Srivastava, H.M., Karlsson, P.W.: Multiple Gaussian Hypergeometric Series. Ellis Horwood, New York (1985)
483. Srivastava, H.M., Karlsson, P.W.: Transformations of multiple q-series with quasi-arbitrary terms. J. Math. Anal. Appl. **231**(1), 241–254 (1999)
484. Srivastava, H.M., Manocha, H.L.: A Treatise on Generating Functions. Ellis Horwood Series: Mathematics and Its Applications. Ellis Horwood Ltd., Chichester; Halsted Press [Wiley, Inc.], New York (1984)
485. Srivastava, H.M., Pintér, A.: Remarks on some relationships between the Bernoulli and Euler polynomials. Appl. Math. Lett. **17**(4), 375–380 (2004)
486. Stanley, R.P.: Enumerative Combinatorics. Vol. I. The Wadsworth & Brooks/Cole Mathematics Series. Wadsworth & Brooks/Cole, Monterey (1986)
487. Stedall, J.: The discovery of wonders: reading between the lines of John Wallis's Arithmetica infinitorum. Arch. Hist. Exact Sci. **56**(1), 1–28 (2001)
488. Steffensen, J.F.: Interpolation. Chelsea, New York (1950)
489. Stephens, E.: The Elementary Theory of Operational Mathematics. McGraw-Hill, New York (1937)
490. Stirling, J.: Methodus Differentialis. London (1730)
491. Stubhaug, A.: Att Våga Sitt Tärningskast Gösta Mittag Leffler 1846–1927. Bokförlaget Atlantis, Stockholm (2007)
492. Swarttouw, R.F.: The contiguous function relations for the basic hypergeometric series. J. Math. Anal. Appl. **149**(1), 151–159 (1990)
493. Szegö, G.: Review of Nørlund N.E., Mémoire sur les polynômes de Bernoulli. Acta Math. **43**, 121–196 (1920). [402], JFM
494. Szegö, G.: Ein Beitrag zur Theorie der Thetafunktionen. Sitz.ber Preuss. Akad. Wiss., Phys.-Math. Kl., 242–252 (1926)
495. Talman, J.: Special Functions: A Group Theoretic Approach. W.A. Benjamin, Inc., New York (1968). Based on Lectures by Eugene P. Wigner. With an Introduction by Eugene P. Wigner
496. Thibaut, B.F.: Grundriss der Allgemeinen Arithmetik. Göttingen (1809)
497. Thomae, J.: Beiträge zur Theorie der durch die Heinische Reihe: $1+((1-q^{\alpha})(1-q^{\beta})/(1-q)(1-q)^{\gamma})x+\cdots$ darstellbaren Functionen. J. Reine Angew. Math. **70**, 258–281 (1869)

498. Thomae, J.: Über die höheren hypergeometrischen Reihen, insbesondere die Reihe $1 + \frac{a_0 a_1 a_2}{1 \cdot b_1 b_2} x + \frac{a_0(a_0+1)a_1(a_1+1)a_2(a_2+1)}{1 \cdot 2 \cdot b_1(b_1+1)b_2(b_2+1)} x^2 + \cdots$. Math. Ann. **2**, 427–444 (1870)
499. Thomae, J.: Les séries Heinéennes supérieures. Ann. Mat. Pura Appl. Bologna II **4**, 105–139 (1871)
500. Thomae, J.: Abriss einer Theorie der Functionen einer Complexen Veränderlichen und der Thetafunctionen, 2nd enlarged edn. Nebert, Halle (1873)
501. Thomae, J.: Über die Funktionen welche durch Reihen von der Form dargestellt werden. J. Reine Angew. Math. **87**, 26–73 (1879)
502. Timoteo, V.S., Lima, C.L.: Effect of q-deformation in the NJL gap equation. Phys. Lett. B **448**(1–2), 1–5 (1999)
503. Toscano, L.: Sul l'iterazione degli operatori xD e Dx. Ist. Lombardo, Rend., II. Ser. **67**, 543–551 (1934)
504. Toscano, L.: Sulla iterazione dell'operatore xD. Rend. Mat. Appl. (V) **8**, 337–350 (1949)
505. Toscano, L.: Numeri di Stirling generalizzati e operatori permutabili di secondo ordine. Matematiche (Catania) **24**, 492–518 (1969)
506. Toscano, L.: Sui polinomi ipergeometrici a più variabili del tipo F_D di Lauricella. Matematiche (Catania) **27**, 219–250 (1972)
507. Toscano, L.: L'operatore xD e i numeri di Bernoulli e di Eulero. Matematiche **31**, 63–89 (1976). 1977
508. Tricomi, F.G.: Vorlesungen über Orthogonalreihen. Die Grundlehren der mathematischen Wissenschaften in Einzeldarstellungen mit besonderer Berücksichtigung der Anwendungsgebiete, vol. LXXVI. Springer, Berlin (1955)
509. Truesdell, C.: A Unified Theory of Special Functions Based Upon the Functional Equation $\frac{\partial}{\partial z} F(z, \alpha) = F(z, \alpha + 1)$. Annals of Math. Studies, vol. 18. Princeton University Press, Princeton. London: Geoffrey Cumberlege. Oxford University Press (1948)
510. Tweddle, I.: James Stirling's Methodus Differentialis. An annotated translation of Stirling's text. Sources and Studies in the History of Mathematics and Physical Sciences. Springer, London (2003)
511. Tweedie, Ch., Stirling, J.: A Sketch of His Life and Works Along with His Scientific Correspondence. Clarendon Press, Oxford (1922)
512. Van der Jeugt, J.: The q-boson operator algebra and q-Hermite polynomials. Lett. Math. Phys. **24**(4), 267–274 (1992)
513. Van der Jeugt, J.: Transformation formula for a double Clausenian hypergeometric series, its q-analogue, and its invariance group. J. Comput. Appl. Math. **139**, 65–73 (2002)
514. Van der Jeugt, J., Srinivasa Rao, K.: Invariance groups of transformations of basic hypergeometric series. J. Math. Phys. **40**(12), 6692–6700 (1999)
515. Vandermonde, A.T.: Abhandlungen aus der Reinen Mathematik. In Deutscher Sprache herausgegeben von Carl Itzigsohn. Springer, Berlin (1888) (German)
516. Vandermonde, A.T.: Mémoire sur des irrationnelles de différents ordre avec une application au cercle. Mém. Acad. R. Sci. Paris 489–498 (1772) (French)
517. Vandiver, H.S.: Simple explicit expressions for generalized Bernoulli numbers of the first order. Duke Math. J. **8**, 575–584 (1941)
518. Vein, R., Dale, P.: Determinants and Their Applications in Mathematical Physics. Applied Mathematical Sciences, vol. 134. Springer, New York (1999)
519. Veneziano, G.: Construction of a crossing-symmetric, Regge-behaved amplitude for linearly rising trajectories. Nuovo Cimento A **57**, 190–197 (1968)
520. Verma, A.: Expansions involving hypergeometric functions of two variables. Math. Comput. **20**, 590–596 (1966)
521. Verma, A.: A quadratic transformation of a basic hypergeometric series. SIAM J. Math. Anal. **11**(3), 425–427 (1980)
522. Vidunas, R.: Degenerate Gauss hypergeometric functions. Kyushu J. Math. **61**(1), 109–135 (2007)
523. Vilenkin, N.J., Klimyk, A.U.: Representations of Lie Groups and Special Functions, vol. 3. Kluwer Academic Publishers, Dordrecht (1992)

524. Vivanti, G.: Lezioni sulla teoria delle funzioni ellittiche tenute nella. Universita di Messina, Messina (1900)
525. von Ettingshausen, A.: Die Combinatorische Analysis. Vienna (1826)
526. von Grüson, J.F., Neuer analytischer Lehrsatz, Abhandlungen der Berliner Academie 1814–1815
527. Walker, E.: A Study of the Traité des Indivisibles of Gilles Persone de Roberval. Bureau of Publications, Teachers College, Columbia University, New York (1932)
528. Wallis, J.: Arithmetica Infinitorum. Oxford (1655)
529. Wallisser, R.: Über ganze Funktionen, die in einer geometrischen Folge ganze Werte annehmen. Monatshefte Math. **100**, 329–335 (1985)
530. Wallisser, R.: On entire functions assuming integer values in a geometric sequence. In: Théorie des Nombres, Quebec, PQ, 1987, pp. 981–989. de Gruyter, Berlin (1989)
531. Ward, M.: A calculus of sequences. Am. J. Math. **58**, 255–266 (1936)
532. Warnaar, O.: On the q-analogue of the sum of cubes. Electron. J. Comb. **11** (2004)
533. Watson, G.N.: The continuations of functions defined by generalised hypergeometric series (x). Camb. Philos. Soc. Trans. **21**, 281–299 (1910)
534. Watson, G.N.: A note on generalized hypergeometric series. Proc. Lond. Math. Soc. (2) **23**, XIII–XV (1925)
535. Watson, G.N.: A new proof of the Rogers-Ramanujan identities. J. Lond. Math. Soc. **4**, 4–9 (1929)
536. Watson, G.N.: Notes on generating functions of polynomials: (1) Laguerre polynomials. J. Lond. Math. Soc. **8**, 189–192 (1933)
537. Watson, G.N.: Notes on generating functions of polynomials: (2) Hermite polynomials. J. Lond. Math. Soc. **8**, 194–199 (1933)
538. Watson, G.N.: A note on the polynomials of Hermite and Laguerre. J. Lond. Math. Soc. **13**, 29–32 (1938)
539. Weierstraß, K.: Über die Theorie der analytischen Facultäten. J. Reine Angew. Math. **51**, 1–60 (1856)
540. Westfall, R.S.: Never at Rest. A Biography of Isaac Newton. Cambridge University Press, Cambridge (1980)
541. Whipple, F.J.W.: A group of generalized hypergeometric series: relations between 120 allied series of the type ${}_3F_2(a, b, c; d, e)$. Proc. Lond. Math. Soc. (2) **23**, 104–114 (1925)
542. Whipple, F.J.W.: On well-poised series, generalized hypergeometric series having parameters in pairs, each pair with the same sum. Proc. Lond. Math. Soc. (2) **24**, 247–263 (1926)
543. Whipple, F.J.W.: Well-poised series and other generalized hypergeometric series. Proc. Lond. Math. Soc. (2) **25**, 525–544 (1926)
544. Whipple, F.J.W.: Some transformations of generalized hypergeometric series. Proc. Lond. Math. Soc. (2) **26**, 257–272 (1927)
545. Worpitzky, J.: Studien über die Bernoullischen und Eulerschen Zahlen. J. Reine Angew. Math. **94**, 203–232 (1883)
546. Yang, K.-W.: Matrix q-hypergeometric series. Discrete Math. **146**(1–3), 271–284 (1995)
547. Zappa, G., Casadio, G.: L'attivita matematica di Francesco Faà di Bruno tra il 1850 e il 1859. Mem. Accad. Sci. Torino Cl. Sci. Fis. Mat. Natur. (5) **16**(1–4) (1992)
548. Zeng, J.: The q-Stirling numbers, continued fractions and the q-Charlier and q-Laguerre polynomials. J. Comput. Appl. Math. **57**(3), 413–424 (1995)
549. Zhang, C.: Sur la fonction q-gamma de Jackson. Aequ. Math. **62**(1–2), 60–78 (2001)
550. Zhang, Y., Chen, W.: q-Triplicate inverse series relations with applications to q-series. Rocky Mt. J. Math. **35**(4), 1407–1427 (2005)
551. Zhe, C., Yan, H.: Quantum group-theoretic approach to vibrating and rotating diatomic molecules. Phys. Lett. A **158**(5), 242–246 (1991)
552. Zhe, C., Yan, H.: The $SU_q(2)$ quantum group symmetry and diatomic molecules. Phys. Lett. A **154**(5–6), 254–258 (1991)
553. Zhe, C., Guo, H.Y., Yan, H.: The q-deformed oscillator model and the vibrational spectra of diatomic molecules. Phys. Lett. A **156**(3–4), 192–196 (1991)

Index before 1900

T. Ernst, *A Comprehensive Treatment of q-Calculus*,
DOI 10.1007/978-3-0348-0431-8, © Springer Basel 2012

Index after 1900

T. Ernst, *A Comprehensive Treatment of q-Calculus*,
DOI 10.1007/978-3-0348-0431-8, © Springer Basel 2012

Physics index

T. Ernst, *A Comprehensive Treatment of q-Calculus*,
DOI 10.1007/978-3-0348-0431-8, © Springer Basel 2012

Name index before 1900

DOI 10.1007/978-3-0348-0431-8,

Name index after 1900

Name index Physics

T. Ernst, *A Comprehensive Treatment of q-Calculus*,
DOI 10.1007/978-3-0348-0431-8, © Springer Basel 2012

Notation index Chapter 1, 2, 6–9

T. Ernst, *A Comprehensive Treatment of q-Calculus*,
DOI 10.1007/978-3-0348-0431-8, © Springer Basel 2012

Notation index Chapter 3

T. Ernst, *A Comprehensive Treatment of q-Calculus*,
DOI 10.1007/978-3-0348-0431-8, © Springer Basel 2012

Notation index Chapter 4, 5

$\triangle_{\omega \text{NWA},2,q}$, 153
$\triangle^n_{\overline{m}_{1g},\ldots,\overline{m}_{nq}\text{NWA},q}$, 112
$(-1)^{\overline{n}_q}$, 110
$(-1)^{\tilde{n}_q}$, 110
$(A,\star)$, 98
$(\boxplus_{q,l=0}^{\infty} a_l x^l)^k$, 110
$(\oplus_{q,l=0}^{\infty} a_l x^l)^k$, 110
$(\tilde{n}_q)^k$, 109
$(x)_{k,q}$, 171
$[\alpha]$, 99
$\Phi^{(1)}_{\nu,q}(x)$, 114
$\Phi^{(n)}_{\nu,q}$, 114
$\Phi^{(n)}_{\nu,q}(x)$, 114
α, β, 98
$\alpha \nsim \beta$, 99
$\alpha \sim \beta$, 99
$\triangle_{\tilde{n}_q \text{JHC},q}$, 105
$\nabla_{\omega \text{JHC},q}$, 106
$\nabla_{\omega \text{NWA},q}$, 106
$\nabla^n_{\omega_1,\ldots,\omega_n \text{JHC},q}$, 113, 133
$\nabla^n_{\omega_1,\ldots,\omega_n \text{NWA},2,q}$, 154
$\nabla^n_{\omega_1,\ldots,\omega_n \text{NWA},q}$, 112
$\triangle_{\omega \text{JHC},q}$, 106
$\triangle_{\omega \text{NWA},2,q}$, 153
$\triangle_{\omega \text{NWA},q}$, 106
$\triangle_{\overline{n}_q \text{NWA},q}$, 105
$\triangle^n_{\tilde{m}_{1g},\ldots,\tilde{m}_{nq}\text{JHC},q}$, 113
$\triangle^n_{\omega_1,\ldots,\omega_n \text{JHC},q}$, 113
$\triangle^n_{\omega_1,\ldots,\omega_n \text{NWA},2,q}$, 165
$\triangle^n_{\omega_1,\ldots,\omega_n \text{NWA},q}$, 112
$\beta^{(n)}_{\nu,q}(x)$, 116
$\eta^{(n)}_{\nu,q}(x)$, 128
$\frac{1}{(x\oplus_q y;q)_a}$, 104
$\gamma^{(n)}_{\nu,q}(x)$, 123
γ_q, 122
$\mathbb{R}_q$, 98
$\mathcal{B}_q(n)$, 181
$\nabla_{\text{JHC},q}$, 106
$\nabla^n_{\text{JHC},q}$, 113
$\nabla_{\text{NWA},2,q}$, 153
$\nabla_{\text{NWA},q}$, 106
$\nabla^n_{\text{NWA},q}$, 112
$\overline{n}_q$, 100
$\sigma_{\text{JHC},m,q}(n)$, 135
$\sigma_{\text{NWA},m,q}(n)$, 131
$\tau_{\text{NWA},m,q}(n)$, 160
$\text{B}^{(-n)}_{\text{JHC},\nu,q}(x)$, 143
$\text{B}^{(n)}_{\text{JHC},\nu,q}(x|\omega_1,\ldots,\omega_n)$, 124
$\text{B}^{(n)}_{\text{JHC},\nu,q}(x)$, 124
$\text{B}^{(n+p)}_{\text{JHC},\nu,q}(x \oplus_q y|\omega_1,\ldots,\omega_{n+p})$, 142
$\text{B}_{\text{JHC},n,q}$, 124

T. Ernst, *A Comprehensive Treatment of q-Calculus*,
DOI 10.1007/978-3-0348-0431-8, © Springer Basel 2012

$\mathrm{B}^{(-n)}_{\mathrm{NWA},\nu,q}$, 137
$\mathrm{B}^{(n)}_{\mathrm{NWA},\nu,q}(x)$, 117
$\mathrm{B}^{(n+p)}_{\mathrm{NWA},\nu,q}(x \oplus_q y|\omega_1,\ldots,\omega_{n+p})$, 136
$\mathrm{B}_{\mathrm{NWA},n,q}$, 118
$\mathrm{B}_{\mathrm{NWA},q}$, 100
$\mathrm{B}^{(n)}_{\mathrm{NWA},\nu,q}(x|\omega_1,\ldots,\omega_n)$, 117
$\mathrm{D}_{\oplus_q}$, 102
$\mathrm{E}(\boxplus_q)$, 104
$\mathrm{E}(\oplus_q)$, 104
$\mathrm{E}(\oplus_q)^{\omega}$, 104
$\mathrm{E}^{l}_{\mathrm{C},q}$, 171
$\mathrm{F}^{(n)}_{\mathrm{JHC},\nu,q}(x)$, 133
$\mathrm{F}^{(-n)}_{\mathrm{JHC},\nu,q}$, 143
$\mathrm{F}^{(n)}_{\mathrm{JHC},\nu,q}$, 134
$\mathrm{F}^{(n)}_{\mathrm{JHC},\nu,q}(x)$, 133
$\mathrm{F}^{(n)}_{\mathrm{JHC},\nu,q}(x|\omega_1,\ldots,\omega_n)$, 133
$\mathrm{F}_{\mathrm{JHC},k,q}$, 133
$\mathrm{F}_{\mathrm{NWA},\nu,q}(x)$, 130
$\mathrm{F}^{(-n)}_{\mathrm{NWA},\nu,q}$, 137
$\mathrm{F}^{(n)}_{\mathrm{NWA},\nu,q}(x|\omega_1,\ldots,\omega_n)$, 129
$\mathrm{F}^{(n)}_{\mathrm{NWA},\nu,q}(x)$, 131
$\mathrm{F}_{\mathrm{NWA},\nu,q}$, 130
$\mathrm{F}_{\mathrm{NWA},k,q}$, 130
$\mathrm{G}^{(-n)}_{\mathrm{NWA},\nu,q}$, 162
$\mathrm{G}^{(-n)}_{\mathrm{NWA},\nu,q}(x)$, 162
$\mathrm{G}^{(n)}_{\mathrm{NWA},\nu,q}(x)$, 158
$\mathrm{G}^{(n)}_{\mathrm{NWA},\nu,q}(x|\omega_1,\ldots,\omega_n)$, 159
$\mathrm{J}_{\mathrm{B,J},q}$, 127
$\mathrm{J}_{\mathrm{B,N},q}$, 122
$\mathrm{J}_{\mathrm{L,N},q}$, 158
$\mathrm{L}^{(-n)}_{\mathrm{NWA},\nu,q}$, 162
$\mathrm{L}^{(-n)}_{\mathrm{NWA},\nu,q}(x)$, 162
$\mathrm{L}^{(n)}_{\mathrm{NWA},\nu,q}(x)$, 154
$\mathrm{L}^{(n)}_{\mathrm{NWA},\nu,q}(x,\omega_1,\ldots,\omega_n)$, 154
$\mathrm{L}^{1}_{q}(I)$, 111
$\mathrm{S}(n,k)_q$, 173
$\mathrm{S}^{n}_{\mathrm{B,J},q}$, 125
$\mathrm{S}^{n}_{\mathrm{B,N},q}$, 119
$\mathrm{S}_{\mathrm{C},m,q}(n)$, 171, 182
$\mathrm{S}^{n}_{\mathrm{L,N},q}$, 155
e_k, 175
h_k, 176
$\mathrm{s}(n,k)_q$, 173
θ, 99
$\theta^{(n)}_{\nu,q}(x)$, 132
θ_q, 179
$\tilde{n}_q$, 100
$\triangle_{\mathrm{CG},q}$, 170, 183
$\triangle_{\mathrm{H},q}$, 171
$\triangle_{\mathrm{JHC},q}$, 105
$\triangle_{\mathrm{J},x,q}$, 191
$\triangle_{\mathrm{NWA},2,q}$, 153
$\triangle_{\mathrm{NWA},q}$, 107
$\triangle^{-1}_{\mathrm{CG},q}$, 188
$f(\overline{k}_q)$, 110
$f(\tilde{k}_q)$, 110
$f(x \oplus_q y \boxminus_q z)$, 106
$s_{\mathrm{JHC},m,q}(n)$, 127
$s_{\mathrm{NWA},m,q}(n)$, 120
$t_{\mathrm{NWA},m,q}(n)$, 157
ν, 98
$\Phi^{(0)}_{\nu,q}(x)$, 114

Notation index Chapter 10–11

$(-1)^{\vec{k}}$, 359
${}_r\Phi_s(a_1,\ldots,a_r;b_1,\ldots,b_s|q;z)$, 428
${}_r\Psi_s(a_i;b_j|q;u_p)$, 430
$\nabla_q(h)$, 401
$\triangle_q(h)$, 401
$[\epsilon_j f](x_1,\ldots,x_n)$, 427
$\frac{1}{(\vec{x};\vec{q})_{\vec{\beta}}}$, 359
$[\partial_{x_j,q_j} f](x_1,\ldots,x_n)$, 427
$\Phi_1(a;b,b';c|q;x_1,x_2)$, 365
$\Phi_2(a;b,b';c,c'|q;x_1,x_2)$, 365
$\Phi_3(a,a';b,b';c|q;x_1,x_2)$, 365
$\Phi_4(a;b;c,c'|q;x_1,x_2)$, 361, 365
$\Phi_{\mathrm{A}}^{(n)}(a,\vec{b};\vec{c}|q;\vec{x})$, 437
$\Phi_{\mathrm{B}}^{(n)}(\vec{a},\vec{b};c|q;\vec{x})$, 437
$\Phi_{\mathrm{C}}^{(n)}(a,b;\vec{c}|q;\vec{x})$, 437
$\Phi_{\mathrm{D}}^{(n)}(a,\vec{b};c|q;\vec{x})$, 387, 399, 420
$\theta_{q,j}$, 366
$\triangle_p^+$, 430
$\triangle_p^-$, 430
$\{\vec{\alpha}\}_{\vec{m},\vec{q}}$, 359
$[\vec{a},\vec{b}]$, 360
$\{\vec{l}\}_{\vec{q}}\,|\cdot$, 359
$\vec{q}^{\binom{\vec{k}}{2}}$, 359
$\vec{x}^{\vec{\alpha}}$, 359
$\vec{xy}$, 420
$\mathrm{D}_{\vec{q},\vec{x}}^{\vec{l}} F(\vec{x},\vec{q})$, 360
$(\mathrm{D}_{\vec{q},\vec{x}}\vec{\epsilon}^{\vec{t}})^{\vec{l}} F(\vec{x},\vec{q})$, 360
$\mathrm{F}_1(a;b,b';c;x_1,x_2)$, 364
$\mathrm{F}_2(a;b,b';c,c';x_1,x_2)$, 364
$\mathrm{F}_3(a,a';b,b';c;x_1,x_2)$, 364
$\mathrm{F}_4(a;b;c,c';x_1,x_2)$, 364
$F(\vec{x}\boxplus_{\vec{q}}\vec{y})$, 361
$F(\vec{x}\boxplus_{\vec{q}}\vec{y})$, 362
$F(\vec{x}\oplus_{\vec{q},t}\vec{y})$, 361
$F(\vec{x}\oplus_{\vec{q}}\vec{y})$, 361
$F(\vec{x}\oplus_{\vec{q}}\vec{y})$, 362
$F(x\boxplus_q y)$, 362
$F(x\oplus_q y)$, 362
$\langle\vec{\alpha};\vec{q}\rangle_{\vec{k}}$, 359
$[M_{x_j} f](x_1,\ldots,x_n)$, 427
$\oplus_{q,\star}$, 371
$\mathrm{P}_{\vec{k},\vec{q}}(\vec{x},\vec{y})$, 359

T. Ernst, *A Comprehensive Treatment of q-Calculus*,
DOI 10.1007/978-3-0348-0431-8, © Springer Basel 2012

Notation index Chapter 12

T. Ernst, *A Comprehensive Treatment of q-Calculus*,
DOI 10.1007/978-3-0348-0431-8, © Springer Basel 2012

Printed by Publishers' Graphics LLC
BT20130401.09.07.119